Klaus Buchholz, Volker Kasche,
Uwe Theo Bornscheuer

बाइोकाटिलिस्ट
और

Biocatalysts and Enzyme Technology

टैकनोलोजी

БИОКАТАЛИСТ
АНD(И)
β ЭНЗАЙМ ТЕКНОЛОГИ

Клауз Бухолз
Волкер Каше
Уве Тхео Борншхер

WILEY-VCH Verlag GmbH & Co. KGaA

Authors

Prof. Dr. Klaus Buchholz
Technical University of Braunschweig
Technology of Carbohydrates
Hans-Sommer-Strasse 10
38106 Braunschweig
Germany

Prof. Dr. Volker Kasche
Technical University of Hamburg-Harburg
AB Biotechnology II
Denickestrasse 15
21071 Hamburg
Germany

Prof. Dr. Uwe Theo Bornscheuer
University of Greifswald
Institute for Chemistry and Biochemistry
Soldmannstrasse 16
17487 Greifswald
Germany

Cover illustration:
Bioreactor with kind permission from Bioengineering, Switzerland.

Library of Congress Card No.: applied for

British Library Cataloguing-in-Publication Data
A catalogue record for this book is available from the British Library.

Bibliographic information published by Die Deutsche Bibliothek
Die Deutsche Bibliothek lists this publication in the Deutsche Nationalbibliografie; detailed bibliographic data is available in the Internet at <http://dnd.ddb.de>.

Printed in the Federal Republic of Germany
Printed on acid-free and low chlorine paper

Cover Design Grafig-Design SCHULZ, Fußgönheim
Typesetting Fotosatz Detzner, Speyer
Printing Strauss GmbH, Mörlenbach
Binding Litges & Dopf Buchbinderei GmbH, Heppenheim

ISBN-13: 978-3-527-30497-4
ISBN-10: 3-527-30497-5

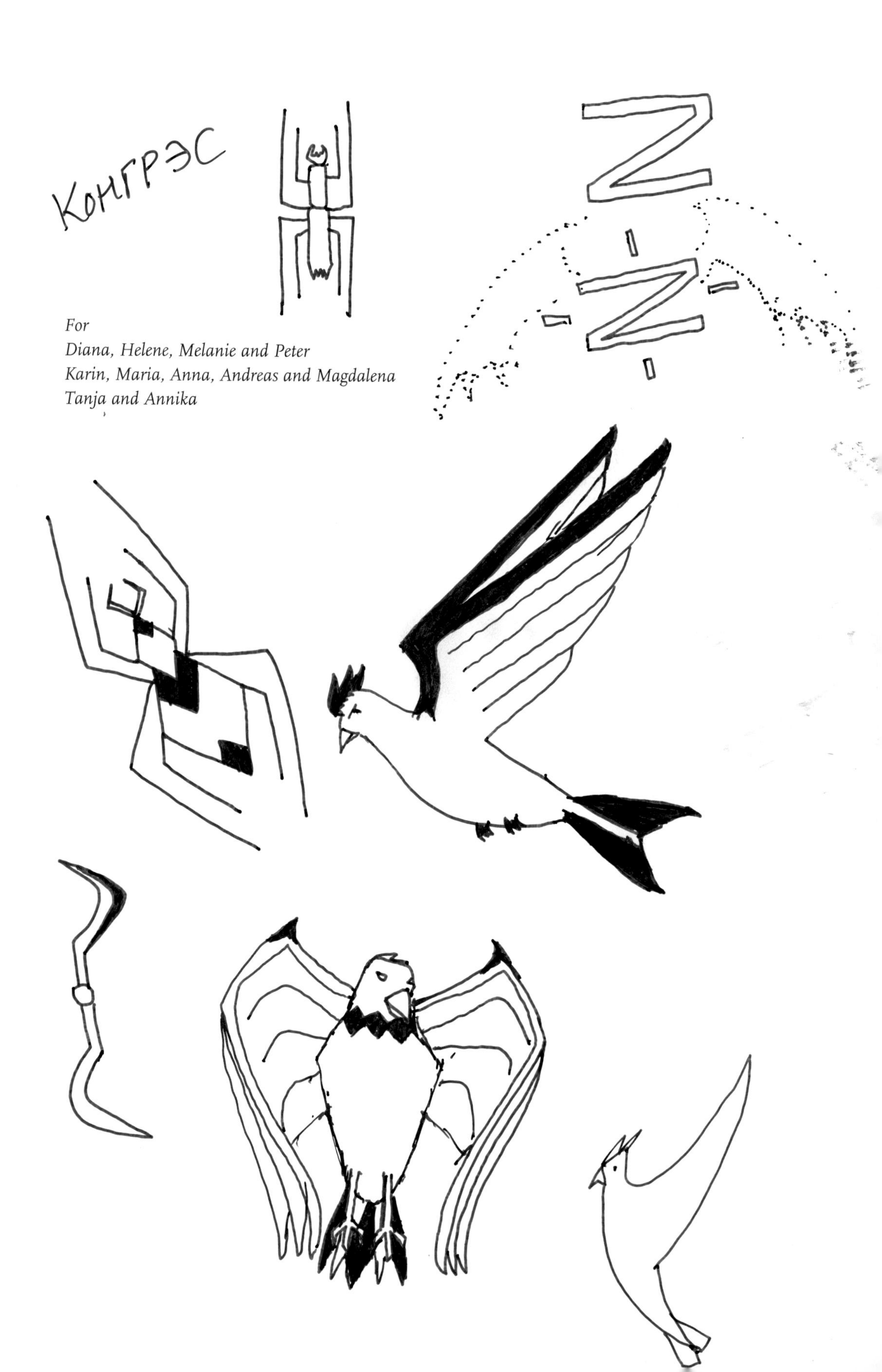

For
Diana, Helene, Melanie and Peter
Karin, Maria, Anna, Andreas and Magdalena
Tanja and Annika

Немецкий

Preface

To the First German Edition

Biotechnology is the technical application of biological systems or parts thereof to provide products and services to meet human needs. It can, besides other techniques, contribute towards doing this in a sustainable manner. Since, in the majority of cases, renewable raw materials and biological systems are used in biotechnological processes, these processes can – and should – be performed practically without waste, as all of the byproducts can be recycled.

The development of natural and engineering science fundamentals for the design of such processes remains a challenge to biotechnology – a field that originated from the overlapping areas of biology, chemistry, and process engineering.

The requisite education for a career in biotechnology consists, in addition to a basic knowledge of each of these fields, of further biotechnological aspects which must provide an overview over the entire field and a deeper insight into different areas of biotechnology. The biotechnological production of various materials is performed either in fermenters using living cells (technical microbiology), or with enzymes – either in an isolated form or contained in cells – as biocatalysts. Indeed, the latter aspect has developed during recent years to form that area of biotechnology known as enzyme technology, or applied biocatalysis.

The aim of the present textbook is to provide a deeper insight into the fundamentals of enzyme technology and applied biocatalysis. It especially stresses the following inter-relationships: A thorough understanding of enzymes as biocatalysts and the integration of knowledge of the natural sciences of biology (especially biochemistry), cell and molecular biology; physico-chemical aspects of catalysis and molecular interactions in solutions; heterogeneous systems and interphase boundaries; and the physics of mass transfer processes. The same applies to the inter-relations between enzyme technology and chemical and process engineering, which are based on the above natural sciences.

In less than a century since the start of industrial enzyme production, enzyme technology and its products have steadily gained increasing importance. In the industrial production of materials to meet the demands of everyday life, enzymes play an important role – and one which is often barely recognized. Their application ranges from the production of processed foods such as bread, cheese, juice and beer,

Biocatalysts and Enzyme Technology. K. Buchholz, V. Kasche, U. T. Bornscheuer

ISBN: 3-527-30497-5

to pharmaceuticals and fine chemicals, to the processing of leather and textiles, as process aids in detergents, and also in environmental engineering.

Meeting the demand for these new products – which increasingly include newly developed and/or sterically pure pharmaceuticals and fine chemicals – has become an important incentive for the further development of biocatalysts and enzyme technology. Of similar importance is the development of new sustainable production processes for existing products, and this is detailed in Chapter 1, which forms an Introduction.

Enzymes as catalysts are of key importance in biotechnology, similar to the role of nucleic acids as carriers of genetic information. Their application as isolated catalysts justifies detailed examination of the fundamentals of enzymes as biocatalysts, and this topic is covered in Chapter 2. Enzymes can also be analyzed on a molecular level, and their kinetics described mathematically. This is essential for an analytical description and the rational design of enzyme processes. Enzymes can also catalyze a reaction in both directions – a property which may be applied in enzyme technology to achieve a reaction end-point both rapidly and with a high product yield. The thermodynamics of the catalyzed reaction must also be considered, as well as the properties of the enzyme. The amount of enzyme required for a given conversion of substrate per unit time must be calculated in order to estimate enzyme costs, and in turn the economic feasibility of a process. Thus, the quantitative treatment of biocatalysis is also highlighted in Chapter 2.

When the enzyme costs are too high, they can be reduced by improving the production of enzymes, and this subject is reviewed in Chapter 3 (*Chapter 4 in the present book*).

In Chapter 4 (*here Chapter 5*), applied biocatalysis with free enzymes is described, together with examples of relevant enzyme processes. When single enzyme use is economically unfavorable, the enzymes can be either reused or used for continuous processes in membrane reactors (Chapter 4; *here Chapter 5*) or by immobilization (Chapters 5 and 6; *here Chapters 6 and 7*). The immobilization of isolated enzymes is described in detail in Chapter 5, while the immobilization of microorganisms and cells, with special reference to environmental technology, is detailed in Chapter 6.

In order to describe analytically the processes associated with immobilized biocatalysts that are required for rational process design, the coupling of reaction and diffusion in these systems must be considered. To characterize immobilized biocatalysts, methods which were developed previously for analogous biological and process engineering (heterogeneous catalysis) systems can be used (Chapter 7; *here Chapter 8*).

Details of reactors and process engineering techniques in enzyme technology are provided in Chapter 8 (*here Chapter 9*), while the analytical applications of free and immobilized enzymes is treated in Chapter 10 (*not covered in the present book*).

Within each chapter an introductory survey is provided, together with exercises and references to more general literature and original papers citing or relating the content of that chapter.

This textbook is designed to address both advanced and graduate students in biology, chemistry and biochemical, chemical and process engineering, as well as scientists in industry, research institutes and universities. It should provide a solid foundation that covers all relevant aspects of research and development in applied biocatalysis/enzyme technology. It should be remembered that these topics are not of equal importance in all cases, and therefore selective use of the book – depending on the individual reader's requirements – might be the best approach to its use.

In addition to a balanced methodological basis, we have also tried to present extensive data and examples of new processes, in order to stress the relevance of these in industrial practice.

From our point of view it is also important to stress the interactions, which exist beyond the scientific and engineering context within our society and environment. The importance and necessity of these interactions for a sustainable development has been realized during the past two decades, and this has resulted in new economic and political boundary conditions for scientific and engineering development. Problems such as allergic responses to enzymes in detergents and, more recently, to enzymes produced in recombinant organisms, have direct influences on enzyme technology/applied biocatalysis. Therefore, an integrated process design must also consider its environmental impact, from the supply and efficient use of the raw materials to the minimization and recycling of the byproducts and waste. Political boundary conditions derived from the concept of sustainability, when expressed in laws and other regulations, necessitates due consideration in research and development. The design of sustainable processes is therefore an important challenge for applied biocatalysis/enzyme technology. Ethical aspects must also be considered when gene technology is applied, and this is an increasing consideration in the production of technical and pharmaceutical enzymes. The many interactions between research and development and economic and political boundary conditions must be considered for all applications of natural and engineering sciences. Most importantly, this must be appreciated during the early phases of any development, with subsequent evaluation and selection of the best alternative production processes to meet a variety of human needs, as is illustrated in the following scheme:

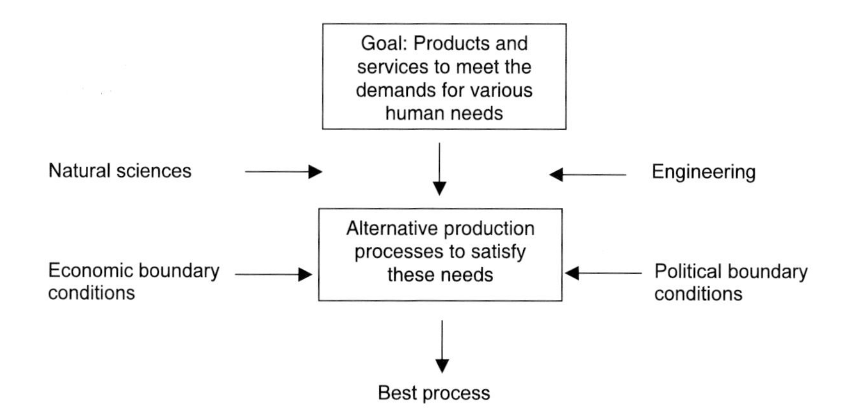

This book has been developed from our lecture notes and materials, and we also thank all those who provided valuable help and recommendations for the book's production. In particular, we thank Dipl.-Ing. Klaus Gollembiewsky, Dr. Lieker, Dr. Noll-Borchers, and Dipl. Chem. André Rieks.

Klaus Buchholz
Volker Kasche

To the First English Edition

The basic philosophy of the previous German edition is retained, but the contents have been revised and updated to account for the considerable development in enzyme technology/applied biocatalysis since the German edition was prepared some 10 years ago. Hence, a new chapter (Chapter 3) has been added to account for the increasing importance of enzymes as biocatalysts in organic chemistry. Recent progress in protein design (by rational means and directed evolution) has been considerably expanded in Chapter 2. The final chapter has been amended with more detailed case studies to illustrate the problems that must be solved in the design of enzyme processes. An appendix on information retrieval using library and internet resources has also been added, and we thank Thomas Hapke (Subject Librarian for Chemical Engineering at the Library of the Technical University Hamburg-Harburg) for help in the preparation of this material. The chapter on enzymes for analytical purposes has been removed in this English edition as it now is beyond the scope of this textbook.

We thank Prof. Dr. L. Jaenicke and Prof. Dr. J.K.P. Weder for their very constructive suggestions for corrections and improvements of the German edition.

The authors of this edition thank Prof. Dr. Andreas Bommarius, Dr. Aurelio Hidalgo, Dr. Janne Kerovuo, Dr. Tanja Kummer, Dr. Dieter Krämer, Dr. Brian Morgan, Sven Pedersen, Poul Poulsen, Prof. Dr. Peter Reilly, Dr. Klaus Sauber, Dr. Wilhelm Tischer, and Dr. David Weiner for valuable discussions, revisions and suggestions while preparing this book.

January 2005

Klaus Buchholz
Volker Kasche
Uwe T. Bornscheuer

Contents

Biocatalysts and Enzyme Technology. K. Buchholz, V. Kasche, U.T. Bornscheuer

ISBN: 3-527-30497-5

1 Introduction to Enzyme Technology

1.1 Introduction

Biotechnology offers an increasing potential for the production of goods to meet various human needs. In enzyme technology – a sub-field of biotechnology – new processes have been and are being developed to manufacture both bulk and high added-value products utilizing enzymes as biocatalysts, in order to meet needs such as food (e.g., bread, cheese, beer, vinegar), fine chemicals (e.g., amino acids, vitamins), and pharmaceuticals. Enzymes are also used to provide services, as in washing and environmental processes, or for analytical and diagnostic purposes. The driving force in the development of enzyme technology, both in academia and industry, has been and will continue to be:

- the development of new and better products, processes and services to meet these needs; and/or
- the improvement of processes to produce existing products from new raw materials as biomass.

The goal of these approaches is to design innovative products and processes that are not only competitive but also meet criteria of sustainability. The concept of sustainability was introduced by the World Commission on Environment and Development (WCED, 1987) with the aim to promote a necessary "... development that meets the needs of the present without compromising the ability of future generations to meet their own needs". To determine the sustainability of a process, criteria that evaluate its economic, environmental and social impact must be used (Gram et al., 2001; Raven, 2002; Clark and Dickson, 2003). A positive effect in all these three fields is required for a sustainable process. Criteria for the quantitative evaluation of the economic and environmental impact are in contrast with the criteria for the social impact, easy to formulate. In order to be economically and environmentally more sustainable than an existing processes, a new process must be designed to reduce not only the consumption of resources (e.g., raw materials, energy, air, water), waste production and environmental impact, but also to increase the recycling of waste per kilogram of product.

Biocatalysts and Enzyme Technology. K. Buchholz, V. Kasche, U.T. Bornscheuer

ISBN: 3-527-30497-5

1.1.1
What are *Biocatalysts*?

Biocatalysts are either proteins (*enzymes*) or, in a few cases, they may be nucleic acids (*ribozymes*; some RNA molecules can catalyze the hydrolysis of RNA. These ribozymes were detected in the 1980s and will not be dealt with here; Cech, 1993). Today, we know that enzymes are necessary in all living systems, to catalyze all chemical reactions required for their survival and reproduction – rapidly, selectively and efficiently. Isolated enzymes can also catalyze these reactions. In the case of enzymes however, the question whether they can also act as catalysts outside living systems had been a point of controversy among biochemists in the beginning of the twentieth century. It was shown at an early stage however that enzymes could indeed be used as catalysts outside living cells, and several processes in which they were applied as biocatalysts have been patented (see Section 1.3).

These excellent properties of enzymes are utilized in enzyme technology. For example, they can be used as biocatalysts to catalyze chemical reactions on an industrial scale in a sustainable manner. Their application covers the production of desired products for all human material needs (e.g., food, animal feed, pharmaceuticals, fine and bulk chemicals, fibers, hygiene, and environmental technology), as well as in a wide range of analytical purposes, especially in diagnostics. In fact, during the past 50 years the rapid increase in our knowledge of enzymes – as well as their biosynthesis and molecular biology – now allows their rational use as biocatalysts in many processes, and in addition their modification and optimization for new synthetic schemes and the solution of analytical problems.

This introductory chapter outlines the technical and economic potential of enzyme technology as part of biotechnology. Briefly, it describes the historical background of enzymes, as well as their advantages and disadvantages, and compares these to alternative production processes. In addition, the current and potential importance – and the problems to consider in the rational design of enzyme processes – are also outlined.

1.2
Goals and Potential of Biotechnological Production Processes

Biomass – that is, renewable raw materials – has been and will continue to be a sustainable resource which is required to meet a variety of human material needs. In developed countries such as Germany, biomass covers ≈30 % of the raw material need – equivalent to ~7000 kg per person per year. The distribution of biomass across different human demands is shown schematically in Figure 1.1. This distribution of the consumption is representative for a developed country in the regions that have a high energy consumption during the winter. However, the consumption of energy (expressed as tons of coal equivalent per capita in 1999) showed a wide range, from 11.4 in the United States, to 5.5 in Germany and the UK, 0.8 in China, and 0.43 in India (United Nations, 2002). This is mainly due to differences in energy use for housing,

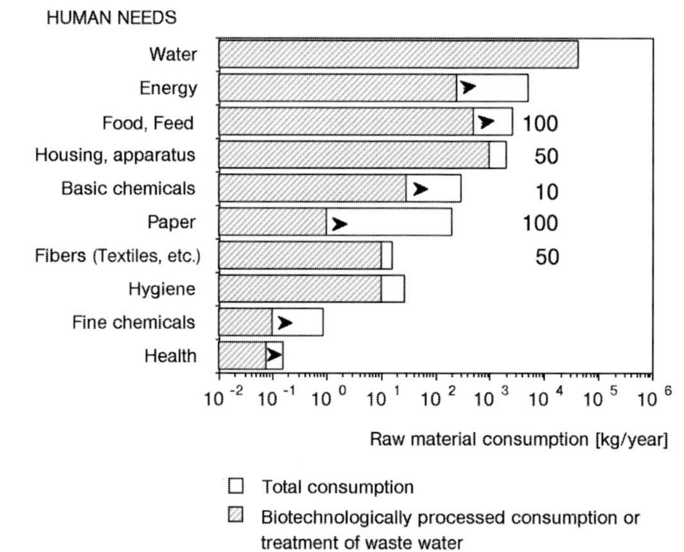

Fig. 1.1 Consumption of raw materials for various human needs per person and year in Germany 1992. The water consumption is only for household use. The arrowheads indicate the current increase in biotechnological processing of the products for different demands. For food and animal feed, only renewable raw materials (biomass) can be used, the figures to the right give the percentage biomass of the raw materials used for the production. For hygiene, fine chemicals and health products, 0–100 % of the raw materials can be biomass, depending on the product. After the use of the products the unavoidable waste must be recycled in a sustainable manner. Besides waste water, this results in about 1000 kg of solid waste (soil, building materials, plastics, sludge, etc.) that must be recycled in suitable environmental biotechnology processes. Energy is measured in coal equivalents.

transport and the production of other material needs. In less-developed countries, although the fraction of biomass as raw material to meet human demands is higher than in the developed countries, the total consumption is smaller.

Biomass – in contrast to non-renewable raw materials such as metals, coal, and oil – is renewable in a sustainable manner when the following criteria are fulfilled:

- the C-, N-, O-, and salt-cycles in the biosphere are conserved; and
- the conditions for a sustainable biomass production through photosynthesis and biological turnover of biomass in soil and aqueous systems are conserved.

Currently, these criteria are not fulfilled on a global level, one example being the imbalance between the CO_2 production to meet energy requirements and its consumption by photosynthesis in the presently decreasing areas of rain forests. This leads to global warming and other consequences that further violate these criteria. International treaties – for example, the Kyoto convention – have been introduced in an attempt to counteract these developments and to reach a goal that fulfills the above criteria.

When the above sustainability criteria are fulfilled, biomass can be used as raw material to meet the human demands illustrated in Figure 1.1. The needs for human food and animal feed must be met completely by biomass, though when these needs of highest priority are met, biomass can be used to fulfill the other demands shown in Figure 1.1. This applies especially to those areas with lower total raw material consumption than for food. From this point, it also follows that a large consumption of biomass to meet energy demands is only possible in countries with a low population density and a high biomass production.

By definition, biotechnological processes are especially suited to the production of compounds from biomass as the raw material (Fig. 1.2). The economic importance of such processes is detailed in Table 1.1. This also involves the development of suitable equipment to obtain more sustainable processes. From the information

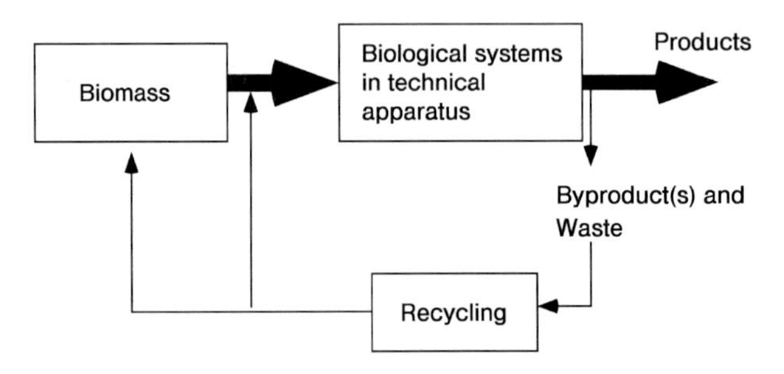

Fig. 1.2 Schematic view of an ideal sustainable biotechnological production process. Biomass as a regenerable resource is converted into desired products with minimal waste and byproduct production. The waste and byproducts must be completely recycled.

Table 1.1 Yearly production and value of biotechnologically produced products to meet human needs. All production data are from 1999–2003, values are only given where sources gives these or when they can be estimated, based on prices in the European Union (EU).

Human need	*Product*	*World production (tons year^{-1})*	*Value (10^9 Euro year^{-1})*	*Production method*	
				Biotechnology	*Chemical*
Food and feed	Beer/Wine	170 000 000	200	F,E	
	Cheese	13 000 000	100	F,E	
	Baker's yeast (1992)	1 800 000	?	F	
	Vegetable oils (refined)	25 000 000	?	E	+
	Acetic acid (10%)	1 200 000	?	F	
Fine chemicals	Amino acids	>1 000 000	>5	F,E	+
	Glucose-Fructose syrup	12 000 000	5	E	
	Vitamin C	>100 000	>2	F,E	+
	Aspartame (Dipeptide)	10 000	?	F,E	+
	Citric acid	800 000	1	F	
	Herbicides, insecticides	2 200 000	?	E	+
	Enzymes	>10 000	2	F	

Table 1.1 Continued

Human need	***Product***	***World production (tons year^{-1})***	***Value (10^9 Euro year^{-1})***	***Production method***	
				Biotechnology	***Chemical***
Basic chemicals	Products from starch (Ethanol, etc.)	35 000 000	10	E,F	
	Acrylamide	400 000	?	E*	+
	Acetic acid	3 400 000	?		+
Fibers for textiles	Cotton	17 000 000		E	+
	Wool	1 000 000		E	+
	Linen	70 000		F,E	+
Paper		300 000 000		E	+
Hygienics/ detergents	Biotensides	?	?	E,F	+
	Washing powder	22 000 000		F	+
Therapeuticals	Antibiotics	>60 000	≈60	F,E	+
	Insulin	≈10	3	F,E	
	Other peptide hormones	?	?	E,F	+
	Recombinant proteins (Factor VIII, Interferons, tPA, growth factors, monoclonal antibodies, etc.)	?	20	F,E	
Diagnostics	Monoclonal antibodies	?	>1	F,E	+
	DNA/Protein-Chips	?	?	F,E	+
	Enzyme-based	<1	>1	F,E	+
Environment	Clean water (only Germany)	13 000 000 000	13	F	+
	clean air/soil (soil remediation)	?	?	F	+
Comparison					
Chemical industry	All products	≈1 000 000 000	≈2 000		
	Chemical catalysts	180 000	8		

F = fermentation; E = enzyme technology; * 25 % by enzyme technology

Sources: UN (2003); Hassan and Richter (2002); ifok (2003); C & EN/Datamonitor (2001); Novozymes (2004).

provided in Figure 1.1 it also follows that biotechnology has a major potential in the development of sustainable processes to meet all human needs.

Enzyme technology is a part of biotechnology that the European Federation of Biotechnology 1989 defined as:

> *"Biotechnology is the integration of natural sciences and engineering sciences in order to achieve the application of organisms, cells, parts thereof and molecular analogues for products and services."*

However, the following amendment was added:

> *"It is a clear understanding that Biotechnology is directed to the benefit of mankind by obeying biological principles."*

Some of these principles have been outlined above. Here, we add the requirement that traditional classical – as well as new biotechnological processes – must be improved and/or developed in order to be sustainable (Fig. 1.2). The fundamentals needed for the development of such processes in the interdisciplinary field of biotechnology require the close cooperation of biologists, chemists, bioengineers, and chemical engineers.

1.3
Historical Highlights of Enzyme Technology/Applied Biocatalysis

1.3.1
Early Developments

Applied biocatalysis has its roots in the ancient manufacture and preservation of food and alcoholic drinks, as can be seen in old Egyptian pictures. Cheese making has always involved the use of enzymes, and as far back as about 400 BC, Homer's Iliad mentions the use of a kid's stomach for making cheese.

With the development of modern natural science during the 18th and 19th centuries, applied biocatalysis began to develop a more scientific basis. In 1833, Payen and Persoz investigated the action of extracts of germinating barley in the hydrolysis of starch to yield dextrin and sugar, and went on to formulate some basic principles of enzyme action (Payen and Persoz, 1833):

- small amounts of the preparation were able to liquify large amounts of starch,
- the material was thermolabile,
- the active substance could be precipitated from aqueous solution by alcohol, and thus be concentrated and purified. This active substance was called *diastase* (a mixture of amylases).

In 1835, the hydrolysis of starch by diastase was acknowledged as a catalytic reaction by Berzelius. In 1839, he also interpreted fermentation as being caused by a catalytic force, and postulated that a body – by its mere presence – could, by affinity to the fermentable substance, cause its rearrangement to the products (Hoffmann-Ostenhof, 1954).

The application of diastase was a major issue from the 1830s onwards, and the enzyme was used to produce dextrin that was used mainly in France in bakeries, and also in the production of beer and wines from fruits. The process was described in more detail, including its applications and economic calculations, by Payen (1874) (Fig. 1.3). Indeed, it was demonstrated that the use of malt in this hydrolytic process was more economic than that of sulfuric acid.

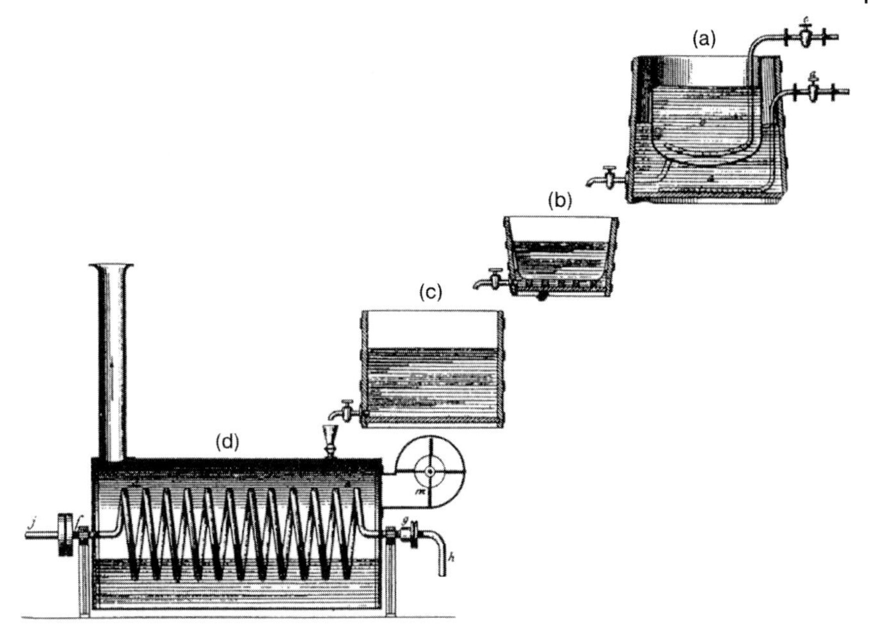

Fig. 1.3 Process for dextrin production, with reaction vessel (a), filter (b), reservoir (c), and concentration unit (d).

Lab preparations were also used to produce cheese (Knapp, 1847), and Berzelius later reported that 1 part of lab ferment preparation coagulated 1800 parts of milk, and that only 0.06 parts of the ferment was lost. This provided further evidence for Berzelius' hypothesis that ferments were indeed catalysts.

About two decades later, the distinction of organized and unorganized ferments was proposed (Wagner, 1857), and further developed by Payen (1874). These investigators noted that fermentation appeared to be a contact (catalytic) process of a degradation or addition process (with water), and could be carried out by two substances or bodies:

- A nitrogen-containing organic (unorganized) substance, such as protein material undergoing degradation.
- A living (organized) body, a lower-class plant or an "infusorium", an example being the production of alcohol by fermentation.

It is likely that the effect is the same, insofar as the ferment of the organized class produces a body of the unorganized class – and perhaps a large number of singular ferments. Consequently, in 1878, Kühne named the latter class of substances, *enzymes.*

Progress in the knowledge of soluble ferments (enzymes) remained slow until the 1890s, mainly due to a scientific discussion where leading scientists such as Pasteur denied the existence of "unorganized soluble ferments" that had no chemical identity. Consequently, the subject of enzymatic catalysis remained obscure, and was considered only to be associated with processes in living systems. In the theory of fermentation, a degree of mystery still played a role: Some *vital factor, "le principe vital"*, which differed from chemical forces, was considered to be an important principle in the chemical processes associated with the synthesis of materials isolated from living matter. But Liebing and his school took an opposite view, and considered fermentation simply to be a decay process.

In 1874, in Copenhagen, Denmark, Christian Hansen started the first company (Christian Hansen's Laboratory) for the marketing of standardized enzyme preparations, namely rennet for cheese making (Buchholz and Poulson, 2000).

1.3.2
Scientific Progress Since 1890: The Biochemical Paradigm; Growing Success in Application

From about 1894 onwards, Emil Fischer elaborated on the essential aspects of enzyme catalysis. The first aspect was *specificity*, and in a series of experiments Fischer investigated the action of different enzymes using several glycosides and oligosaccharides. For this investigation he compared invertin and emulsin. He extracted invertin from yeast – a normal procedure – and showed that it hydrolyzed the α-, but not the β-methyl-D-glucoside. In contrast, emulsin – a commercial preparation from Merck – hydrolyzed the β-, but not the α-methyl-D-glucoside. Fischer therefore deduced the famous picture of a "lock and key", which he considered a precondition for the potential of an enzyme to have a chemical effect on the substrate. In this way he assumed that the "geometrical form of the (enzyme) molecule concerning its asymmetry, corresponds to that of the natural hexoses" (sugars) (Fischer, 1909).

The second aspect referred to the protein nature of enzymes. In 1894, Fischer stated that amongst the agents which serve the living cell, the proteins are the most important. He was convinced that enzymes were proteins, but it took more than 20 years until the chemical nature of enzymes was acknowledged. Indeed, Willstätter, as late as 1927, still denied that enzymes were proteins (Fruton, 1979).

A few years after Fischer's initial investigations, Eduard Buchner published a series of papers (1897, 1898) which signalled a breakthrough in fermentation and enzymology. In his first paper on alcoholic fermentation without yeast cells, he stated, in a remarkably short and precise manner, that "... a separation of the (alcoholic) fermentation from the living yeast cells was not successful up to now". In subsequent reports he described a process which solved this problem (Buchner, 1897), and provided experimental details for the preparation of a cell-free pressed juice from yeast cells, that transformed sugar into alcohol and carbon dioxide. Buchner presented the proof that (alcoholic) fermentation did not require the presence of "... such a complex apparatus as is the yeast cell". The agent was in fact a soluble substance – without doubt a protein body – which he called *zymase* (Buchner, 1897). In referring to the deep controversy on his findings and theory, and in contradiction to the ideas of Pasteur (see above), Buchner insisted that his new experimental findings could not be disproved by older theories.

After a prolonged initial period of about a hundred years, during which time a number of alternative and mysterious theories were proposed, Buchner's elaborate results brought about a new biochemical paradigm. It stated – in strict contrast to the theories of Pasteur – that enzyme catalysis, including complex phenomena such as alcoholic fermentation, was a chemical process not necessarily linked to the presence and action of living cells, nor requiring a vital force – a *vis vitalis*. With this, the technical development of enzymatic processes was provided with a new, scientific basis on which to proceed in a rational manner.

The activity in scientific research on enzymes increased significantly due to this new guidance, and was reflected in a pronounced increase in the number of papers published on the subject of soluble ferments from the mid-1880s onwards (Buchholz and Poulson, 2000). Further important findings followed within a somewhat short time. In 1898, Croft-Hill performed the first enzymatic synthesis – of isomaltose – by allowing a yeast extract (α-glycosidase) to act on a 40 % glucose solution (Sumner and Somers, 1953). In 1900, Kastle and Loevenhart showed that the hydrolysis of fat and other esters by lipases was a reversible reaction, and that enzymatic synthesis could occur in a dilute mixture of alcohol and acid (Sumner and Myrbäck, 1950). This principle was subsequently utilized in the synthesis of numerous glycosides by Fischer and coworkers in 1902, and by Bourquelot and coworkers in 1913 (Wallenfels and Diekmann, 1966). In 1897, Bertrand observed that certain enzymes required dialysable substances to exert catalytic activity, and these he termed *coenzymes*.

The final proof that enzymes were in fact proteins was the crystallization of urease by Sumner in 1926, and of further enzymes (e.g., trypsin) by Northrup and Kunitz in 1930–1931. In all known cases, the pure enzyme crystals turned out to be proteins (Sumner and Myrbäck, 1950).

Despite these advances, the number of new applications of enzymes remained very small. In the USA, J. Takamine began isolating bacterial amylases in the 1890s, in what was later to become known as Miles Laboratories. In 1895, Boidin discovered a new process for the manufacture of alcohol, termed the "Amyloprocess". This comprised cooking of the cereals, inoculation with a mold which formed saccharifying enzymes, and subsequent fermentation with yeast (Uhlig, 1998). The early applications and patents on enzyme applications in the food industry (which numbered about 10 until 1911) have been reviewed by Neidleman (1991).

At the beginning of the 20th century, plant lipases were produced and utilized for the production of fatty acids from oils and fats, typically in the scale of 10 tons per week (Ullmann, 1914). Likewise, in the chill-proofing of beer, proteolytic enzymes have been used successfully since 1911 in the USA (Tauber, 1949). Lintner, as early as 1890, noted that wheat diastase interacts in dough making, and studied the effect extensively. As a result, the addition of malt extract came into practice, and in 1922 American bakers used 30 million pounds (13.5×10^6 kg) of malt extract valued at US$ 2.5 million (Tauber, 1949).

The use of isolated enzymes in the manufacture of leather played a major role in their industrial scale production. For the preparation of hides and skins for tanning, the early tanners kept the dehaired skins in a warm suspension of the dungs of dogs and birds. In 1898, Wood was the first to show that the bating action of the dung was caused by the enzymes (pepsin, trypsin, lipase) which it contained. In the context of Wood's investigations the first commercial bate, called Erodine, was prepared from cultures of *Bacillus erodiens*, based on a German patent granted to Popp and Becker in 1896. In order to produce Erodine, bacterial cultures were adsorbed onto wood meal and mixed with ammonium chloride (Tauber, 1949).

In 1907, Röhm patented the application of a mixture of pancreatic extract and ammonium salts as a bating agent (Tauber, 1949). Röhm's motivation as a chemist was to find an alternative to the unpleasant bating practices using dungs. Although the first tests with solutions of only ammonia failed, Röhm was aware of Buchner's studies on enzymes. He came to assume that enzymes might be the active principle in the dung, and so began to seek sources of enzymes that were technically feasible. His tests with pancreas extract were successful, and on this basis in 1907 he founded his company, which successfully entered the market and expanded rapidly. In 1908, the company sold 10 tons of a product with the tradename Oropon, followed by 53 tons and 150 tons in the subsequent years. In 1913, the company (Fig. 1.4) was employing 22 chemists, 30 other employees, and 48 workers (Trommsdorf, 1976). The US-based subsidiary – today's Röhm and Haas Company – was founded in 1911. The example of Rohm's company illustrates that although the market for this new product was an important factor, knowledge of the principles of enzyme action was equally important in providing an economically and technically feasible solution.

This success of the enzymatic bating process was followed by new applications of pancreas proteases, including substitution therapy in maldigestion, desizing of textile fibers, wound treatment, and the removal of protein clots in large-scale washing procedures. During the 1920s, however, when insulin was discovered in pancreas, the pancreatic tissue became used as a source of insulin in the treatment of diabetes. Consequently, other enzyme sources were needed in order to provide existing enzy-

Fig. 1.4 The factory of the Röhm & Haas Company, Darmstadt, Germany, 1911.

matic processes with the necessary biocatalysts, and the search successfully turned to microorganisms such as bacteria, fungi, and yeasts.

1.3.3
Developments Since 1950

Between 1950 and 1970, a combination of new scientific and technical knowledge, market demands for enzymes for use in washing processes, starch processing and cheaper raw materials for sweeteners and optically pure amino acids, stimulated the further development of enzyme technology. As a result, an increasing number of enzymes that could be used for enzyme processes were found, purified and characterized. Among these were penicillin amidase (or acylase), used for the hydrolysis of penicillin and first identified in 1950, followed some years later by glucose isomerase, which is used to isomerize glucose to the sweeter molecule, fructose. With the new techniques of enzyme immobilization, enzymes could be reused and their costs in enzyme processes reduced. Although sucrose obtained from sugar-cane or sugar beet was the main sweetener used, an alternative raw material was that of starch, which is produced in large quantities from corn (mainly in the USA). Starch can be hydrolyzed to glucose, but on a weight basis glucose is less sweet than sucrose or its hydrolysis products, glucose and fructose. As glucose isomerase can isomerize glucose to fructose, starch became an alternative sweetener source. The process was patented in 1960, but it lasted almost 15 years until the enzyme process to convert starch to glucose-fructose syrups became industrialized. This was in part due to an increase in sucrose prices and to the introduction of immobilized glucose isomerase as a biocatalyst. Although European engineers and scientists had contributed strongly to the development of this process, it is applied only minimally in Europe (~1 % of world production) due to the protection of sucrose production from sugar beet.

Enzyme immobilization was first introduced to enable the reuse of costly enzymes. Some of the initial attempts to do this were described during the early parts of the last century (Hedin, 1915), but the enzymes when adsorbed to charcoal proved to be very unstable. Around the time of 1950, several groups began to immobilize enzymes on other supports (Michel and Evers, 1947; Grubhofer and Schleith, 1954; Manecke 1955, cited in Silman and Katchalski, 1966). Georg Manecke was one of the first to succeed in making relatively stable immobilized systems of proteins on polymer carriers, and although he was granted a patent on his method he could not convince industry of the importance to further develop this invention. Rather, it was a group of chemists working with Ephraim Katchalski-Katzir in Israel who opened the eyes of industry to the world of immobilized enzymes (among Katzir's co-workers were Klaus Mosbach and Malcolm Lilly who later made important contributions to establish enzyme technology). The first industrial applications of immobilized enzymes, besides the isomerization of glucose to fructose to produce high-fructose corn syrups (HFCS), were in the production of optically pure amino acids (Tosa et al., 1969) and the hydrolysis of penicillin G (Carleysmith and Lilly, 1979, together with Beecham Pharmaceuticals, UK, and G. Schmidt-Kastner, Bayer, Germany).

Even today, the largest immobilized enzyme product in terms of volume is immobilized glucose isomerase. As these products were introduced they became more efficient, and stable biocatalysts were developed that were cheaper and easy to use. As a result, the productivity of commercial immobilized glucose isomerase increased from ~500 kg HFCS kg^{-1} immobilized enzyme product (in 1975) to ~15 000 kg kg^{-1} (in 1997) (Buchholz and Poulson 2000).

During the 1960s, enzyme production gained speed only in modest proportions, as reflected by the growing sales of bacterial amylases and proteases. Indeed, the annual turnover of the enzyme division of Novo Industri (now Novozymes), the leading enzyme manufacturer at the time, did not exceed $1 million until 1965. However, with the appearance of the detergent proteases, the use of enzymes increased dramatically, and during the late 1960s, everybody wanted Biotex, the protease-containing detergent. At the same time, an acid/enzyme process to produce dextrose using glucoamylase was used increasingly in starch processing. As a consequence, by 1969 – within only a four-year period – Novo's enzyme turnover exceeded US$ 50 million annually, and in 2003 Novozymes' turnover was approximately US$ 1000 million. The present global market is estimated to be around 2 billion € (Novozymes, 2004) (Fig. 1.5a), and this has been reflected in the increased employment within the enzyme producing industry (Fig. 1.5b).

The main industrial enzyme processes with free or immobilized enzymes as biocatalysts are listed in Table 1.2.

The introduction of gene technology during the 1970s provided a strong impetus for both improved and cheaper biocatalysts, and also widened the scope of application. Productivity by recombinant microorganisms was dramatically improved, as was enzyme stability, and this led to a considerable lowering of prices and improvements in the economics of enzyme applications. Today, most of the enzymes used as biocatalysts in enzyme processes – except for food processing – are recombinant.

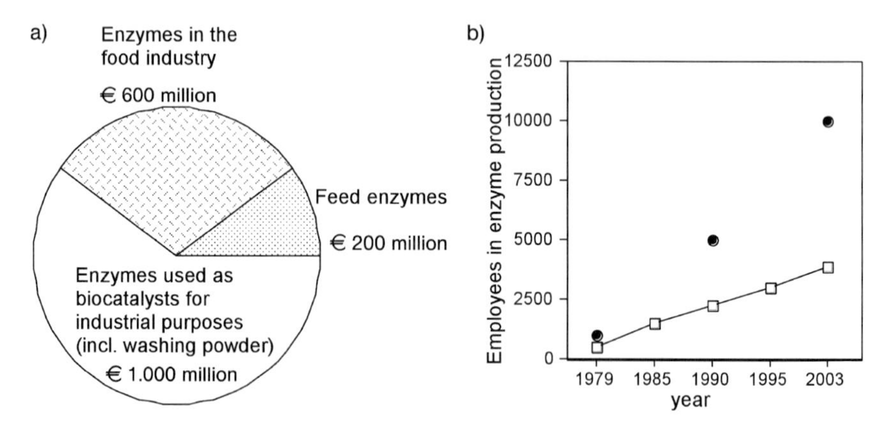

Fig. 1.5 Current market for enzymes for different purposes (a) and the increase in the application of enzymes reflected in the number of employees in the enzyme-producing industry (b) (for Novozymes (□) from yearly reports; worldwide (●) estimated).

Table 1.2 Products produced in quantities larger than 1000 t year^{-1} by different companies with enzymes as biocatalysts. Some other products produced in the range 10 to 1000 t year^{-1} in recently developed enzyme processes are also included.

Product	***Enzyme***	***Free or immobilized enzyme***	***Companies***
> 10 000 000 t a^{-1}			
HFCS	amylase	free	several
	glucoamylase	free	
	glucose isomerase	immobilized	
Ethanol	amylase,	free	several
(gasoline additive)	glucoamylase	free	
> 10 000 t a^{-1}			
Acrylamide	nitrilase	immobilized cells	Nitto, DSM
6-Aminopenicillanic acid (6-APA)	penicillin amidase	immobilized	several
Cacao butter	lipase	immobilized	Fuji Oil, Unilever
Isomaltulose	sucrose mutase	in immobilized cells	Südzucker
Lactose-free milk or whey	β-galactosidase	free or immobilized	several
> 1000 t a^{-1}			
7-Aminocephalo-	(*R*)-amino oxidase	immobilized	several
sporanic acid (7-ACA)	glutaryl amidase	immobilized	
7-Aminodesacetoxy-	glutaryl amidase	immobilized	DSM
cephalosporanic acid (7-ADCA)	(modified (?))		
(*S*)-Aspartic acid	aspartase	immobilized (?)	Tanabe
Aspartame	thermolysisn	immobilized	Toso, DSM
(*S*)-Methoxyisopropyl amine	lipase	immobilized	BASF
(*R*)-Pantothenic acid	aldolactonase		Fuji chem. Ind.
(*R*)-Phenylglycine	hydantoinase,	immobilized	several
	carbamoylase		
(*S*)-Amino acids	aminoacylase	free	Degussa, Tanabe
1000 > 10 t a^{-1}			
Amoxicillin	penicillin amidase	immobilized	DSM
Cephalexin	penicillin amidase	immobilized	DSM
(*S*)-DOPA	β-tyrosinase	immobilized	Ajinomoto
Human insulin	carboxypeptidase A	free	Aventis
	lysyl endopeptidase	free	several
	trypsin	free	BASF
Sterically pure alcohols			
and amines	lipase	immobilized	BASF
(*R*)-Mandelic acid	nitrilase	immobilized	BASF

The recent development of techniques of site-directed mutagenesis, gene shuffling and directed evolution have opened the perspective of modifying the selectivity and specificity of enzymes (see Chapter 2, Section 2.11, and Chapter 3).

More detailed accounts on the scientific and technological development can be found in articles by Sumner and Myrbäck (1950), Sumner and Somers (1953), Ullmann (1914), Tauber (1949), Neidleman (1991), by Turner (Roberts et al., 1995), and Buchholz and Poulsen (2000). A profound analysis of the background of Biotechnology and "Zymotechnica" has been presented by Bud (1992, 1993).

1.4 Biotechnological Processes: The Use of Isolated or Intracellular Enzymes as Biocatalysts

Biotechnological processes use one or more enzymes with or without cofactors or cosubstrates as biocatalysts (Fig. 1.6). When more enzymes and cosubstrate regeneration are required, fermentation processes with living cells are more effective than processes with isolated enzymes. These processes will not be dealt with here, except for their application in environmental biotechnology (see Chapter 7). For enzyme processes which utilize few enzymes (≤3) without any cosubstrate (ATP, NADH) regeneration, those with isolated enzymes or enzymes in either dead or living cells have the following advantages compared with fermentations:

- Higher space-time yields can be obtained than with living cells; smaller reactors can then be used, reducing processing costs.
- The risk that a desired product is converted by other enzymes in the cells can be reduced.
- The increased stabilty and reuse of immobilized biocatalysts allows continuous processing for up to several months.

For such enzyme processes:

1. The required intra- or extracellular enzyme must be produced in sufficient quantities and purity (free from other disturbing enzymes and other compounds).
2. Cells without intracellular enzymes that may disturb the enzyme process must be selected.
3. The enzyme costs must be less than 5–10 % of the total product value.

Tables 1.1 and 1.2 provide information about important enzyme technological products and the processes in which they are produced. In many cases, enzyme and chemical processes are combined to obtain these products. This will also be the case in the future when both processes fulfill the economic and sustainability criteria listed above. When this is not the case, new processes must be developed that fulfill these criteria better. This is illustrated by the first two large-scale enzyme processes, namely the hydrolysis of penicillin (Fig. 1.7) and the production of glucose-fructose syrup from starch (Fig. 1.8). Both processes were developed some 30 years ago, and both replaced purely chemical processes that were economically unfavorable, and not sustainable.

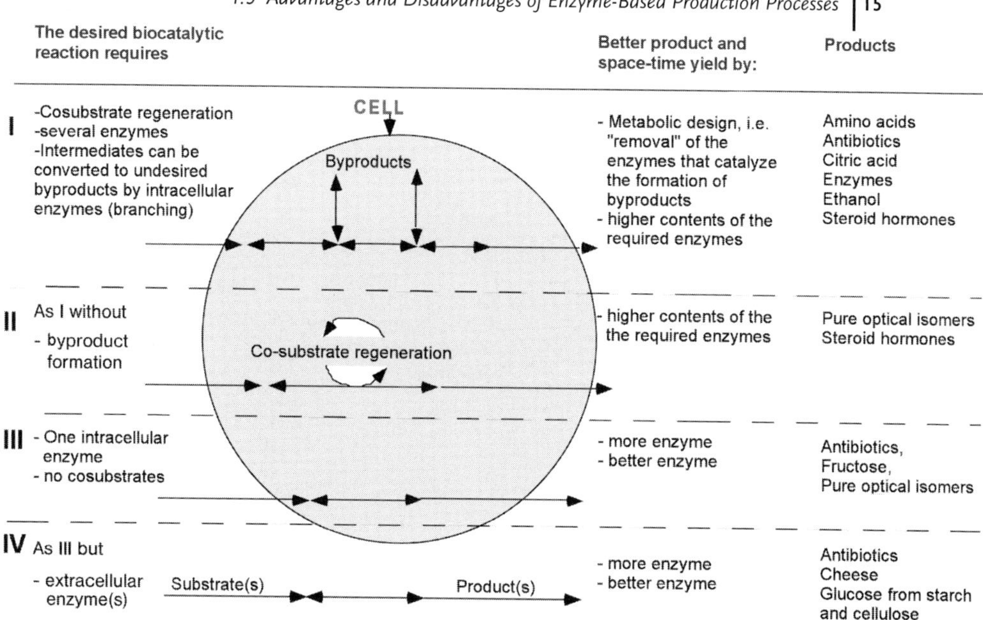

Fig. 1.6 Classification of biocatalytic processes with enzymes as biocatalysts. I and II must be performed with enzymes in living cells; III can be performed with enzymes in dead cells or as IV with isolated enzymes. Processes II–IV will be covered in this book. Process I will only be treated in connection with waste water and exhaust air treatment with immobilized cells (see Chapter 7).

1.5
Advantages and Disadvantages of Enzyme-Based Production Processes

In Figures 1.7 and 1.8, the enzyme processes for the hydrolysis of penicillin and the production of glucose-fructose syrup are compared with previously used procedures that had the same aims. In the case of penicillin hydrolysis, the chemical process uses environmentally problematic solvents and toxic compounds, leading to toxic wastes that are difficult to recycle. The process is, therefore, not sustainable. The enzyme process is more sustainable than the previous process, and leads to a considerable reduction in waste which in turn reduces the processing costs. In this process the product yield could also be increased to >95 %. The hydrolysis of starch and isomerization of glucose cannot be performed chemically at reasonable cost, as each would result in lower yields, unwanted byproducts, and the considerable production of waste acids. These processes illustrate some of the advantages of enzyme processes compared with alternative processes (Box 1.1). However, it must be remembered that the use of enzymes as biocatalysts may be limited by their biological and chemical properties (see Chapter 2).

Benzylpenicillin, Penicillin G

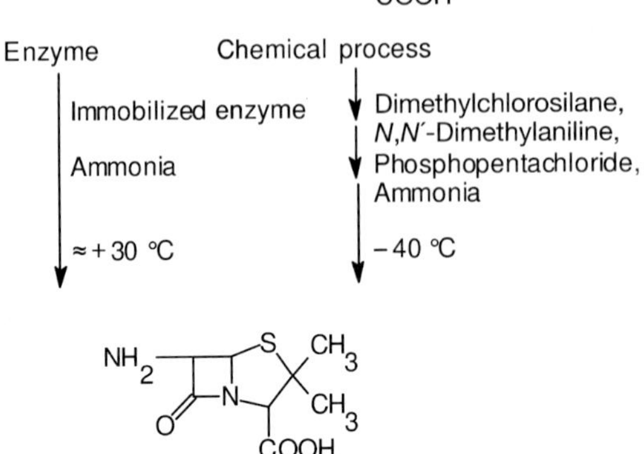

6-Aminopenicillanic acid (6-APA) + Phenylacetic acid

Required for the production of 500 t 6-APA

For hydrolysis:

Enzyme process		Chemical process	
1000 t	penicillin G	1000 t	penicillin G
45 t	ammonia	300 t	dimethylchlorosilane
≈ 1 t	immobilized enzyme	800 t	*N,N*-Dimethylaniline
10000 m^3	water	600 t	phosphopentachloride
		160 t	ammonia
		4200 m^3	dichloromethane
		4200 m^3	*n*-butanol

For downstream processing:

Enzyme process	Chemical process
acetone	hydrochloric acid
ammonium bicarbonate	butylacetate
	acetone

Fig. 1.7 Comparison of the old (chemical) and the new (enzyme) process for the hydrolysis of penicillin G. The product, 6-aminopenicillinanic acid (6-APA), is used for the synthesis of semisynthetic penicillins with side chains other than phenylacetic acid. In the enzyme process, the byproduct phenylacetic acid can be recycled in the production of penicillin by fermentation (from Tischer, 1990).

Enzymes are proteins that are essential for living systems and, in the right place, they catalyze all chemical conversions required for the system's survival and reproduction. However, in the wrong place they can be harmful to an organism. Peptidases from the pancreas are normally transported into the intestine where they are necessary for the digestion of proteins to amino acids. The amino acids are transported into blood vessels and distributed to different cells, where they are used for the synthesis of new proteins. Under shock situations or pancreas insufficiencies, these peptidases may be transported from the pancreas directly into the bloodstream

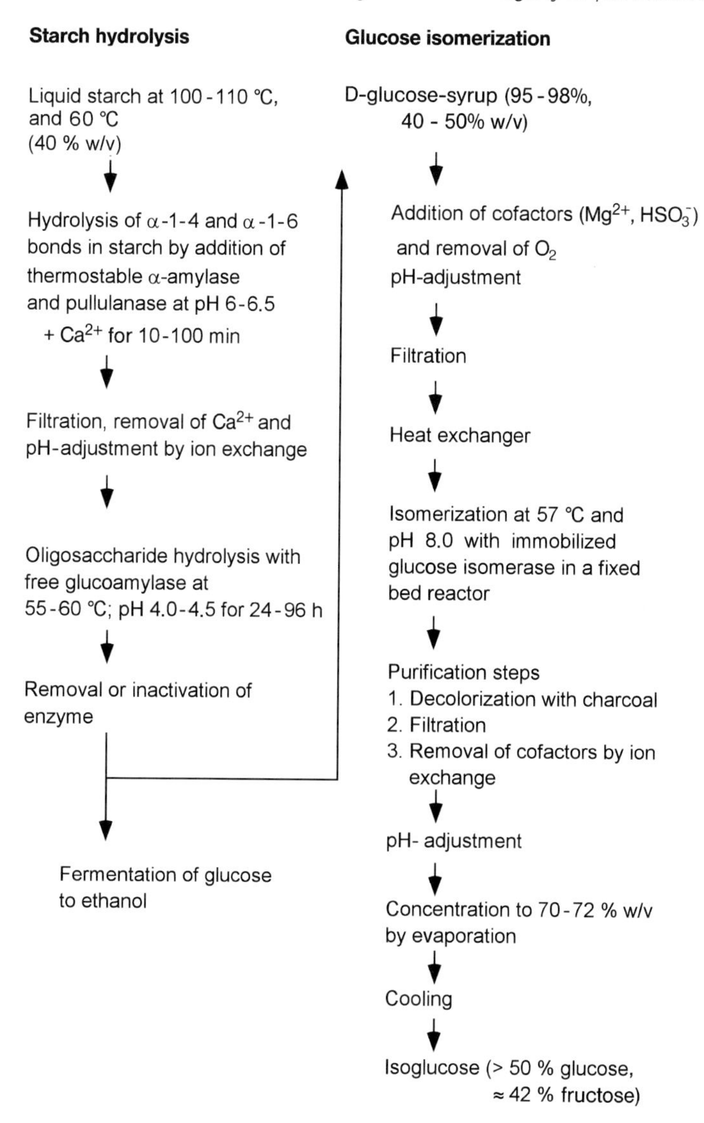

Fig. 1.8 The enzyme process for the hydrolysis of starch to glucose, and the isomerization of glucose to fructose. (w/v = weight per volume) (see Chapters 5, 6, and 9).

where they may cause harmful blood clotting. To prevent this from occurring, the blood contains inhibitors for pancreatic peptidases.

Enzymes are normal constituents of food and, as with all orally ingested proteins, they are hydrolyzed in the stomach and intestine. However, if enzymes or other proteins are inhaled as small particles or aerosols in the lungs, they can be transferred directly into the bloodstream. There, they are recognized as foreign proteins and induce an immune reaction – that is, the production of antibodies against them. This may also lead to enzyme or protein allergies. These risks must be considered in the production and use of enzymes and other proteins, and simple measures can be taken to minimize them. For example, enzymes used in washing powders are dried as large particles covered with a wax layer; consequently, the size of the particles is so large (>100 μm) that they cannot be inhaled into the lungs. Likewise, when enzymes are used in the liquid phase, aerosol formation must be prevented.

Some proteins (enzymes) can also be transferred from the digestive tract into the bloodstream and cause allergies. This applies to proteins that are digested very slowly in the stomach and intestine (Fuchs and Astwood, 1996; Jank and Haslberger, 2003). The slow digestion has been correlated with a high thermal stability, and enzymes used on an industrial scale as biocatalysts should therefore be rapidly hydrolyzed by peptidases in the digestive system in order to minimize the allergy risk. This applies especially to enzymes that cannot easily be used in closed systems, and particularly those used in food processing.

1.6 Goals and Essential System Properties for New or Improved Enzyme Processes

1.6.1 Goals

The advantages detailed in Box 1.1 are not sufficient alone for the industrial use of enzyme processes. Sustainability goals, derived from the criteria outlined in Section 1.1, must also be considered (Table 1.3).

Enzyme processes have become competitive and have been introduced into industry when they attain these goals better than alternative processes. This, however, also requires that these goals are quantified such that the amount of product and by-products (or waste) produced with a given amount of enzyme in a given time must be determined. For this aim, enzyme processes – as with all catalyzed chemical processes – can be divided into two categories (Fig. 1.9):

1. *Equilibrium-controlled processes*: the desired product concentration or property has a maximum at the end-point of the process (B in Fig. 1.9); the chemical equilibrium is independent of the properties of the catalyst (enzyme), but is dependent upon on pH and temperature.

2. *Kinetically controlled processes*: the desired product concentration or property (such as fiber length or smoothness in textiles or paper) reaches a maximum (A in Fig. 1.9), the concentration or properties of which depend on the properties of the catalyst (enzyme, see Chapter 2), pH, and temperature. The process must be stopped when the maximum is reached.

In both cases the time to reach the maximum product concentration or property depends on the properties and amount of enzyme used, and of the catalyzed process

Box 1.1 Advantages and disadvantages of cells and enzymes as biocatalysts in comparison with chemical catalysts.	
Advantages:	• Stereo- and regioselective • Low temperatures (0–110 °C) required • Low energy consumption • Active at pH 2–12 • Less byproducts • Non-toxic when correctly used • Can be reused (immobilized) • Can be degraded biologically • Can be produced in unlimited quantities
Disadvantages:	• Cells and enzymes are – unstable at high temperatures – unstable at extreme pH-values – unstable in aggressive solvents – inhibited by some metal ions – hydrolyzed by peptidases • Some enzymes – are still very expensive – require expensive cosubstrates • When inhaled or ingested enzymes are, as all foreign proteins, potential allergens

(endo- or exothermal, pH- and temperature dependence of equilibrium constants, solubility and stability of substrates, products, etc.). This must be considered in the rational design of enzyme processes. Another difference to consider is that in these processes the enzymes are used at substrate concentrations (up to 1 M) that are much higher than those in living systems (≤0.01 M). At enzyme concentrations used in enzyme technology, the formation of undesired byproducts in uncatalyzed bimolecular reactions cannot be neglected.

1.6.2
Essential System Properties for Rational Design of an Enzyme Process

The steps to be considered for the design of enzyme processes that are within the scope of this book may be illustrated based on the equilibrium-controlled hydrolysis of the substrate lactose to the products glucose and galactose, as shown in Figure 1.9. The stages are summarized in Figure 1.10.

A high substrate content is favorable in order to reduce downstream processing costs. In milk, the lactose content cannot be changed, but in whey it can be increased by nanofiltration. The upper limit is given by the solubility (150–200 g L^{-1}), which is lower than for other disaccharides such as sucrose. As both substrates and products have no basic or acidic functional groups, the equilibrium constant should not depend on pH, but on the temperature. This dependence must be known in order to select a suitable process temperature (T), though the selection also depends on the

Table 1.3 Economic and environmental sustainability goals that can be realized in enzyme processes (modified from Uhlig, 1998).

Goals	*Means to achieve the goals*	*Products/Processes*
Cost reduction	Yield increase	Penicillin-Cephalosporin C hydrolysis
	Biocatalyst reuse and increased productivity by immobilization	Glucose isomerization
	Better utilization of the raw material	Isomaltulose production Juice and wine production
	Reduction of process costs for • filtration • energy • desizing of fibers • cheese ripening • malting in beer production	 Sterile filtration of plant extracts; Low temperature washing powder Desizing with enzymes Increase rate of process with enzymes
	Reduction of residence time in starch processing	
Improvement of biological properties and quality	Produce only isomers with the desired biological property	Racemate resolution
	Improved preservation of foods	Juice concentrates
	Improvement of technical properties	Protein modification, flour for baking, transesterification of vegetable oils, biodiesel
	Improved taste (sweeteness)	Glucose isomerization to glucose-fructose syrup
Utilization of new regenerable sources of raw materials	Utilization of wastes from food and wood industry (whey, filter cakes with starch and protein from vegetable oil production, cellulose)	Drinks from whey Ethanol, biodiesel Animal feed
Reduction of environmental impact	Reduction of non-recyclable waste	Penicillin-, Cephalosporin C hydrolysis, leather production, paper bleaching
	Waste recycling	Utilization of whey

properties of the biocatalyst. Its selectivity (ratio of hydrolysis to synthesis rates) must be high in order to minimize the formation of byproducts (oligosaccharides). In addition, its catalytic properties and stability as a function of pH and temperature must also be known in order to calculate the amount of biocatalyst required to reach the endpoint of the process within a given time. When other constraints have been identi-

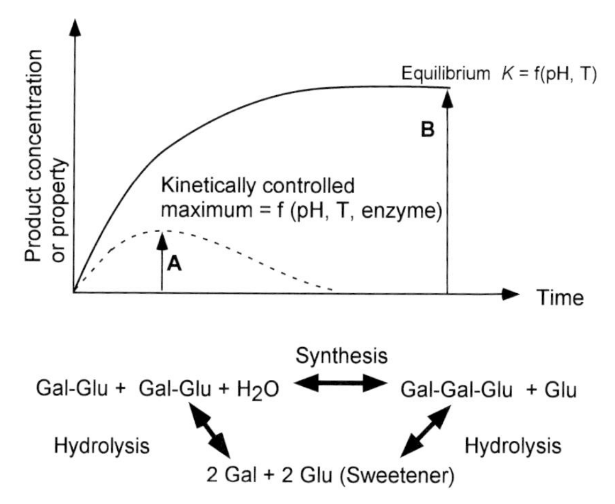

Fig. 1.9 Time-dependence (progress curves) of equilibrium- (solid line) and kinetically (broken line) controlled processes catalyzed by enzymes. The suitable end-points of these processes are those where the maximum product concentration or property is achieved – that is, A for the kinetically and B for the equilibrium-controlled process. Such processes are illustrated with the hydrolysis of lactose in milk or whey. [Whey is an inevitable byproduct in cheese production, where mainly protein and fats are precipitated in the milk by addition of a 'coagulating' enzyme (chymosin or rennin, a carboxyl acid peptidase, EC 3.4.23.4). The remaining liquid phase (whey) contains ~5 % sugars (mainly lactose, a disaccharide galactosyl-glucose), 1 % protein, 1 % amino acids, and 1 % ions (Ca^{2+}, Na^{+}, phosphate ions, etc.). Previously, with mainly small dairies, whey was used as a feed, or condensed to various sweet local products. Today, in large cheese-producing dairies, up to 10^7 tons of whey are formed each year that cannot be used as before. The main content lactose cannot be used as a sweetener as a part of the population cannot tolerate lactose (this fraction is higher in parts of Asia and Africa). When it is hydrolyzed, its sweetness is increased. The enzyme β-glucosidase that catalyzes the hydrolysis of lactose also catalyzes the kinetically controlled synthesis of tri- and tetra-saccharides. When lactose is consumed, these oligosaccharides are hydrolyzed. This also illustrates the formation of undesired byproducts in enzyme-catalyzed processes. The oligosaccharides are byproducts in the equilibrium-controlled hydrolysis of lactose. On the other hand, in the kinetically controlled synthesis of the oligosaccharide that can be used as prebiotics, byproduct formation is due to the hydrolysis to monosaccharides (Illanes et al., 1999; Bruins et al., 2003). The formation of the byproducts must be minimized by selecting suitable process conditions and biocatalysts.

fied, a process window in a pH-*T*-plane can be found where it can be carried out with optimal yield and minimal biocatalyst costs. The maximal yield as a function of pH and *T* is only defined by the catalyzed reaction. When this maximum is outside the process window, the enzyme process can only be improved by screening for a better biocatalyst or changing its properties by recombinant methods, so that the process can be carried out at pH- and *T*-values where this maximum can be reached.

In order to reduce the enzyme costs, the enzyme production can be improved (see Chapter 4) or the enzymes used in a re-usable form. This can be achieved by their immobilization to porous particles that can easily be filtered off at the end of the process (see Chapters 6 and 7). In these systems, the kinetics differ from those of

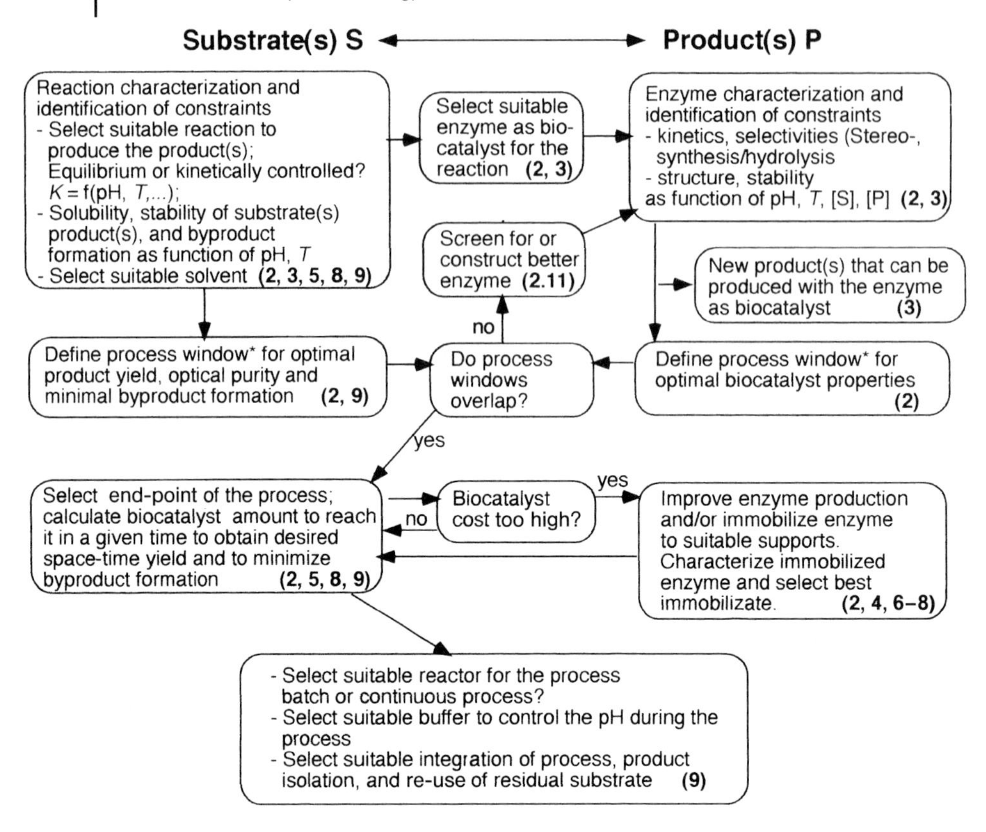

Fig. 1.10 Steps to consider in the design of an enzyme process to produce existing or new products (bold numbers refer to chapters in this book). *Process window = the range in a pH-T- (or pH-[S]-, pH-[P]-, T-[S]-, T-[P]-) plane where the reaction can be carried out with a given yield or optical purity, and where the properties (activity, selectivity, stability) of the biocatalyst are optimal.

systems with free enzymes, as the mass transfer inside and to and from the particles with the biocatalyst causes the formation of concentration and pH-gradients that influence rates and yields (this topic is dealt with in Chapter 8).

Once the process conditions and its end-point have been chosen in the process window, the enzyme costs per kg product are influenced by the type of reactor (batch, or continuous stirred tank or fixed-bed reactor) selected to carry out the process (see Chapter 9). There, the procedure to design an enzyme process (summarized in Fig. 1.10), will be illustrated in more detail as case studies for the design of a classical (HFCS production) and newer (7-ACA production) enzyme process.

1.6.3
Current Use and Potential of Enzyme Technology

Current amounts of products obtained with industrial biotechnical and enzyme processes are detailed in Tables 1.1 and 1.2, respectively. The value of the enzymes used as biocatalysts for different applications are shown in Figure 1.5. Besides industrial uses, many enzymes are used for analytical purposes, mainly in diagnostics, though on a weight basis less than 1 ‰ of all produced enzymes are used for these applications. Some enzymes are produced in increasing amounts for therapeutic purposes; this applies especially to recombinant enzymes such as factor VIII, tPA and urokinase that cannot be produced in sufficient amounts from natural sources (blood serum or urine). Another advantage of the recombinant production of these enzymes is that possible contamination with pathogenic human viruses (HIV, herpes) can be avoided.

The large number of new enzyme processes (>100) introduced during the past 30 years has recently been reviewed in detail (Liese et al., 2000). The type of process used, the compounds produced, and the enzymes used for these processes have been analyzed statistically (Straathof et al., 2002). These data show that hydrolases, lyases and oxidoreductases are used in two-thirds of all processes, while only about 1 % of the about 3000 known enzymes are used in larger amounts for enzyme technological and therapeutic purposes. During the past 10 years, the three-dimensional structures and detailed mechanisms of the reactions that they catalyze have been determined for many of the enzymes seen to be important in enzyme technology. This information allows a more rational improvement of their properties that is essential for their application. Based on the above discussion and on the information shown in Figure 1.5B, the number of new enzyme processes is expected to increase further during the next few decades. The rational and sustainable design of these processes – and the improvement of existing processes – requires the interdisciplinary cooperation of (bio)chemists, micro- and molecular biologists and (bio)chemical engineers. The (bio)chemist must determine the mechanism and properties of the catalyzed process, the kinetics of the enzyme-catalyzed process and other relevant properties of substrate, product and free and immobilized enzyme (stability, solubility, pH- and temperature dependence of equilibrium constants, selectivities), and select the suitable support for the immobilization together with the engineer. This also provides information about the properties of the enzyme that should be improved (specificity, selectivity, pH-optimum, stability, metal ion requirement, yield in the fermentation), and this is a task for the micro- or molecular biologist. The methods by which this problem may be solved is either to screen for better enzymes in nature, or to promote molecular *in vitro* evolution (see Chapter 2, Section 2.11). Finally, the engineer must use this knowledge to scale up the process to the production scale. In improving the latter procedure, however, the engineer will also identify problems that must be solved by the (bio)chemists and micro- and molecular biologists, and this is illustrated in Figure 1.8. The number of processing steps can be reduced when the enzymes used function at the same pH-value and have the same requirements for metal ions, but this problem has not yet been sufficiently solved.

Fields where large amounts of enzymes will be required in order to realize more sustainable new enzyme processes to meet human needs include:

- The production of optically pure therapeutics and fine chemicals. It is expected that, in future, only the isomer with desirable biological activity will be approved for use by regulatory authorities. Many pharmaceuticals and fine chemicals are still provided only as racemates, the resolution of which for any process has a maximal yield of 50 %. For a sustainable process, the other 50 % must be racemized, and to solve this problem the rational integration of chemical and enzyme processes is required in the development of dynamic kinetic resolution processes or asymmetric synthesis (Collins et al., 1992; Breuer et al., 2004).
- The synthesis of antibiotics (Bruggink, 2002).
- Paper production or recycling to reduce waste and energy consumption (Bajpai, 1998).
- The regio- and stereoselective synthesis of oligosaccharides for food and pharmaceutical purposes.
- The selective glycosylation of peptides, proteins and other drugs.
- Environmental biotechnology.

1.7
Exercises

1. How was it shown that enzymes can act as catalysts outside living cells? In which enzyme process was this knowledge first applied?
2. How can the process in Figure 1.8 be improved by a reduction in the number of processing steps? What must be done to achieve this?
3. Explain the relevance of Figure 1.9 for enzyme technology. Which system properties must be known in addition to the properties of the biocatalyst to improve the yields of these processes?
4. Which properties of the enzyme and the catalyzed process must be known to minimize byproduct formation in the production of oligosaccharides from lactose, as shown in Figure 1.9? (Hint: Use Figure 1.10 to answer this question).
5. Test whether Figure 1.1 is in agreement with your consumption pattern.
6. How can the allergic and toxic risks due to enzymes be avoided in enzyme technology?

Literature

Overview on Enzyme Technology

The following books give an overview on enzyme technology from the point of view of the biotechnological and chemical industry (enzyme producers and users). Besides the established and new applications of free and immobilized enzymes, some also cover health, legal and economic aspects of enzyme technology:

Atkinson, B. Mavituna, F., *Biochemical Engineering and Biotechnology Handbook*, 2nd Ed., Stockton Press, New York, 1991

Bommarius, A.S., Riebel, B., *Biocatalysis*, Wiley-VCh, Weinheim, 2004

Collins, A.N., Sheldrake, G.N., Crosby, J., (Eds.), *Chirality in Industry*, J. Wiley & Sons, Chichester, 1992

Godfrey, T., Reichelt, J., (Eds.), *Industrial Enzymology*, 2nd Ed., pp. 435–482, Stockton Press, New York, 1996

Kirst, H.A., Yeh, W.-K., Zmijewski, Jr., M.J., (Eds.), *Enzyme Technologies for Pharmaceutical and Biotechnological Applications.*, Marcel Dekker, New York, 2001

Liese, A., Seelbach, K., Wandrey, C., *Industrial Biotransformations*, Wiley-VCH, Weinheim, 2000

Tanaka, A., Tosa, T., Kobayashi, T., (Eds.), *Industrial Application of Immobilized Biocatalysts*, Marcel Dekker, New York, 1993

Straathof, A., Adlercreutz, P., (Eds.), *Applied Biocatalysis*, Harwood Academic Publishers, Amsterdam, 2000

Uhlig, H., *Industrial enzymes and their applications*, John Wiley & Sons, Inc., New York, 1998

Whitaker, J.R., Voragen, A.G.J., Wong, D.W.S. (Eds.), *Handbook of food enzymology*, Marcel Dekker, New York, 2003

Historical Development

These articles and books cover the historical development of biotechnology and enzyme technology:

Bud, R., The zymotechnic roots of biotechnology, *Br. J. Hist. Sci.*, **1992**, *25*, 127–144

Bud, R., *The Uses of Life, A History of Biotechnology*, Cambridge University Press, 1993

McLaren, A.D., Packer, L., Some aspects of enzyme reactions in heterogeneous systems, *Adv. Enzymol.* **1970**, *33*, 245–303

Mosbach, K., (Ed.), *Immobilized enzymes*, Methods Enzymol, Vol. 44, Academic Press, New York 1976

Mosbach, K., (Ed.), *Immobilized enzymes and cells*, Methods Enzymol. Vol. 135–137, Acad.Press, New York, 1987

Silman, I.M., Katchalski, E., Water insoluble derivatives of enzymes, antigens and antibodies, *Annu.Rev. Biochem.*, **1966**, *35*, 873–908

Sumner, J.B., Myrbäck, K., In: *The Enzymes*, 1, Part 1, 1–27, 1950

References

Bajpai, P., Applications of enzymes in the pulp and paper industry, *Biotechnol. Prog.*, **1999**, *15*, 147–157

Bruggink, A., (Ed.), *Synthesis of β-lactam antibiotics*, Kluwer Acad. Publ., Dordrecht, 2001

Bruins, M.E., Strubel, M., van Lieshout, J.F.T., Janssen, A.E.M., Boom, R.M., Oligosaccharide snthesis by the hyperthermostable β-glucosidase from *Pyrococcus furiosus*: kinetics and modelling, *Enzyme. Microb. Technol.*, **2003**, *33*, 3–11

Buchholz, K., Poulson, P.B., *Overview of History of Applied Biocatalysis*, in: Applied Biocatalysis, 1–15, A.J.J. Straathof, P. Adlercreutz, (Eds.), Harwood Academic Publishers, Amsterdam, 2000

Buchner, E., Alkoholische Gährung ohne Hefezellen. *Ber. D. Chem. Ges.*, **1897**, *30*, 117–124

Buchner, E., Über zellfreie Gährung, *Ber. D. Chem. Ges.*, **1898**, *31*, 568–574

Carleysmith, S. W., Lilly, M.D., Deacylation of benzylpenicillin by immobilised Penicillin acylase in a continuous four-stage stirred-tank reactor, *Biotechnol. Bioeng.*, **1979**, *21*, 1057–1073

Cech, T.R., Catalytic RNA: Structure and mechanism, *Biochem. Soc. Trans.*, **1993**, *21*, 229–234

Clark, W.C., Dickson, N.M., Sustainable science: The emerging research programm, *Proc. Natl. Acad. Sci. USA*, **2003**, *100*, 8059–8061

C & EN, *Chemical engineering news*, Datamonitor, 2001

Fischer, E., Untersuchungen über Kohlenhydrate und Fermente. Springer, Berlin, 1999

Fruton, J. S., in: *The Origins of Modern Biochemistry*, Srinivasan, P.R., Fruton, J.S. and Edsall, J.T. (Eds.), pp. 1–18, New York Academy of Sciences, New York, 1979

Fuchs, R.L., Astwood, J., Allergenicity Assessment of foods derived from genetically modified plants, *Food Technol.*, **1996**, 83–88

Gram, A., Treffenfeldt, W., Lange, U., McIntyre, T., Wolf, O., *The application of Biotechnology to Industrial Sustainability*, OECD Publications Service, Paris, 2001

Hassan, A., Richter, S., Closed loop management of spent catalysts in the chemical industry. *Chem. Eng. Technol.*, **2002**, *25*, 1141–1148

Hedin, S. G., *Grundzüge der physikalischen Chemie in ihrer Beziehung zur Biologie*, Kap. 4, J. F. Bergmann Verlag, Wiesbaden, 1915

Hoffmann-Ostenhof, O., *Enzymologie*, Springer, Wien, 1954

Illanes, A., Wilson, L., Raiman, L., Design of immobilized enzyme reactors for the continuous production of fructose sirup from whey permeate, *Bioprocess. Eng.*, **1999**, *21*, 509–551

Ifok, Institut für Katalyseforschung, Rostock, Germany, 2003

Jank, B., Haslberger, A.G., Improved evaluation of potential allergens in GM food, *Trends Biotechnol.*, **2003**, *21*, 249–250

Knapp, F., *Lehrbuch der chemischen Technologie*, F. Vieweg und Sohn, Braunschweig, 1847

Neidleman, S.L., Enzymes in the food industry: a backward glance, *Food Technology*, **1991**, 45, 88–91

Novozymes, *Annual report for 2003*, Novozymes, Copenhagen, 2004 (www.NOVOZYMES. com)

Payen, A., *Handbuch der technischen Chemie*, in: F. Stohmann and C. Engler, (Eds.), Vol. II, p.127 E. Schweizerbartsche Verlagsbuchhandlung, Stuttgart, 1874

Payen, A., Persoz, J.F., Mémoire sur la diastase, les principaux produits de ses réactions, et leurs applications aux arts industriels, *Annales de Chimie et de Physique*, **1833**, 2me Série *53*, 73–92

Raven, P.H., Science, sustainability, and the Human Prospect, *Science*, **2002**, *297*, 954–958

Roberts, S.M., Turner, N.J., Willets, A.J. and Turner, M.K., *Biocatalysis*, p. 1, Cambridge University Press, Cambridge, 1995

Straathof, A., Panke, S., Schmid, A., The production of fine chemicals by biotransformations, *Curr. Opin. Biotechnol.*, **2002**, *13*, 548–556

Sumner, J.B. and Somers, G. F., *Chemistry and Methods of Enzymes*, Academic Press, New York, XIII-XVI, 1953

Tauber, H., *The Chemistry and Technology of Enzymes*. Wiley, New York, 1949

Tischer, W., *Umweltschutz durch technische Biokatalysatoren, in* Symposium Umweltschutz durch Biotechnik, Boehringer Mannheim GmbH, (Ed.), Boehringer Mannheim, 1990

Tosa, T., Mori, T., Fuse, N., Chibata, I., Studies on continuous enzyme reactions 6. Enzymatic properties of DEAE-Sepharose Aminoacylase complex, *Agr. Biol. Chem.*, **1969**, *33*, 1047–1056

Trommsdorf, E. *Dr., Otto Röhm – Chemiker und Unternehmer*, Econ, Düsseldorf, 1976

Ullmann, F., *Enzyklopädie der technischen Chemie*, Vol 5, p. 445, Urban und Schwarzenberg, Berlin, 1914

UN, *Energy Statistics Yearbook for 1999*, United Nations, New York, 2002

UN, *Industrial commodity statistics yearbook 2001*, United nations, New York, 2003

Wagner, R., *Die chemische Technologie*, O. Wiegand, Leipzig, 1857

Wallenfels, K., Diekmann, H., in: *Hoppe-Seyler*, **1996**, *6B*, 1156–1210

WCED – World Commission on Environment and Development, *Our common future*, Oxford Univ. Press, Oxford, 1987

Internet resources for enzyme technology
see Appendix I (p. 419)

2
Basics of Enzymes as Biocatalysts

For the development of a new enzyme process the following questions must be answered:	To do this the following must be known or performed:
How can a suitable enzyme be selected?	Enzyme classification (Section 2.2).
What structural properties of enzymes are important for their application in enzyme technology?	Structure (primary, secondary, tertiary, quarternary); amino acid residues with functional groups on the enzyme surface or clefts in it (Section 2.3).
How can the biological function of enzymes as biocatalysts be described and applied for equilibrium and kinetically controlled processes? Mechanism? Quantitative measures for the biological function of enzymes? Variations in these for enzymes with the same function from different sources?	The enzyme function that can be described by: • Substrate binding, characterized by a dissociation constant K_m. • A monomolecular catalytic reaction between the enzyme and the part of the bound substrate that is changed in the reaction, characterized by a first order rate constant (turnover number or k_{cat}). • Finally the dissociation of the product(s) from the enzyme When they are bound they cause product inhibition, characterized by a dissociation constant K_i. This also determines the substrate- and stereospecificity of enzymes (Sections 2.4 –2.6).

Biocatalysts and Enzyme Technology. K. Buchholz, V. Kasche, U.T. Bornscheuer

ISBN: 3-527-30497-5

For the development of a new enzyme process the following questions must be answered:	To do this the following must be known or performed:
How can the substrate and stereospecificity be determined from enzyme kinetic properties? How do the enzyme kinetic properties depend on pH, *T*, ionic strength, inhibitors and primary structure?	The determination of k_{cat}, K_m and K_i from initial rate measurements $v = f(k_{cat}, K_m, K_i$ [substrate] [product]) as function of pH, *T*, ionic strength, [inhibitor] (Section 2.7).
What system properties determine the end-point, with maximal product yield, of enzyme processes? How much enzyme is required to reach this end-point in a given time?	How pH, *T*, ionic strength, inhibitors, activators, k_{cat}, K_m, K_i, stereoselectivity, influence the equilibrium and steric purity at the end-points, and the time to reach these in equilibrium controlled and kinetically controlled processes (Section 2.8).
How can enzyme processes be carried out at substrate concentrations up to ≈1 M, even with substrates that are slightly soluble in aqueous solutions?	Solubilities of substrates and products should be known. Carry out the enzyme process either in aqueous suspensions or emulsions, or in organic solvents, ionic liquids or supercritical gases (Section 2.9).
What factors influence the stability of enzymes in enzyme processes?	Factors that influence enzyme de- and renaturation (Section 2.10).
How can an enzyme with improved properties as biocatalyst in a given enzyme process be found or constructed?	Screen for enzymes with improved properties among: • existing enzymes with the same function in other organisms or • new enzymes formed by mutations in living organisms or by directed or random evolution in vitro or • existing enzymes that have been redesigned by site directed mutagenesis (Section 2.11).

2.1 Introduction

The biochemical basis of enzymology is fundamental for the successful development and application of enzyme processes. The subject is decribed in detail in several excellent textbooks (see literature list), but the following sections include a summary of some general aspects and principles such as the classification, structure, function (binding of substrate(s) and catalysis), and substrate-, stereo-, and regiospecificity of enzymes, as well as the mechanisms and kinetics of enzyme-catalyzed reactions.

Some aspects of enzymes and enzyme-catalyzed processes that are important for enzyme technology are, however, often not treated in the biochemically oriented literature. These include the points:

- enzymes can be used as biocatalysts for equilibrium-controlled processes in both directions, or to obtain non-equilibrium concentrations of products in kinetically controlled synthesis;
- enzymes are not strictly stereospecific;
- the end-points of enzyme processes and the amount of enzyme required to reach these in a given time;
- the physico-chemical properties of the catalyzed process and its process window for optimal yields;
- enzymes can be used to catalyze reactions with non-natural substrates also in aqueous supensions and organic solvents, supercritical gases and ionic liquids;
- stability of enzymes, substrates and products;
- the possibility of changing the properties of existing enzymes to enable their use in the process window for the optimal yield of the catalyzed process by:
  - screening natural enzyme variants,
  - genetic engineering (site-directed mutagenesis, random or directed evolution).

Thus, these topics will be treated in greater detail in this chapter.

2.2 Enzyme Classification

The system which has long been used by the Enzyme Commission to classify known enzymes is shown in Table 2.1. Every enzyme is given four numbers after the abbreviation EC (www.chem.qmul.ac.uk/iubmb/enzyme). The first number indicates one of the six possible reaction types that the enzyme can catalyze; the second number defines the chemical structures that are changed in this process; the third defines the properties of the enzyme involved in the catalytic reaction or further characteristics of the catalyzed reaction; and the fourth number is a running number. This classification system now covers more than 3000 enzymes. It emphasizes the function of the enzyme as a catalyst of a process in one direction, as is mostly the case in living systems. As a catalyst the enzyme can, however, also catalyze the reverse reaction, and this property is often used in enzyme technology. This point has

Table 2.1 The enzyme classification system developed by the Enzyme Commission (EC), a commission of IUPAC (International Union of Pure and Applied Chemistry). The classification based on the second number is incomplete here, the third (enzyme property) and fourth (running number) EC number are not covered.

Enzyme classes and sub-classes (Functions)	*Remarks*
1 **Oxidoreductases** (Oxidation-reduction reactions)	Cosubstrate required Two-substrate reactions
1.1 at -CH–OH 1.2 at -C=O 1.3 at -C=C-	
2 **Transferases** (Group transfer reactions)	Two-substrate reactions, one substrate must be activated
2.1 C1-groups, 2.2 Aldehyde- or ketogroups 2.3 Acyl-groups 2.4 Glycosyl-groups	
3 **Hydrolases** (strictly transferases that transfer groups to H_2O, i.e., hydrolysis reactions)	Two-substrate reactions, one of these is H_2O
3.1 Ester bonds 3.2 Glycoside bonds 3.3 Ether bonds 3.4 Peptide bonds 3.5 Amide bonds	
4 **Lyases** (non-hydrolytic bond-breaking reactions)	One-substrate reactions $\rightarrow$ bond breaking Two-substrate reactions$\leftarrow$ bond formation
4.1 C–C 4.2 C–O	
5 **Isomerases** (Isomerization reactions)	One-substrate reactions
5.1 Racemizations 5.2 *Cis-trans*-isomerizations 5.3 Intramolecular oxidoreductases	
6 **Ligases** (Bond-formation reactions)	Require ATP as cosubstrate Two-substrate reactions
6.1 C–O 6.2 C–S 6.3 C–N 6.4 C–C	

already been made in the list of the enzymes that are currently applied in enzyme processes (see Table 1.2), where enzymes with the same function but from different sources have the same EC number. The quantitative properties with which the enzyme carries out its function (kinetics) varies with the source (organism) of the enzyme. This is due to the fact that the primary structure of the enzyme with the same EC number can differ much from source to source (see Sections 2.3 and 2.4). Thus, in addition to the EC number, the source of the enzyme must always be stated.

The above-described EC classification system is based on the biochemical function of enzymes in living systems. This database contains no quantitative data on the properties of the enzymes. Newer enzyme databases (a list with description and evaluation of internet databases of importance for enzyme technology is given in Appendix I) provide information on:

- new classifications of enzymes based on their three-dimensional (3-D) structure and function (www.biochem.ucl.ac.uk/bsm/cath_new)
- their properties (www.brenda.uni-koeln.de)
- the biotransformations that they can catalyze (www.genome.ad.jp/kegg/ligand.html)

With the further development of bioinformatics it will be possible to combine these databases in order to select the enzymes that can catalyze a given reaction.

Once an enzyme has been identified as a suitable biocatalyst for a process, its development also involves the screening of other sources (microorganisms) for the same enzyme with properties that are better suited to this purpose than the original enzyme. Besides the catalytic properties of the enzyme, the end-points of the enzyme-catalyzed process, the selectivity and stability of the enzyme under process conditions, and the properties of the source microorganism (enzyme yield, safety class) must be considered in selecting the optimal biocatalyst. Recently, this method to find better enzymes has been amended by rational enzyme design, directed evolution or random mutagenesis of already available enzymes (see Section 2.11).

2.3 Enzyme Synthesis and Structure

The *primary structure* – that is, the amino acid sequence – is defined by the base sequence in the structural gene (DNA) and the genetic code (Fig. 2.1). Only the information in one strand of the double-stranded DNA, the sense strand, is used here. It is first transcribed into mRNA (*transcription*), the information from which is then translated in the synthesis of the polypeptide with the correct primary structure (*translation*). The synthesis direction is always $NH_3^+ \rightarrow COO^-$ – that is, the polypeptide chain starts with an amino group.

For intracellular proteins, the main *secondary structure* elements, the α-helix and the β-sheet, are formed spontaneously either co- or post-translationally, and are stabilized by hydrogen bonds, or hydrophobic interactions between amino acid residues, respectively. These secondary structures are folded spontaneously into

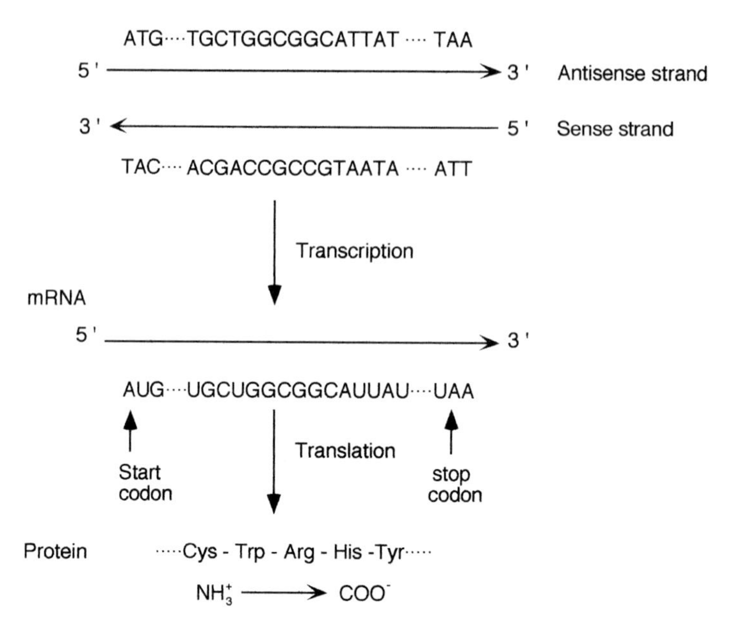

Fig. 2.1 During transcription (DNA→mRNA), only one strand of DNA (sense strand) is used for the synthesis of complementary mRNA. During ribosomal protein synthesis the latter is translated into an amino acid sequence or the primary protein structure. The anti-sense DNA strand contains information for a different primary structure.

domains, stabilized by hydrophobic, charge–charge, charge–dipole and dipole–dipole interactions. The *tertiary* or *3-dimensional structure* of an enzyme often consists of more than one domain, and is stabilized by such interactions between amino acid residues far away from each other along the primary structure. The domains are separated by linker polypeptides without a secondary structure; this gives the tertiary structure a flexibility that is essential for the function of the enzyme. The spontaneous folding of the translation product competes with bimolecular aggregation reactions between partly folded translation products. The latter reaction cannot be neglected at the high protein content in the cell, and leads to inactive enzyme products. To prevent this loss, living systems have developed a special class of proteins (chaperones) that can act as folding catalysts and prevent the aggregation reaction of intracellular enzymes.

The spontaneous folding of some extracellular enzymes must be inhibited in the cell by other chaperones in order to allow translocation of the unfolded polypeptide through the cell membrane. The folding then occurs spontaneously, but extracellularly.

Many enzymes that are used in enzyme technology consist of oligomers with identical or non-identical subunits. This *quaternary structure* is formed spontaneously by non-covalent interactions between the subunits.

The folded translation product of a large number of enzymes is not catalytically active. This applies generally for extracellular and periplasmic enzymes. Different post-translational modifications are required to activate the enzyme precursors (pro-enzymes). These are:

- Non-covalent binding of cofactors (metal ions, pyridoxal phosphate, biotin, etc.).
- Covalent binding of cofactors and other molecules (hemes, mono- and oligosaccharides (only for extracellular enzymes in eucaryotic cells), phosphate groups, etc.).
- Proteolytic processing of the polypeptide chain. For extracellular enzymes, this applies first for the N-terminal signal peptide required for the binding of the pre-pro-enzyme (pre stands for the signal peptide) to the cytoplasmic part of the membrane protein translocation system. The signal peptide is hydrolyzed by peptidases that belong to this system and are localized on the outer membrane side (Dalbey and Robinson, 1999). The formed pro-enzyme is then activated by proteolytic processing reactions (hydrolysis of one or more peptide bonds in the pro-enzyme) in or outside the membrane. The first step has for some amidases and peptidases (pro-calpain, pro-penicillin amidase, pepsinogen, the first peptidases in processing chains of physiological importance as apoptosis, blood clotting and its reversal, activation of digestive enzymes, etc.), recently been shown to be an intra- or intermolecular autoproteolytic reaction (Kasche et al., 1999). Many hydrolases that are used in large amounts in enzyme technology (amylases, galactosidases, lipases, glutaryl and penicillin amidases, peptidases as subtilisins, trypsins, etc.) are proteolytically processed extracellular enzymes. Crude preparations of these enzymes generally contain several proteolysis products with the same enzyme function, but different activities. This must be considered when they are applied to enzyme processes where consistent properties of the enzyme used are required. Enzyme kinetic constants must be determined for pure enzyme preparations, and this is especially important when the properties of the same enzyme from different sources or mutant enzymes are compared.

For all post-translational modifications the surface of the folded translation product is important. The functional groups involved in the covalent and non-covalent binding of cofactors, and the peptide bonds hydrolyzed in proteolytic processing, must be localized on this surface (Table 2.2). About 40–50 % of the amino acid residues on the surface of an enzyme are hydrophobic. This is important for the oligomerization of subunit enzymes, and is applied in the purification of enzymes with hydrophobic adsorbents (see Chapter 4).

The binding of substrates, inorganic ions or organic molecules to enzymes may lead to large changes in the tertiary and quaternary structure (Fig. 2.2), when the enzyme's properties may be markedly changed. Hexokinase without bound glucose is a weak hydrolase (ATPase), but with glucose it becomes a transferase with a markedly reduced hydrolase function. This illustrates that a flexible tertiary and quaternary structure is important for the function of an enzyme. When the enzyme is immobilized (see Chapter 6), this flexibility must be retained.

Table 2.2 Average content of amino acids with functional groups, and their frequency as active site residues in enzymes. A large fraction of these are localized on or in clefts of the protein surface and may interact with molecules in solution (Creighton, 1993; Bartlett et al., 2002).

Amino acid	*Content [%]*	*Functional group*	*Frequency in active site [%]*
Ser	7.8	-OH	4
Lys	7	$-NH_2$	9
Thr	6.5	-OH	3
Asp	4.8	-COOH	15
Glu	4.8	-COOH	11
Arg	3.8	$-NH_2$	11
Tyr	3.4	-OH	5
Cys	3.4	-SH	6
His	2.2	-N-	18
Met	1.6	-S-	<1
Trp	1.2	-N-	1

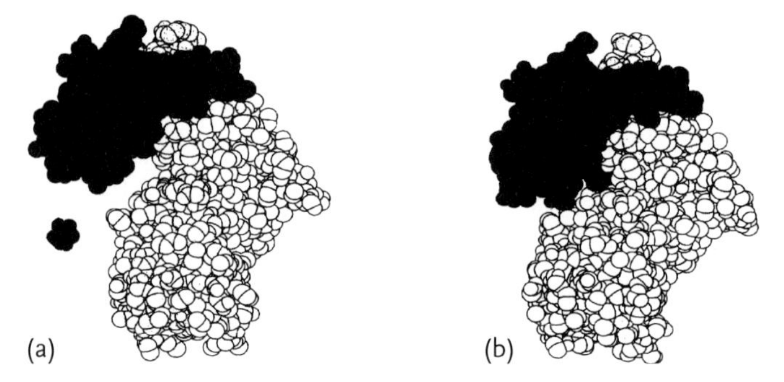

Fig. 2.2 Change in the tertiary structure of an enzyme due to substrate binding, illustrated for the binding of glucose to hexokinase. (a) Without glucose, (b) after binding of glucose (from Darnell et al., 1986; Copyright 1986 Scientific American Books; with permission by W.H. Freeman and Co.).

2.4
Enzyme Function and its General Mechanism

Enzyme function is defined by the reactions in which that enzyme acts as a catalyst. This generally involves three steps, each of which can be subdivided into different reactions:

1. Non-covalent binding of one or two substrates to functional groups in the *binding subsites* S_i before and S'_i after the chemical bond that is changed in the enzyme-catalyzed process, on the enzyme surface. The numbering i starts from 1 in both directions from this bond. This convention applied for peptides is shown in Figure 2.3, but it can also be applied for other substrates.

2. When the substrate is correctly bound to the binding subsites, the functional groups in the *catalytically active amino acid residues* C in clefts on the enzyme surface can interact with the part of the substrate leading to its chemical transformation.

3. Dissociation of the products formed in step (2) from the binding subsites.

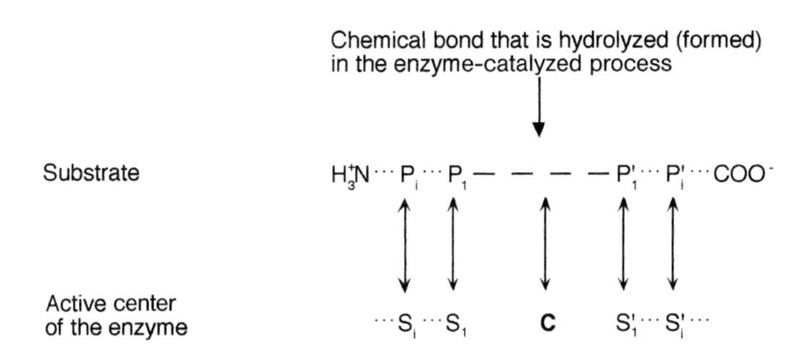

C: catalytically active amino acid residues on the enzyme surface

S, S´: binding sub-sites on the enzyme surface numbered by subscripts as shown

Fig. 2.3 Conventions used to describe the substrate specificity of enzymes and the binding (S-, S′-) and catalytically active (C) subsites of the active site of the enzyme, illustrated for peptidases (Schechter and Berger, 1967). This can be applied to all enzymes that catalyze the formation or breaking of a covalent bond. The number of subsites involved depends on the size of the molecules involved.

The *active site* of the enzyme is the sum of the binding subsites and the catalytically active amino acid residues (catalytic subsite).

The active sites of many industrially important enzymes, such as amylase, carboxypeptidase A, α-chymotrypsin, glycoamylase, glucose isomerase, glutaryl amidase, lipase, penicillin amidase, and trypsin, have been mapped. The detailed mechanisms with which carboxypeptidase A, α-chymotrypsin and penicillin amidase carry out the first two steps of its enzymatic function given above are shown in Figures 2.4 and 2.5.

In the binding step, both charged and uncharged functional groups in the binding subsites are involved. The amino acid residues in these sites are moved to allow for a better interaction with the substrate (induced fit). This illustrates the importance of a flexible tertiary structure stressed above. In the binding step, the enzyme can discriminate between charged and uncharged substrates. Due to the positively charged Arg_{145} in the S'_1-subsite of carboxypeptidase A, it can bind and hydrolyze the negatively charged C-terminal amino acid in a polypeptide. Hence, it is an exopeptidase. The endo-peptidase α-chymotrypsin has a negative charge in the S'_1-subsite and therefore cannot hydrolyse such peptides; rather, it can only hydrolyze peptide bonds in a polypeptide before neutral or positively charged amino acid residues.

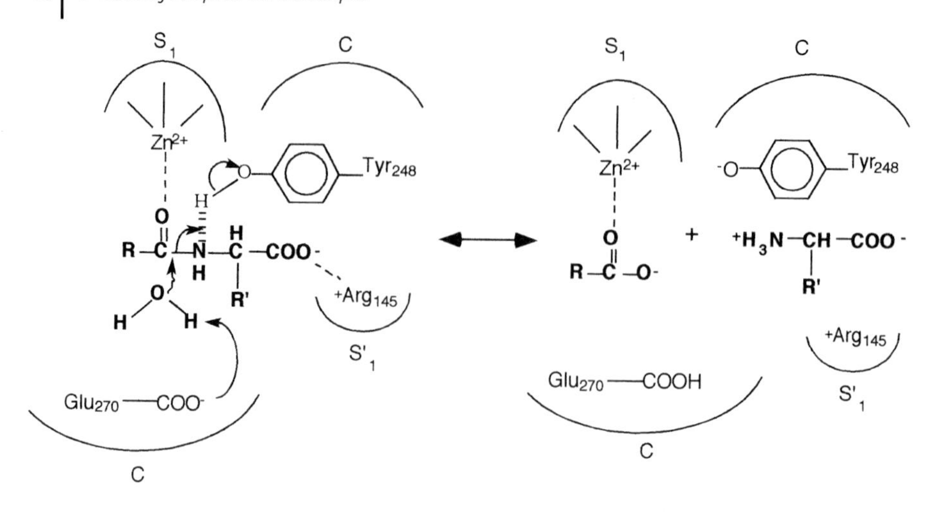

Fig. 2.4 The active center of the enzyme carboxypeptidase A (EC 3.4.17.1), an exopeptidase that catalyzes the hydrolysis of peptide bonds at the carboxy-terminal end of peptides. This enzyme is used in the production of human insulin and for peptide synthesis. The amino acid residues in the binding subsites S_1 and S'_1 and the catalytically active subsite are also shown (from Walsh, 1979).

In this step and in subsequent catalytic reactions, the enzymes can also discriminate between stereoisomers. This requires at least two binding interactions with groups bound to asymmetric C-atoms in the substrate. This is illustrated for the mechanism in Figure 2.4. When the C-terminal R′ and carboxyl group of an (*S*)-amino acid is bound correctly to functional groups that bind R′, and Arg_{145} in the S'_1-subsites of carboxypeptidase A, the N–H group in the amide bond that is hydrolyzed is correctly oriented with respect to the catalytically active amino acid residues C. For a C-terminal (*R*)-amino acid, only two of these three interactions can occur simultaneously. Carboxypeptidase is thus stereospecific for C-terminal (*S*)-amino acid residues. The interactions between the functional groups and R′ can discriminate between different amino acid residues in this position, and demonstrate the substrate specificity of enzymes.

The catalytic step, hydrolysis in Figure 2.4 and acylation and deacylation in Figure 2.5, starts with a proton transfer from or to the catalytically amino acid residues. The withdrawal of a proton from water by Glu_{270} leads to a nucleophilic attack of OH^- on the peptide bond that is hydrolyzed without the formation of a covalent acyl-enzyme intermediate. In Figure 2.5a (p. 37), the charge and proton transport in a catalytic triad Asp_{102}, His_{57} and Ser_{195}, and in Figure 2.5b (p. 38) in the N-terminal Ser_{B1}, activates the O in Ser for a nucleophilic attack on the amide bond. A mechanism similar to Figure 5a applies for lipases (EC 3.1.1.3) that also have a catalytic triad (see Fig. 3.20, p. 127). (Such proton transfer mechanisms start the catalytic step in many enzyme catalyzed reactions, not only for hydrolytic enzymes; Dodson and Wlowad-

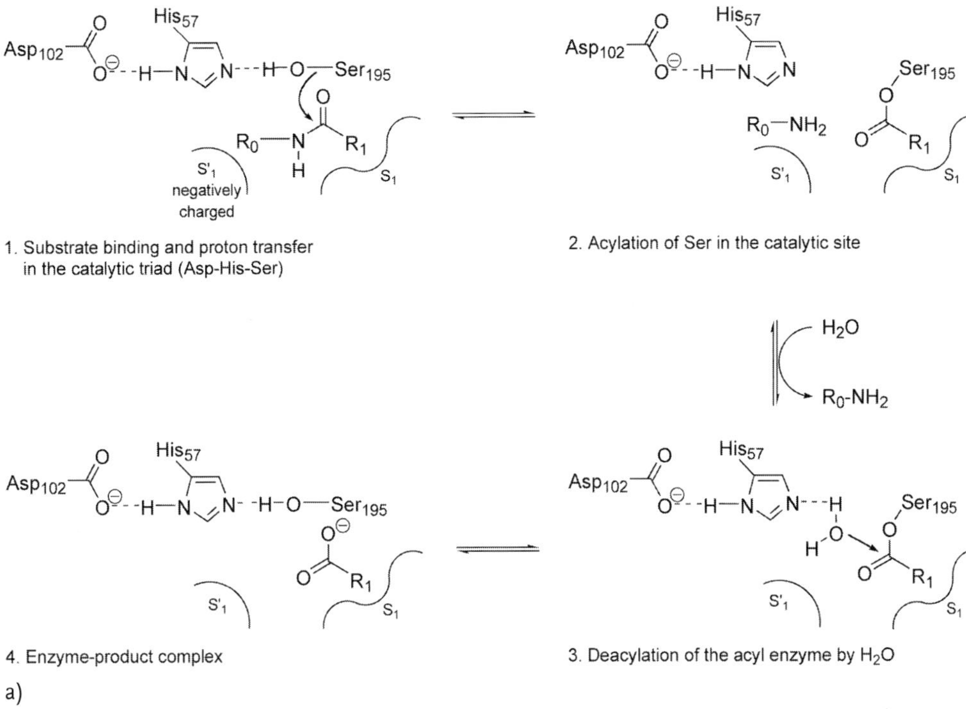

1. Substrate binding and proton transfer in the catalytic triad (Asp-His-Ser)

2. Acylation of Ser in the catalytic site

3. Deacylation of the acyl enzyme by H_2O

4. Enzyme-product complex

a)

Fig. 2.5 Mechanism of the enzyme-catalyzed hydrolysis of peptide and amide bonds involving the formation of an acyl-enzyme intermediate. The catalytic sites are known, whereas the binding subsites S_1 and S_1' are not yet completely identified.
(a) Hydrolysis of peptide bond with the serine endo-peptidase α-chymotrypsin (EC 3.4.21.1). This enzyme catalyzes the hydrolysis of peptide bonds in the polypeptide chain where the S_1 and S_1' binding subsites preferably bind aromatic amino acid residues (Tyr, Trp, Phe), and Arg and Leu, respectively. The serine peptidases have the catalytic triad Asp-His-Ser in the catalytic subsite. The reaction starts with a charge and proton transfer in this triad that results in a nucleophilic attack on the carboxyl group of the substrate (from Walsh, 1979).

er, 1998; Brannigan et al., 1995). This leads to an acylation of Ser – that is, the formation of a covalent acyl-enzyme – and the formation of a C-terminal peptide or amino acid. In both cases the involved His- and Ser-residues must be uncharged to allow for the proton transport. This shows that the rate of enzyme-catalyzed processes depend on pH. Below the p*K*-values of this His-residue, the rate of the catalytic step will decrease. Following acylation, the acyl-enzyme is deacylated by a nucleophilic attack of H_2O.

Finally, the product(s) must dissociate from the binding subsites. As they are part of the substrate their binding to the active site is weaker. The binding of the product in the active site inhibits the binding of a substrate. This leads to *product inhibition* that cannot be avoided in enzyme processes. This point is important to consider in enzyme processes, as the product concentration increases during the reaction towards the desired end-point. In these processes much larger substrate concentra-

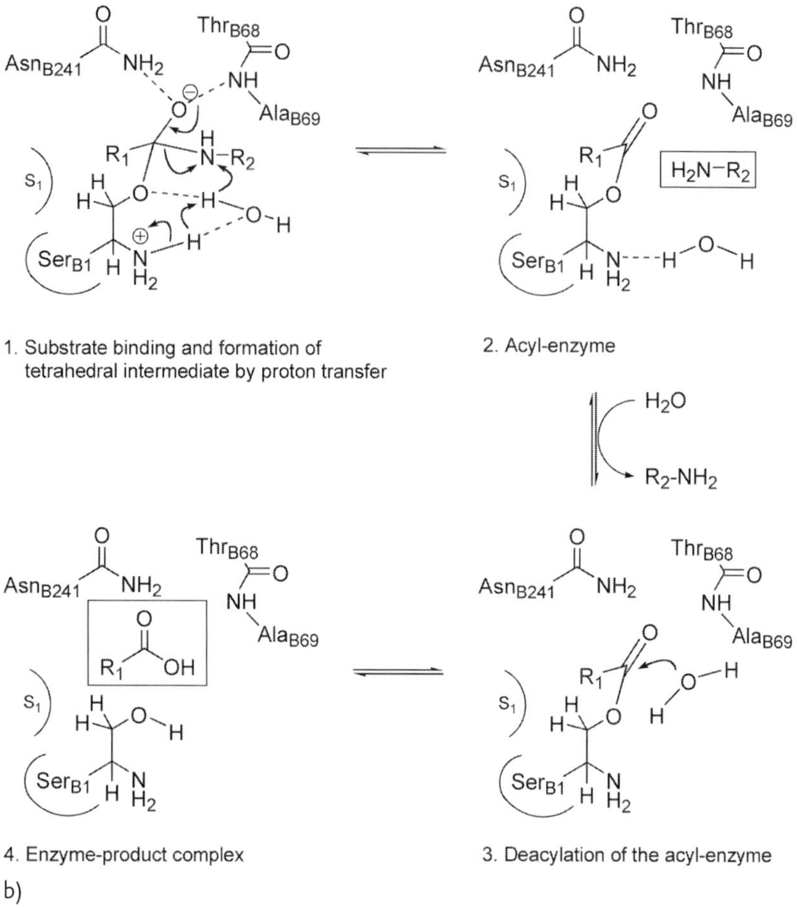

1. Substrate binding and formation of tetrahedral intermediate by proton transfer

2. Acyl-enzyme

3. Deacylation of the acyl-enzyme

4. Enzyme-product complex

b)

Fig. 2.5 (b) Hydrolysis of amide bonds with penicillin amidase (EC 3.5.1.11) from *E. coli* (from Duggleby et al., 1995). In this enzyme the charge and proton transfer mechanism that starts the catalytic reaction is performed by the N-terminal Ser of the B chain and an adjacent water molecule.

tions are also used (up to 2 M, as in the isomerization of glucose) compared to living systems (≤10 mM).

As shown in Figures 2.4 and 2.5, the amino acid residues in the active center are located far from each other in the primary structure, but must be near each other in the tertiary structure. The primary structure of enzymes with the same function and EC number from different sources can differ by up to 70 %. As the function is conserved, it can be assumed that the structure of the active site and its sterical orientation is conserved in these natural enzyme variants. This has been verified for some enzymes studied in detail. By the occurrence of spontaneous *mutations* (changes in the bases in the structural gene in the DNA), the cytochrome gene from *Paracoccus denitrificans* has been changed during evolution so that the same gene from tuna fish encodes for a cytochrome with 70 % change in amino acids in the primary structure (Salemme, 1977). The same has recently been shown to apply for enzymes of technological interest as such as glycoamylase (Sauer et al., 2000), glucose isomerase (Hartley et al., 2000), lipases (Kazlauskas, 1994), and penicillin amidases (Brannigan et al., 1995; McDonough et al., 1999; Kasche et al., 2003) (Fig. 2.6). The spontaneous mutation frequency in living systems is 10^{-10} to 10^{-9} wrongly incorporated nucleotides per base pair and cell division. For an enzyme with a molecular weight of 10^5 Da, this implies that one enzyme mutant (with one amino acid exchange) is formed after about 10^4 to 10^5 cell divisions (Lewin, 1997). Only those cells in which the enzyme function is conserved after so-called neutral mutations survive. Mutations that lead to changes in the active site and cause loss of enzyme function are lethal for the organism, and enzymes with these properties disappear during evolution. These results also show that most spontaneous mutations in enzyme genes are neutral (Kimura and Ohta, 1973). They cause the large differences observed in the primary structure of the same enzyme from different sources, and result in the natural enzyme variants with different properties of interest for their application. For a specific enzyme process, the optimal enzyme must be found among the natural variants formed by neutral mutations during evolution. Gene technology is now an alternative for this screening for enzyme variants in living systems. The gene from an existing enzyme can be mutated by DNA-replications in a test tube, where much

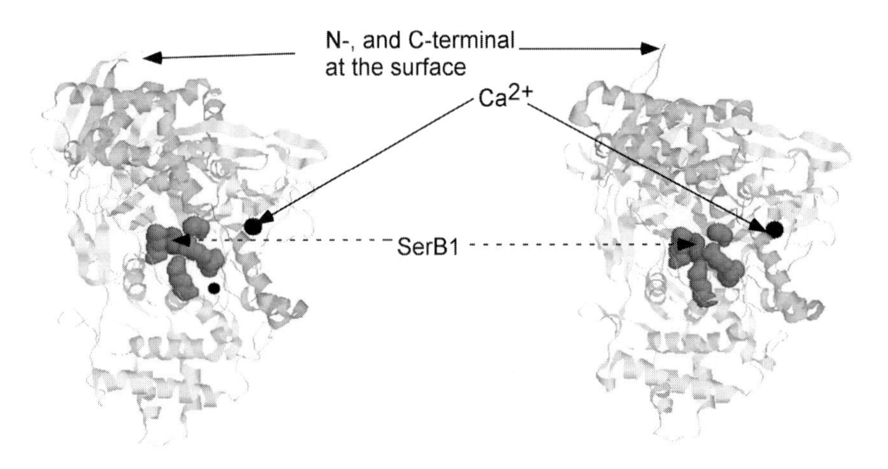

Fig. 2.6 Three-dimensional ribbon structures of the penicillin amidase from *E. coli* (left) and *P. rettgeri* (right). The active enzyme consists of a A- (darker) and B-(lighter) polypeptide chains. The residues of the active site (the catalytically active residue Ser B1 is given by broken arrows) that are conserved in these and the penicillin amidases from *A. faecalis*, *A. viscosus*, *B. megaterium* and *K. citrophila* are given as dark gray balls. Their orientation is conserved in the known structures. Other conserved residues important for the properties of these enzymes, especially for the intramolecular autoproteolysis and the tight binding of a Ca^{2+}-ion (black ball shown by the arrows) that holds the A- and B-chains together, are not shown (Brannigan et al., 1995; McDonough et al., 1999; Kasche et al., 2003).

larger mutation frequencies can be obtained than in living systems. The mutated enzymes with improved properties for the specific application are then selected by suitable screening methods (see Section 2.11).

2.5
Free Energy Changes and the Specificity of Enzyme-Catalyzed Reactions

Once an enzyme has bound substrate(s) in the active site, the catalytically active functional groups in the amino acid residues and cofactors can act as nucleophiles or electrophiles on the groups on the substrate(s) that are transformed in the catalyzed reaction. The binding interactions are both electrostatic and hydrophobic. In the enzyme–substrate(s) complex, both the structure of the substrate(s) and the enzyme active center are changed so that the functional groups are spatially oriented near the groups on the substrate(s) with which they interact with a precision within 10^{-2} nm. This leads to a reduction in activation energies in comparison with the uncatalyzed reaction that increases the rate of the catalyzed reaction by factors of up to more than 10^{10}. This is shown schematically in the free energy diagrams (Figs. 2.7 and 2.8). In comparison with the uncatalyzed reaction, the catalyzed reaction requires at least two reaction steps – binding, followed by the catalyzed chemical reaction.

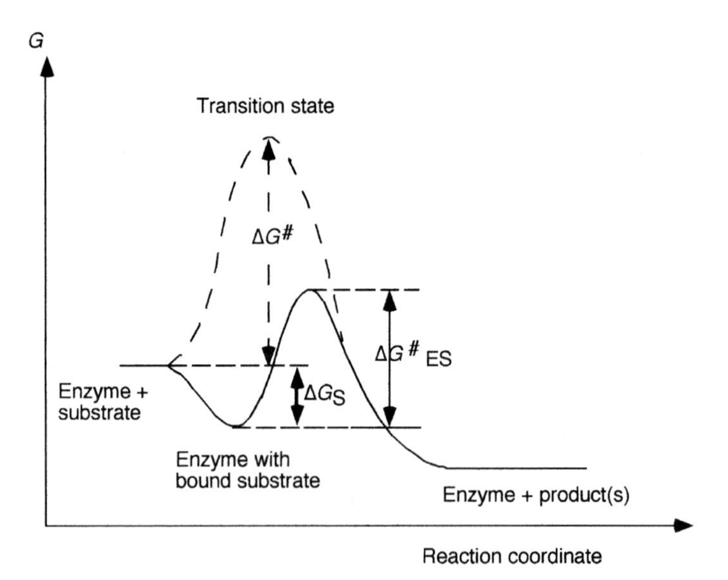

Fig. 2.7 Free energy diagram for an uncatalyzed (----) and enzyme-catalyzed (———) reaction involving one binding step and one chemical reaction (see Fig. 2.4). The apparent activation energy of the latter ($\Delta G^{\#}_{E,S} - \Delta G_S$) is much smaller than for the uncatalyzed reaction ($\Delta G^{\#}$); ΔG_S free energy change due to substrate binding.

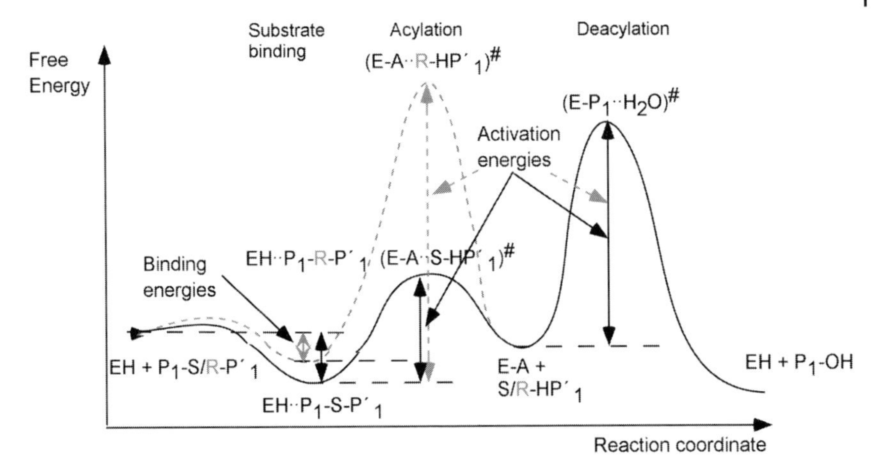

Fig. 2.8 Free energy diagram for an enzyme-catalyzed reaction involving the formation of an acyl-enzyme intermediate, consisting of one binding process and two chemical reactions as for the peptide or amide bond hydrolysis shown in Figure 2.5 with either a *S*-(—) or *R*-(--) amino acid in the P_1'-position, respectively (see Fig. 2.3). The acyl-enzyme E-A is formed by acylation by the amino acid in the P_1-position.

A detailed reaction mechanism is required to analyze all factors that may influence the kinetics of an enzyme-catalyzed process. The free energy diagram in Figure 2.7 can be used for a thermodynamic discussion of the peptide bond hydrolysis catalyzed by carboxypeptidase A shown in Figure 2.4. Where this hydrolysis involves the formation of covalent acyl-enzyme intermediates (see Fig. 2.5), this diagram is insufficient. In this case, the diagram in Figure 2.8 must be used, where both the acylation and deacylation reactions, but not product binding are included.

The free energy changes for substrate binding and the activation energies for the chemical reactions in Figures 2.7 and 2.8 can be related to dissociation and rate constants for the substrate binding and catalytic steps of an enzyme-catalyzed reaction. The free energy change ΔG_S defines a dissociation constant K_S from which the commonly used Michaelis–Menten constant K_m can be derived, and is used as a quantitative measure for the substrate binding (see Section 2.7.1). On a molecular level, it is the sum of the free energies for the binding interactions between the substrate and functional groups in the active site of the enzyme. The activation energy $\Delta G^{\#}_{ES}$ defines the first-order rate constant k_{cat} that is a quantitative measure of the catalytic steps(s) in an enzyme-catalyzed reaction. It is also called the *turnover number*. When the chemical step involves two reactions (Fig. 2.8), the turnover number depends on the first-order rate constants for both reactions, as will be derived in Section 2.7.1.

One enzyme can catalyze the conversion of different substrates. The interactions with the functional groups in the active site leads to different free energy changes (Figs. 2.7 and 2.8). Therefore, K_m and k_{cat} vary from substrate to substrate. This is the thermodynamic basis for the *substrate specificity* of enzymes. Changes in K_m and k_{cat} are also expected for reactions with the same substrate and enzymes with the same function from different sources.

The main cause for the acceleration of an enzyme-catalyzed versus an uncatalyzed reaction of up to more than a factor 10^{10}, was until recently considered to be mainly due to the decrease in the activation energy. The enzyme structure in the transition states in Figures 2.7 and 2.8 was considered to be complementary to the transition state of the substrate. This causes a stabilization of this state that leads to a reduction in the activation energy. The structure of the enzyme is, however, already changed when it binds the substrate (induced fit, as shown in Fig. 2.2) (Done et al., 1998). For the evolution of enzymes, it is favorable when the binding energy ΔG_S also contributes to the reduction of the activation energy. Therefore, both the substrate binding and catalytic reaction contribute to the substrate specificity of an enzyme and the acceleration of the rate compared with the uncatalyzed reaction (Menger, 1992; Schowen, 2003; Garcia-Viloca et al., 2004).

In Figure 2.8 the free energy diagram for the hydrolysis of a dipeptide or a N-acylated amino acid where the C-terminal amino acid is either a (*R*) or (*S*)-enantiomer is shown for an enzyme-catalyzed reaction involving acyl-enzyme intermediates. The corresponding reaction mechanisms are given in Figure 2.5. Due to a different orientation of the residues of the (*R*)- and (*S*)-substrate relative to the functional groups in the active site of the enzyme, both the binding and activation energies differ for the two substrates. In the case shown, this applies only for the binding and acylation step for a (*S*)-specific enzyme. For a N-terminal (*R, S*)-amino acid in a dipeptide, differences in the activation energy are also expected for the deacylation step. This shows that enzymes are *stereospecific*, but not always strictly so as both substrates can be hydrolyzed by the enzyme. Both substrate binding and catalysis contribute to the stereospecificity. From the above discussion, it can also be concluded that the stereospecificity should differ for enzymes with the same function from different sources. This discussion on the stereospecificty should also apply for the *regiospecificity* of enzymes.

Although the above analysis on the thermodynamic basis for the substrate- and stereospecificity of enzymes was derived for hydrolases, it is equally applicable to all types of enzymes. For a more quantitative analysis of all the factors that influence the kinetics and yields of enzyme-catalyzed reactions, the kinetic constants K_m and k_{cat} must be determined (see Section 2.7).

2.6 Equilibrium- and Kinetically Controlled Reactions Catalyzed by Enzymes

When enzymes catalyze equilibrium-controlled reactions (see Figs. 2.4 and 2.5), they accelerate the rate to reach the equilibrium, but do not influence the equilibrium constant or the end-point of the reaction (Fig. 2.9). In enzyme technology, high product yields are essential in order to obtain a process that is competitive. The equilibrium (thermodynamic) yield is only determined by the initial substrate concentra-

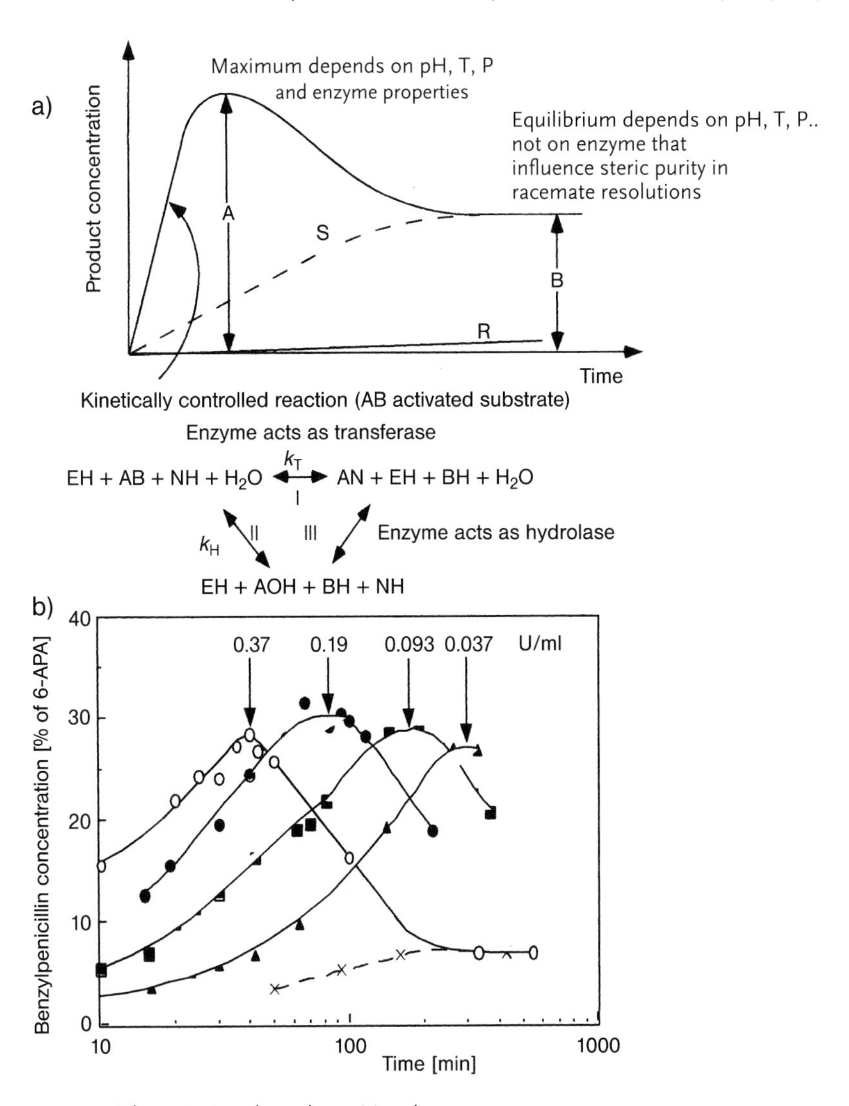

Fig. 2.9 Schematic time-dependence (a) and experimental results (b) for kinetically and equilibrium-controlled reactions catalyzed by hydrolases with acyl-enzyme and other covalent intermediates.
(a) This also shows that the steric purity of the product in equilibrium-controlled racemate resolutions depends on the stereoselectivity of the used enzyme.
(b) Kinetically (—) and equilibrium-controlled (---) synthesis of penicillin G (AN) from equal concentrations of phenylacetyl glycine (AB) or phenylacetic acid (AOH) and 6-aminopenicillamic acid (NH) at different enzyme (penicillin amidase from *E. coli*) concentrations given in U ml^{-1} (pH 6.0, 25 °C). The kinetically controlled maximum is (within experimental error) independent on the concentration of the enzyme, and is much larger than the equilibrium concentration.

tion, the solubilities of substrate and product and the equilibrium constant that can be influenced by the system properties pH, T, P, etc. From this, a process window in the latter properties can be derived where a given yield (as >90 %) can be obtained. Among the enzymes that can catalyze this process, the best enzyme is the one with the best properties (activity, stability) inside this process window. This demonstrates that in the rational design of an enzyme process one must consider the properties of both the enzyme and the catalyzed process.

For an equilibrium-controlled, enzyme-catalyzed resolution of racemates, the equilibrium yield of the desired enantiomer is not influenced by the properties of the enzyme. In this case, however, the steric purity of the desired enantiomer must also be considered. This is influenced by the stereospecificity of the enzyme (Figs. 2.8 and 2.9a). Thus, for these processes this property must also be considered in the selection of the optimal enzyme.

A detailed analysis of the reaction mechanisms in Figure 2.5 shows that H_2O acts as a nucleophile in the deacylation step. However, might other molecules (R–OH or R–NH_2) also act as nucleophiles here? When this is the case, the hydrolases could also act as transferases, transferring acyl-groups to R–OH or R–NH_2. This has been shown to apply for hydrolases that form covalent intermediates, such as acyl-enzymes (Fig. 2.5), glycosyl-enzymes or intermediates as cyclic phosphates observed in RNase-catalyzed reactions (Kasche, 1986). These reactions require the use of an activated substrate (such as an ester or amide). The general reaction scheme of such enzyme-catalyzed, *kinetically controlled reactions*, and experimental data for the synthesis of benzylpenicillin are shown in Figure 2.9b. In these processes, the nucleophiles H_2O and NH (6-aminopenicillamic acid in Fig. 2.9b) compete in the deacylation reaction of the acyl-enzyme (see Fig. 2.5b). In the former reaction, the enzyme acts as a hydrolase with an apparent hydrolase rate constant k_H. In the latter reaction, where the enzyme acts as a transferase with an apparent transferase rate constant k_T, a condensation product is formed that can also be hydrolyzed by the enzyme. The concentration of the latter increases until its synthesis rate equals the hydrolysis rate, where the maximal product concentration is observed. The final condensation product concentration is given by the equilibrium constant for the reaction:

$$AN + H_2O \leftrightarrow AOH + NH$$

Thus, in kinetically controlled reactions much larger concentrations of the product AN than for an equilibrium controlled reaction can be obtained. The maximum product concentration depends on the selectivity $(k_T/k_H)_{app}$ and the rate with which AN is hydrolyzed by the enzyme. Thus, in contrast to the equilibrium-controlled process the maximal product yield in the kinetically controlled process depends on the properties of the enzyme. The relationship between $(k_T/k_H)_{app}$ and intrinsic rate constants of the enzyme can be derived from Scheme 2.1.

The ratio $(k_T/k_H)_{app}$ is determined from the initial rates of the formation of the condensation AN (v_T) and hydrolysis product AOH (v_H). From Figure 2.9 and Scheme 2.1 the following relationship is derived:

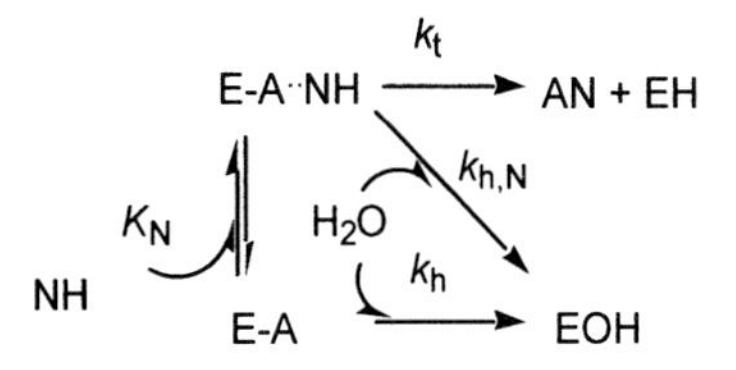

Scheme 2.1

$$(v_T/v_H) = (k_T/k_H)_{app}\,[NH]/[H_2O] = \\ = k_t\,[E\text{-}A\cdot\cdot NH]/(k_h\,[H_2O][E\text{-}A] + k_{h,N}[H_2O][E\text{-}A\cdot\cdot NH]) \tag{2.1}$$

assuming that equilibrium is established in the nucleophile binding, i.e.,

$$K_N = [E\text{-}A][NH]/[E\text{-}A\cdot\cdot NH] \tag{2.2}$$

then

$$(k_T/k_H)_{app} = k_t\,/(k_h\,K_N + k_{h,N}\,[NH]) \tag{2.3}$$

i.e., $(k_T/k_H)_{app}$ depends on both intrinsic enzyme properties and [NH].

Such kinetically controlled processes can be applied in enzyme technology for the production of condensation products catalyzed by hydrolases that also can act as transferases (Kasche, 1986, 2001). Some processes have already been realized on an industrial scale (production of insulin and other peptides such as the sweetener aspartame; synthesis of semisynthetic cephalosporins and penicillins; kinetic resolutions of racemates, etc.) (Bornscheuer and Kazlauskas, 1999; Bruggink, 2001; Kasche, 2001; Liese et al., 2000; Zmiejewski et al., 1991). Hydrolases that can be used as biocatalysts for such processes and their apparent transferase to hydrolase ratio $(k_T/k_H)_{app}$ derived from initial rate measurements are listed in Table 2.3. The ratio $(k_T/k_H)_{app}$ is also a measure for the P_1'-specificity of peptidases and amidases (see Fig. 2.3) and can be used to map these (Kasche, 2001; Galunsky and Kasche, 2002).

The concentrations of condensation products (biopolymers as nucleic acids, polysaccharides, and proteins) in living cells are much higher than those that exist at thermodynamic equilibrium in these systems. The synthesis of these compounds at concentrations much higher than can exist at equilibrium, occurs in kinetically controlled processes from activated substrates (trinucleotides such as ATP, disaccharides, adenylated amino acids). The transferases that catalyze these reactions, have been optimized during evolution so that their transferase activity is much larger than the hydrolase activity (Table 2.3). These enzymes and their activated substrates are still too expensive for use in enzyme technology, although exceptions here are dextran- and levansucrases that can synthetize polysaccharides from the activated substrate sucrose (Buchholz and Monsan, 2003).

Table 2.3 Transferases and hydrolases, that can be used as transferases in kinetically controlled synthesis of condensation products and their selectivities $(k_T/k_H)_{app}$ for different nucleophiles (Kasche, 1986).

Enzyme	*Nucleophile*	$(k_T/k_H)_{app}$
2. Transferases		
DNA-Polymerase	DNA	10^7
Hexokinase	Glucose	10^6
Dextransucrase	Glucose	
3. Hydrolases		
Lipase[a)]	Acids	
	Alcohols	10
	Amines	
Alkaline Phosphatase	TRIS	10^2–10^3
RNase I	Alcohols	10
	Nucleoside	10^2–10^3
Glycosidases	Alcohols	10^2
β-Galactosidase	Lactose	10^2–10^3
Serine-, Thiol-	(*S*)-Amino acids	10–10^2
Peptidases	(*S*)-Amino acid esters	10^2–10^4
	(*S*)-Amino acid amides	10^2–10^5
	(*R*)-Amino acid amides	10–10^3
	Alcohols	–10^2
	TRIS	–10^2
Amidases		
Penicillin amidase	6-Aminopenicillanic acid	10^3–10^4
	(*S*)-Amino acids	–10^4
	Alcohols	1

a) For reactions with lipases (esterification are often carried out in almost water-free systems) the above data apply for aqueous systems.

It can be seen from the data in Table 2.3 that many hydrolases may be used as biocatalysts for the kinetically controlled synthesis of condensation products. However, in order to obtain large condensation product yields, much higher nucleophile concentrations (up to 1 M) must be used compared to those in living systems ($<10^{-2}$ M). These concentrations must be chosen so that $(k_T/k_H)_{app})([NH]/[H_2O]) \gg 1$ (see Table 2.3 and Eq. 2.1).

2.7
Kinetics of Enzyme-Catalyzed Reactions

Enzymes are used as catalysts to decrease the time required to reach the end-point of equilibrium- and kinetically controlled processes (see Fig. 2.9). This time is a function of the enzyme concentration and enzyme properties (binding of substrate and product, activation energy of the catalytic reactions), as discussed in Section 2.5. In order to calculate this time, these properties must be determined based on a kinetic analysis of the enzyme-catalyzed reaction (see Section 2.7.1). The end-point of an enzyme-catalyzed process must be selected to give the highest possible product yield. When the end-point is the equilibrium state of the reaction, the yield cannot be influenced by the properties of the enzyme. The latter, however, influence the yield at the product maximum in a kinetically controlled process. The product yield depends also on initial substrate concentrations pH, temperature and ionic strength. This implies that the selection of the process conditions for an enzyme process, depend on the pH-, ionic strength-, and temperature-dependence of the catalyzed process (see Section 2.8). These can be unfavorable for the enzyme used as biocatalyst, when it is unstable at the conditions optimal for the process. From this it follows that the pH-, temperature-, and ionic strength dependence of the properties of the enzyme (binding, catalysis, stability) must also be known in order to determine the best process conditions (see Section 2.7.2).

2.7.1 Quantitative Relations for Kinetic Characteristics and Selectivities of Enzyme-Catalyzed Reactions

2.7.1.1 Turnover Number (k_{cat}) and Michaelis–Menten Constant (K_m)

The enzyme-catalyzed reactions discussed in Sections 2.4 to 2.6, as well as for all other enzymes, can be described by one of the following kinetic schemes (Note: the enzyme in Section 2.7, in contrast to the usual abbreviation E, is written EH; this is done to stress that in many enzyme-catalyzed processes, proton transfer steps are directly involved (see Fig. 2.5; Benkovic and Hammes-Schiffer, 2003) and to indicate the pH-dependence of these processes. Later the common E will be used):

1. One-substrate (AB) reactions (only for lyases in one direction and isomerases in both directions).

$$\mathrm{EH + AB} \underset{k_{-1}}{\overset{k_1}{\rightleftharpoons}} \mathrm{(EH{\cdot\cdot}AB)} \begin{cases} \overset{k_2}{\rightleftharpoons} \mathrm{EH + AB'} \quad \text{(Isomerases)} \\ \underset{k_{-2}}{\rightleftharpoons} \mathrm{EH + A + B} \quad \text{(Lyases)} \end{cases} \tag{2.4}$$

2. Two-substrate (AB and NH) reactions (for all other enzymes).

(a) Both substrates cannot be bound simultaneously in the active center.

$$\mathrm{EH + AB} \underset{k_{-1}}{\overset{k_1}{\rightleftharpoons}} \mathrm{(EH{\cdot\cdot}AB)} \underset{k_{-2}}{\overset{k_2}{\rightleftharpoons}} \mathrm{(E{\cdot\cdot}A) + BH} \underset{k_{-3}}{\overset{k_3,\ +\mathrm{NH},\ -\mathrm{BH}}{\rightleftharpoons}} \mathrm{(NH{\cdot\cdot}EA)} \underset{k_{-4}}{\overset{k_4}{\rightleftharpoons}} \mathrm{EH + AN} \tag{2.5}$$

substrate binding catalytic reactions

(b) Both substrates can be bound simultaneously in the active center.

$$\begin{array}{ccc} & (EH\cdot\cdot AB) + NH & \\ \nearrow\!\!\swarrow & & \searrow\!\!\nwarrow \\ EH + AB + NH & & (NH\cdot\cdot EH\cdot\cdot AB) \underset{k_{-2}}{\overset{k_2}{\rightleftharpoons}} (NH\cdot\cdot EA) + BH \underset{k_{-3}}{\overset{k_3}{\rightleftharpoons}} EH + AN + BH \\ \searrow\!\!\nwarrow & & \nearrow\!\!\swarrow \\ & (NH\cdot\cdot EH) + AB & \end{array} \tag{2.6}$$

substrate binding catalytic reactions

Most enzyme-catalyzed reactions can be described by one of these three schemes. In the case where more than one substrate NH participate as nucleophiles, as in kinetically controlled processes, the reaction schemes (2.5) and (2.6) will include more branching reactions. From these schemes, relationships that provide quantitative measures for the enzyme properties, substrate binding and rate of the enzyme-catalyzed reaction, can be derived. They are derived assuming that the following boundary conditions are fulfilled:

1. Substrate concentration >> Enzyme concentration

   $[NH]_0$, $[AB]_0 >> [EH]_0$ (subscript 0 for $t = 0$).

2. Only the initial rate is measured. Then $[NH]_0$ and $[AB]_0$ are practically constant, and the reverse reactions after the binding of the substrate(s) can be neglected.

3. Equilibrium in the binding of the substrates is obtained. This is an assumption that is problematic.

4. Steady-state in the concentration of the intermediates EH¨AB, H¨EH¨AB, NH¨EA, etc.

   (¨ non-covalent interactions, EA = covalent enzyme–substrate intermediate (as an acyl- or glycosyl-enzyme).

With conditions (1) to (4) and mass conservation relationships, a set of linear equations can be derived, from which the reaction rate can be expressed in the rate constants of the schemes (2.4) to (2.6) and the known concentrations. Most enzyme-catalyzed reactions, also in enzyme technology, are two-substrate reactions. They will therefore be considered here[1]. The rate of product formation (or substrate consumption) for the scheme (2.5) is

1) In biochemistry textbooks, generally only one-substrate reactions (Eq. 2.4) are described. The rate of product formation is then $v = k_2[EH\cdot\cdot AB]$. The mass conservation relationship for the enzyme is:
$[EH]_0 = [EH] + [EH\cdot\cdot AB]$
and the steady-state condition for $[EH\cdot\cdot AB]$ gives
$k_1 [EH][AB]_0 = (k_{-1} + k_2) [EH\cdot\cdot AB]$
From these two equations, we obtain
$[EH\cdot\cdot AB] = [EH]_0[AB]_0/((k_{-1} + k_2)/k_1 + [AB]_0)$
The rate of product formation is then
$v = k_2 [EH]_0[AB]_0/(K_m + [AB]_0)$
where the turnover number k_2 and the Michaelis–Menten constant $K_m = (k_{-1} + k_2)/k_1$ are the quantitative measures for the catalytic and substrate binding properties of the enzyme.

$$v = d[AN]/dt = k_4 [EA^{\cdot\cdot}NH] \tag{2.7}$$

From the linear set of equations an expression for $[EH^{\cdot\cdot}AN]$ in the known concentrations $[EH]_0$, $[AB]_0$ and $[NH]_0$ and the rate constants can be derived. After insertion in Eq. (2.7), the following expression in the form of a Michaelis–Menten equation of a one-substrate reaction is obtained:

$$v = \frac{k_{cat,AB} [AB]_0[EH]_0}{(K_{m,AB} + [AB]_0)} \tag{2.8}$$

with the turnover number:

$$k_{cat,AB} = \frac{k_2\, k_3\, k_4\, [NH]_0}{k_3 \cdot (k_4 + k_2)[NH]_0 + k_2\, k_4} \tag{2.9}$$

and the Michaelis–Menten constant:

$$K_{m,AB} = \frac{(k_{-1} + k_2) \cdot k_3 \cdot k_4 \cdot [NH]_0}{(k_3 \cdot (k_4 + k_2) \cdot [NH]_0 + k_2\, k_4)\, k_1} \tag{2.10}$$

For large $[NH]_0$ (as in hydrolysis reactions with $[NH]_0 = H_2O$), Eqs. (2.9) and (2.10) become:

$$k_{cat,AB} = \frac{k_2\, k_4}{k_2 + k_4} \tag{2.11}$$

and

$$K_{m,AB} = \frac{(k_{-1} + k_2)\, k_4}{(k_2 + k_4)\, k_1} \tag{2.12}$$

that is, for the same enzyme these quantities are characteristic for the substrate AB only under these conditions. For smaller $[NH]_0$, they also depend on this concentration. Equation (2.8) was first derived in 1913 by Michaelis and Menten. It is a general expression for the rate of enzyme-catalyzed reactions, and plots of the rate as function of the substrate concentration are called Michaelis–Mentens plots (Fig. 2.10).

The turnover number k_{cat} has the dimension s^{-1}, and is the number of molecules converted per second by one enzyme molecule under substrate saturation conditions. It is only influenced by the rates of the catalytic reactions (k_2, k_4), and is a measure for the activation energy $G^{\#}_{ES}$ in Figures 2.7 and 2.8. The Michaelis–Menten constant K_m has the dimension mol L^{-1} (M), and is the substrate concentration at which 50 % of the maximal velocity V_{max} ($= k_{cat} [EH]_0$) has been reached. It is an apparent equilibrium constant and a measure for the substrate binding energy ΔG_S in

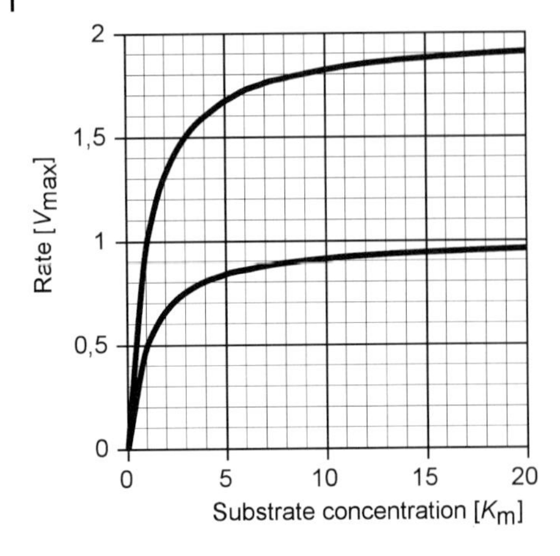

Fig. 2.10 Michaelis–Menten plot – that is, rate v (Eq. 2.8) in units of V_{max} for an enzyme-catalyzed reaction as a function of the substrate concentration (in K_m-units) at enzyme concentrations $[EH]_0$ (lower) and $2[EH]_0$ (upper curve). Note that $v = 0.9\ V_{max}$ requires a substrate content of $\approx 10\ K_m$.

Figures 2.7 and 2.8. It is determined by the equilibrium constant for the substrate binding and rate constants for the catalytic reactions. The smaller the K_m-value, the better is the binding of the substrate to the enzyme. The turnover number k_{cat} and the Michelis–Menten constant K_m are used to characterize the kinetic properties of an enzyme. They are influenced by both the enzyme and the substrate. The ratio k_{cat}/K_m (dimension: $[M^{-1}\ s^{-1}]$), an apparent second-order rate constant, is called *specificity constant*, and is a quantitative measure for the substrate specificity of an enzyme. The specificity constant cannot be larger than the rate of a diffusion-controlled bimolecular reaction ($<10^8 - 10^9\ M^{-1}\ s^{-1}$) (Fersht, 1999). For the determination of the substrate specificity of one enzyme and different substrates or of different enzymes for one substrate, k_{cat} and K_m must be determined. When the enzyme concentration is unknown, as in homogenates, $V_{max} = k_{cat} \cdot [EH]_0$ is determined. The turnover number can only be determined for enzymes for which the active site can be titrated – that is, when the active enzyme concentration $[EH]_0$ can be determined. This is possible for enzymes such as serine peptidases, penicillin amidases, and lipases that can be practically irreversibly covalently acylated in the active site (Fersht, 1999). V_{max} is a measure for the amount of enzyme expressed in activity units. For this, the enzyme unit (U) is still used, although it is not a Système Internationale (SI) unit. One U is the amount of enzyme that catalyzes the conversion of 1 µmol substrate per min under defined conditions (pH, T). One SI unit for enzyme activity (Katal, abbreviated kat) is the amount of enzyme that converts 1 mol of substrate per second, but this is hardly used in enzymological literature. One kat is equivalent to 60×10^6 U.

Different methods have been developed to determine k_{cat} and K_m from the determinations of v at different substrate concentrations. One suitable method was developed by Eadie and Hofstee. Equation (2.8) can be linearized as:

$$\frac{v}{[AB]_0\,[EH]_0} = \frac{k_{cat}}{K_m} - \frac{v}{[EH]_0\,K_m} \tag{2.13}$$

$v/[AB]_0[EH]_0$ as function $v/[EH]_0$ provide a linear plot (the Eadie–Hofstee plot), from which k_{cat} and K_m can be determined from the x- and y-intercepts. Frequently, these quantities are determined by another linear plot ($1/v$ is plotted against $1/[AB]_0$; Lineweaver–Burk plot). It has been shown that the experimental error in the determination of k_{cat} and K_m is larger for the Lineweaver–Burk plot than for the Eadie–Hofstee plot (Deranleau, 1969), but even for the latter plot the errors are not negligible (±10 %) when the error in rate determination is ±2 %.

Values of k_{cat} and K_m for different substrates (same enzyme) and different enzymes (same substrate) are listed in Tables 2.4 and 2.5. These tables provide information about the substrate specificity as the binding of different P_1'- and P_1-residues, hydrolysis of ester or amide bonds – that is, activation of these bonds by the catalytic subsite **C** in Figure 2.3, and the interactions between the active site and the substrate. It is clear from the data in Table 2.4 that the substrate specificity for one en-

Table 2.4 Turnover number k_{cat} and Michaelis–Menten constant K_m for different hydrolases from different sources; NH in Eqs. (2.5) and (2.6) = H_2O; ↓; = bond that is hydrolyzed (from Laidler, 1958 and the laboratory of the author Kasche).

Enzyme	*Substrate* ($P_i..P_1$↓$P_1'..P_1$)	*Temp.* [°C]	*pH*	k_{cat} [s^{-1}]	K_m [mM]	k_{cat}/K_m [M^{-1} s^{-1}]
Trypsin	Benzoyl-Arg↓NH_2	25.5	7.8	27	2.1	13 000
(bovine)	Benzoyl-Arg↓OEt	25.0	8.0	19	0.02	1 000 000
	Z-Lys↓OMe	30	8.2	101	0.23	440 000
	Z-Lys↓Ala	40	8.2	0		≈0
	Z-Lys↓Ala-Ala-Ala	40	8.2	4.6	3.7	1 200
α-Chymotrypsin	Acetyl-Tyr↓OEt	25.0	7.0	160	3.7	42 000
bovine	Acetyl-Tyr↓NH_2	25.0	7.8	0.28	7	40
	Benzoyl-Tyr↓OEt	25.0	7.8	78	4	20 000
	Benzoyl-Phe↓OEt	25.0	7.8	37	6	6 300
	Benzoyl-Met↓OEt	25.0	7.8	0.77	0.8	1 000
	Ac-ProAlaProPhe↓Ala	37	8.0	0		≈0
	Ac-ProAlaProPhe↓AlaAlaNH_2	37	8.0	37	0.83	44 000
Mouse NZB	Acetyl-Tyr↓Oet	25	7.0	250	2.1	120 000
Mouse A/sn	Acetyl-Tyr↓OEt	25	7.0	210	2.1	100 000
Carboxy-Peptidase A	Carbobenzoyl-Gly-↓-Try	25	7.5	89	2	17 000
	Carbobenzoyl-Gly-↓-Leu	25	7.5	10	28	390
Penicillin amidase						
E. coli		25	7.8	48	0.01	4 800 000
A. faecalis	Penicillin G	25	7.8	80	0.008	10 000 000
K. citrophila		25	7.8	60	0.02	3 000 000
Adenosine-triphosphatase	ATP	25	7.0	104	0.012	8 300 000
Urease	Urea	20.8	7.1	20 000	4	5 000 000

Table 2.5 Stereoselectivity in the hydrolysis of the same substrate with different hydrolases (pH 7.5, 25 °C, $I = 0.2$ M) (Michaelis, 1991; Lummer et al., 1999; Galunsky et al., 2002). ↓ indicates the bond that is hydrolyzed.

Substrate P_1-↓-P_1'	K_m [mM]	k_{cat} [s^{-1}]	k_{cat}/K_m [$M^{-1}\,s^{-1}$]	*Stereoselectivity* E_{eq} $(k_{cat}/K_m)_S/(k_{cat}/K_m)_R$
Hydrolysis with penicillin amidase (EC 3.5.1.11 from *E. coli*)				
(*S*)-Phg-↓O-Me	20	11	550	0.5
(*R*)-Phg-↓O-Me	32	35	1100	
N-Acetyl-(*S*)-Phg-↓O-Me			1.3	>13
N-Acetyl-(*R*)-Phg-↓O-Me			<0.1	
(*S*)-Phg-↓-NH_2	11	6.4	580	0.27
(*R*)-Phg-↓-NH_2	17	36	2100	
Hydrolysis with α-chymotrypsin (EC 3.4.21.1 bovine)				
(*S*)-Phg-↓O-Me	50	0.46	9.2	14
(*R*)-Phg-↓O-Me	140	0.08	0.57	
N-Acetyl-(*S*)-Phg-↓O-Me	4	0.8	200	285
N-Acetyl-(*R*)-Phg-↓O-Me	7	0.005	0.7	
Hydrolysis with proteinase K (EC 3.4.21.14 from *Tritirachium album*)				
(*S*)-Phg-↓O-Me			0.6	2
(*R*)-Phg-↓O-Me			0.3	

Phg: phenyl-glycyl

zyme can vary by several orders of magnitude for different substrates, and that the specificity constants for some enzymes approach the rate of a diffusion-controlled reaction $k_{cat}/K_m \approx 10^8$ (M s)$^{-1}$. Trypsin and α-chymotrypsin are better esterases than amidases, and have different P_1-specificities. The difference in esterase and amidase activity is mainly caused by differences in the substrate binding (trypsin) or the catalytic reactions (α-chymotrypsin). These endopeptidases cannot hydrolyze a C-terminal unprotected amino acid as the carboxyl group carries a negative charge. The substrate is repelled by the negatively charged active site of these enzymes (see Fig. 2.5). The data in Table 2.4 for α-chymotrypsin and penicillin amidase also show that the specificity constants for the same substrate vary considerably with the source of the enzyme.

2.7.1.2 Stereoselectivities for Equilibrium- and Kinetically Controlled Reactions

It is clear from the data in Table 2.5 that the same substrate can be hydrolyzed by different enzymes. It is also apparent that the stereoselectivity of enzymes differs, and from this it follows that hydrolases, as well as other enzymes, have no strict stereospecificity. (The stereospecificity gives qualitative information on whether the enzyme prefers (*S*)- or (*R*)-substrates.) Enzymes that act preferably on (*S*)-enantiomers can also convert (*R*)-enantiomers, and *vice versa*. The results in Table 2.5 show that the stereoselectivity of the studied enzymes in all cases is due to changes both in k_{cat} and K_m for the enantiomeric substrates. The binding of the less-specific substrate results in a larger distance between the bond that is changed in the catalytic reaction and the catalytic subsite **C** than for the preferred substrate (Fig. 2.11). This results in different activation energies for the transformation of the (*S*)- and (*R*)-enantiomers (see Fig. 2.8). The data for penicillin amidase also show that the stereospecificity of an enzyme can be different in the S_1- and S_1'- binding subsites. The stereoselectivity is increased and the stereospecificity reversed when an additional binding site is involved in binding of the substrate (Fig. 2.11). This follows from the results for *N*-acetylated substrates in Table 2.5, where the acetyl group is bound in the S_2-subsite.

The stereoselectivity of enzymes is increasingly applied to produce pure enantiomers from prochiral compounds (assymetric synthesis) or racemic mixtures (racemate resolution). An α-keto acid is a prochiral compound that in living cells is transformed to a (*S*)-amino acid in reactions catalyzed by transaminases (transferases) or dehydrogenases (oxidoreductases). One of the first industrial enzyme processes was the production of (*S*)-amino acids from a racemic mixture of *N*-acetylated (*R*, *S*)-amino acids using a (*S*)-specific aminoacylase (a hydrolase) (Tosa et al., 1969). The deacylated (*S*)-amino acid was separated from the *N*-acetyl-(*R*)-amino acid by crystallisation or ion-exchange chromatography. The isolated *N*-acetyl-(*R*)-amino acid was racemized, after which almost all of the racemic mixture could be transformed into the (*S*)-amino acid (see Chapter 3).

The enzyme reactions that can be used for the production of pure enantiomers are shown in Figure 2.12. In the kinetic resolution of racemates II–IV, ≤50 % of the racemic mixture can be transformed to the desired enantiomer. To increase the yield, the latter must be separated from the other enantiomer that can be racemized in either chemical or enzymatic, catalyzed by racemases, processes. This can also be achieved when the kinetic resolution of racemates and the racemization can be carried out simultaneously, and this is referred to as "dynamic kinetic resolution". The application of enzymes for racemate resolutions or enantiomer production from pro-chiral compounds is covered in detail in Chapter 3. The stereoselectivities of the enzymes used in these processes must be known, but can be determined as follows for equilibrium- and kinetically controlled reactions.

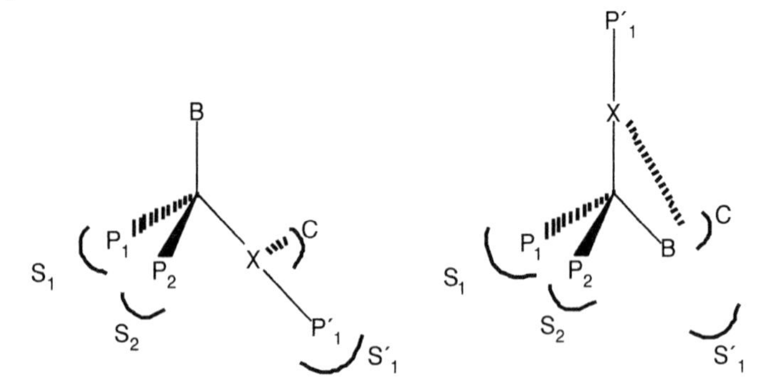

Fig. 2.11 Binding interactions that determine the stereospecificity of an enzyme. They are illustrated for an enzyme that is specific for an (*S*)-enantiomer (with groups P_1, P_2, B, and X-P'_1 bound to the asymmetric C atom) that has binding interactions with the S_1, S_2 and S'_1 binding subsites. The correct binding of this enantiomer is shown to the left. The catalytic subsite (C) is then near the bond changed in the catalytic reaction. For the (*R*)-enantiomer, this and the other three binding interactions cannot occur simultaneously. Only two of the three possible binding interactions are possible. In the case shown to the right one of these cases leads to a large distance between the catalytic site and the bond to be changed. Thus, either the binding of the (*R*)-enantiomer is weaker than for the (*S*)-enantiomer, or the catalytic reaction for the (*R*)-enantiomer is slower as the activation energy will increase (see Fig. 2.8).

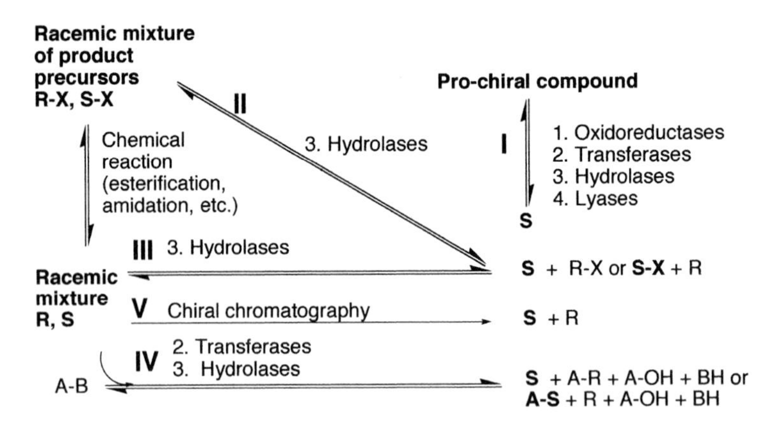

Fig. 2.12 Different enzyme and chromatographic separation processes for the production of pure (*S*)-enantiomers from racemic mixtures by kinetic resolution or asymmetric synthesis from pro-chiral compounds. The class of enzymes that can catalyze the different reactions are given based on the classification given in Section 2.2. The enzyme processes I–IV are equilibrium-controlled, with the exception of process IV, which is catalyzed by hydrolases in aqueous systems that are kinetically controlled reactions. In process I, a 100 % product yield can be obtained, whereas in the other processes (II–V) only 50 % of the racemic mixture can be converted to the desired enantiomer. To obtain 100 % yield, the unwanted enantiomer must be racemized. A similar scheme can be designed for the production of the (*R*)-enantiomer.

The stereoselectivity E of an enzyme-catalyzed reaction is defined as the ratio of the rates of the (*S*)- and (*R*)-enantiomer consumption (Chen et al., 1982):

$$E = \nu_S/\nu_R \tag{2.14}$$

For equilibrium-controlled reactions such as hydrolysis reactions, the stereo- or enantioselectivity has been shown to be:

$$E_{eq} = (k_{cat}/K_m)_S/(k_{cat}/K_m)_R \tag{2.15}$$

that is, the ratio of the specificity constants of the two enantiomers. The intrinsic enzyme property E_{eq} can be calculated from the determination of these constants, based on initial rate measurements, for the isolated enantiomers. Kinetically controlled reactions are only used for the resolution of racemic nucleophile mixtures (see Section 2.6, Scheme 2.1 and Fig. 2.12). The kinetically controlled resolution of a racemic mixture of the activated substrate (*R*,*S*)-A-B with a nucleophile NH, is not a suitable racemate resolution procedure as it gives two products with the desired enantiomer ((*S*)-A-OH and (*S*)-A-N)). Thus, for kinetically controlled racemate resolutions, the relationship for E for only the nucleophile binding subsite (S_1'-stereoselectivity) must be derived using the definition given in Eq. (2.14). The intrinsic (concentration independent) S_1'-stereoselectivity can only be determined from initial rates of the formation of the (*S*)- and (*R*)-enantiomeric product using a racemic nucleophile mixture (Galunsky and Kasche, 2002).[2)] Under the assumption that equilibrium is obtained in nucleophile binding it is (see p. 45):

$$E_{kin} = \nu_{A\text{-}(S)\text{-}N}/\nu_{A\text{-}(R)\text{-}N} = (k_t/K_N)_S/(k_t/K_N)_R \tag{2.16}$$

The relationships for E determined by Eqs. (2.15) and (2.16) can be used for a thermodynamic analysis of the pH-, *T*, and ionic strength dependence of E. This requires a large number of measurements, but allows the determination of E in the range from $\approx 10^{-4}$ to $\approx 10^4$.

A simpler and more rapid, but less accurate, estimation of E is based on the determination of the enantiomer excess of the product ee_P ($ee_P = |[P_S] - [P_R]|/([P_S] + [P_R])$) or substrate ee_S ($ee_S = |[S_S] - [S_R]|/([S_S] + [S_R])$) and the extent of the reaction when the enantiomeric excess is measured

$$c = 1 - ([S_S] + [S_R])/[S_S]_0 + [S_R]_0)$$

where the subscript 0 denotes the initial concentration, for an equilibrium- or kinetically controlled resolution of a racemic mixture from the relationships:

2) Previously, the ratio $(k_T/k_H)_S/(k_T/k_H)_R$, determined from separate measurements with the two enantiomers, was used as a measure for the stereoselectivity in kinetically controlled reactions. This ratio, however, depends on [NH] – that is, it is concentration-dependent. It should therefore not be used as a measure for the stereoselectivity of the enzyme.

$$E = \frac{\ln[1 - c(1 + ee_P)]}{\ln[1 - c(1 - ee_P)]} \quad \text{or} \quad E = \frac{\ln[1 - c(1 - ee_S)]}{\ln[1 - c(1 + ee_S)]} \tag{2.17}$$

E can also be determined from ee_P and ee_S using the relationship:

$$E = \frac{\ln\left(\dfrac{1 - ee_S}{1 + \dfrac{ee_S}{ee_P}}\right)}{\ln\left(\dfrac{1 - ee_S}{1 + \dfrac{ee_S}{ee_P}}\right)} \tag{2.18}$$

The E-values calculated from Eqs. (2.17) or (2.18) equal the values determined from the initial rates of Eqs. (2.15) or (2.16) only when the reverse reaction in equilibrium-controlled processes, product inhibition, and product hydrolysis in kinetically controlled synthesis (Fig. 2.9a) can be neglected. It may be assumed that the different

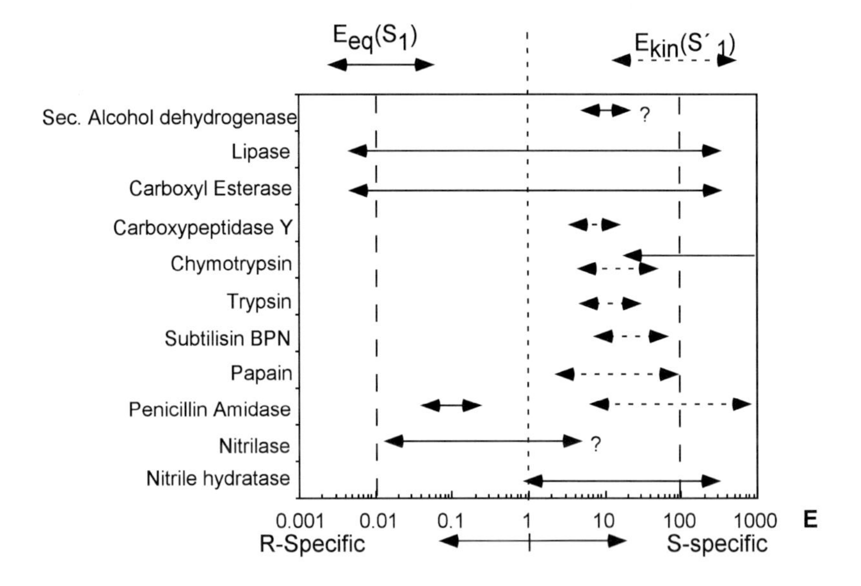

Fig. 2.13 Range of published stereoselectivities E_{eq} (Eq. 2.15) for the S_1-binding subsite (acyl-binding in Fig. 2.3) and E_{kin} (Eq. 2.16) for the S'_1-binding subsite (leaving group in equilibrium-controlled hydrolysis or nucleophile binding subsite in kinetically controlled synthesis) for enzymes (mainly hydrolases). Note that penicillin amidase has different stereospecificities in both binding subsites. The same applies for lipases and esterases, where the binding subsites are the acid and alcohol binding sites.

product enantiomers inhibit the enzyme differently. When this is the case, the steric purity of the desired product enantiomer is also influenced by product inhibition (Straathof and Jongejan, 1997).

The experimental error in determining E from Eqs. (2.17) and (2.18) is much larger than for the E-values determined from Eqs. (2.15) and (2.16), especially at ee_P- or ee_S-values close to 1. Therefore, Eqs. (2.17) and (2.18) can only be used to determine E-values in the range from $\approx 10^{-2}$ to 10^2.

Figure 2.13 provides a summary of the range of published stereoselectivities for enzymes in equilibrium- and kinetically controlled reactions. For applications in enzyme technological kinetic resolution of racemates the stereoselectivity E should be >100 or <0.01 for *S*- and *R*-specific enzymes respectively, in order to obtain products with high steric purity.

2.7.2 Dependency of k_{cat}, K_m and Selectivities on pH, Temperature, Inhibitors, Activators and Ionic Strength in Aqueous Solutions

In Section 2.7.1, quantitative relationships for "undisturbed" enzyme kinetics and stereoselectivity were derived. These were based on a discussion of the reactions occurring when the substrate is bound in the active site. The latter can be influenced either directly or indirectly by other compounds or changes in pH or temperature, and this can result in either an increase (activation) or decrease (inhibition) in the rate of the enzyme-catalyzed process. Quantitative studies on these changes can provide valuable information about the groups that control the enzyme function (which amino acid residues are in the active center, how enzymes can be selectively inhibited, etc.). This information in turn can be applied to the search for better enzymes by screening natural enzyme sources or creating new enzymes by random or directed evolution or rational protein design (see Section 2.11). The quantitative relationship describing the influence of these factors can be derived from Eqs. (2.8), (2.11), (2.12), and (2.14) to (2.18). How these properties are changed in aqueous suspensions or non-conventional solvents (organic solvents, supercritical gases, ionic liquids) will be discussed in Section 2.9.

2.7.2.1 pH Dependency

k_{cat} and K_m

Acidic and basic groups occur on the surface of enzymes, but they can also form part of the active site. Hence, their ionization state can influence the k_{cat} and K_m values of the enzyme. In Section 2.7.1, the pH-dependence of k_{cat} and K_m was not considered, and the values were considered to be apparent (pH-dependent). Here, this situation will be illustrated for α-chymotrypsin (see Fig. 2.5). When the histidine in the active site of α-chymotrypsin is protonated (EH), k_{cat} is 0 and the enzyme function is lost. Only that fraction of the enzyme where the histidine group is uncharged (f_-) contributes to the activity of the enzyme, at which time the apparent k_{cat}-value is:

$$k_{cat} = f\_ \, k^0_{cat} = \frac{k^0_{cat}}{1 + \frac{[H^+]}{K_1}} \tag{2.19}$$

that is, a function of pH where pK_1 is the pK-value for the histidine in the active site.

The same can apply for K_m when another group with $pK = pK_2$ influences the binding of the substrate. We assume that both the charged and uncharged group with this pK can bind the substrate with different K_m values. Then, the apparent K_m-value as function of pH is:

$$K_m = \frac{K'_m}{\left(1 + \frac{K_2}{[H^+]}\right)} + \frac{K''_m}{\left(1 + \frac{[H^+]}{K_2}\right)} \tag{2.20}$$

When Eqs. (2.19) and (2.20) are inserted into Eq. (2.8), we get the pH-dependence of the rate

$$v = \frac{\dfrac{k^0_{cat}\,[EH]_0\,[AB]_0}{\left(1 + \frac{[H^+]}{K_1}\right)}}{\left([AB]_0 + \dfrac{K'_m}{\left(1 + \frac{K_2}{[H^+]}\right)} + \dfrac{K''_m}{\left(1 + \frac{[H^+]}{K_2}\right)}\right)} \tag{2.21}$$

When $[H^+] > K_1$ (or $pH < pK_1$) and $K_1 > K_2$, v will increase with pH ($[H^+]$ decreases). When $pH > pK_1$ and pK_2, $[H^+]/K_1$ is ≈ 0 and the denominator will increase with pH, and v will decrease. From this, it follows that v will have a maximum between pK_1 and pK_2 (pH-Optimum). This is generally observed in enzyme kinetics (Fersht, 1999). From Figure 2.14 pK_1 and pK_2 can be determined. When $k_{cat} = 1/2\, k^0_{cat}$, $[H^+]$ equals K_1, or $pH = pK_1$. The amino acid residues in enzymes have similar pK-values

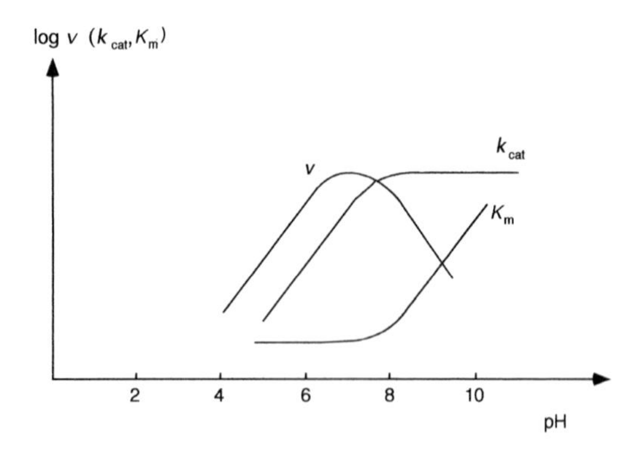

Fig. 2.14 pH-dependence of v, k_{cat} and K_m for α-chymotrypsin-catalyzed hydrolysis of an uncharged substrate (from Bender et al., 1964).

as the isolated amino acids. Thus, curves such as Figure 2.14 can provide qualitative information on the type of amino acids that give rise to the pH-dependence of k_{cat} and K_m. For α-chymotrypsin, such kinetic evidence has been supported by structural studies on the active site (see Fig. 2.5a). The amino acid residue with the pK_1-value ≈ 7 is His_{57}. The K_m-value is influenced by a residue with pK_2 ≈ 9 that must be an amino group. Equation (2.21) shows that a pH-optimum can only occur when two different amino acid residues control the pH-dependence of k_{cat} and K_m. This applies for most enzymes (exception: pepsin in the stomach with pH = 1–2). Quantitative studies to determine k_{cat} and K_m must thus be carried out in buffered systems. Especially when acids or bases are formed or consumed in an enzyme-catalyzed reaction, the buffers used must have a sufficient buffering capacity at the pH where the measurements are carried out. It is also important that the buffers do not participate in the enzyme-catalyzed reaction, so that the use of buffers with amino-groups that may deacylate acyl-enzymes are excluded. Therefore Tris buffers must not be used as a buffer in enzyme kinetic studies (Kasche and Zöllner, 1982). Charged substrates may also influence the pH-dependence of k_{cat} and K_m, either directly or indirectly, as they influence the ionic strength (Lummer et al., 1999).

Selectivities

In kinetically controlled reactions, the reactive group of the nucleophile is often an amino group, and it must be uncharged in order to act as a nucleophile. In the equation for the selectivity of such reactions (Eq. 2.3), k_t depends on the pK-value of this amino group. It decreases with decreasing pH below this value. Kinetically controlled synthesis must therefore be carried out at pH values at least one pH unit above the pK for the amino group of the nucleophile.

The stereoselectivity E_{eq} has been found to depend on pH for alcohol dehydrogenase and penicillin amidase (Lummer et al., 1999). This can occur when the pK of a group that influences k_{cat} is changed differently by the bound (*S*)- or (*R*)-substrate. The observed changes were less than one order of magnitude.

2.7.2.2 Temperature Dependency

k_{cat} and K_m

In Section 2.7.1 it was shown that k_{cat} is a rate constant for a monomolecular reaction and K_m a dissociation constant. From the temperature-dependence of these based on free energy changes (Figs. 2.7 and 2.8), the temperature-dependence of v (Eq. 2.8) can be written as:

$$v = \frac{A\,[EH]_0\,[AB]_0\,e^{-\frac{\Delta G^{\#}_{ES}}{RT}}}{[AB]_0 + B\,e^{-\frac{\Delta G_S}{RT}}} \tag{2.22}$$

where A and B are constants. The activation energy $\Delta G^{\#}_{ES}$ is always > 0, ΔG_S can be < 0 or > 0. This implies that k_{cat} always increases with temperature, whereas K_m can increase or decrease with temperature. $\Delta G^{\#}_{ES}$ and ΔG_S can be determined from the temperature-dependence of k_{cat} and K_m using Arrhenius plots. As $|\Delta G^{\#}_{ES}| > |\Delta G_S|$, v will generally increase with temperature up to a "temperature optimum" (Fig. 2.15), above which v will decrease. Enzymes have a flexible structure that, at higher temperatures, leads to unfolding (denaturation) as well as structural changes in the active center that reduce v. The temperature optimum for enzymes in aqueous solutions varies with their source from about 40–50 °C for enzymes from mesophilic organisms, to up to 100 °C for enzymes from thermophilic organisms. As a rule, enzyme processes should be carried out about 10–20 °C below the "temperature optimum" in order to avoid rapid denaturation of the enzyme. (The word "optimum" is here a misleading designation.)

Selectivities

The stereoselectivity E_{eq} for equilibrium-controlled processes (Eq. 2.15) is important when enzymes are used for the equilibrium-controlled kinetic resolution of racemates. From the thermodynamic quantities in Figures 2.7 and 2.8, the temperature-dependence of k_{cat} and K_m, the following equation for $\ln E_{eq}(T)$ can be derived:

$$\ln E_{eq} = -\frac{\left(\Delta(G^{\#}_{ES})_S - \Delta(\Delta G^{\#}_{ES})_R\right) - \left(\Delta(G_S)_R - \Delta(G_S)_S\right)}{RT} \tag{2.23}$$

that is, the enantioselectivity is, as expected, temperature-dependent – an effect which was first shown experimentally in 1989 (Phillips, 1996). In general, the differences in activation energies ($\Delta G^{\#}$) are larger than the differences in the binding energies (ΔG). Then, E_{eq} should decrease with temperature. This has also been observed in most studies on the temperature-dependence of E_{eq} (Phillips, 1996; Galunsky et al., 1997; Sakai et al., 1997). Figure 2.16 shows that E_{eq} can be increased by up to a factor of 10, or even reversed, by a reduction in temperature.

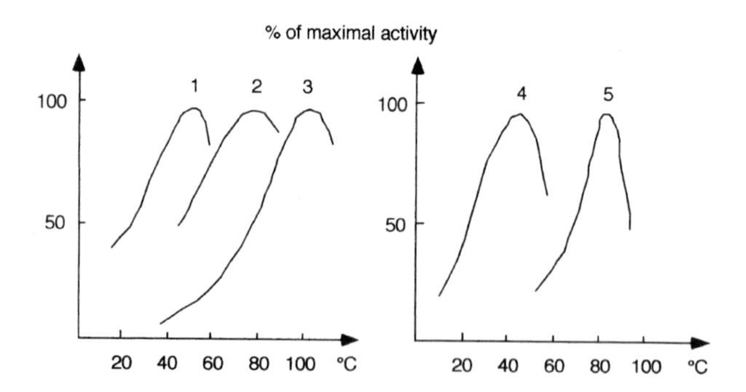

Fig. 2.15 Temperature-dependence of the activity of different enzymes. Amylase from pancreas (1); *Bacillus subtilis* (2), and *Bacillus licheniformis* (3); Peptidase from pancreas (4) and *Bacillus subtilis* (5) (from Godfrey, 1996).

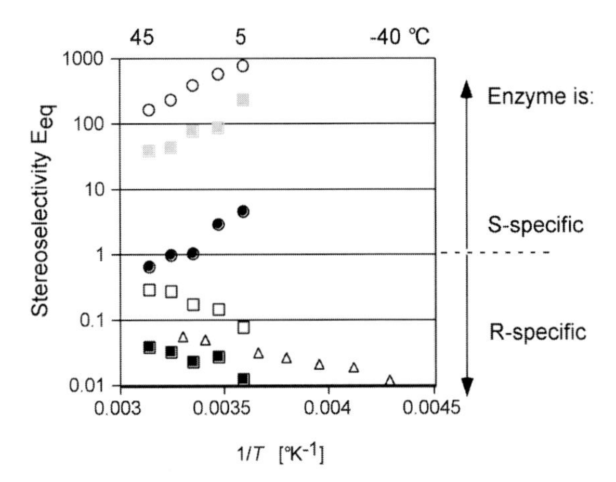

Fig. 2.16 Temperature-dependency of the stereoselectivity E_{eq} for different binding subsites of α-chymotrypsin and penicillin amidase in aqueous solution, and lipase in organic solvent. Bovine α-chymotrypsin: (S_1-selectivity, see Fig. 2.3) hydrolysis of *N*-acetyl-(*R*,*S*)-phenylglycine methyl ester (closed circles) and (*R*,*S*)-phenylglycine methyl ester (open circles) at pH 7.5 (Galunsky et al., 1997). Penicillin amidase from *E. coli*: (S_1-selectivity) hydrolysis of (*R*,*S*)-phenylglycine-amide (open squares), (*R*,*S*)-hydroxy-phenylglycine-ester (filled squares) and N-acetyl-(*R*,*S*)-phenylglycine (gray squares, S_1'-selectivity) at pH 7.5 (Kasche et al., 1996). Lipase from *P. cepacia* (triangles): (nucleophile binding site) esterification of racemic azirine with vinyl acetate in diethylether (Sakai et al., 1997). Note that stereospecificity of penicillin amidase differs in the S_1- and S_1'-subsites.

For kinetically controlled racemate resolutions, similar expressions as Eq. (2.23) for the temperature dependence of E_{kin} can be derived. Where its temperature dependence has been studied it has been found that E_{kin} in most cases decreases with increasing temperature (Kasche et al., 1996, Galunsky et al., 1997). This, and the temperature-dependence of E_{kin}, has been used to increase the yield and steric purity of the product in a kinetically controlled racemate resolution in the production of β-lactam antibiotics that is carried out at 5 °C (Zmiejewski et al., 1991).

2.7.2.3 **Binding of Activator and Inhibitor Molecules**

Every compound that directly or indirectly changes the structure of the active center changes the rate v (Eq. 2.8) of an enzyme-catalyzed reaction. When v increases, the compound is an activator, whereas inhibitors lead to a decrease in v. An enzyme can have binding sites for these also outside the active center (Fig. 2.17).

The enzyme may bind a competitive inhibitor in the active site:

$$EH + I_c \underset{k_d}{\overset{k_a}{\longleftrightarrow}} EH \cdot \cdot I_c \quad K_i = k_d/k_a \tag{2.24}$$

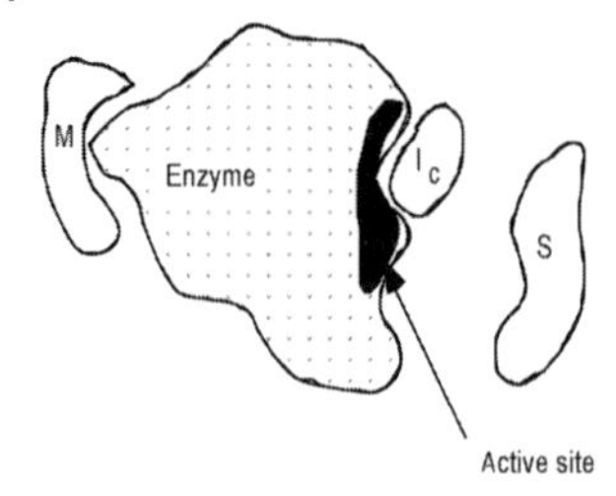

Fig. 2.17 Interactions of enzymes with inhibitors (I_c is a competitive inhibitor), modulators M (activators or non-competitive inhibitors), and substrates S.

When such inhibitors are bound, the enzyme cannot perform its function – that is, it is competitively inhibited. This type of inhibitors can be:

1. Products (product inhibition) such as H^+ for hydrolases when this ion is formed during the reaction, as in the case for His_{57} in α-chymotrypsin can be protonated, or phenylacetic acid for penicillin for penicillin amidase (see Fig. 2.5).
2. Substrates that can acylate enzymes almost irreversibly – that is, they are slowly deacylated. To this sugroup belong neurotoxins, or penicillins in Gram-positive bacteria.
3. Similar effects are caused by chelating compounds (such as EDTA) that can bind metal ions essential for the activity of the enzyme (Ca^{2+} or Zn^{2+}).

When the inhibitor or activator is bound outside the active site, the equilibrium $M + EH \leftrightarrow M \cdot\cdot EH$ must be considered besides the enzyme reactions (Fig. 2.17 and Eqs. 2.4 to 2.6). The conformation (tertiary structure) of $M \cdot\cdot EH$ differs from EH. The active center is not involved, and the enzyme-catalyzed reaction can occur with $M \cdot\cdot EH$. As the structure of the active center is changed in $M \cdot\cdot EH$, k_{cat} and K_m can differ from those for EH. When v increases with the concentration of M, the latter is an *activator* for the enzyme. This frequently applies for metal ions such as Ca^{2+} for trypsin or Mg^{2+} for phosphodiesterase. The opposite case where v decreases with the concentration of M is *non-competitive inhibition*, when M is a non-competitive inhibitor. The decrease in v can either be due to an increase in K_m in $M \cdot\cdot EH$ or that k_{cat} for $M \cdot\cdot EH$ is smaller than for EH. Both cases can occur simultaneously, at which point we have a *mixed inhibition*. Expressions for k_{cat} and K_m in the presence of competitive and non-competitive inhibitors (where $K_{m,\,EH} = K_{m,EH..M}$ and $k_{cat,EH..M} = 0$) are given in Table 2.6. The type of inhibition can be determined by measuring $v = f([S])$ at constant inhibitor concentration and plotting the data using Eq. (2.13) (Fig. 2.18).

Substrates can also be inhibitors, for example when they are bound incorrectly in the active site or bound outside the active site. An example where this applies is for hydrolases that hydrolyze oligosaccharides or the two-substrate reactions (Eqs. 2.5 and 2.6) where v is measured keeping the concentration of one substrate constant while the content of the other substrate is changed. This is frequently observed at high substrate concentrations that are of interest for enzyme technology. In these

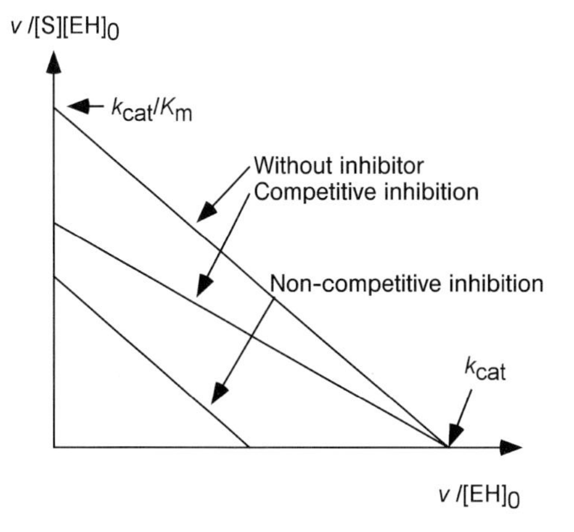

Fig. 2.18 Graphical determination of the type of inhibition from Eadie–Hofstee plots with and without a constant concentration inhibitor. The type of substrate inhibition (competitive or non-competitive) cannot be determined from such curves.

Table 2.6 Influence of competitive and non-competitive inhibitors on k_{cat} and K_m.

	Competitive inhibitor	*Non-competitive inhibitor*
k_{cat}	unchanged	$k_{cat}/(1 + [I]/K_I)$
K_m	$K_m\,(1 + [I]/K_I)$	unchanged

[I] = free inhibitor concentration.

cases, maxima in v as function of the substrate concentration can be observed. For non-competitive substrate inhibition with [S] = [I] = $[AB]_0$ (S for substrate, [NH] = 0,) Table 2.6. provides the following expression for Eq. (2.8) for this case:

$$v = \frac{k_{cat}\,[EH]\,[S]}{K_m + [S]\left(1 + \frac{K_m}{K_i}\right) + \frac{[S]^2}{K_i}} \tag{2.25}$$

2.7.2.4 Influence of Ionic Strength

Enzymes are polyelectrolytes, and the active site may consist of charged amino acid residues. Several substrates of interest in enzyme technology, such as amino acids, peptides, organic acids, and nucleotides, contain acidic or basic functional groups. In enzyme technology, these substrates are used at concentrations of up to ≈1 M, at which the ionic strength I is given by:

$$I = 0.5 \sum c_i z_i^2 \tag{2.26}$$

where c_i = ion concentration (ion i); and z_i = ion charge (ion i).

The ionic strength (I) influences the activity of the ions, the rates and the equilibrium constants. The activity of a charged substrate S is, at ionic strengths up to ≈0.1 M, given by:

$$\{S\} = [S] \cdot \exp\left(-\frac{z_s^2 \cdot A\sqrt{I}}{1 + B\sqrt{I}}\right) = [S]\,\gamma \tag{2.27}$$

where γ is the activity coefficient, z_S the charge, and A and B constants.

The influence of the ionic strength and ion-ion interactions on the rate (k_{cat}) and equilibrium constant (K_m) can be expressed as follows, with A ≈ 0.5 and B ≈ 1 in Eq. (2.27) (Martinek et al., 1971; Dale and White, 1982):

$$k_{cat} = k_{cat}^{o} \cdot \exp\left(z_S z_E \sqrt{I}/(1 + \sqrt{I})\right) \tag{2.28}$$

and

$$K_m = K_m^{o} \cdot \exp - \left(z_S z_E \sqrt{I}(1 + \sqrt{I})\right) \tag{2.29}$$

where the superscript o applies to the ionic strength-independent constants.

This ionic strength dependence is shown graphically in Figure 2.19, and shows that ionic strength effects cannot be neglected at concentrations >0.1 M. As I also influences the pK-groups that control the pH dependency of the enzyme, it may also influence the pH-optimum. Few data are available on the influence of I on k_{cat}, K_m and selectivities, though some are provided in the references to this section. The selectivity and yield in the kinetically controlled synthesis of semisynthetic β-lactam antibiotics (ampicillin, cephalexin) have been found to decrease with ionic strength

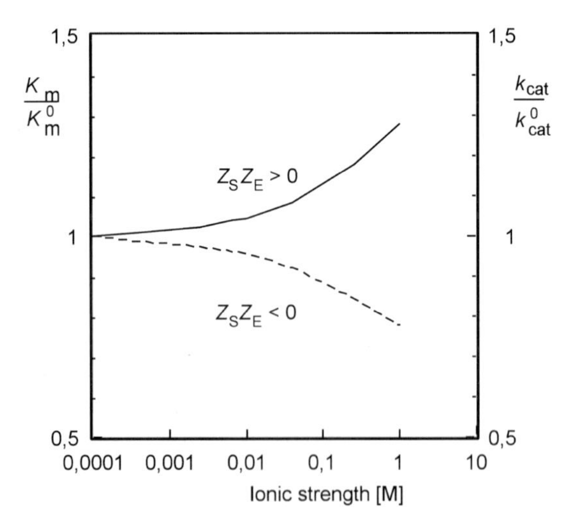

Fig. 2.19 Ionic strength-dependence of k_{cat} and K_m of an enzyme-catalyzed reaction for different charges of the substrate and the active site $|Z_E| = |Z_S| = 1$.

that reduces the binding of the nucleophile to the charged binding site on the enzyme (Kasche, 1986, and unpublished data).

2.8 End-points of Enzyme Processes and Amount of Enzyme Required to Reach the End-point in a Given Time

The time dependency and possible end-points of enzyme-catalyzed reactions are shown in Figures 1.9 and 2.9. At the end-point, a maximal product yield or optimal product quality is desired. In equilibrium-controlled processes, the enzyme cannot influence the yield, but it can influence the time to reach the end-point. In kinetically controlled synthesis (Figs. 1.9 and 2.9), or in processes where fibers are treated by enzymes either to reach an optimal fiber length or to polish the fibers (treatment of recycled paper or to improve cotton or wool fibers for textile production), the product concentration or product quality has a maximum at which the process must be stopped by removing the enzyme. This maximum, is (in contrast to the equilibrium-controlled process) influenced by the properties of the enzyme. In kinetic racemate resolutions and asymmetric synthesis, it is not only the yield that is important in the selection of a suitable end-point, but rather the steric purity of the product. The selection of end-points for such processes will be discussed in the end of this section.

In a technological process the enzyme costs should be less than 5–10 % of the total process costs. Hence, in order to estimate costs, the amount of enzyme required to reach the end-point within a given time must be determined. For equilibrium-controlled processes this can be done as follows.

From the given time *t* and the desired change in substrate concentration, from the initial $[S]_0$ to the final concentration at the selected end-point $[S]_t$ the required space-time yield (*STY*) is:

$$STY = \frac{[S]_0 - [S]_t}{t} \tag{2.30}$$

The *STY* has the dimensions [mol substrate per unit time and reactor volume], and is an important quantity in enzyme technology as it determines the reactor volume. It is a given quantity that is determined by previous processing steps. When all penicillin G produced per day by fermentation (see Chapter 1) must be hydrolyzed to 6-APA, this gives the minimal *STY* required in the enzyme process. The equations derived in Section 2.7.1 are insufficient to calculate the enzyme amount required to obtain the necessary *STY* as they are based on initial rates where only a fraction of the substrate is converted to product. The reverse reactions and product inhibition must also be considered. Therefore, the analytical description of initial rate enzyme kinetics, which may be found in biochemically oriented textbooks on enzymology, must be amended for enzyme technological purposes. The enzyme amount required to reach the end-point in a given time can be derived as follows. The rate of the change in the substrate content is:

$$\frac{d[S]}{dt} = -\frac{V_{max}(t)}{f([S],[P])} \tag{2.31}$$

where $V_{max}(t)$ is given by the following relationship:

$$V_{max}(t) = V_{max,0}\, e^{-k_i t} \tag{2.32}$$

assuming that the enzyme is inactivated in a first-order reaction with the rate constant k_i, and that $V_{max,0}$ is the initial enzyme concentration. The differential equation (Eq. 2.31) can be solved after variable separation, and provides the following relationship for $V_{max,0}$

$$V_{max,0} = \frac{\int_{[S]_t}^{[S]_0} f([S],[P])\, d[S]}{(1 - e^{-k_i t})/k_i} \tag{2.33}$$

where the denominator tends to t when k_d approaches 0 – that is, enzyme inactivation can be neglected. The function f([S], [P]) for a process where the reverse reaction can be neglected can be derived from Eq. (2.8) and the relationships in Table 2.6. It is:

$$\frac{K_m + [S]}{[S]} \tag{2.34}$$

when substrate and product inhibition can be neglected,

$$\frac{K_m + (1 + K_m/K_i)[S] + [S]^2/K_i}{[S]} \tag{2.35}$$

for non-competitive substrate inhibition, and

$$\frac{K_m(1 + [S]_0/K_i) + (1 - K_m/K_i)[S]}{[S]} \tag{2.36}$$

for competitive product inhibition.

In many important equilibrium-controlled enzyme processes the theoretical yield is below 100 % substrate conversion. This applies for:

- the isomerization of glucose, with ca. 50 %,
- the hydrolysis of penicillin and cephalosporin C, with ca. 90–95 %,
- the hydrolysis of maltose, with ≈95 %

substrate conversion. For these processes, 100 % substrate conversion can only be obtained at pH-values and temperatures far outside the range where the enzymes used as biocatalysts are stable. For these processes, the back reaction (see Section 2.7.1) cannot be neglected, and Eqs. (2.33) to (2.36) can only be used for rough estimates, as Eqs. (2.8) and (2.33) do not include the reverse reaction. For one-substrate reactions, such as the isomerization of glucose (Glu) to fructose (Fru), the following Michaelis–Menten equation can be derived:

$$v = \frac{\frac{k_{cat}}{K_m}[\text{Glu}] - \frac{k'_{cat}}{K'_m}[\text{Fru}]}{1 + \frac{[\text{Glu}]}{K_m} + \frac{[\text{Fru}]}{K'_m}}[\text{EH}]_0 \tag{2.37}$$

where the prime applies for the reverse reaction. Corresponding equations for the hydrolysis of penicillin can be found in the literature (Spiess et al., 1999).

In the kinetically controlled synthesis, the suitable end-point is the maximum product concentration $[\text{AN}]_{max}$ (see Fig. 2.9). It is:

$$[\text{AN}]_{max} = \frac{[\text{AB}]_0 [\text{NH}]_0}{(k_H/k_T)_{app}[\text{H}_2\text{O}] + [\text{NH}]_0} \tag{2.38}$$

when the rate of hydrolysis of AN can be neglected, or

$$[\text{AN}]_{max} = \left(\frac{(k_T/k_H)_{app}}{(k_T/k_H)_{app}[\text{NH}] + [\text{H}_2\text{O}]}\right)\frac{\left(\frac{k_{cat}}{K_m}\right)_{\text{AB, NH}}}{\left(\frac{k_{cat}}{K_m}\right)_{\text{AN, NH}}}[\text{NH}][\text{AB}] \tag{2.39}$$

when the rate of hydrolysis of AN cannot be neglected (Kasche, 2001). From these relationships it follows that $[\text{AN}]_{max}$ depends on the properties of the enzyme. Analytical equations from which the amount of enzyme required to reach these concentrations in a given time can be determined have not yet been derived.

Optimal yields at the end-points for equilibrium- and kinetically controlled processes are generally obtained at pH- and temperature-values, outside the range where the enzymes used as biocatalysts have their optimal properties. From this, it follows that *in the design of enzyme processes, both the properties of the enzyme and of the enzyme-catalyzed process must be considered.* This will be illustrated for the influence of temperature and pH.

2.8.1
Temperature-Dependency of the Product Yield

For equilibrium-controlled processes, thermodynamic data are required to analyze the temperature-dependency of the product yield, but unfortunately these are still missing for many important enzyme processes. Some data are provided in Table 2.7 (Tewari, 1990).

For kinetically controlled processes, it has been observed that the yield decreases with temperature (Kasche, 1986). Few data exist relating to investigations of whether the reactions are either endo- or exothermal (Michaelis, 1991).

Table 2.7 Thermodynamic data for enzyme catalyzed reactions (end-point = equilibrium) (Goldberg and Tewari, 1995; http://xpdb.nist.gov/enzyme_thermodynamics/enzyme1.pl).

Process	*Enthalpy change* $[kJ\ mol^{-1}]$ [a]	*Process is*	*Change in yield with increasing temperature*
Reduction of formate	–15	exothermal	smaller
Hydrolysis of			
Penicillin G	20	endothermal	larger
Penicillin V	<0	exothermal	smaller
Saccharose	–15	exothermal	smaller
Lactose	0.5	?	?
Maltose	–4	exothermal	smaller
Isomaltose	>0	endothermal	larger
Cellulose	<0	exothermal	smaller
Starch	<0	exothermal	smaller
Lipids	?	?	?
Peptide bond	–5 [b]	exothermal	smaller
Formation of Asp from fumarate and NH_3	–24	exothermal	smaller
Isomerization of glucose	3	endothermal	larger

[a] At pH 7–8, 25–40 °C.
[b] Between protected amino acids (also in polypeptides).

2.8.2
pH-Dependency of the Yield at the End-point

For equilibrium-controlled processes, where acid or bases are produced or consumed, the equilibrium constant and the product yield is a function of pH. This applies for the hydrolysis and synthesis of peptides, peptide antibiotics (such as cephalosporins and penicillins), lipids, and esters. Calculation of the equilibrium as function of pH can be performed when the pK-values of the substrates and products are known. The apparent association constant K_{app} for an equilibrium condensation of a base and an acid is, when all activity coefficients (Eq. 2.27) are 1,

$$K_{app} = \frac{(\Sigma\,[\text{condensation product}])\,[H_2O]}{(\Sigma\,[\text{acid}])\,(\Sigma\,[\text{base}])} = K_{ref}\,\frac{\left(\dfrac{\Sigma\,[\text{condensation product}]}{[\text{reference condensation product}]}\right)}{\left(\dfrac{(\Sigma\,[\text{acid}]}{[\text{reference acid}]}\right)\left(\dfrac{\Sigma\,[\text{base}]}{[\text{reference base}]}\right)} \tag{2.40}$$

where the sums are over all proteolytic states of the substrates and the product K_{ref} is a reference association constant applying to the condensation of only one of the different protonated states of the acid and base. K_{ref} applies to a reference pH-value and is thus independent of pH. The ratios in the nominator and denominator in Eq. (2.40) can be determined from dissociation schemes, as shown in Eq. (2.41).

$$H_3^+N\text{–}R_i\text{–}COOH \overset{K_1}{\longleftrightarrow} H_3^+N\text{–}R_i\text{–}COO^- + H^+ \overset{K_2}{\longleftrightarrow} H_2N\text{–}R_i\text{–}COO^- + 2H^+ \quad (2.41)$$

In Figure 2.20, K_{app}/K_{ref} has been calculated as a function of pH for some condensation reactions. The yields increase with K_{app}. For the hydrolysis of the condensation products the yields increase with decreasing K_{app}. For the synthesis of peptides, the highest yields are obtained at neutral pH-values, where the serine peptidases that can be used as biocatalysts have their pH optimum (see Fig. 2.14). For the synthesis of penicillins with penicillin amidase and esters with serine peptidases, the highest yields are obtained at pH-values much below the pH optima of these enzymes. Penicillin amidases also have a low pH-stability at pH 5 (Fig. 2.28), and the association

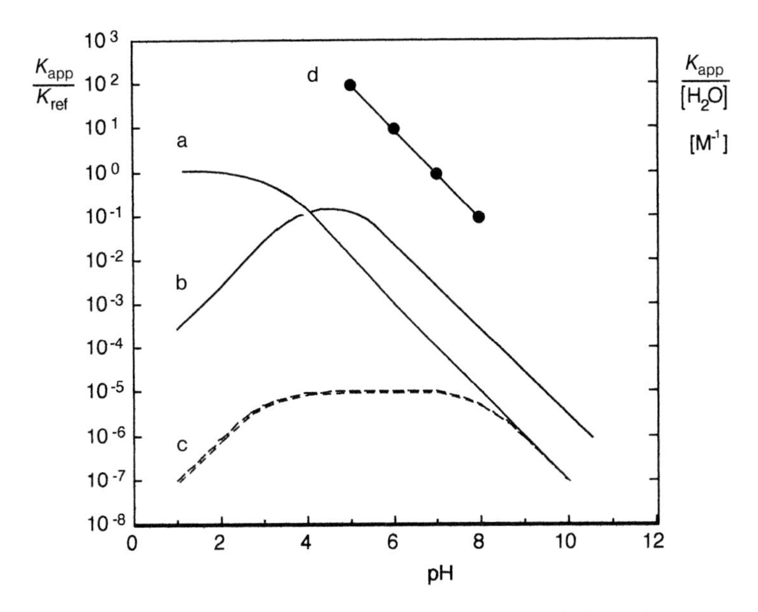

Fig. 2.20 Calculated K_{app}/K_{ref} from Eq. (2.40) and experimental values ($K_{app}/[H_2O]$) pH-dependence of the association constants for the formation of esters (a) peptide antibiotics (b, d) and peptides (c). (a) Ester synthesis from a carboxylic acid ($pK_1' = 3$), reference states: the uncharged compounds. (b) Synthesis of penicillin G ($pK_1''' = 2.9$) from phenylacetic acid ($pK_1' = 4.9$) and 6-aminopenicillanic acid (6-APA, $pK_1'' = 2.9$, $pK_2'' = 4.6$), reference states: uncharged product and neutral acid and positively charged 6-APA. (c) Synthesis of a dipeptide from an amino-protected ($pK_1' = 3.0$) and carboxyl-protected ($pK_2'' = 7.8$) amino acid, reference states: the uncharged compounds. (d) Experimental data for the synthesis of penicillin G at 25 °C ($I = 0.2$ M). ′, ″, and ‴, denote the dissociation constants for the acid, base and condensation product, respectively; Subscript 1 denotes the acid dissociation constant subscript 2 the base dissociation constant.

constants are too low to obtain high yields. Thus, these enzymes are not suitable as biocatalysts for the equilibrium-controlled synthesis of these condensation products. The opposite applies when they are used as biocatalysts for the hydrolysis of the condensation products. From Figure 2.20, it follows that the hydrolysis of penicillin with a high product yield should be carried out at pH > 9. At this pH, however, the possible biocatalyst – penicillin amidase from *E. coli* – has a low pH-stability (Fig. 2.28). The process with this enzyme must therefore be carried out under suboptimal conditions. This again demonstrates that both the properties of the process and the enzyme must be considered to select the optimal condition for an enzyme process.

For kinetically controlled processes, the pH-dependency of the maximum product concentration can be derived from Eqs. (2.38) and (2.39). This is, however, difficult to perform as the pH-dependency of the rate constants in these equations cannot easily be determined. These processes must be carried out at pH-values where the nucleophile is uncharged – that is, above the pK-values of these. Only the uncharged nucleophiles can deacylate the acyl-enzyme. Therefore, penicillin synthesis is carried out at pH-values >5 and peptide synthesis at pH >9 (Kasche, 1986).

2.8.3
End-points for Kinetic Racemate Resolutions

In these processes not only high product yields but also a high steric purity of the product is essential. The steric purity is given by the enantiomeric excess of the product (ee_P) or substrate (ee_S), the yield of the desired enantiomer (product or remaining substrate) is less than 50 % of the initial racemate concentration (see Fig. 2.12, p. 54 and Chapter 3). These experimental conditions must also be chosen so that more than 90 % of the desired enantiomer is formed in the equilibrium- or kinetically controlled process, respectively. For equilibrium-controlled processes this requires that the equilibrium constant and its pH- and *T*-dependence is known (see above). For kinetically controlled processes, this can be achieved when the concentration ratio $[AB]_0/[NH]_0$ and the selectivity v_T/v_H in Eq. (2.1) is larger than 1 (Kasche, 1986).

Selection of the end-point based on steric purity can be made based on plots of ee_P and ee_S as function of the extent of the reaction (Fig. 2.21). From this figure it fol-

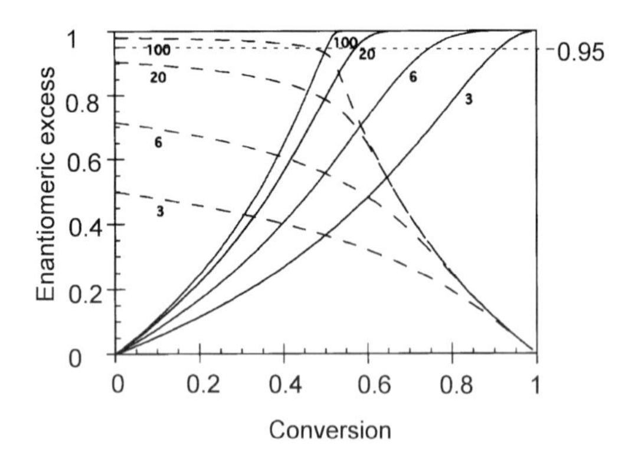

Fig. 2.21 The enantiomeric excess ee_p of the product (----) and ee_S for the remaining substrate (——) as a function of the extent of the reaction for kinetic racemate resolutions catalyzed by (*S*)- or (*R*)-specific enzymes with different E-values (the E-values for the (*R*)-specific enzymes are are the inverse of the given E-values). The horizontal dotted line (100) shows that a product yield of ≥45 % with ee ≥0.95 requires a (*S*)-specific enzyme with E ≥100.

lows that for $ee_P \geq 0.95$ and product yields ≥45 % of the initial racemate concentration, (*S*)-specific enzymes with $E \geq 100$ or (*R*)-specific enzymes with $E \leq 0.01$ must be used. The amount of enzyme required to reach this end-point for equilibrium-controlled processes can be calculated from Eq. (2.33) when product inhibition, and the competitive inhibition of the undesired enantiomer is considered.

2.9
Enzyme-catalyzed Processes with Slightly Soluble Products and Substrates

Enzymes have been evolved over a billion years to secure the survival of living systems where their biological function is carried out in aqueous systems. Enzymes catalyze reactions in these systems where the substrate concentrations are comparatively small (< 10 mM). In enzyme processes, at least one order of magnitude higher concentrations are desired to reduce processing costs. This may be above the solubility (in water) of many hydrophobic natural compounds or organic molecules that are substrates or products in enzyme processes. This section covers methods to solve this problem. They are:

1. The enzyme reactions can be carried out in aqueous suspensions (or emulsions) with precipitated substrate and/or product (see Section 2.9.1). This is possible when the rate of mass transfer (solid to liquid, or emulsion to liquid) is not rate-limiting (Fig. 2.22). This requires that the size of the solid particles or droplets in the emulsion are small (Eq. 2.42), at which point high *STY* can be achieved (Fig. 2.23).

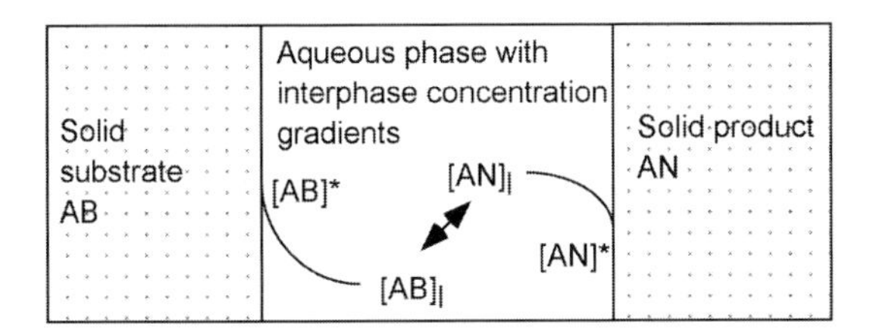

Fig. 2.22 The concentration gradients at the phase boundaries in enzyme-catalyzed processes in aqueous suspensions involving insoluble substrates or products. The asterisk denotes the concentration at the phase boundary at equilibrium between the solid and aqueous phases.

2. When method 1 is not possible, the reaction can be carried out in different one- or two-phase systems with solvents in which the substrates and products are soluble (Table 2.8). Besides traditional organic solvents, supercritical gases such as CO_2, or ionic liquids can be used here (Fig. 2.24). The organic solvents and ionic liquids used must be non-toxic and easy to recycle. In such systems equilibria that are unfavorable in water can be shifted in the desired direction (Eq. 2.43 and Fig. 2.25).

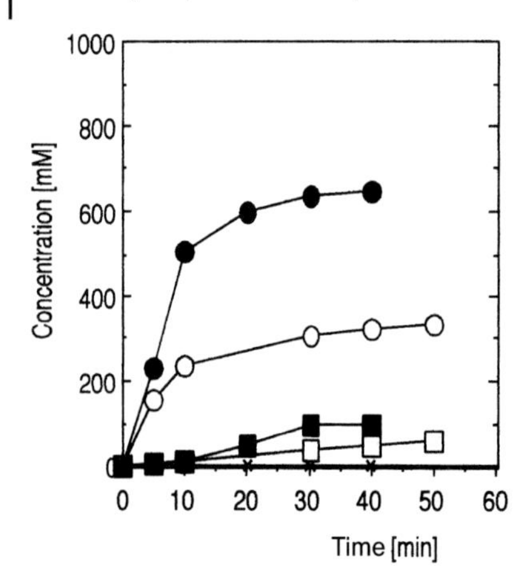

Fig. 2.23 Kinetically controlled synthesis of soluble *N*-acetyl-(*S*)-Tyr-(*S*)-Arg-NH_2 (ATAA, circles), a precursor of the dipeptide (Tyr-Arg) kyotorphin, from *N*-acetyl-(*S*)-Tyr-OEt (ATEE) as a suspension and soluble (*S*)-Arg-NH_2 with the peptidase α-chymotrypsin (CT). *Conditions:* enzyme content 10 µg/ml, 25 °C; pH 9.0 (carbonate buffer, I = 0.2 M), the pH is kept constant during the reaction. *Squares:* the soluble hydrolysis product *N*-acetyl-(*S*)-Tyr. *Closed symbols:* starting concentrations: 800 mM (*S*)-Arg-NH_2 and 750 mM (total content) ATEE as a suspension. *Open symbols:* starting concentrations: 400 mM (*S*)-Arg-NH_2 and 400 mM (total content) ATEE as a suspension. The dissolved ATEE concentration in the aqueous system during the reaction was <1 mM (Kasche and Galunsky, 1995).

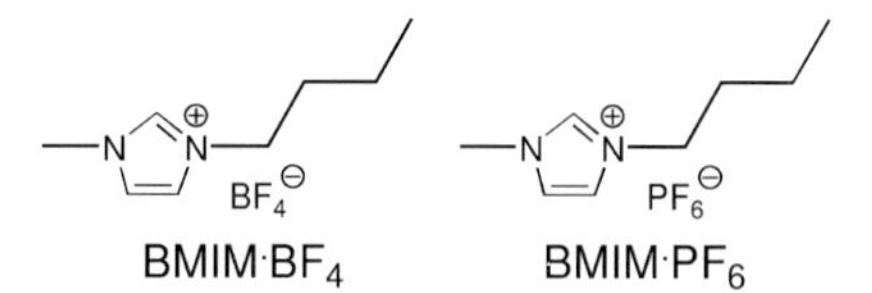

Fig. 2.24 Structures of the two most often used ionic liquids 1-butyl-3-methylimidazolium tetrafluoroborate (BMIM-BF_4) or its hexafluorophosphate (BMIM-PF_6).

In many processes one of the substrates (such as alcohols) can be used as a solvent. Such systems are especially used in enzyme processes where lipase (EC 3.1.1.3) is used as a biocatalyst. This enzyme catalyzes the conversion of slightly soluble hydrophobic substrates at organic–aqueous interphases also in living systems. In these "dry" systems, water is required for optimal enzyme activity (Fig. 2.26), and a minimal number of water molecules must be bound to the enzyme to provide it with the flexibility required to carry out its function.

2.9.1 Enzyme-catalyzed Processes in Aqueous Suspensions

In an aqueous suspension with a slightly soluble substrate that is converted by an enzyme in the aqueous phase, a concentration gradient is formed in the interphase between the solvent and the solid phase substrate (Fig. 2.22). This causes a mass transfer of the substrate from the solid to liquid phase. The rate of dissolution of the substrate is:

$$k_{L,AB}\, a_{AB}\, ([AB]^{*} - [AB]) = \frac{Sh\, D_{AB}\, 3\, \alpha}{2\, (r_{AB})^{2}}\, ([AB]^{*} - [AB]_{l}) \qquad (2.42)$$

where $k_{L,AB}$ and a_{AB} are the mass transfer coefficient and surface area per volume unit, respectively; Sh is the Sherwood number, D_{AB} is the diffusion coefficient of the substrate, r_{AB} is the average radius of the solid substrate particles, and α is the volume fraction of the solid particles in the suspension. This rate can be estimated for a 10 % suspension ($\alpha = 0.1$) of a substrate with MW = 500 Da ($D_{AB} \approx 5 \times 10^{-6}$ cm^2/s) Sh = 10, $r_{AB} = 10^{-2}$ cm, and a concentration gradient of 1 mM. This gives a rate of dissolution of (Eq. 2.42):

$$75 \times 10^{-6}\ \text{M/s} = 4500\ \text{U/l}$$

With a stationary substrate concentration in the aqueous phase > K_m, and an enzyme content of 10 µM, the rate of of the enzymatic reaction equals the rate of dissolution when $k_{cat} \geq 8.3\ s^{-1}$. From this and Table 2.4, it follows that many enzymes have turnover numbers above this value. Therefore it should be possible to carry out enzyme processes in suspensions (emulsions) of slightly soluble substrates with total concentrations of more than 1 M successfully and with high space-time yield. This has been demonstrated in some studies (Fig. 2.23) (Kasche and Galunsky, 1995; Cao et al., 1997; Vulfson et al., 2001).

One reason why such processes have not yet been studied in more detail or applied in enzyme technology might be that slightly soluble products can be formed. This may be favorable in equilibrium-controlled processes with slightly soluble products, where the equilibrium is shifted in the direction of the product. However, in enzyme processes carried out in suspensions, the enzyme must be removed once the end-point has been reached. This is difficult with soluble enzymes, but it is possible when the enzyme is immobilized (see Chapters 6 to 8) in non-porous or porous supports that can be separated from the insoluble substrate or product. This separation is especially easy with magnetic supports (Bozhinova et al., 2004).

For an insoluble product, precipitation of the product in the pores of a porous support must be avoided to allow reuse of the immobilized enzyme. The solubility is inversely proportional to r_{AB} (Freundlich, 1909). Thus, precipitation can be avoided when supports with either small or no pores are used (Kasche and Galunsky, 1995; Bozhinova et al., 2004), see also Section 8.8.

2.9.1.1 Changes in Rates, k_{cat}, K_m, and Selectivities in these Systems Compared with Homogeneous Aqueous Solutions

The enzyme-catalyzed reaction occurs in the aqueous phase. Thus, k_{cat}, K_m and the stereoselectivities are not changed. This also applies for the rate and the selectivity $(k_T/k_H)_{app}$, when the rate of dissolution is not limiting (Eq. 2.42), for the same bulk concentration. When the water content is low in these systems it may be difficult to control the pH for reactions where acids or bases are either consumed or formed. This may reduce the rates and change $(k_T/k_H)_{app}$ for kinetically controlled processes (Vulfson et al., 2001).

2.9.2
Enzyme-catalyzed Processes in Non-conventional Solvents Where Products and Substrates Are Dissolved (and the Enzyme Suspended)

Biological systems are generally aqueous solutions with up to 20 % soluble and some proportion of slightly soluble components such as membranes, fats, waxes, structural proteins (hair, wool), and cellulose – that is, they are heterogeneous systems. The slightly soluble fraction consists of hydrophobic, low molecular weight molecules (lipids) or biopolymers. Many biologically active lipids, vitamins, polycyclic or peptide hormones are slightly soluble in water, and in biological systems the concentrations at which they are present and biologically active are below their solubility (≤1 μM).

Initially, the enzymatic technological production of hydrophobic compounds was considered to be difficult due to their low solubility. That this does not apply for aqueous suspension systems, was outlined in Section 2.9.1. The possibility of using one- or two-phase systems with solvents to increase the solubility of hydrophobic compounds in enzyme processes was demonstrated more than 100 years ago, and has also been extensively studied during the past 20 years (Halling, 1994, 2000; Vulfson et al., 2001). Such solvents, where the enzyme-catalyzed reactions occur in the aqueous phase or in the phase with suspended enzyme, include (Table 2.8)

- water-miscible solvents (methanol, ethanol, acetone, DMSO, dimethylformamide, etc.);
- solvents that are not miscible with water (liquid alkanes, higher alcohols, ethyl acetate, chlorinated hydrocarbons, etc.);
- supercritical solvents such as CO_2 (with the limitation that many substances exhibit a low solubility in these); and
- ionic liquids (salts that are liquid at room temperature).

The use of organic solvents in industrial processes is problematic due to their dangerous properties (toxicity, flammability, etc.), and this has led to strict regulations for their use and containment. In many countries regulations derived from laws for the protection of the environment require the use of less dangerous materials in the processing industry when this is possible. This search for environmentally more friendly compounds must be documented. This applies to solvents such as halogenated hydrocarbons that destroy the ozone layer in the atmosphere, and these have therefore been replaced by other solvents, and must not be used for enzyme technological purposes. Other organic solvents must be recycled in the production plant, as their emission is (or must be) limited by environmental regulations. Supercritical solvents such as CO_2 are also environmentally friendly alternatives, but whether this also applies to ionic liquids – which have been used only relatively recently in enzyme technology – requires further study, notably with regard to their influences on biological systems (Poliakoff et al., 2002; Park and Kazlauskas, 2003). Ionic liquids (ILs) are low-melting point (<100 °C) salts that represent a new class of non-aqueous solvents. In contrast to conventional organic solvents, they do not possess a vapor pressure, so their reuse should be facilitated. As they do not evaporate, they are also considered as "green solvents" (Seddon, 2003), but few data are currently available regarding their toxicity and biodegradability.

The most often-used ILs in biocatalysis are 1-alkyl-3-methylimidazolium salts (Fig. 2.24). The solvent properties of these can be modulated considerably by changing the cation or anion. Thus, ionic liquids of different polarity are available which makes them either water-miscible (e.g., BMIM-BF_4) or immiscible (e.g., BMIM-PF_6). Recent studies have also revealed that ILs not only replace organic solvents, but that the performance of an enzyme is also influenced in ILs. Ionic liquids have in some cases been shown to increase regio- and enantioselectivity of enzymatic reactions, activity and stability (Kragl et al., 2002; Park and Kazlauskas, 2003; van Rantwijk et al., 2003). Proteases, lipases, esterases, glycosidases and oxidoreductases were each found to be active in ILs. As yet, the widespread use of these solvents – especially in industrial processes – has been hampered by their very high price (typically 800-fold more than conventional solvents), along with a need for more efficient methods for their reuse and/or product isolation.

Enzyme processes with these solvents can be carried out with the one- or two-phase systems, as shown in Table 2.8. The additional solvents can influence the enzyme or the enzyme-catalyzed reaction either *directly* or *indirectly*.

The direct influence is caused by the binding of solvent molecules to the hydrophobic parts of the surface, which results in structural changes to the active site. The bound solvent changes the kinetic properties of the enzyme. This can be studied in system IV in Table 2.8, where indirect effects are excluded. Existing data show that bound organic solvents (hexane, octane, hexanol, ethyl acetate) influence enzymes in different ways. For example, the secondary and tertiary structures of α-chymotrypsin, penicillin amidase and proteinase K were changed, as shown by their fluorescence and circular dichroism spectra in buffers saturated with the organic sol-

Table 2.8 Systems in which enzyme-catalyzed processes can be carried out in the presence of organic solvents or ionic liquids.

Two-phase system			***One-phase system***	
I Solvent not miscible with water[a)]	II Enzyme with water suspended in organic solvents or Ionic liquids	III Immobilized enzyme suspended in organic solvent or ionic liquids	IV Buffer saturated with organic solvent	V Buffer with water miscible organic solvent

a) The aqueous phase is usually small (≤1 %).

vents – that is, by the binding of organic molecules to the enzymes (Michaelis, 1991). This caused different changes in the S-, S′-binding and catalytic sites. For α-chymotrypsin and penicillin amidase, the specificity constants were hardly influenced. However, the selectivity (stereo- and synthesis/hydrolysis ratio in kinetically controlled synthesis) in the S_1'-binding site was changed by almost an order of magnitude for α-chymotrypsin, but unchanged for penicillin amidase in the presence of bound organic molecules. For proteinase K, the specificity constant for the hydrolysis of *N*-acetyl-L-tyrosine-ethylester increased by almost an order of magnitude in the presence of bound organic solvent (Kasche et al., 1991).

The indirect effect of the solvent is caused by its influence on the end-point due to the partition of substrates and products in the two-phase systems (II, III) in Table 2.8. The solvents change the equilibrium in equilibrium-controlled processes (this is shown schematically in Fig. 2.25). The apparent equilibrium constant for the two-phase system K_{biph} is given by the following relationship:

$$K_{biph} = K_{aq} \frac{(1 + \alpha p_c)\,(1 + \alpha)}{(1 + \alpha p_A)\,(1 + \alpha p_B)} \tag{2.43}$$

where α = the volume ratio V_{org}/V_{aq}. From this equation, it follows that higher yields in C are only obtained when $p_C > p_A \approx p_B$ – that is, when the product is much more hydrophobic than the substrates.

The situation is different when the organic solvent itself is one of the substrates such as a liquid ester, alcohol, lipid, etc., in which the other substrates can be dissolved. This is the case for many biotransfomations that can be catalyzed by lipase, an enzyme which has evolved for the bioconversion of hydrophobic compounds. This enzyme has a wide substrate specificity, and is used in enzyme processes such

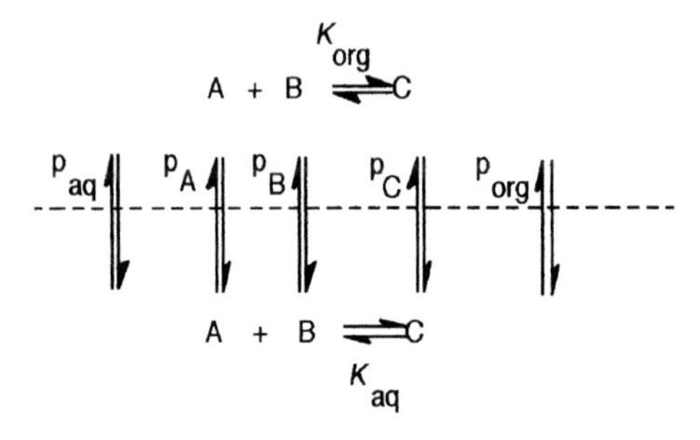

Fig. 2.25 Enzyme-catalyzed process in a two-phase system. The enzyme-catalyzed process can only occur in the aqueous phase or at the interphase. K_{aq} and K_{org} are the equilibrium constants in the aqueous and organic phases, respectively; p_A, p_B and p_C (=c_{org}/c_{aq}) are partition coefficients; p_{aq} and p_{org} are the partition coefficients for the solvents.

as the transesterfication of lipids, racemate resolution of hydrophobic compounds used as fine chemicals or pharmaceuticals (see Chapter 3).

2.9.2.1 Changes in Rates, k_{cat}, K_m, and Selectivities in These Systems Compared with Homogeneous Aqueous Solutions

k_{cat}, K_m and Rates

In the two-phase systems I and, the one-phase systems IV and V in Table 2.8, k_{cat}, and K_m are changed due to bound organic molecules. Whether this also applies for ionic liquids that are not miscible with water has not yet been studied. The rates are reduced as the concentration of the substrate is lower than its solubility in water.

In the other two-phase systems (II and III), the dependence of the rate on the water content is shown diagrammatically in Figure 2.26. The studies on biocatalysis in these solvents have also raised the question: How much water is required to maintain the catalytic function of an enzyme? Enzymes have been evolved for catalysis in the presence of water. Therefore, enzymes should not function as biocatalysts in the absence of water. This has been confirmed in studies on the systems I and III in Table 2.8. The results can be summarized as shown in Figure 2.26. Observations from the direct effects of bound organic solvents show that the kinetic properties of the enzymes are changed by less than an order of magnitude. So, how can the maximum in Figure 2.26 be explained? This must be caused by at least two different effects: At low water content, the solvent and the enzyme compete for the water molecules dissolved in the organic solvent. The enzyme activity increases until it has bound the water molecules that are required for its optimal function. When the water amount is increased further two phases are formed (to system I), and the enzyme that is located in the aqueous phase is diluted. This causes the decrease in the rate, as shown in Figure 2.26. Also, the substrate content in the aqueous phase is re-

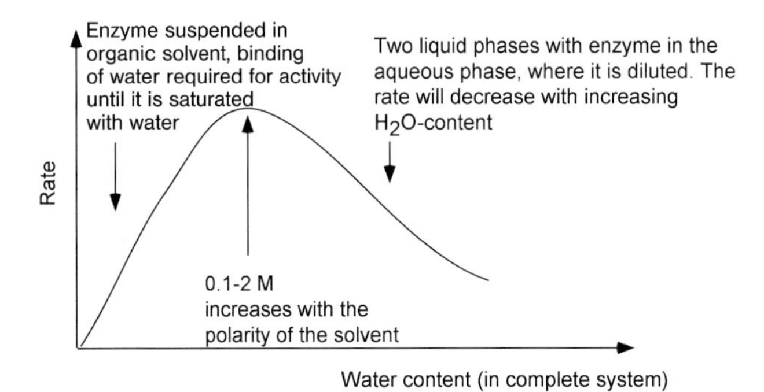

Fig. 2.26 The rate of an enzyme-catalyzed reaction with low (systems II and III in Table 2.8) and high (system I in Table 2.8) water content as a function of total water content. The total enzyme amount in the system is kept constant.

duced. At the phase boundary it will always be smaller than its solubility in water – that is, lower than at the phase boundary shown in Figure 2.22. Thus, the mass transfer between the phases, and the reduced flexibility of the enzyme (the conformation is not changed in non-aqueous solvents), cause a reduction in the rates in such two-phase systems up to a factor of 10^{-3}, compared to aqueous solutions or suspensions with dissolved or suspended substrates (Halling, 1994; Kasche and Galunsky, 1995; Schmitke et al., 1996). From Figure 2.25 it follows that the enzyme processes should be carried out at the water content with the maximum rate – that is, generally system II or III in Table 2.8. However, in these systems the suspended free enzymes tend to aggregate, leading to an insufficient usage of the enzymes' active sites. In order to avoid any reduction in rate caused by this effect, the enzymes should be used in a disaggregated form, and this can be achieved either by immobilization or by using covalent modifications to reduce the formation of enzyme aggregates.

Stereoselectivity

Most racemate resolutions with organic solvents or ionic liquids are carried out in the two-phase systems II or III in Table 2.8. In these, both direct and indirect effects cause the observed changes in stereoselectivity. Most published data show that the changes are larger than in aqueous solution (Carrea et al., 1995; Halling, 2000). Although no conclusive explanation for this has yet been given, it has been shown that it is strongly influenced by how the enzyme was incubated (pH) and lyophilized before being added to the organic solvent, as well as by the presence of metal ions that also influence the rates (Halling, 2000).

Selectivity in Kinetically Controlled Processes

Due to the low water content, hydrolysis of products is not expected for kinetically controlled processes, such as transesterfications, in systems II and III. These reactions can therefore be treated as equilibrium-controlled processes.

The conclusion here is that enzyme processes involving slightly soluble substrates or products can be carried out in aqueous suspensions or emulsions with higher space-time yields than in the systems with other solvents shown in Table 2.8. The use of organic solvents to solubilize substrates for such processes is favorable in cases where the solvent does not have difficult properties (e.g., supercritical CO_2; Mesiano et al., 1999), or when one of the substrates can be used as the solvent. This is often the case in lipase-catalyzed processes.

2.10 Stability, Denaturation, and Renaturation of Enzymes

In enzyme processes it is desired to use the biocatalyst for long periods in order to reduce enzyme costs. In living systems, enzymes have not evolved for long-term use as their biological half-life is generally in the range of minutes to 100 days. Optimal use of the enzyme requires that factors influencing its stability be known, with stability being reduced as little as possible under the process conditions.

The biologically active tertiary and quaternary structure of enzymes is formed spontaneously – that is, the native structure is thermodynamically stable. However, this stability can be reduced when interactions that stabilize the native structure (e.g., hydrophobic bonds, hydrogen bonds, binding ion–ion interactions such as $NH^+ \cdots COO^-$) are weakened by increased temperature or pH-values that lead to unfolding, loss of function and, ultimately, to denaturation of the enzyme (Fig. 2.27).

The denaturation is reversible when there are no covalent changes in the primary structure of the enzyme. The stability of an enzyme is also influenced when substrates and organic solvent molecules are bound to the enzyme (see Section 2.9). It has been shown that substrates and organic solvents immiscible with water stabilize enzymes, whereas solvents that are miscible with water destabilize enzymes. In supercritical CO_2 the stability of enzymes is reduced when the water content in the supercritical gas increases (Kasche et al., 1988). Enzymes can be stabilized when destabilizing interactions such as repulsive ion–ion interactions are weakened at high ionic strengths, or in the presence of doubly charged "crosslinking" ions such as Ca^{2+} or HPO_4^{2-}. The stability of an enzyme is also a function of its primary structure (Fig. 2.28). For penicillin amidase it has also been shown that pH-stability can be increased by covalent immobilization in porous particles (see Chapter 6).

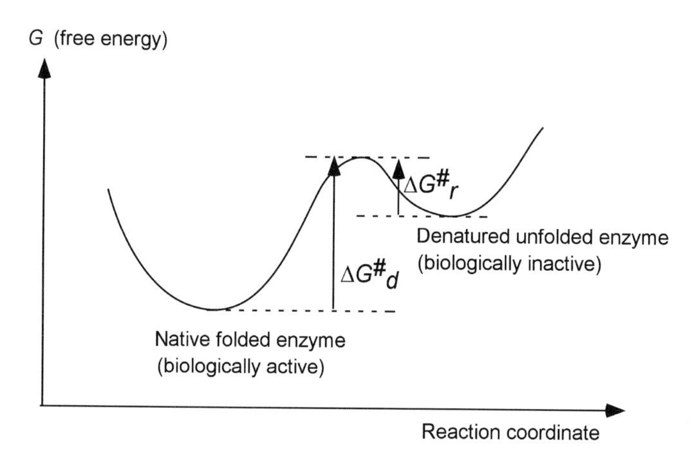

Fig. 2.27 Free energy diagram for the reversible denaturation of an enzyme with the activation energies for denaturation ($\Delta G^{\#}_{d.}$) and renaturation ($\Delta G^{\#}_{r.}$). The native structure is formed spontaneously at normal temperatures. At higher temperatures or extreme pH-values, the equilibrium is shifted to the unfolded (denatured) enzyme. The denatured enzyme can be renatured by slowly changing the temperature or pH to the original value. A rapid change "freezes" the denatured state, due to the slow rate of refolding. The renaturation must also be carried out at low enzyme concentrations to avoid aggregation of the denatured enzymes. A comparison with Figure 2.7 shows that enzymes are stabilized when they bind substrates.

Table 2.9 Denaturation of an enzyme by factors that influence rates and yields in enzyme processes, and how this can be minimized.

Factor/purpose	*Denaturation by*	*Reduce denaturation by*
Temperature increase/ higher rates and yields	Unfolding (reversible) Chemical modification (irreversible)	Use temperatures ≥20 °C below temperature optimum; use enzyme with higher temperature optimum
Temperature decrease/ higher yields	Dissociation of oligomeric enzymes (reversible)	Crosslink the oligomeric enzyme
Shear forces/ increase mass transfer	Unfolding (reversible)	Crosslink or immobilize the enzyme
Increase or decrease pH/ higher rates and yields	Unfolding (reversible) when $pH \gg$ or $\ll pI$	Crosslink or immobilize enzyme, or use enzyme with better pH-stability
O_2/increase rate with oxidases H_2O_2/bleaching agent in washing powder	Oxidation of –SH or methionine (irreversible)	Find or construct more stable enzymes
High substrate content/ increase yields reduce product purification costs	Chemical modification (irreversible) Glucose can react with amino groups of lysine	Find or construct more stable enzymes
Organic solvents	Unfolding (reversible)	Immobilize or crosslink enzyme
Peptidase in enzyme preparation	Hydrolysis of peptide bond (irreversible)	Use a more pure enzyme or immobilize the enzyme
Intramolecular autoproteolysis	Hydrolysis of peptide bond (irreversible)	Find or construct a more stable enzyme
Heavy metal ions as impurities in water, buffer compounds	Irreversible inhibition of -SH	Use water and buffer with minimal heavy metal ion content
UV or ionizing radiation/ sterilization	Chemical modification (irreversible)	Immobilize enzyme

Table 2.10 Rate constant for the irreversible denaturation and chemical changes in Lysozyme at 100 °C (Ahern and Klibanov, 1988).

Process	*First-order rate constant [h^{-1}] at pH*		
	4	*6*	*8*
All processes	0.49	4.1	50
Deamidation of Asn	0.45	4.1	18
Hydrolysis of Asp-X peptide bonds	0.12	0	0
Destruction of Cys	0	0	6
Formation of "wrong" structures (reversible process (?))	0	0	32

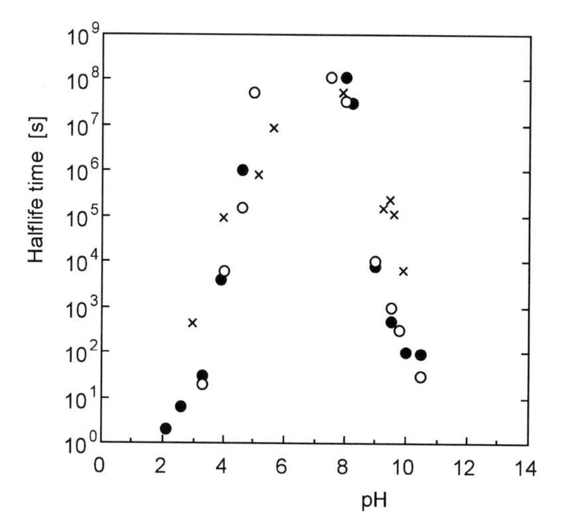

Fig. 2.28 The pH-stability of different penicillin amidase forms from *E. coli* ($PA_{7.0}$ ●, $PA_{6.7}$ ○; subscripts indicate their isoelectric point) and *Alcaligenes faecalis* (x) at 25 °C. The half-life time of the monomolecular denaturation was measured in buffer with ionic strength 1 M by activity determinations (Wiesemann, 1991; A. Rieks and V. Kasche, unpublished data).

Different factors that may cause reversible or irreversible (leading to covalent changes in the primary structure) enzyme denaturation during enzyme-catalyzed processes are listed in Table 2.9. Intramolecular autoproteolysis is a newly discovered process in which a peptide bond hydrolysis within a polypeptide chain is catalyzed by amino acid residues, in the same polypeptide chain, near this bond (Kasche et al., 1999). These processes can also cause irreversible enzyme denaturation.

The half-lives of the enzymes range from days (lipase, β-glucosidase) to >100 days (penicillin amidase) under optimal process conditions, though the reason for these widely ranging stabilities of enzymes remains incompletely understood. Rates for irreversible thermal denaturation are listed in Table 2.10. One possible way to avoid denaturation caused by factors listed in Table 2.9 and to reduce enzyme costs would be to screen for more stable enzymes in living systems, or to develop and construct better enzymes.

2.11 Better Enzymes by Natural Evolution, *in vitro* Evolution, or Rational Enzyme Engineering

One of the main problems in enzyme technology is to identify or design an optimal enzyme for a specific enzyme process. These processes can only be realized on an industrial scale when the enzyme costs can be reduced to less than 5 % of the costs to transform the substrate(s) to the desired product(s). As shown in Section 2.8, a process window for optimal yields in a pH-T-plane for an equilibrium-controlled process can be constructed based on thermodynamic data for the process. From Eq. (2.31), it follows that the amount of enzyme required to reach this yield in a given time depends on:

1. The enzyme kinetic constants (k_{cat}, K_m, and K_i).
2. The stability (chemical, pH- or temperature) of the enzyme under process conditions, given by the half-life, determined from the apparent first-order rate constant k_i (Eq. 2.32).

For equilibrium-controlled racemate resolutions the steric purity of the product(s) also depends on:

3. The enantioselectivity E_{eq} of the enzyme (Eq. 2.15) is determined by the enzyme kinetic constants in 1. This also applies for kinetically controlled racemate resolutions. In this case, however, the enantioselectivity E_{kin} (Eq. 2.16) depends also on other rate and equilibrium constants than for equilibrium-controlled processes.

For other kinetically controlled processes the product yield is additionally influenced by:

4. The selectivity for the synthesis and hydrolysis reactions $(k_T/k_H)_{app}$ (Eq. 2.3).

For a given enzyme process, the biocatalyst costs can be reduced when a better enzyme, with respect to the properties 1 to 4 above, can be found among natural variants or constructed by gene technological methods.

That neutral (or silent) mutations cause variations in the primary structure, and the properties 1 to 4 of natural enzyme variants of the same enzyme was outlined in Section 2.4. These mutations are very important for the evolution of living systems as they can result in a better adaptation of organisms in changing environments (higher or lower pH, temperature, substrate or salt content, new substrates, etc.). On the enzyme level, this leads to the large variation in the primary structure observed for the same enzyme from different organisms that differ in properties 1 to 4. For enzyme technology, this implies that better enzymes for a process can be found by screening for the enzyme in different living organisms or for the enzyme gene in DNA from environmental samples ("metagenomes"; Lorenz et al., 2002). This allows screening for enzymes in the more than 99 % of all microorganisms that have not yet been – or cannot be – cultivated. The metagenome is a mixture of DNA from living and dead cells found in environmental samples that contain genes for enzymes; these genes can be expressed in suitable host organisms. Suitable environments to screen for an enzyme or its gene are those with pH- and temperature-values in the optimal process window, and with high contents of the substrates or analogous compounds that are used in the process.

In this context, it is of interest to analyze the ranges within which properties 1 to 4 can differ for the same enzyme due to natural evolution. This will be covered in Section 2.11.1, where the question is raised as to whether these properties can be improved more than observed in living systems by evolution *in vitro* (Section 2.11.2) or by rational enzyme engineering (Section 2.11.3) (Bornscheuer and Pohl, 2001).

2.11.1 Changes in Enzyme Properties by Natural Evolution

k_{cat} and K_m

The protein (including enzymes) content in a living cell cannot be increased above the value of ≈20 % observed in procaryotic and eucaryotic cells (Atkinson, 1969). New metabolic reactions that require additional enzymes and their genes can therefore only be obtained when enzymes and their genes required for other metabolic reactions are:

- lost during the evolution; and/or
- improved during the evolution so that less enzyme is required to catalyze these metabolic reactions.

Both cases are observed during the evolution of living systems. Eucaryotic cells such as mammalian cells have lost genes for the enzymes required to synthesize certain amino acids, to allow the incorporation of new genes that code for enzymes that catalyze new metabolic reactions. The synthesis of enzymes requires much energy (substrate). An organism that synthesizes an enzyme with improved k_{cat}-value requires less enzyme and energy to catalyze the metabolic reaction than organisms with a less active enzyme. For organisms living at the same ambient temperature, the one with the higher k_{cat} has an evolutionary advantage, and this leads to variations in turnover number at a constant temperature for the same enzyme from different organisms. The changes in k_{cat} during evolution are, however, limited by a maximum value of the specificity constant (k_{cat}/K_m) of ~10^8 $(M\ s)^{-1}$ (this is discussed in Section 2.7.1), and the substrate concentrations in living systems. It is not advantageous for an organism to synthesize enzyme molecules that cost energy, when only a fraction of those molecules will catalyze metabolic reactions. This occurs when $[S] \ll K_m$. Therefore, enzymes in living systems have been selected during evolution with K_m-values in the range of the substrate concentrations in order to minimize energy costs for their synthesis (Fersht, 1999; Hochachka and Somero, 2002). This situation is shown graphically in Figure 2.29. From the boundary condition for the substrate specificity, it follows that the upper limit of k_{cat} obtainable by natural evolution is $\approx 10^8\ [S]\ s^{-1}$ – that is, it is limited by the substrate content in living systems and their environment.

For organisms that can survive over a large temperature range (e.g., procaryotes, fishes), the temperature-dependence of k_{cat} and K_m is also of evolutionary importance. For organisms living at low temperatures (psychrophiles), the enzyme that catalyzes a metabolic reaction must be evolved so that its k_{cat}-value is higher at this temperature than for an enzyme that catalyzes the same reaction in an organism living at a higher temperature (meso- or thermophiles). When this is not the case, the psychrophilic organism must produce more of the enzyme, and this is an evolutionary disadvantage. The variation in the k_{cat}-values observed at one temperature for lac-

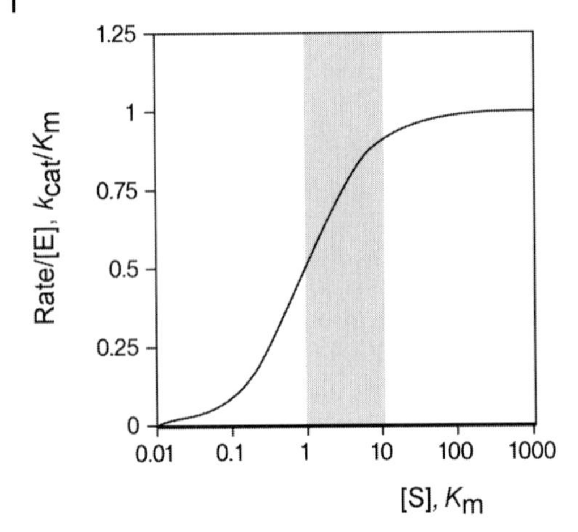

Fig. 2.29 The specific rate of an enzyme calculated from Eq. (2.8) and expressed in k_{cat}/K_m units as a function of the substrate content [S] in the living cell or its environment expressed in K_m-units. It is an evolutionary advantage to use as much as possible of the synthesized enzyme for catalytic conversion of the substrate. This occurs when [S] is 1–10 K_m (shaded area). At higher ratios, the specific rate is independent of [S]; this is unfavorable for metabolic regulation.

tate dehydrogenase from psychro-, meso- and thermophilic organisms has been found to differ by one order of magnitude (Fields and Somero, 1998). As these organisms can live at different temperatures, the rates of the reaction must be regulated to become less dependent on temperature. This requires enzymes for which the K_m-values increase with temperature. This applies to lactate dehydrogenases, as discussed above, with K_m-values at one temperature having also been found to vary by up to one order of magnitude (Fields and Somero, 1998).

For non-natural substrates the k_{cat}, K_m, and specificity constants have been found to differ by up to one order of magnitude for the same enzyme from different organisms (Table 2.11). These changes did not differ markedly for specific and less-specific substrates (see also Table 2.4).

In enzyme technology, much higher substrate contents (up to 1 M) are employed than in living systems, and this is generally much higher than the K_m-values of the enzymes used. In these cases the enzyme activity under process conditions can only be increased by using enzyme variants with higher k_{cat}-values. From the above analysis it follows that similar changes in k_{cat} as observed for natural variants should be expected for enzymes improved by *in vitro* evolution or rational enzyme engineering.

Enzyme Stability

During evolution, organisms have been selected that can live in environments of different pH (range 1 to 11) and temperature (range 0 to >100 °C). This implies that enzymes in organisms have evolved to achieve a better biological activity at pH- and temperature-values in the environment of these organisms, than the same enzymes in organisms that survive better under other conditions (pH, *T*). Organisms produce both intra- and extracellular enzymes, but due to intracellular proteolysis (protein turnover) the intracellular enzymes have biological half-lives that are not limited by their pH- and temperature stability (see Chapter 4.3). From this, it follows that intracellular enzymes have not been thoroughly selected for pH- and temperature stability during evolution.

Table. 2.11 Kinetic constants and selectivities of penicillin amidases (PA) from different sources and two genetically constructed hybrid enzymes with the A- and B-chains from different sources (Hybrid I $A_{Kc}B_{Ec}$ and Hybrid II $A_{Ec}B_{Kc}$; Ec = *E. coli*, Kc = *K. citrophila*): k_{cat}, K_m and specificity constant for the hydrolysis of benzylpenicillin (BP), 2-nitro-5-phenylacetamidobenzoic acid (NIPAB), phenylglycine amide ($PhgNH_2$) and phenylglycine methyl ester (PhgOMe) at pH 7.5 ($I = 0.2$) and 25 °C.

Substrate	BP			NIPAB			(R)-PhgNH₂			(R)-PhgOMe		
PA from	k_{cat} [s^{-1}]	K_m [μM]	k_{cat}/K_m [$M^{-1}s^{-1}$]	k_{cat} [s^{-1}]	K_m [μM]	k_{cat}/K_m [$M^{-1}s^{-1}$]	k_{cat} [s^{-1}]	K_m [mM]	k_{cat}/K_m [$M^{-1}s^{-1}$]	k_{cat} [s^{-1}]	K_m [mM]	k_{cat}/K_m [$M^{-1}s^{-1}$]
E. coli	50	10	5.0×10^6	20	20	1.0×10^6	16	16	1.0×10^3	16	15	1.1×10^3
A. faecalis	80	8	1.0×10^7	82	12	6.8×10^6	7	6	1.2×10^3	43	6	7.2×10^3
K. citrophila	60	18	3.3×10^6	30	20	1.5×10^6	15	20	7.5×10^2	13	13	1.0×10^3
A. viscosus	140	20	7.0×10^6	5	130	3.8×10^4	45	480	9.4×10^1	42	13	3.2×10^3
Hybrid I	100	12	8.3×10^6	40	20	2.0×10^6	15	21	7.1×10^2	18	21	8.6×10^2
Hybrid II	115	18	6.4×10^6	6	20	3.0×10^5	14	25	5.6×10^2	–	–	–

Stereoselectivity E_{eq} for the hydrolysis of (*R,S*)-$PhgNH_2$, (*R,S*)-$pOHPhgNH_2$ (S_1-stereoselectivity) and PAA-(*R,S*)-Phe (S_1'- stereoselectivity) at pH 7.5 ($I = 0.2$) and 25 °C; Selectivity $(k_T/k_H)_{app}$ in kinetically controlled synthesis of amoxycillin from 20 mM (*R*)-$pOHPhgNH_2$ and 20 mM 6-aminopenicillanic acid at pH 6.5 ($I = 0.2$) and 25 °C, and cephalexin from 200 mM (*R*)-$PhgNH_2$ and 200 mM 7-aminodesacetoxycephalosporanic acid at pH 7.5 ($I = 0.2$) and 25 °C.

Reaction	(R,S)-PhgNH₂ hydrolysis	(R,S)-pOHPhgNH₂ hydrolysis	PAA-(R,S)-Phe hydrolysis	Amoxycillin synthesis	Cephalexin synthesis
PA from	E_{eq}	E_{eq}	E_{eq}	$(k_T/k_H)_{app}$	$(k_T/k_H)_{app}$
E. coli	0.5	0.03	1000	3200	3300
A. faecalis	0.4	n.d.	250	2500	800
K. citrophila	0.3	0.2	n.d	1900	n.d.
A. viscosus	0.7	n.d	n.d	n.d.	n.d.
Hybrid I	0.4	0.2	n.d	1600	n.d.
Hybrid II	n.d.	n.d.	n.d	2500	n.d.

n.d. = not determined.

Extracellular enzymes are produced to hydrolyze biopolymers and to transform organic compounds into products that can be taken up and metabolized in the cells. They can also be used to detoxify compounds outside the cells that are toxic to the cells. For the cells, it is an evolutionary advantage when the pH- and temperature stability of these enzymes are increased as they then have to produce less enzyme to transform extracellular compounds into molecules required for their survival. One enzyme process where pH-stability of the enzyme used is essential is in the hydrolysis of penicillin G (see Sections 2.7, 2.10 and Exercise 14, Section 2.12). The enzyme first used in this process was the extracellular (periplasmic) penicillin amidase from *E. coli*. Due to its limited pH-stability, it could only be used up to pH ≈ 8, but at this pH <95 % of penicillin is hydrolyzed in the equilibrium-controlled process. By introducing the same enzyme from *A. faecalis* that could be used at pH ≈ 9 due to its better pH-stability (Fig. 2.28), the hydrolysis yield was increased to ≈99% (Verhaert et al., 1997; Ignatova et al., 1998).

From this it follows that the pH- and temperature-stability for intracellular enzymes could be improved by *in vitro* evolution or by rational protein design more than for those extracellular enzymes selected for better pH- and temperature-stability.

The chemical stability of an enzyme is a measure of its chemical modification due to reactions with metabolites and other organic molecules in the environment. Hydroxy groups of sugars can react with amino groups on enzymes, while O_2 or H_2O_2 can oxidize Met or Cys, etc. In natural systems the concentration of these compounds is much less than is used in enzyme processes, and therefore the rate of enzyme modification by such bimolecular reactions is much higher in the latter situation. In this case, chemical stability is a measure of the covalent modification of the enzyme by the substrates and products of the enzyme process. Thus, it could be assumed that during evolution enzymes have not been selected for optimal chemical stability, and that the same enzyme from different organisms can vary widely with regard to its chemical stability. Although this point has not yet been studied in detail, it follows that this property of enzymes could be improved by *in vitro* evolution or by rational protein design.

Stereoselectivity

With the exception of amino acids, hydroxy acids, carbohydrates and some other compounds, optically active molecules are only present as one enantiomer in living systems. This is also reflected in the list of enzymes that catalyze racemization reactions (EC 5.1). The enzymes involved in the biosynthesis of sterically pure compounds from racemic mixtures of the above compounds must therefore have been selected for high stereoselectivity during evolution. This applies to the synthesis of oligomers and polymers from monomers. Ribosomal peptide (protein) synthesis involves only (*S*)-amino acids, whereas peptide antibiotics that are synthesized non-ribosomally contain both (*S*)- and (*R*)- amino acids (β-lactam antibiotics as penicillin, gramisidin S, cyclosporin, tyrocidine, etc.) (Kleinkauf, von Döhren, 1990). The latter are mainly produced in procaryotes. Some ribosomal antibacterial peptides from eucaryotes (e.g., dermorphin from frog skin) contain (*R*)-amino acids that are formed by racemization reactions after the peptide synthesis from (*S*)-amino acids. In natural systems these oligomers and polymers are hydrolyzed by hydrolases. As the monomers can be racemized in procaryotes, the hydrolases in these are not selected for high stereoselectivities as the transferases that catalyze the polymerization reactions. Therefore many peptidases can be used for the kinetically controlled synthesis of peptide bonds between (*S*)- and (*R*)-amino acids (Kasche, 2001; see also Fig. 2.12).

In living systems, sterically pure compounds are synthesized from pro-chiral compounds containing double bonds or are keto-acids. The enzymes catalyzing these reactions, such as dehydratases (EC 4.2.1) and oxygenases (EC 1.13) for additions to double bonds, or transaminases (EC 2.6.1) and dehydrogenases (EC 1), must have been selected for high stereoselectivity during evolution.

Enzymes that in natural systems mainly catalyze reactions involving optically pure substrates or substrates that are optically inactive, are not expected to be selected for high stereoselectivity during evolution. Lipases have a wide range of stereoselectivity among the natural enzyme variants (see Fig. 2.13). The stereoselectivity of penicillin amidases, which still have unknown biological functions, have both (*S*)- and (*R*)-specific binding subsites, and the stereoselectivity is known to vary by up to one order of magnitude for natural enzyme variants (see Table 2.11).

As with the stability of enzymes, it is possible to increase the stereoselectivity of enzymes by *in vitro* evolution or by rational enzyme engineering (see Sections 2.11.2 and 2.11.3).

Selectivity in Kinetically Controlled Synthesis of Condensation Products

In natural systems, processes are catalyzed by transferases that have been selected for high $(k_T/k_H)_{app}$-values during evolution (see Table 2.3), whereas in enzyme processes the hydrolases are mainly used as biocatalysts. This property varies up to a factor of 4 for the same hydrolase from different sources for the same reaction (see Table 2.11). These enzymes have not been selected with regard to this property during evolution.

2.11.2
Methods to Improve the Properties of Enzymes by *in vitro* Evolution

The traditional method to identify new enzymes is based on screening, for example of soil samples or of strain collections by enrichment cultures. Once a suitable biocatalyst has been identified, strain improvement as well as cloning and expression of the encoding gene(s) enables production on a large scale. Unfortunately, not all microorganisms are culturable by using common fermentation technology, and the number of accessible microorganisms from a sample is estimated as 0.001–1 %, depending on their origin (Lorenz et al., 2003; Miller, 2000). In addition, not all enzymes found in nature are suitable for a certain synthetic problem – for example, the enzyme's activity, stability, substrate specificity, and enantioselectivity are not always satisfactory. Until recently, these limitations were often overcome by rational protein design when, if the gene encoding the enzyme and its three-dimensional structure were available, mutation sites can be identified followed by the introduction of appropriate amino acids by site-directed mutagenesis (see Section 2.11.3).

An alternative which emerged during the mid-1990s was the strategy of in-vitro evolution (also called directed or molecular evolution). This "evolution in the test tube" comprises essentially two steps: (1) random mutagenesis of the gene encoding the enzyme; and (2) identification of desired biocatalyst variants within these mutant libraries by screening or selection (Fig. 2.30).

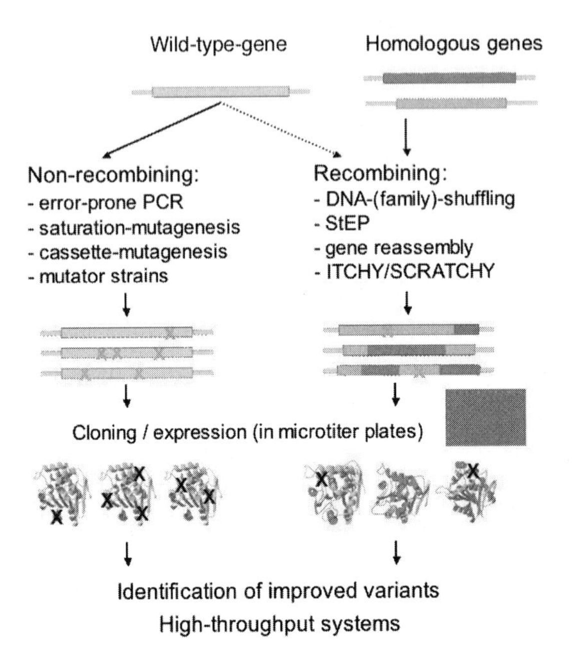

Fig. 2.30 Principle of directed evolution. The gene(s) encoding the wild-type or homologous enzyme(s) are subjected to random mutagenesis using non-recombining or recombining methods. The resulting mutant libraries are then cloned and expressed (often in microtiter plates). The desired improved variants are identified by high-throughput screening systems, usually using microtiter plate-based assays or selection (e.g., using agar plate assays; not shown).

Prerequisites for in-vitro evolution are the availability of the gene(s) encoding the enzyme(s) of interest, a suitable (usually microbial) expression system, an effective method to create mutant libraries, and a suitable screening or selection system. Many detailed protocols for this are available, including several books (Arnold and Georgiou, 2003a,b; Brakmann and Johnsson, 2002; Brakmann and Schwienhorst, 2004) and reviews (Bornscheuer, 2001; Neylon, 2004; Reetz, 2004; Turner, 2003).

2.11.2.1 **Methods to Create Mutant Libraries**

A broad range of methods has been developed to create mutant libraries. These can be divided into two approaches: (1) a non-recombining mutagenesis, in which one parent gene is subjected to random mutagenesis leading to variants with point mutations; or (2) recombining methods, in which several parental genes (usually showing high sequence homology) are randomized. This results in a library of chimeras rather than an accumulation of point mutations.

One challenge in directed evolution experiments is the coverage of a sufficiently large sequence space – that is, the creation of as many variants as possible. When considering a protein (enzyme) consisting of 200 amino acids, the number of possible variants of a protein by introduction of M substitutions in N amino acids can be calculated from the formula $19^{M}[N!/(N-M)!M!]$. (From the 20 proteinogenic amino acids, one is already present in the wild-type enzyme; therefore the value 19.) Thus, for two random mutations already more than seven million variants are possible, with three or more substitutions, the creation and screening of a library becomes very challenging (Table 2.12)

The most prominent methods for the creation of libraries is the error-prone polymerase chain reaction (epPCR) in which conditions are used that lead to the introduction of approximately one mutation per 1000 base pairs (Cadwell and Joyce, 1992). This is achieved by changing the reaction conditions – that is, to use Mn^{2+} salts instead of Mg^{2+} salts (the polymerase is magnesium-dependent), the use of *Taq* polymerase from *Thermomyces aquaticus*, and variations in the concentrations of the four desoxynucleotides (dNTPs). It should be noted, that due to the non-biased exchange of nucleotides and the degenerated genetic code, only an average of 6 of all possible 19 amino acid substitutions are accessible by this method. Another approach utilizes mutator strains, for example the *E. coli* derivative *Epicurian coli* XL1-Red, which lacks DNA repair mechanisms (Bornscheuer et al., 1998; Greener et al., 1996). The introduction of a plasmid bearing the gene encoding the protein of interest leads to mutations during replication. Both methods introduce point mutations, and several iterative rounds of mutation followed by identification of best variants are usually required to obtain a biocatalyst with the desired properties.

Alternatively, methods of recombination (also referred to as sexual mutagenesis) can be used. The first example was the DNA (or gene-) shuffling developed by Stemmer, in which DNAse degrades the gene followed by recombination of the fragments using PCR with and without primers (Stemmer, 1994a,b). This process mimics natural recombination and has been proven in various examples as a very effec-

Table 2.12 Sequence space of possible variants for a protein consisting of 200 amino acids at a given number of substitutions.

Substitutions [M]	*Number of variants [sequence length N = 200]*
1	3800
2	7183 900
3	9 008 610 600
4	8 429 807 368 950
5	6 278 520 528 393 960

tive tool to create desired enzymes. More recently, this method was further refined and termed "DNA family shuffling" or "molecular breeding", enabling the creation of chimeric libraries from a family of genes.

The Arnold laboratory developed two variations of DNA shuffling: The staggered extension process (StEP) is based on a modified PCR protocol using a set of primers and short reaction times for annealing and polymerization. Truncated oligomers dissociate from the template and anneal randomly to different templates leading to recombination. Several repetitions allow the formation of full-length genes (Zhao, 1998). Other methods are ITCHY (incremental truncation of chimeric hybrid enzymes) and related approaches (Lutz et al., 2001; Ostermeier et al., 1999). Table 2.13 provides an overview of methods, while more details and comparisons of different strategies for the creation of mutant libraries can be found in reviews (Kurtzman et al., 2001; Neylon, 2004).

Table 2.13 Selected methods to create mutant libraries for directed evolution (Kurtzman et al., 2001; Neylon, 2004).

Method	*Pros*	*Cons*
Error-prone PCR	Easy to perform, mutation rate adjustable	Non-biased amino acid substitutions Only point mutations accessible
Mutator strains	Easy to perform	Entire organism/plasmid is mutated Only point mutations accessible
DNA-shuffling	Modest sequence homology sufficient Several parent genes can be used Creation of chimeras possible Useful mutations are combined, harmful ones are lost	Requires sequence homology
StEP	Similar to DNA-shuffling, more simple No fragment purification necessary	Requires sequence homology PCR protocol must be specifically adapted
SHIPREC	No sequence homology required	Low diversity library in single round (might be repeated) Limited to two parents of similar length Deletions/duplications possible
ITCHY	Similar to SHIPREC	Similar to SHIPREC
THIO-ITCHY	Similar to ITCHY, but more efficient/easier	Similar to ITCHY
GSSM	All single amino acid substitutions are covered	Technically out of reach for most researchers

2.11.2.2 **Assay Systems**

The major challenge in directed evolution is the identification of desired variants within the mutant libraries. Suitable assay methods should enable a fast, very accurate and targeted identification of desired biocatalysts out of libraries comprising 10^4–10^6 mutants. In principle, two different approaches can be applied, namely screening or selection.

Selection

Selection-based systems have been used traditionally to enrich certain microorganisms. For in-vitro evolution, selection methods are less frequently used as they usually can only be applied to enzymatic reactions that occur in the metabolism of the host strain. On the other hand, selection-based systems allow a considerably higher throughput compared to screening systems (see below). Often, selection is performed as a complementation – that is, an essential metabolite is produced only by a mutated enzyme variant. For instance, a growth assay was used to identify monomeric chorismate mutases. Libraries were screened using media lacking L-tyrosine and L-phenylalanine (MacBeath et al., 1998). In a similar manner, complementation of biochemical pathways has also been used to identify mutants of an enzyme involved in tryptophan biosynthesis. HisA and TrpF (isomerases involved in the biosynthesis of histidine and tryptophan, respectively) have a similar tertiary structure, and the aminoaldose substrates (ProFAR and PRA) used are very similar except for a different residue at the amino functionality. Using random mutagenesis and selection by complementation on media lacking tryptophan, several HisA variants that catalyze the TrpF reaction both *in vivo* and *in vitro* were identified (Juergens et al., 2000). One of these variants also retained significant HisA activity (Fig. 2.31).

Stemmer's group subjected four genes of cephalosporinases from *Enterobacter, Yersinia, Citrobacter,* and *Klebsiella* species to error-prone PCR or DNA-shuffling. Libraries from four generations (a total of 50 000 colonies) were assayed by selection on agar plates with increasing concentrations of Moxalactam (a β-lactam antibiotic). Only those clones could survive, which were able to hydrolyze the β-lactam antibiotic. The best variants from epPCR gave only an 8-fold increased activity, but the best chimeras from multiple gene-shuffling showed 270- to 540-fold resistance to Moxalactame (Crameri et al., 1998). Sequencing of a mutant revealed low homology compared to the parental genes (Fig. 2.32) and a total of 33 amino acid substitutions and seven crossovers were found. These changes would have been rather impossible to achieve using epPCR and single-gene shuffling only; thus, these investigations demonstrate the power of DNA-shuffling.

Mutants of an esterase from *Pseudomonas fluorescens* (PFE) produced by directed evolution using the mutator strain *Epicurian coli* XL1-Red were assayed for altered substrate specificity using a selection procedure (Bornscheuer et al., 1999). The key to the identification of improved variants acting on a sterically-hindered 3-hydroxy ester – which was not hydrolyzed by the wild-type esterase – was an agar plate assay system based on pH-indicators, thus leading to a change in color upon hydrolysis of the ethyl ester. Parallel assaying of replica-plated colonies on agar plates supplemented with the glycerol derivative of the 3-hydroxy ester was used to refine the

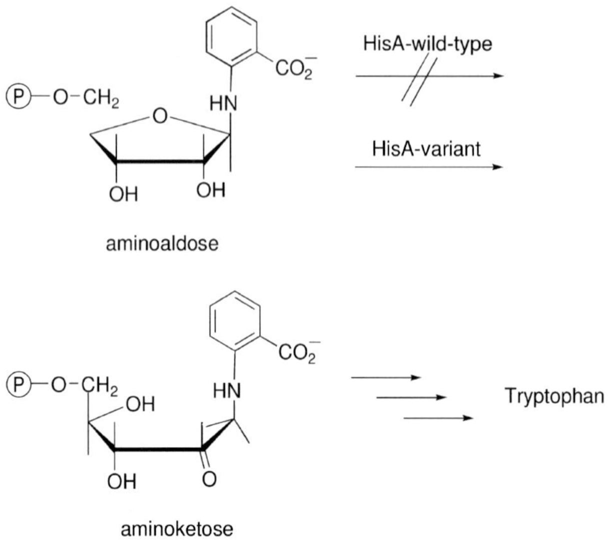

HisA-wild-type

HisA-variant

aminoaldose

aminoketose

Tryptophan

Fig. 2.31 A growth selection assay was used to identify variants of HisA capable of tryptophan precursor synthesis.

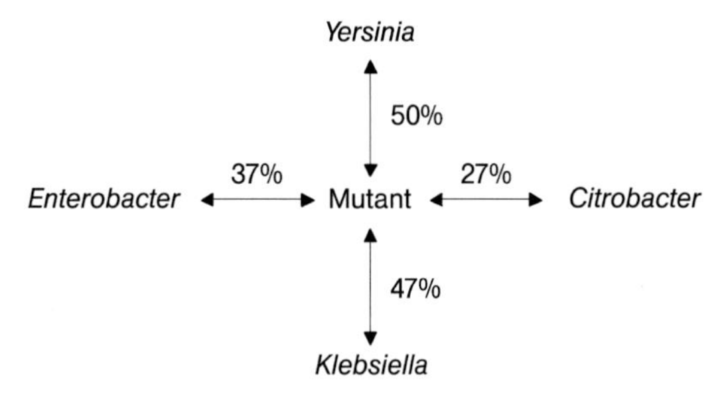

Fig. 2.32 Homology between a mutant of a cephalosporinase gene obtained by DNA-shuffling of parental genes from four different species (Crameri et al., 1998).

identification, because only *E. coli* colonies producing active esterases had access to the carbon source glycerol, thus leading to enhanced growth and in turn larger colonies. By this strategy, a double mutant was identified, which efficiently catalyzed hydrolysis.

Screening

Screening-based systems (not to be confused with the use of the term "screening" for the identification of microorganisms) are much more frequently used. Due to the very high number of variants generated by directed evolution, common analytical tools such as gas chromatography and HPLC are less useful, as they are usually

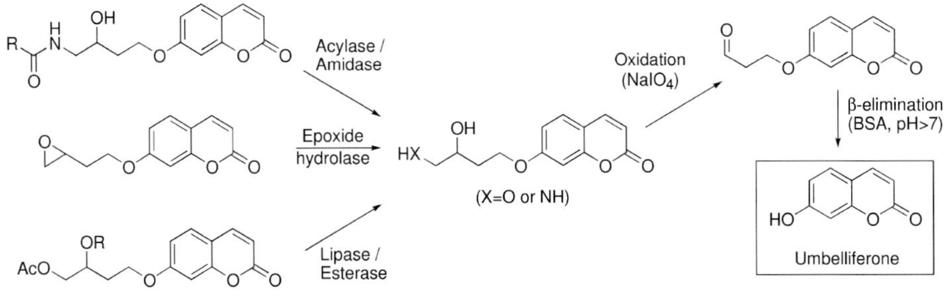

Fig. 2.33 Fluorogenic assay based on umbelliferone derivatives. Enzyme activity yields a product which, upon oxidation with sodium periodate and treatment with bovine serum albumin (BSA), yields umbelliferone (Reymond and Wahler, 2002).

too time-consuming. High-throughput GC-MS or NMR techniques have also been described, but these require the availability of rather expensive equipment and, in the case of screening for enantioselective biocatalysts, the use of deuterated substrates. Phage display, ribosome display and fluorescense-activated cell sorting (FACS) have also been used to screen within mutant libraries containing on the order of $>10^6$ variants, but they are not generally applicable.

The most frequently used methods are based on photometric and fluorimetric assays performed in microtiter plate (MTP)-based formats in combination with high-throughput robot-assistance. These allow a rather accurate screening of several tens of thousands of variants within a reasonable time, and provide information about the enzymes investigated, notably their activity by determining inital rates or end-points and their stereoselectivity by using both enantiomers of the compound of interest. One versatile example is the use of umbelliferone derivatives (Fig. 2.33). Esters or amides of umbelliferone are rather unstable, especially at extreme pH and at elevated temperatures. The ether derivatives shown in Figure 2.33 are very stable as the fluorophore is linked to the substrate via an ether bond. Only after enzymatic reaction and treatment with sodium periodate and bovine serum albumin (BSA), is the fluorophore released (Reymond and Wahler, 2002). Other techniques are outlined in the following paragraphs.

Examples

Reetz and coworkers turned a non-enantioselective (2 % ee, E = 1.1) lipase from *Pseudomonas aeruginosa* PAO1 into a variant with very good selectivity (E >51, >95 % ee) in the kinetic resolution of 2-methyldecanoate. The identification of variants was based on optically pure (*R*)- and (*S*)-*p*-nitrophenyl esters of 2-methyldecanoate in a spectrophotometric screening. In the first step, the wild-type lipase gene was subjected to several rounds of random mutagenesis by epPCR, leading to a variant with an E = 11 (81 % ee) followed by saturation mutagenesis (E = 25). Key to further doubling of enantioselectivity was a combination of DNA-shuffling, combinatorial cassette mutagenesis and saturation mutagenesis which led to a maximal recombina-

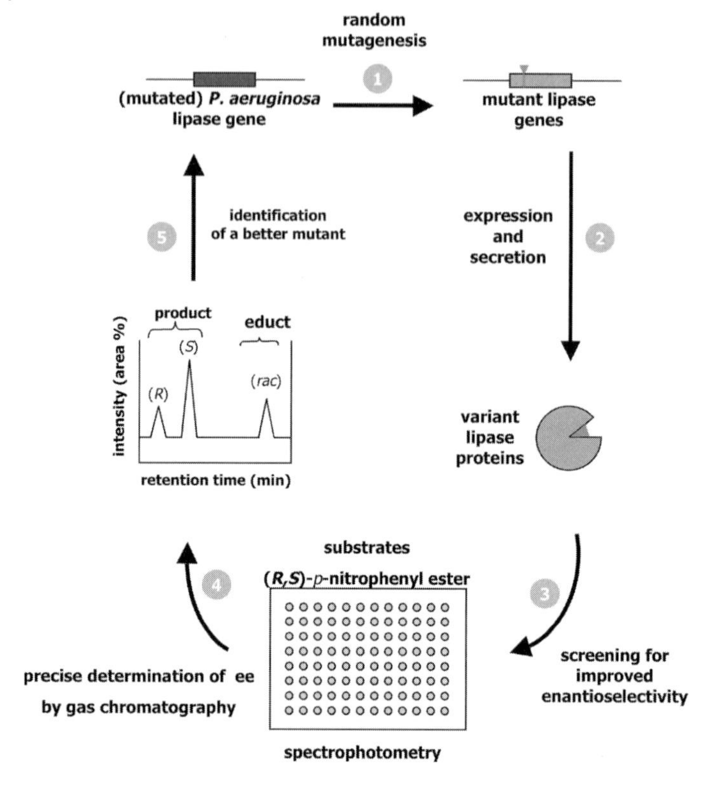

Fig. 2.34 Overview of the directed evolution of a lipase from *Pseudomonas aeruginosa* for the enantioselective resolution of 2-methyl decanoate. In the first step (1), the lipase gene was subjected to random mutagenesis; next the mutated genes were expressed and secreted (2). Screening for improved enantioselectivity was based on a spectrophotometric assay using optically pure (*R*)- or (*S*)-*p*-nitrophenyl esters of the substrate (3). Hit mutants with improved enantioselectivity were then verified by gas chromatography (4). The cycle was repeated several times to identify the best mutants (5) (Reetz et al., 2001).

tion of best variants. The best mutant (E >51) contained six amino acid substitutions, and a total of approximately 40 000 variants was screened (Reetz et al., 2001). The overall strategy is illustrated in Figure 2.34, and the overall changes in enantioselectivity using the combination of different approaches for random mutagenesis are summarized in Figure 2.35.

The Arnold group reported the inversion of enantioselectivity of a hydantoinase from D-selectivity (40 % ee) to moderate L-preference (20 % ee at 30 % conversion) by a combination of epPCR and saturation mutagenesis. Only one amino acid substitution was sufficient to invert enantioselectivity. Thus production of L-methionine from (*R*,*S*)-5-(2-methylthioethyl)hydantoin in a whole-cell system of recombinant *E. coli* containing also a L-carbamoylase and a racemase at high conversion became feasible (May et al., 2000; see also Chapter 3, Section 3.2.2.6).

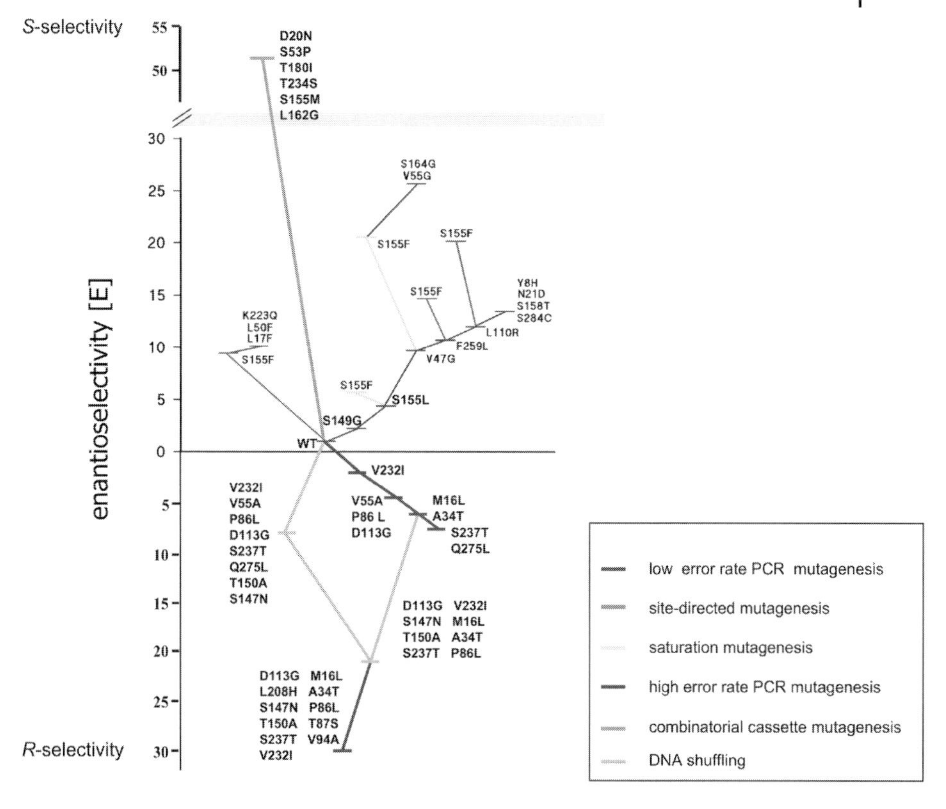

Fig. 2.35 Changes in enantioselectivity of a lipase from *Ps. aeruginosa* using methods of directed evolution. Starting from the non-selective wild-type (WT, E = 1.1), the combination of various genetic tools led to the creation and identification of variants with high (*S*)-selectivity (E = 51) and with good (*R*)-selectivity (E = 30) (Reetz et al., 2001).

Even if a biocatalyst with proper substrate specificity (and stereoselectivity) has already been identified, the requirements for a cost-effective process are not always fulfilled. Enzyme properties such as pH-, temperature- and solvent-stability are very difficult to improve using "classical" methods such as immobilization techniques or site-directed mutagenesis. Again, directed evolution has been shown to be a versatile tool to meet this challenge.

For example, an esterase from *Bacillus subtilis* hydrolyzes the *p*-nitrobenzyl ester of Loracarbef, a cephalosporin antibiotic. Unfortunately, the wild-type enzyme was only weakly active in the presence of dimethylformamide (DMF), which must be added to dissolve the substrate. A combination of epPCR and DNA-shuffling led to the generation of a variant with 150 times higher activity compared to the wild-type in 15 % DMF (Moore and Arnold, 1996). Later, the thermostability of this esterase could also be increased by ~14 °C by directed evolution. In a similar manner, performance of subtilisin E in DMF was improved 470-fold.

Researchers at Novozymes (Denmark) subjected a heme peroxidase from *Coprinus cinereus* (CiP) to multiple rounds of directed evolution in an effort to produce a mutant suitable for use as a dye-transfer inhibitor in laundry detergent (Cherry et al., 1999). Mutants obtained by epPCR and site-directed mutagenesis were screened for improved stability by measuring residual activity after incubation under conditions mimicking those in a washing machine (e.g, pH 10.5, 50 °C, 5–10 mM peroxide). Subsequent in-vivo shuffling led to dramatic improvements in oxidative stability, yielding a mutant with 174 times higher thermal stability and 100 times improved oxidative stability of wild-type CiP.

It is often assumed, that improving a biocatalyst in one direction is countered with a loss of other enzyme characteristics. However, it was shown recently that it is possible to increase the thermostability of a cold-adapted protease to 60 °C while maintaining high activity at 10 °C (Miyazaki et al., 2000). The best psychrophilic subtilisin S41 variant contained only seven amino acid substitutions resembling only a tiny fraction of the usual 30–80 % sequence difference found between psychrophilic enzymes and mesophilic counterparts.

In another example, researchers at Maxygen (USA) and Novozymes (Denmark) simultaneously screened for four properties in a library of family-shuffled subtilisins: activity at 23 °C; thermostability; organic-solvent tolerance; and pH-profile, and reported variants with considerably improved characteristics for all parameters (Ness et al., 1999).

These and some other selected examples are summarized in Table 2.14.

Table 2.14 Selected examples for the directed evolution of enzymes (Bornscheuer, 2001).

Enzyme	***Target***	***Mutagenesis method***	***Assay***	***Main result***
Pseudomonas fluorescens esterase	Altered substrate specificity	MS	Growth/ pH-indicator	Active double mutant
Pseudomonas fluorescens esterase	Increased enantioselectivity	MS/epPCR	MTP w/ chiral resorufin esters	E increased from 3.5 to 6.5
Pseudomonas aeruginosa lipase	Increased enantioselectivity	epPCR/SM/ cassette mutagenesis	MTP w/ chiral *p*-nitrophenyl esters	E increased from 1 to 51
Bacillus subtilis esterase	High activity in DMF towards pNB-ester of Loracarbef	epPCR/SM/SH	MTP w/*p*-nitrophenyl ester	Activity increased ca. 150 fold
Coprinus cinereus heme peroxidase	Stability and activity under laundry conditions	epPCR/SM/SH	Incubation in MTP at high pH, w/H_2O_2, measuring residual activity	100-fold oxidative/170-fold thermal stability

2.11.3 Rational Enzyme Engineering

When the German edition of this book was written during the early 1990s, the dominant method used here was site-directed mutagenesis – that is, the genetic exchange of one amino acid residue with another to change the enzyme's properties. With the exception of an improved stability of glucose isomerase (Hartley et al., 2000) and subtilisin (Bryan, 2000), amino acid exchanges in the active site of enzymes were mainly studied, and there were few applications to enzyme technology. Nonetheless, these studies were very important in elucidating the detailed mechanism of enzyme-catalyzed reactions, and today extensive data are available on enzymes that are used widely in enzyme technology. In most cases the mutations led to enzyme variants with lost enzymatic function or with reduced or marginal changes in specificity constants for the same substrate, while larger changes were observed for substrate specificity (Hartley, 2000; Sauer et al., 2000). With our present knowledge this is expected, as most of the amino acid residues in the active site that interact directly with the substrate are highly conserved in the same enzyme from different species. Since 1993, a large number of studies have been conducted sucessfully to improve properties 1 to 4 in Section 2.11.1 for important enzymes in enzyme technology using either random or directed evolution (see Section 2.11.2). However, in many cases the improvements were due to exchanges of less-conserved amino acid residues that were located far from the active site (Arnold, 2001). This indicated that less-conserved amino acid residues in, around and far from the active site, might be function-modulating – that is, they are not essential for enzyme function but alter the quantitative measures with which the function is carried out (k_{cat}, K_m, E, and $(k_T/k_H)_{app}$). These results, the amino acid alignments of the same enzyme from different sources, and elucidation of the 3-D-structures of these enzymes can provide us with new tools for a more rational enzyme engineering, when possible function-modulating residues can be identified from these data. Indeed, this is an important task in the developing field of bioinformatics (Chen, 2001).

The goal of rational protein design is to identify the amino acid residues that might be function-modulating or might limit the stability of the enzyme in a given enzyme process. These residues could then be mutated using site-directed mutagenesis to select the mutant enzymes with improved properties. However, as shown above, no general rules have yet been formulated to identify these residues.

Some results of rational enzyme engineering based on 3-D-structures, multiple alignments and other data to improve properties 1 to 4 (see Section 2.11.1) of enzymes used as biocatalysts in enzyme technology are listed in Table 2.15. These include enzymes used for the following processes:

- Starch conversion by amylases (Nielsen, Borchert, 2000), glucoamylases (Sauer et al., 2000) and glucose isomerases (Hartley et al., 2000).
- Protein hydrolysis during washing by subtilisin (Bryan, 2000).
- Hydrolysis and synthesis of β-lactam antibiotics by glutaryl amidase (Sio et al., 2002; Oh et al., 2003) and penicillin amidase (Van Der Laan et al., 2000; Kasche et al., 2003).

Table 2.15 Summary of the results of rational design to improve the property of an enzyme important for an enzyme process, based on the 3-D structure, multiple alignments by site-directed mutagenesis (SDM).

Enzyme	*Enzyme property that must be improved to increase the yield and the rate of an enzyme process*	*Selection of amino acid residue that should be changed by SDM based on*	*Observed change in enzyme property that should be improved*	*Comments*
α-amylase	Specificity constant	3-D structure, SDM of active site residues	Loss of activity	Tightly bound Ca^{2+} required (binding constant in nM-range) for activity, but this ion is not involved in the enzyme function
	Temperature stability in the hydrolysis of starch	3-D structure, properties of hybrid enzymes, Ca^{2+}-binding sites and alignment, remove amide containing side chains to increase temperature stability (see Table 2.10)	Temperature stability could be increased by a factor of 10 at 90 °C	
Gluco-amylase	Specificity constant	3-D structure, SDM of active site residues	Reductions	
	Temperature stability in maltose and malto-tetraose hydrolysis	3-D structure, SDM to form new S-S bonds	Temperature optimum increased by 4 °C	
Glucose isomerase	Thermal stability to reduce activity loss due to reaction of glucose and Lys amino groups, in glucose isomerization; Specificity constant	3-D structure and alignment of primary sequence of class I enzyme from 11 sources, SDM of 17 conserved amino acid residues	Increased by a factor of almost 3 (no activity change) by exchanging 3 Lys by Arg; Amino acid exchange in 14 of these led to almost complete loss of activity, the other had up to 90 % of the wt value	Binding of Mg^{2+}, Mn^{2+}, or Co^{2+} (binding constant in μM range) required for activity and stability, these ions are not involved in the enzyme function
Glutaryl amidase	Substrate specificity to reduce enzyme costs in the hydrolysis of adipoyl-7-ADCA	3-D structure and selection of 3 amino acid residues suitable for SDM from evolution *in vitro*. They are in or near the active site.	Increase by a factor of 6	Based on alignment similar to the penicillin amidase below (but no bound Ca^{2+})
	Increase the specificity constant for the hydrolysis of cephalosporin C (see case study, Chapter 9)	3-D structure and selection of residues for SDM that interact with cephalosporin C	Increase by a factor of 6 due to mutation of two amino acid residues	

Table 2.15 Continued

Enzyme	*Enzyme property that must be improved to increase the yield and the rate of an enzyme process*	*Selection of amino acid residue that should be changed by SDM based on*	*Observed change in enzyme property that should be improved*	*Comments*
Penicillin amidase	pH-stability and turnover number to increase product yield and reduced enzyme costs in the hydrolysis of penicillin G	Part of pro-peptide activates enzyme from *A. faecalis* that has a higher k_{cat} and better pH-stability than the *E. coli* enzyme (Fig. 2.28) SDM to avoid complete hydrolysis of pro-peptide	k_{cat} could be increased by a factor of 3	Pro-peptide assists folding to active structure A tightly bound Ca^{2+} is required for membrane transport and maturation of the pro-enzyme. It is probably also required for activity and stability, but is not involved in the enzyme function
	Stereoselectivity and (k_T/k_H) in racemate resolution or kinetically controlled synthesis of β-lactam antibiotics	3-D structure (*E. coli* enzyme), alignment of enzyme from five sources and molecular modeling, SDM of residues in and near the active site of	Decreased specificity constants, 6-fold increase in stereoselectivity and equal or decreased (k_T/k_H) for 8 mutants of *A. faecalis* enzyme	
Subtilisin	Thermal and chemical stability and reduced oxidation of Met by H_2O_2 in washing powder	3-D structure, SDM to form new stabilizing S–S bonds or increase new stabilizing side chain interactions, and replace Met	3 of 18 new S–S bonds increased thermal stability by up to a factor of 100 (enzyme from *B. amyloliquefaciens*) or 20 (*B. subtilis* enzyme); chemical stability increased by a factor of 100 by replacing Met with Ser (specificity constant decreased by a factor of 10)	More than 50 % of all amino acid have been replaced by SDM Pro-peptide assists folding to active structure Tightly bound Ca^{2+} required (binding constant in nM-range) for activity, but is not involved in the enzyme function.

Based on the property changes listed in Table 2.15, there seems to be good agreement with the conclusions on the magnitude of differences in the properties of natural enzyme variants outlined in Section 2.11.1.

2.11.4
New Enzymes by Biosynthesis (Catalytic Antibodies) or Chemical Synthesis (Synzymes)

Catalytic Antibodies

Different monoclonal antibodies against compounds with structures similar to the transition states of substrates can be produced in cell cultures using hybridoma technology. The hybridomas have binding subsites as present in the active site of enzymes. Among the different monoclonal antibodies formed after immunization, some may have catalytic subsites near these binding sites. These catalytic antibodies could be used as biocatalysts for the conversion of substrates, and can be selected from different monoclonal antibodies by using screening techniques outlined in Section 2.11.2. Catalytic antibodies have been under development for about 20 years, but although their specificity constants have increased, they remain many orders of magnitude below those of the enzymes able to convert the same substrates (Wentworth, 2002).

Synzymes

Known active site structures determined with X-ray crystallography have been used by polymer chemists to mimic the sites in chemically synthesized polymers. Despite these studies having been conducted for over 30 years, the best specificity constants obtained remain many orders of magnitude lower than those of the enzyme of which the active site is being mimicked.

2.12
Exercises

1. For the ionization of TRIS (for formula, see Merck Index) and phosphoric acid the heat of ionization is 40 kJ mol^{-1}, and $\approx$0 kJ mol^{-1}, respectively. In which buffer would you study the pH- and temperature-dependence of enzyme-catalyzed reactions? What other reasons can you give for the advice to avoid the use of TRIS as buffer when you study enzyme kinetics? ($pK_{TRIS} \approx 8.5$) (see Table 2.3).

2. Why do oxidoreductases require co-substrates (and which)?

3. Explain Figures 2.4 and 2.5. Why does carboxypeptidase A hydrolyze at the C-end and α-chymotrypsin not? (Hint: the latter enzyme has a negative charge in the S_1'-subsite).
   What influence has the ionic strength here? In what steps can the enzyme function be divided? Can product inhibition be avoided?

4. What other similarities, besides the conserved active site structure, do you find in Figure 2.6?

5. Derive the Michaelis–Menten equation (see Section 2.8).

6. A pyrophosphatase from potatoes catalyzes the hydrolysis of inorganic pyrophosphate at pH 5.3. V_{max} was determined at different temperatures:

Temperature (°C)	15	25	35	40
V_{max} (μM min^{-1})	6.53	10.47	16.79	20.65

   Calculate the activation energy for the enzyme-catalyzed process and compare it with that for the uncatalyzed reaction (= 121 kJ mol^{-1}). What activation energy is determined here? (Hint: see Fig. 2.7).

7. Define k_{cat}, K_m, and k_{cat}/K_m. What information do they give? Why does the last quantity have an upper limit?

8. What information do Tables 2.4 and 2.5 provide about the P_1-, P'_1- substrate specificity and stereoselectivity of the studied enzymes? What is the esterase/amidase activity ratio of the peptidases?

9. Glycolipids are biodegradable detergents. Which enzymes could catalyze the synthesis and hydrolysis of such compounds?

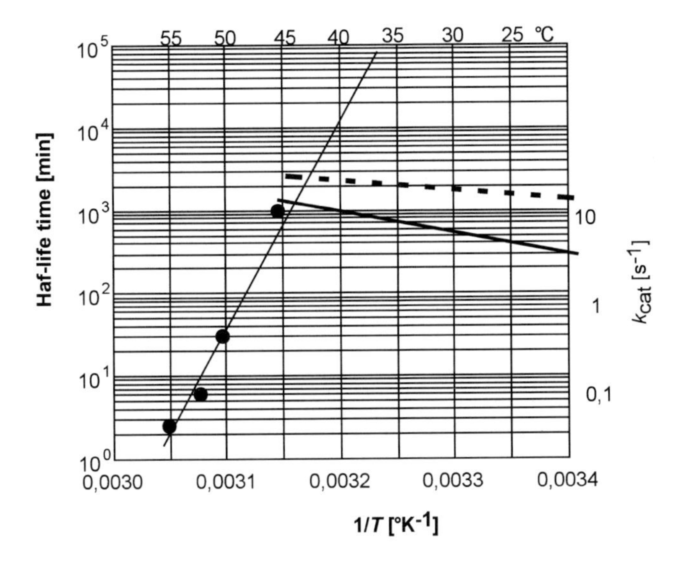

10. The above figure gives the Arrhenius plots for the thermal denaturation at pH 7.5 (closed circles, left scale) of penicillin amidase from *E. coli*, and k_{cat} (right scale) for the hydrolysis of (*R*)-(broken line) and (*S*)-(thick line)-phenylglycine amide by the same enzyme. Calculate and compare the activation energies for the different processes. What activation energy is determined in Figure 2.8?

11. Transglutaminase (EC 2.3.2.13) catalyzes the formation of a peptide bond between polypeptide chains between glutamine (acyl-donor) and the amino group of lysine (acyl-acceptor) (Aeschlimann, Paulsson, 1994). In this reaction ammonia is formed. Try to formulate a reaction mechanism and discuss where such a peptide can be formed in or on a protein. Can this enzyme be used to cat-

alyze reactions where the enzyme is labeled with fluorescent dyes or other ligands, the immobilization of proteins on surfaces? What reactive groups must these ligands or surfaces have? For what purposes can such labeled proteins be used? Is this an equilibrium- or kinetically controlled process? What is the optimal pH for this reaction?
What other applications can you propose for this enzyme in enzyme technology?

12. Why is substrate and product inhibition important in enzyme technology?

13. What properties should organic solvents used in enzyme technology have?

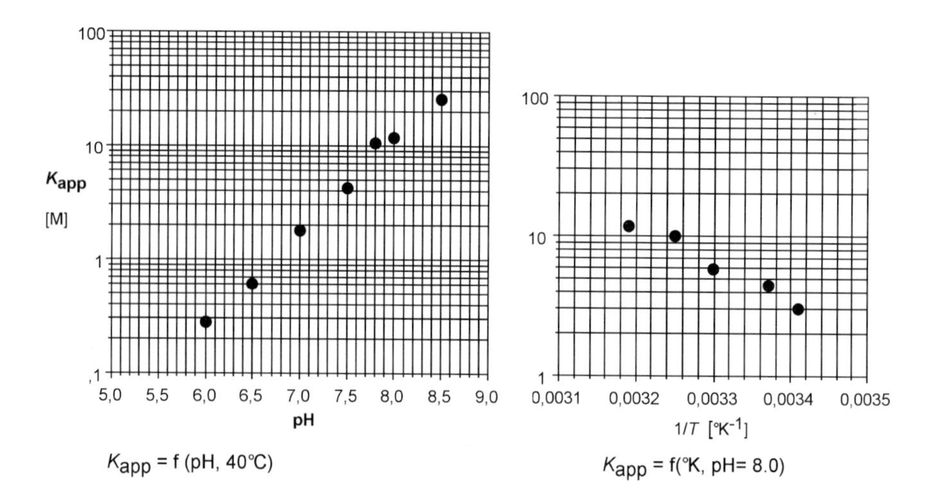

14. The above figure gives the pH- and temperature-dependence of K_{app} (as dissociation constant = [6-aminopenicillamic acid][phenylacetic acid]|[penicillin G]) for the hydrolysis of penicillin G. Determine the process window in a pH-T plane for 95 and 99 % yield for the hydrolysis of 300 mM penicillin G. Which enzyme in Figure 2.28 should be used as the biocatalyst? What other factors can limit the size of this process window? (See also Exercise 10). What buffers are suitable to control the pH during the hydrolysis (Hint: a buffer with a high buffer capacity at the optimal pH for the process, and that does not increase waste water treatment costs.)

15. How much does the temperature change during the hydrolysis of 300 mM penicillin G or 1 M saccharose in an adiabatic reactor (Hint: see Table 2.7). What are the consequences for enzyme technology?

16. Enzyme processes with slightly soluble substrates can be carried out in aqueous suspensions with up to ≥90 % solid substrates. When H^+ is formed or consumed in the process, the pH will change during the reaction. Discuss the possibility of keeping the pH constant (and the product yield high) by adding base or acid at different solid contents. At what solid contents would you carry out such a reaction?

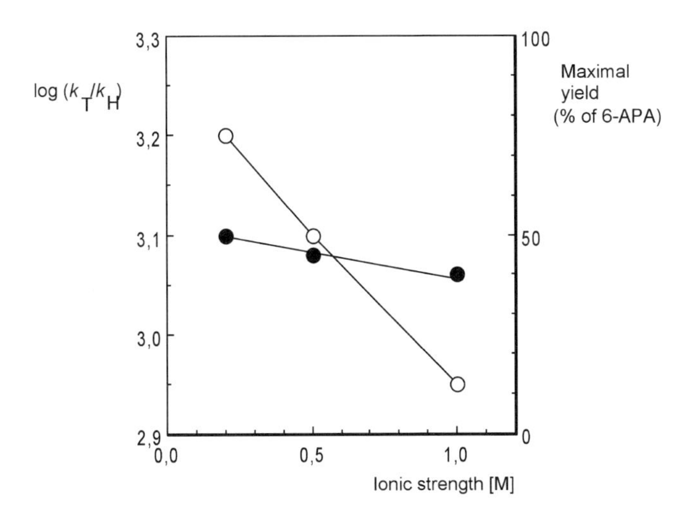

17. The above graph provides the ionic strength dependence of the selectivity (open symbols), and the maximal yield (closed symbols) in the kinetically controlled synthesis of the β-lactam antibiotic ampicillin from 200 mM *R*-phenylglycine-amide and 6-aminopenicillamic acid (6-APA) at pH 7.5 and 25 °C. The hydrolysis of the amide does not depend on *I*. Try to explain the cause for this ionic strength dependence. (Hints: p*K*s in Fig. 2.20, enzyme binding site charge, and how does 6-APA inhibit the hydrolysis of the amide.)

18. Derive and integrate Eq. (2.33) for competitive product inhibition.

19. How can a better enzyme be found or constructed for a given enzyme process?

20. Analyze Table 2.15.

21. Pyruvate decarboxylase can catalyze the carboxylation of acetaldehyde to pyruvate. At what pH and in what solvents would you carry out this reaction to obtain a high product yield? (Hints: What is the p*K*-value for pyruvic acid, CO_2 is used for carboxylation) (Miyazaki et al., 2001).

22. Peptide hormones or neuropeptides are synthesized as parts of ribosomically produced polypeptides. Some of the biologically active peptides are amidated at the C-terminal end. To these belong the following peptide hormones: oxytocin (stimulates contraction of smooth muscles)

    Cys-Tyr-Ile-Gln-Asn-Cys-Pro-Leu-Gly-NH_2

    and vasopressin (antidiuretic)

    Cys-Tyr-Phe-Gln-Asn-Cys-Pro-Arg-Gly-NH_2
    where the two Cys form an S–S bond.

    a) Discuss how this amidated peptide can be produced in an enzyme process from the non-amidated peptide. what properties must the enzyme used as biocatalyst have? (Hints: Figs. 2.3 to 2.5, and Table 2.4).

    b) Find out more about the biosynthesis of these peptide using Internet resources, and suggest processes for their biotechnological production.

Literature

Textbooks

These textbooks cover the bio-, organic-, and physical chemical basics, kinetics and structure function analysis of enzyme catalyzed reactions.

Bisswanger, H., *Enzyme kinetics*, 2nd Ed., Wiley-VCH, Weinheim, 2002
Creighton, T.E., *Proteins: Structures and Molecular Properties*, W.H. Freeman, New York, 1993
Fersht, A.R., *Structure and Mechanism in Protein Science: A Guide to Enzyme Catalysis and Protein Folding*, W.H. Freeman, New York, 1999
Laidler, K.J., *The Chemical Kinetics of Enzyme Action*, The Clarendon Press, Oxford, 1958
Walsh, C., *Enzymatic Reaction Mechanisms*, W.H. Freeman, San Fransisco, 1979

Handbooks

Schomburg, D., Salzmann, I., *Springer Handbook of Enzymes*, 2nd rev. Ed., Springer, Berlin, Heidelberg, New York, 2001
Webb, E.C., *Enzyme Nomenclature*, Academic Press, San Diego, 1992

References

Ahern, T.J., Klibanov, A.M., Analysis of processes causing thermal inactivation of enzymes, *Meth. Biochem. Anal.*, **1988**, *33*, 91–127
Arnold, F. H., Combinatorial and computational challenges for biocatalyst design, *Nature*, **2001**, *409*, 253–257
Arnold, F.H., Georgiou, G. (Eds), *Directed Enzyme Evolution: Screening and Selection Methods*, Humana Press, Totawa 2003a.
Arnold, F.H., Georgiou, G. (Eds), *Directed Evolution Library Creation: Methods and Protocols*, Humana Press, Totawa 2003b.
Atkinson, D.E., Limitation of metabolite concentrations and the conservation of solvent capacity in the living cell, *Current Topics in Cellular Regulation*, **1969**, *1*, 29–43
Bartlett, G.J., Porter, C.T., Borkakoti, N., Thornton, J.M., Analysis of catalytic residues in enzyme active sites, *J. Mol. Biol.*, **2002**, *324*, 105–121
Bender, M.L., Clement, G.E., Kézdy, F.J., D'A Heck, H., The correlation of the pH (pD) dependence and the stepwise mechanism of α-chymotrypsin-catalyzed reactions, *J. Am. Chem. Soc.*, **1964**, *86*, 3680–3689
Benkovic, S.J., Hammes-Schiffer, S., A perspective on enzyme catalysis, *Science*, **2003**, *301*, 1196–1202
Bornscheuer, U.T., Altenbuchner, J., Meyer, H.H., Directed evolution of an esterase for the stereoselective resolution of a key intermediate in the synthesis of Epothilones, *Biotechnol. Bioeng.*, **1998**, *58*, 554–559.
Bornscheuer, U.T., Altenbuchner, J., Meyer, H.H., Directed evolution of an esterase: screening of enzyme libraries based on pH-indicators and a growth assay, *Bioorg. Med. Chem*, **1999**, *7*, 2169–2173.
Bornscheuer, U.T., Directed evolution of enzymes for biocatalytic applications, *Biocat. Biotransf.*, **2001**, *19*, 84–96.
Bornscheuer, U. T., Pohl, M., Improved biocatalysts by directed evolution and rational protein design, *Curr. Opin. Chem. Biol.*, **2001**, *5*, 137–143
Bozhinova, D., Galunsky, B., Yueping, G., Franzreb, M., Köster, R., Kasche, V., Evaluation of magnetic polymer micro-beads as carriers for immobilized biocatalysts for selective and stereoselective transformations, *Biotechnol. Letters*, **2004**, *26*, 343–350
Brakmann, S., Johnsson, K. (Eds), *Directed molecular evolution of proteins*, Wiley-VCH, Weinheim 2002
Brakmann, S., Schwienhorst, A. (Eds), *Evolutionary Methods in Biotechnology: Clever Tricks for Directed Evolution*, Wiley-VCH, Weinheim 2004
Brannigan, J.A., Dodson, G., Duggleby, H.J., Moody, P.C.E., Smith, J.L., Tomchik, D.R., Murzin, A.G. A protein catalytic framework with an N-terminal nucleophile is capable of self-activation, *Nature*, **1995**, *378*, 416–419
Bruggink, A., (Ed.), *Synthesis of β-lactam antibiotics*, Kluwer Academic Publishers, Dordrecht, 2001
Bryan, P.N., Protein engineering of subtilisin. *Biochim. Biophys. Acta*, **2000**, *1543*, 203–222
Buchholz, K., Monsan, P., Dextransucrase, in: *Handbook of Food Enzymology*, J.R. Whitaker, A.G.J. Voragen, D.W.S. Wong (Eds), M. Dekker, New York, 1990

Cadwell, R.C., Joyce, G.F., Randomization of genes by PCR mutagenesis, *PCR Methods Appl.*, **1992**, *2*, 28–33.

Cao, L., Fischer, A., Bornscheuer, U. W., Schmid, R.D., Lipase-catalyzed solid phase synthesis of sugar fatty acid esters. *Biocatal. Biotrans.*, **1997**, *14*, 269–283

Carrea, G., Ottolina, G., Riva, S., Role of solvents in the control of enzyme selectivity in organic media, *Trends Biotechnol.*, **1995**, *13*, 63–70

Chen, C.S., Fujimoto, Y., Girdaukas, G., Shih, C.J., Quantitative analysis of biochemical kinetic resolution of enantiomers, *J. Am. Chem. Soc.*, **1982**, *104*, 7294–7299

Chen, R, Enzyme engineering: rational redesign versus directed evolution, *Trends Biotechnol.*, **2001**, *19*, 13—14

Cherry, J.R., Lamsa, M.H., Schneider, P., Vind, J., Svendsen, A., Jones, A., Pedersen, A.H., Directed evolution of a fungal peroxidase, *Nat. Biotechnol.*, **1999**, *17*, 379–384

Crameri, A., Raillard, S.A., Bermudez, E., Stemmer, W.P., DNA shuffling of a family of genes from diverse species accelerates directed evolution., *Nature*, **1998**, *391*, 288–291

Dalbey, R, Robinson, C., Protein translocation into and across the bacterial plasma membrane and the plant thylakoid membrane, *Trends Biochem. Sci.*, **1999**, *24*, 17–22

Dale, B.E., White, D.H., Ionic strength: a neglected variable in enzyme technology, *Enzyme Microbiol. Technol.*, **1982**, *5*, 227–229

Darnell, J., Lodish, H., Baltimore, D., *Molecular Cell Biology*, Scientific American Books, New York, **1986**, p. 72

Deranleau, D.A., Theory of the measurement of weak molecular complexes, I. General Considerations, *J. Am. Chem. Soc.*, **1969**, *91*, 4044–4049

Dodson. G., Wlowader, A., Catalytic triades and their relatives, *Trends Biochem.*, **1998**, *23*, 347–352

Done, S.H., Brannigan, J.A., Moody, P.C.E., Hubbard, R.E., Ligand-induced conformational change in penicillin acylase, *J. Mol. Biol.*, **1998**, *284*, 463–475

Duggleby, H.J., Tolley, S.P., Hill, C.P., Dodson, E.J., Dodson, G.G., Moody, P.C., Penicillin acylase has a single amino acid catalytic centre, *Nature*, **1995**, *373*, 264–268

Fields, P.A., Somero, G.N., Hot spots in cold adaptation: Localized increases in conformational fexibility in lactate dehydrogenase A_4 orthologs of Antarctic notothenioid fisches, *Proc. Natl. Acad. Sci. USA*, **1998**, *95*, 11476–11481

Freundlich, H., *Kapillarchemie*, Akad. Verlagsgesellschaft, Leipzig, 1909

Galunsky, B., Ignatova, Z., Kasche, V., Temperature effects on S_1- and S_1'-enantionselectivity of α-Chymotrypsin, *Biochim. Biophys. Acta*, **1997**, *1343*, 130–138

Galunsky, B., Kasche, V., Determination of the enantioselectivity for kinetically controlled condensations catalysed by amidases and peptidases, *Adv. Synth. Catal.* **2002**, *2002*, 1115–1119

Garcia-Viloca, M., Gao, J., Karplus, M., Truhlar, D.G., How enzymes work: analysis by modern rate theory and computer simulations, *Science*, **2004**, *303*, 186–195

Godfrey, T., (Eds), in: *Industrial Enzymology*, Godfrey, T., Reichelt, J., (Eds), 2nd Edn., pp. 435–482, Stockton Press, New York, 1996

Goldberg, R.N., Tewari, Y.B., Thermodynamics of enzyme-catalyzed reactions: Part 5, *J. Phys. Chem. Ref. Data*, **1995**, *24*, 1765

Greener, A., Callahan, M., Jerpseth, B., An efficient random mutagenesis technique using *E. coli* mutator strain, *Methods Mol. Biol.*, **1996**, *57*, 375–385.

Halling, P.J., Thermodynamic predictions for biocatalysis in nonconventional media: Theory, tests, and recommendations for experimental design and analysis, *Enzyme Microbiol. Technol.*, **1994**, *16*, 178–205

Halling, P.J., Biocatalysis in low-water media: understanding effects of reaction condition, *Curr. Opin. Chem. Biol.*, **2000**, *4*, 74–80

Hamada, J. S., Deamidation of food proteins to improve functionality. *Crit. Rev. Food Sci. Nutr.*, **1994**, *34*, 283–292

Hartley, B.S., Hanlon, N., Jackson, R.J., Rangarajan, M., Glucose isomerase: insight into protein engineering for increased stability, *Biochim. Biophys. Acta*, **2000**, *1543*, 294–335

Hochachka, P.W., Somero, G.N,. *Biochemical adaptation: Mechanism and Process in Physiological Evolution, Oxford Univ. Press, Oxford,* **2002**

Ignatova, Z., Stoeva, S., Galunsky, B., Hörnle, C., Nurk, A., Piotraschke, E., Voelter, W.,

Kasche, V., Proteolytic processing of penicillin amidase from *A. faecalis* in *E. coli* yields several active forms, *Biotechnol. Lett.*, **1998**, *20*, 977–982

Juergens, C., Strom, A., Wegener, D., Hettwer, S., Wilmanns, M., Sterner, R., Directed evolution of a (beta-alpha)8-barrel enzyme to catalyze related reactions in two different metabolic pathways, *Proc. Natl. Acad. Sci. USA*, **2000**, *97*, 9925–9930

Kasche, V., Zöllner, R., Tris, (hydroxymethyl) methylamine is acylated when it reacts with acyl-chymotrypsin, *Hoppe Seyler's Z. Phys. Biol. Chem.*, **1982**, *363*, 531–534

Kasche, V., Mechanism and yields in enzyme catalyzed equilibrium and kinetically controlled synthesis of β-lactam antibiotics, peptides and other condensation products, *Enzym. Microbiol. Technol.* **1986**, *8*, 4–16

Kasche, V., Schlothauer, R., Brunner, G., Enzyme denaturation in supercritical CO_2: Stabilizing effect of S-S bonds during the depressurization step, *Biotechnol. Lett.*, **1988**, *10*, 569–574

Kasche, V. Galunsky, B., Michaelis, G., Binding of organic solvent molecules influences the P_1'-P_2' stereo- and sequence specificity of α-chymotrypsin in kinetically controlled peptide synthesis, *Biotechnol. Lett.*, **1991**, *13*, 75–80

Kasche, V., Galunsky, B., Enzyme catalyzed biotransformations in aqueous two-phase systems with precipitated substrate and/or product, *Biotechnol. Bioeng.*, **1995**, *45*, 261–267

Kasche, V., Galunsky, B., Nurk, A., Piotraschke, E., Rieks, A., The dependency of the stereoselectivity of penicillin amidases – enzymes with *R*-specific S_1- and *S*-specific S_1'-subsites – on temperature and primary structure. *Biotechnol. Letters*, **1996**, *18*, 455–460

Kasche, V., Lummer, K., Nurk, A., Piotraschke, E., Riecks, A., Stoeva, S., Voelter, W., Intramolecular autoproteolysis initiates the maturation of penicillin amidase from *E. coli*, *Biochim. Biophys. Acta*, **1999**, *1433*, 76–86

Kasche, V., Proteases in peptide synthesis. In: *Proteolytic enzymes.* A practical approach, Beynon, M., Bond., J. (Eds), pp. 265–292, Oxford Univ. Press, Oxford, **2001**

Kasche, V., Galunsky, B., Ignatova, Z., Fragments of pro-peptide activate mature penicillin amidase of *Alcaligenes faecalis*, *Eur. J. Biochem.*, **2003**, *270*, 4721–4728

Kazlauskas, R.J., Elucidating structure–mechanism relationships in lipases: prospects for predicting and engineering catalytic properties, *Trends Biotechnol.*, **1994**, *12*, 464–472

Kimura, M., Ohta, T., Mutation and evolution at the molecular level, *Genetics*, **1973**, *72*, 19–35

Kleinkauf, H., von Döhren, H., Bioactive peptide analogues: In vivo and in vitro production, *Prog. Drug Res.*, **1990**, *34*, 287–317

Kragl, U., Eckstein, M., Kraftzik, N., Enzyme catalysis in ionic liquids, *Curr. Opin. Biotechnol.*, **2002**, *13*, 565–571

Kurtzman, A.L., Govindarajan, S., Vahle, K., Jones, J.T., Heinrichs, V., Patten, P.A., Advances in directed protein evolution by recursive genetic recombination: applications to therapeutic proteins, *Curr. Opin. Biotechnol.*, **2001**, *12*, 361– 370

Lewin, B., *Genes*, Oxford Univ. Press, Oxford, **1997**, p. 94

Lorenz, P., Liebeton, K., Niehaus, F., Eck, J., Screening for novel enzymes for biocatalytic processes; accessing the metagenome as a resource of novel functional sequence space, *Curr. Opin. Biotechnol.*, **2002**, *13*, 572–577

Lorenz, P., Liebeton, K., Niehaus, F., Schleper, C., Eck, J., The impact of non-cultivated biodiversity on enzyme discovery and evolution, *Biocat. Biotransf.*, **2003**, *21*, 87–91

Lummer, K., Riecks, A., Galunsky, B., Kasche, V., pH-dependence of penicillin amidase enantioselectivity for charged substrates, *Biochim. Biophys. Acta*, **1999**, *1433*, 327–334

Lutz, S., Ostermeier, M., Benkovic, S.J., Rapid generation of incremental trunction libraries for protein engineering using a-phosphothioate nucleotides, *Nucleic Acids Res.*, **2001**, *29*, 1–7

MacBeath, G., Kast, P., Hilvert, D., Redesigning enzyme topology by directed evolution, *Science*, **1998**, *279*, 1958–1961

May, O., Nguyen, P.T., Arnold, F.H., Inverting enantioselectivity by directed evolution of hydantoinase for improved production of L-methionine, *Nat. Biotechnol.*, **2000**, *18*, 317–320

Moore, J.C., Arnold, F.H., Directed evolution of a para-nitrobenzyl esterase for aqueous-organic solvents, *Nat. Biotechnol.*, **1996**, *14*, 458–467

Martinek, K., Yatsimirski, A.K., Berezin, I.V., Effect of ionic strength on the steady state kinetics of α-chymotrypsin catalyzed reactions, *Molec. Biol. (USSR)*, **1971**, *5*, 96–109

McDonough, M.A., Klei, H.E., Kelly, J.A., Crystal structure of penicillin G acylase from the Bro 1 mutant strain of *Providencia rettgeri*, *Protein Sci.*, **1999**, *8*, 1971–1981

Mesiano, A.J., Beckman, E.J., Russel, A.J., Supercritical biocatalysis, *Chem. Rev.*, **1999**, *99*, 623–634

Menger, F.M., Analysis of ground-state and transition-state effects in enzyme catalysis, *Biochem.*,**1992**, *31*, 5368–5373

Michaelis, G., *Peptidase-katalysierte Peptidsynthese*, Thesis, TU Hamburg-Harburg, 1991

Miller, C.A., Advances in enzyme discovery, *Inform*, **2000**, *11*, 489–495

Mitchison, C., Wells, J.A., Protein engineering of disulfide bonds in subtilisin BPN', *Biochemistry*, **1989**, *28*, 4807–4815

Miyazaki, K., Wintrode, P.L., Grayling, R.A., Rubingh, D.N., Arnold, F.H., Directed evolution study of temperature adaptation in a psychrophilic enzyme, *J. Mol. Biol.*, **2000**, *297*, 1015–1026

Miyazaki, M., Shibue, M., Ogino, K., Nakamura, H, Maeda, H., Enzymatic synthesis of pyruvic acid from acetaldehyde and carbon dioxide, *Chem. Commun.*, **2001**, *21*, 1800–1801

Nielsen, J.E., Borchert, T.V., Protein engineering of bacterial α-amylases, *Biochim. Biophys. Acta*, **2000**, *1543*, 253–274

Ness, J.E., Welch, M., Giver, L., Bueno, M., Cherry, J.R., Borchert, T.V., Stemmer, W.P., Minshull, J., DNA shuffling of subgenomic sequences of subtilisin, *Nat. Biotechnol.*, **1999**, *17*, 893–896

Neylon, C., Chemical and biochemical strategies for the randomization of protein encoding DNA sequences: library construction methods for directed evolution, *Nucleic Acids Res.*, **2004**, *32*, 1448–1459

Oh, B., Kim., M., Yoon, J., Chung, K., Shin, Y., Lee, D., Kim, Y., Deacylation activity of cephalosporin acylase to cephalosporin C is improved by changing the side-chain conformations of active site residues, *Biochem. Biophys. Res. Commun.*, **2003**, *310*, 19–27

Ostermeier, M., Nixon, A.E., Benkovic, S.J., Incremental truncation as a strategy in the engineering of novel biocatalysts, *Bioorg. Med. Chem.*, **1999**, *7*, 2139–2144.

Park, S., Kazlauskas, R.J., Biocatalysis in ionic liquids – advantages beyond green technology, *Curr. Opin. Biotechnol.*, **2003**, *14*, 432–437

Phillips, R.S., Temperature modulation of the stereochemistry of enzymatic catalysis: Prospects for exploitation, *Trends Biotechnol.*, **1996**, *14*, 13–16

Poliakoff, M., Fitzpatrick, J.M., Farren, T.R., Anastas, P.T., Green Chemistry: Science and Politics of change, *Science*, **2002**, *297*, 807–810.

Reetz, M.T., Wilensek, S., Zha, D., Jaeger, K.-E., Directed evolution of an enantioselective enzyme through combinatorial multiple-cassette mutagenesis, *Angew. Chem. Int. Ed.*, **2001**, *40*, 3589–3591

Reetz, M.T., Controlling the enantioselectivity of enzymes by directed evolution: Practical and theoretical ramifications, *Proc. Natl. Acad. Sci. USA*, **2004**, *101*, 5716–5722

Reymond, J.L., Wahler, D., Substrate arrays as enzyme fingerprinting tools, *Chem.Bio.Chem.*, **2002**, *3*, 701–708

Sakai, T., Kawabata, I., Kishimoto, T., Ema, T., Utaka, M., Enhancement of the enantioselectivity in lipase-catalyzed kinetic resolution of phenyl-2H-azirine-2-mehanol by lowering the temperauture, *J. Org. Chem.*, **1997**, *62*, 4906–4907

Salemme, F.R., Structure and function of cytochrome C, *Annu. Rev. Biochem.* **1977**, *46*, 299–329

Sauer, J., Sigurskjold, B.W., Christensen, U., Frandsen, T.P., Mirgodskaya, E., Harrison, M., Roepstorff, P., Svensson, B., Glucoamylase: structure/function relationships, and protein engineering, *Biochim. Biophys. Acta*, **2000**, *1543*, 275–293

Schechter, I. Berger, A., On size of active site in protease, I. Papain, *Biochem. Biophys. Res. Commun.*, **1967**, *27*, 157–162

Scherer, L.J., Rossi, J.J., Approaches for the sequence-specific knockdown of mRNA, *Nat. Biotechnol.*, **2003**, *21*, 1457–1465

Schmitke, J.L., Wescott, C.R., Klibanov, A.M., The mechanistic dissection of the plunge in enzymatic activity upon transition from water to anhydrous solvents, *J. Am. Chem. Soc.* **1996**, *118*, 3360–3365

Schowen, R.L., How an enzyme surmounts the activation energy barrier, *Proc. Natl. Acad. Sci. USA*, **2003**, *100*, 11931–11932

Seddon, K.R., Ionic liquids. A taste of the future, *Nature Materials*, **2003**, *2*, 1–2

Sio, C.F., Riemen, A.M., van der Laan, J.M., Verhaert, R.M.D., Quax, W.J., Directed evolution of a glutaryl acylase into an adipyl acylase, *Eur. J. Biochem.*, **2002**, *269*, 4495–4504

Spiess, A., Schlothauer, R., Hinrichs, J., Scheidat, B., Kasche, V., pH gradients in heterogeneous biocatalysts and their influence on rates and yields of the catalysed processes. *Biotechnol. Bioeng.*, **1999**, *62*, 267–277

Stemmer, W.P., Rapid evolution of a protein in vitro by DNA shuffling, *Nature*, **1994**, *370*, 389–391

Stemmer, W.P.C., DNA shuffling by random fragmentation and reassembly: *In vitro* recombination for molecular evolution, *Proc. Natl. Acad. Sci. USA*, **1994**, *91*, 10747–10751

Straaathof, A.J.J., Jongejan, J.A., The enantiomeric ratio: Origin, determination and prediction, *Enzyme Microbiol. Technol.* **1997**, *21*, 559–671

Tosa, T., Mori, T., Fuse, N., Chibata, I., Studies on continuous enzyme reactions. 6. Enzymatic properties of DEAE-Sepharose Aminoacylase complex, *Agric. Biol. Chem.*, **1969**, *33*, 1047–1056

Turner, N.J., Directed evolution of enzymes for applied biocatalysis, *Trends Biotechnol.*, **2003**, *21*, 474–478.

van Rantwijk, F., Madeira Lau, R., Sheldon, R.A., Biocatalytic transformations in ionic liquids, *Trends Biotechnol.*, **2003**, *21*, 131–138

Van der Laan, J.M., Riemens, A.D., Quax, W.J., Mutated penicillin acylase genes, *US Patent 6.033*, **2000**, 823

Verhaert, R.M.D., Riemens, A.M., van der Laan, J.M., van Duin, J., Quax, W.J., Molecular cloning and analysis of the gene encoding the thermostable penicillin G acylase from *Alcaligenes faecalis*, *Appl. Env. Microbiol.*, **1997**, *63*, 3412–3418

Vulfson, E.N., Halling, P.J., Holland, H.L., (Eds), *Enzymes in nonaqueous solvents: methods and protocols*, Humana Press, Totowa, New Jersey, **2001**

Wentworth, Jr., P., Antibody design by man and Nature, *Science*, **2002**, *296*, 224–226

Wiesemann, T., *Enzymmodifikation für analytische und präparative Zwecke: natürliche und künstliche Penicillinamidase-Varianten*, Dissertation, TU Hamburg-Harburg, 1991

Zhao, H., Giver, L., Shao, Z., Affholter, J.A., Arnold, F.H., Molecular evolution by staggered extension process (StEP) in vitro recombination., *Nat. Biotechnol.*, **1998**, *16*, 258–261

Zmiejewski, Jr., M.J., Briggs, B.S., Thompson, A.R., Wright, I.G., Enantioselective acylation of a beta-lactam intermediate in the synthesis of Loracarbef using penicillin G amidase, *Tetrahedron Lett.*, **1991**, *32*, 1621–1622

3
Enzymes in Organic Chemistry

Why enzymes in organic synthesis?	Alternative to chemical methods: • high regio- and stereoselectivity • milder reaction conditions • environmentally more friendly
Which enzyme(s)?	Oxidoreductases (Section 3.2.1) Hydrolases (Section 3.2.2) Lyases (Section 3.2.3) Isomerases/Racemases (Section 3.2.4)
Which reaction system? (Section 2.9)	Aqueous Aqueous and water-miscible organic solvent Aqueous and water-immiscible organic solvent Pure organic solvent Other solvents (supercritical fluids, ionic liquids)
Which 'chemistry'	(Dynamic) kinetic resolution vs. asymmetric synthesis? Cofactor free or cofactor-dependent enzymes

3.1
Introduction

The first applications of enzymes in organic chemistry date back almost a century. As early as 1908, Rosenthaler used a hydroxynitrile lyase-containing extract for the preparation of (*R*)-mandelonitrile from benzaldehyde and hydrogen cyanide (HCN). Since then, an increasing number of enzymes have been identified, and their use in organic chemistry has steadily increased in parallel. In particular, since the mid-1970s the number of reports of enzyme utilization – which at present stands at more

Biocatalysts and Enzyme Technology. K. Buchholz, V. Kasche, U.T. Bornscheuer

ISBN: 3-527-30497-5

than 13 000 – as well as the number of industrialized enzyme-related processes has increased substantially.

Several reasons can be identified for this development, including:

- More organic chemists accept the use of biocatalysts.
- Biocatalysis may save additional reaction steps compared to organic synthesis.
- Enzymes are often highly chemo-, regio-, and stereospecific.
- Biocatalysis is a safer and "greener" technology.
- There is a substantially increased demand for optically pure compounds, especially for pharmaceutical applications.
- The production of biocatalysts has been made easier due to the development of recombinant expression systems.
- Many enzymes are available commercially.

The most important application of enzymes in organic chemistry is in the synthesis of optically active compounds. This is due to the excellent enantio- and stereospecificity shown by many enzymes, which makes them attractive alternatives to asymmetric organic syntheses or reactions starting from the chiral pool.

In addition, enzymes are used for the synthesis of chemicals lacking a chiral center. Prominent examples are the production of acrylamide of about 100 000 tons per annum scale and of nicotinamide (both using a nitrile hydratase), and the synthesis of esters, for use in cosmetic applications (see also Table 1.3, p. 13).

In the following examples, the applications of biocatalysts in organic chemistry are organized based on the division suggested by the Enzyme Commission. Thus, a brief introduction, the reaction principle, and selected examples are provided for each enzyme class. Note that with a few exceptions only those reactions using isolated enzymes are included.

During the past two decades, the application of enzymes in organic synthesis has emerged as an extremely broad field, and a wide variety of types of enzymes and examples of their use in chemoenzymatic syntheses has been described. Within this chapter it would be impossible to provide sufficient coverage of all developments – for example, the book *Enzyme Catalysis in Organic Synthesis*, by K. Drauz and H. Waldmann, utilizes almost 1500 pages in three volumes for this area alone! Hence, only selected enzymes are included in this chapter, the intention being to provide a basic introduction to the subject. Thus, readers are encouraged to consult the broad range of excellent books and reviews cited in the literature list, as many of these provide a much more in-depth coverage of the uses of enzymes in organic chemistry.

3.1.1
Kinetic Resolution or Asymmetric Synthesis

Enzymatic syntheses of optically active compounds can begin from the kinetic resolution of racemic mixtures, or can be performed via an asymmetric synthesis (Fig. 3.1). Kinetic resolution will only lead to a maximum yield of 50 % unless the unwanted enantiomer is racemized. This can be achieved using a racemase, or by chemical racemization. If kinetic resolution and racemization are performed simul-

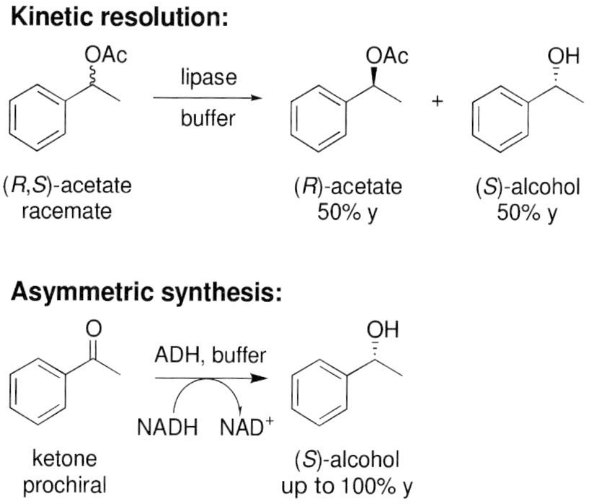

Kinetic resolution:

OAc — lipase, buffer → OAc + OH

(R,S)-acetate racemate — (R)-acetate 50% y — (S)-alcohol 50% y

Asymmetric synthesis:

O — ADH, buffer; NADH → NAD^+ → OH

ketone prochiral — (S)-alcohol up to 100% y

Fig. 3.1 Kinetic resolution of racemates versus desymmetrization of a prochiral compound.

taneously, the process is named dynamic kinetic resolution (DKR), and is exemplified in detail in Section 3.2.2.1. Asymmetric synthesis allows the production of one enantiomer at up to 100 % yield. Examples of this include the desymmetrization of prostereogenic compounds using, for example an alcohol dehydrogenase in the reduction of a ketone (Section 3.2.1.1) or the formation of a chiral compound by, for example, C–C-bond formation using a lyase (see Section 3.2.3).

The enantioselectivity (enantiomeric ratio, E-value) is frequently used to estimate the selectivity of an enzyme in a kinetic resolution (Chen et al., 1982, 1987). The theoretical background and equations to calculate this value are outlined in Chapter 2, Section 2.7.1.

The (*R*)/(*S*)-nomenclature is recommended for the assignment of absolute configurations. In the scientific literature on amino acids and sugars the D,L-nomenclature is still the common practice and is therefore used by us in Chapters 3 to 6, where the biocatalysis of these compounds are treated in greater detail.

3.2 Examples

3.2.1 Oxidoreductases (EC 1)

Although it is estimated that about 25 % of all presently known enzymes are oxidoreductases, the most useful enzymes for preparative applications are dehydrogenases or reductases. The mono- and dioxygenases, oxidases and peroxidases also belong to this class. All of these enzymes require NADH or NADPH as a cofactor and, due to this cofactor-dependency, recycling is necessary in order to conduct cost-effective processes, unless the reaction is performed in a whole-cell system. Currently, most oxidoreductases are not used as isolated biocatalysts and are therefore not covered extensively in the following sections.

3.2.1.1 Dehydrogenases (EC 1.1.1.-, EC 1.2.1.-, EC 1.4.1.-)

Synthesis of Alcohols

The most important application of oxidoreductases in organic chemistry is in the reduction mode, as this yields chiral compounds such as alcohols, hydroxy acids, or amino acids. In the case of alcohol dehydrogenases (ADH), a hydrogen and two electrons are transferred from the reduced nicotinamide moiety to an acceptor molecule such as a ketone or an α-keto acid. In many cases, this reaction is highly stereoselective and the hydride is delivered by the dehydrogenase either from the *re-* or the *si-*face of the carbonyl, yielding the corresponding (*R*)- or (*S*)-products (Fig. 3.2). Many ADHs were found to obey the Prelog rule – which is based on the size of substituents and allows to predict which enantiomer will be produced – although exceptions have also been described (Anti-Prelog). An overview of different ADHs and their selectivities is provided in Table 3.1.

In the literature, most examples for the use of ADHs deal with whole-cell systems, such as baker's yeast. However, more recently ADHs have also been used as isolated enzymes coupled with efficient cofactor regeneration systems which do not require a separate biocatalyst. Instead, NAD(P)H is directly recycled by the ADH used to produce the optically pure alcohol in the presence of isopropanol. The key to suc-

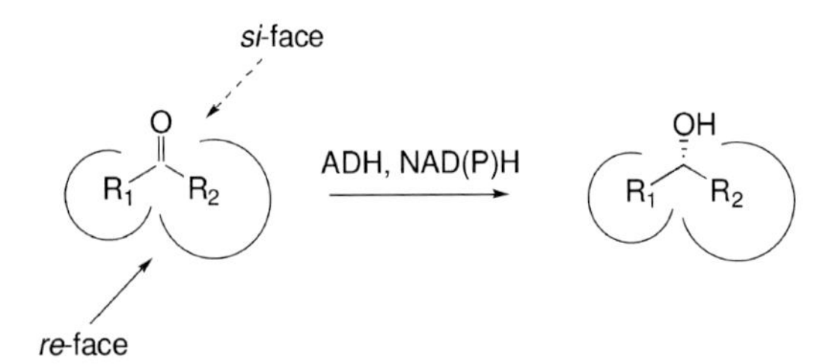

Fig. 3.2 According to the Prelog-rule, the size of the substituents R_1 and R_2 (here: $R_1 < R_2$) determines whether the carbonyl of a ketone is attacked by the hydride either from the *re-* or the *si-*face. In the example shown above, the (*S*)-product is formed, if a sequence rule of $R_2 > R_1$ is assumed.

Table 3.1 Alcohol dehydrogenases from various sources and their selectivities.

ADH	*Attack from*	*Configuration*
Yeast	*Re*	Prelog
Horse liver	*Re*	Prelog
Thermoanaerobium brockii	*Re*	Prelog
Curvularia falcata	*Re*	Prelog
Pseudomonas sp.	*Si*	Anti-Prelog
Lactobacillus kefir	*Si*	Anti-Prelog
Mucor javanicus	*Si*	Anti-Prelog

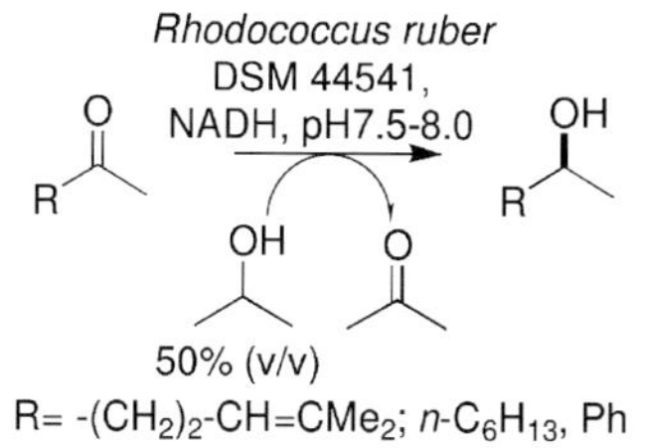

Fig. 3.3 Example of the synthesis of optically active alcohols using an alcohol dehydrogenase. The cofactor NADH is regenerated directly from isopropanol (Stampfer et al., 2002).

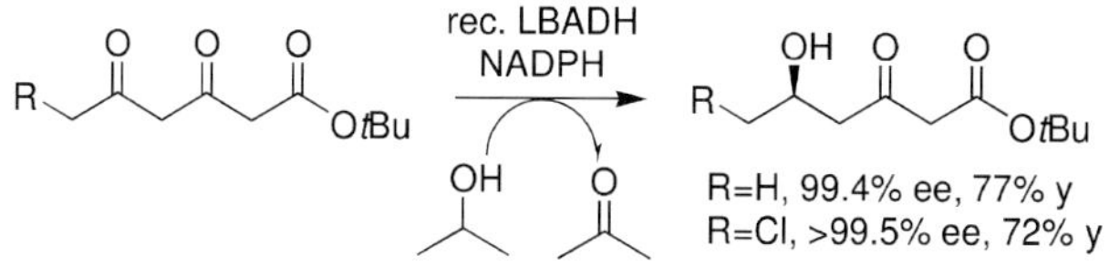

Fig. 3.4 Example of the synthesis of optically active 3-oxo-5-hydroxy carboxylic acids using a recombinant alcohol dehydrogenase from *Lactobacillus brevis* exemplifying the regio- *and* stereoselectivity of this ADH. The cofactor NADH is regenerated directly from isopropanol (Wolberg et al., 2000).

cess was the high stability of the ADH at high concentrations of isopropanol and acetone (Figs. 3.3 and 3.4).

Synthesis of Amino Acids

Amino acids (α-amino carboxylic acids) are widely used in nutrition, medical applications, and organic synthesis. They can be produced by one of the following routes:

- Cultivation of microorganisms, especially overproducers of certain amino acids.
- Extraction from protein hydrolysates.
- Enzymatic synthesis from prochiral precursors or via kinetic resolution of racemates.
- Chemical synthesis.

The first two methods provide access only to natural L-amino acids, whereas the last two approaches allow for the synthesis of non-natural amino acids and the D-enantiomers. L-glutamate (>1 000 000 tons) and L-lysine (>200 000 tons) are amino acids obtained by cultivation with *Corynebacterium glutamicum* as the main producer. Protein hydrolysates are used for the isolation of L-cysteine, L-leucine, L-asparagine, L-arginine, and L-tyrosine.

The enzymatic synthesis can start from racemic amino acids obtained by chemical synthesis (i.e., the Strecker method) followed by kinetic resolution using esterases (i.e., enantioselective hydrolysis of amino acid carboxylic acid esters; see Section 3.2.2.2) or acylases/amidases (i.e., enantioselective hydrolysis of amides; see Section 3.2.2.3) or using hydantoinases (see Section 3.2.2.6). Alternatively, prochiral precursors can be subjected to reductive amination using amino acid dehydrogenases (AADH). With a few exceptions, only L-amino acids are accessible. Although a varie-

Table 3.2 Amino acid dehydrogenases from various sources and their selectivities.

AADH	*Coenzyme*	*Source*
Alanine-DH	NADH	Bacteria
Glutamate-DH	NAD(P)H	Bacteria, Yeast, Fungi, Plants
Serine-DH	NADH	Plant
Valine-DH	NAD(P)H	Bacteria, Plant
Lysine-DH	NADH	Bacteria, Human
Phenylalanine-DH	NADH	Bacteria

ty of AADHs have been desribed in the literature (Table 3.2) (Bommarius, 2002), only a limited number are currently used. More than 20 gene and protein sequences of AADHs have been identified, and several 3D-crystal structures have been resolved. Although sequences vary considerably between AADHs, the residues involved in catalysis and nicotinamide cofactor binding are highly conserved.

The catalytic mechanism of AADHs has been proposed to follow via the formation of an imidine. As shown for leucine dehydrogenase (LeuDH) (Fig. 3.5), the α-keto acid and ammonia react in the amination reaction with stabilization via hydrogen bonding between the carboxylic group and the oxyanion by two lysine residues (Lys_{68} and Lys_{80}). After formation of the imine, a water molecule hydrogen-bonded to Lys_{80} and hydrid transfer from NADH yield the α-amino acid and NAD^+ (Ohshima and Soda, 2000).

Leucine and phenylalanine dehydrogenases (Leu-DH, Phe-DH) are the most important AADHs, especially as they are not restricted to their natural substrates, accept a broad range of other precursors, and efficiently allow the synthesis of non-proteinogenic α-amino acids from the corresponding prostereogenic α-keto acids (Fig. 3.6). Also, the production of isotopically labeled α-amino acids has been described (Kragl et al., 1993).

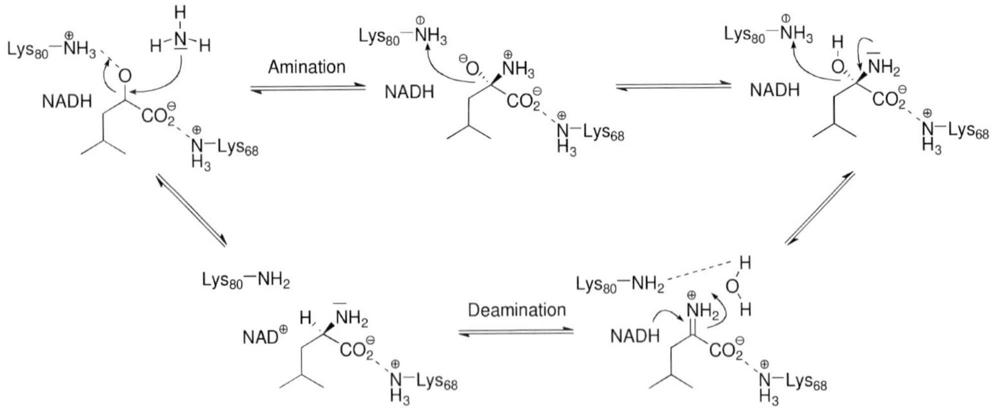

Fig. 3.5 Proposed reaction mechanism for LeuDH (adapted from Ohshima and Soda, 2000).

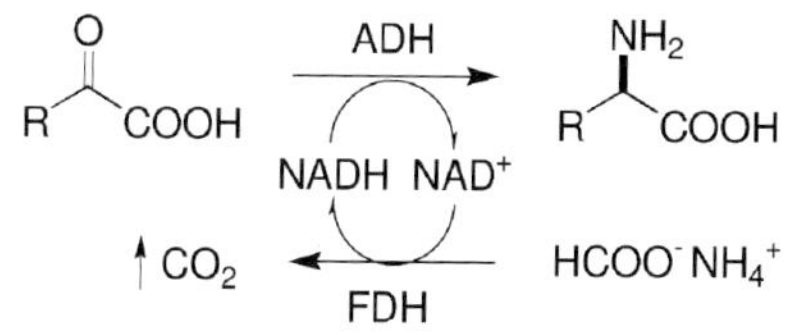

R–CO–COOH → (ADH; NADH → NAD^+) → R–CH(NH_2)–COOH

CO_2 ← (FDH; NAD^+ → NADH) ← $HCOO^- NH_4^+$

Fig. 3.6 Example of the synthesis of optically active L-amino acids using an amino acid dehydrogenase (AADH). The cofactor NADH is regenerated using a formate dehydrogenase from *Candida boidinii*. The equilibrium is shifted towards NADH by using ammonium formate yielding carbon dioxide as by-product.

As pointed out above, cofactor regeneration is required for the cost-effective application of dehydrogenases. An elegant solution for the required recycling of the cofactor was developed by Kula and Wandrey, based on the use of formate dehydrogenase (FDH), for example from the yeast *Candida boidinii* using ammonium formate as co-substrate (Fig. 3.6). Carbon dioxide formed as by-product is highly volatile, and this leads to a favorable shift of the equilibrium. This allows for total turnover numbers (moles of product per mole cofactor) of up to 600000, as demonstrated for the continuous synthesis of phenylalanine (Hummel et al., 1987). The overall reaction is best performed in an enzyme membrane reactor, in which the AADH and the FDH are retained in the reactor compartment using ultrafiltration membranes (see also Chapter 5.4). Covalent coupling of the cofactor NADH to polyethylene glycol (PEG) can be used to avoid leakage through the membrane. An overview of amino acids obtained by this approach is provided in Table 3.3, while further examples and more details can be found in reviews (Bommarius, 2002; Ohshima and Soda, 2000).

From the examples shown in Table 3.3, L-*tert*-leucine (L-^{t}Leu) is the only non-proteinogenic amino acid. L-^{t}Leu is an important building block for a range of pharmaceuticals, as peptide bonds involving this amino acid are only slowly hydrolyzed by

Table 3.3 Examples of the synthesis of L-amino acids using the principle shown in Fig. 3.6 (adapted from Bommarius, 2002).

ADH [a]	*Product*	*Product-conc. [mmol L^{-1}]*	*Conv. [%]*	*STY* [b] *[g $L^{-1} \cdot d^{-1}$]*	*Enzyme consumption* [c] *[U kg^{-1}]*
Leucine-DH	L-Leu	80	80	250	300 (300)
Leucine-DH	L-Leu	70	70	72	730 (350)
Leucine-DH	L-Met	240	60	143	n.d.
Leucine-DH	L-^{t}Leu	425	85	640	1000 (2000)
Alanine-DH	L-Ala	184	46	134	4700 (2600)
Phe-DH	L-Phe	114	95	456	1500 (150)

[a] AADH, amino acid dehydrogenase.
[b] *STY*, space-time-yield.
[c] Values in parentheses refer to the consumption of the cofactor-regenerating enzyme.
n.d., not determined.

peptidases. (Note: According to the Enzyme Commission, proteases should now be named "peptidases", and this term is used throughout the book.) As biocatalytic production using acylase, amidase or hydantoinase/carbamoylase routes failed, reductive amination using LeuDH in combination with the FDH for cofactor recycling was the method of choice for large-scale production (Bommarius et al., 1992). Other non-natural amino acids accessible by this route include L-neopentylglycine, L-β-hydroxyvaline, and 6-hydroxy-L-norleucine.

3.2.1.2 Oxygenases

Oxygenases are enzymes, which introduce either one (monooxygenases) or two (dioxygenases) oxygen atoms into their substrates. Typically, NADH or NADPH serve as reduction equivalents via electron-transfer proteins such as reductases. The major interest in these enzymes for organic synthesis is due to their high regio- and stereoselectivity. Moreover, many of these reactions are difficult to perform by chemical methods, especially if non-activated hydrocarbon moieties need to be transformed. Despite the fact that numerous oxygenases are known, their application in organic synthesis is limited due to a number of problems. These include limited availability of sufficient amounts of enzyme, insufficient stability and often very low specific activity, requirement of costly cofactors and the presence of a reductase. Many enzymes are also membrane-bound, which further limits their application. Some of these problems were overcome by the use of whole-cell systems, preferentially with overexpression of the oxygenase. Dioxygenases have been only used in whole-cell systems, and are not covered here.

P450-Monooxygenases (EC 1.14.13.-)

P450-monooxygenases (also named cytochrome P450 enzymes) are widely distributed in nature, and play a key role in various steps of primary and secondary metabolism, as well as in the detoxification of xenobiotic compounds. A range of reactions is catalyzed by these enzymes (Fig. 3.7), which all include the transfer of molecular oxygen to non-activated aliphatic or aromatic X–H bonds (X: –C, –N, –S) (Goldstein, 1993). Furthermore, a remarkable number of P450 enzymes are able to epoxidize –C=C– double bonds (Lewis, 1996). For these oxygenation reactions, P450 enzymes require cofactors such as NADPH or NADH as reduction equivalents. P450 enzymes show characteristic spectral properties, notably the maximum absorption at 450 nm in the differential spectrum of carbon monoxide, which gave them their name (Omura, 1964).

The P450 superfamily is one of the largest and oldest gene families (Nelson et al., 1993). In 2004, the number of P450 encoding sequences was estimated at over 1000 (*http://www.icgeb.trieste.it/~p450srv/new/p450.html*). For example, the genome project of the plant *Arabidopsis thaliana* has led to the identification of more than 270 putative P450 genes (*http://drnelson.utmem.edu/Arablinks.html*) (Nelson, 1999). The classification of P450 genes is based on primary sequence homologies. P450 genes are identified by the abbreviation (CYP, cytochrome P) followed by a number denoting the family, a letter designating the subfamily (when two or more exist), and a nu-

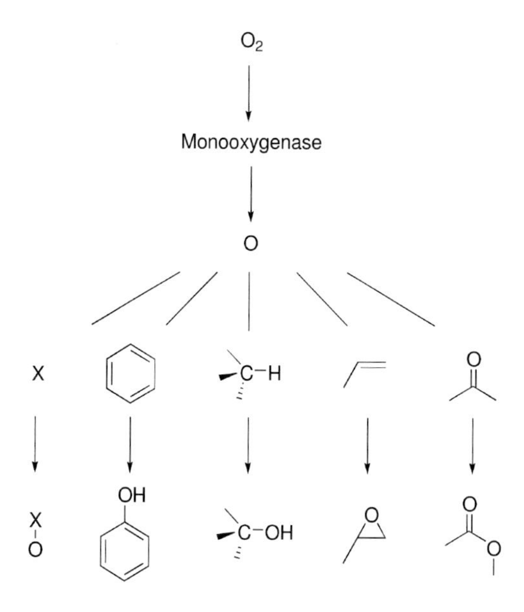

Fig. 3.7 Overview of reactions catalyzed by P450-monooxygenases.

meral representing the individual gene within the subfamily. For example, *CYP*4A1 represents the *first* gene in the P450 subfamily *A* of the P450 gene family *4*.

Depending on the mechanism for the electron transfer system (reductase system), P450 enzymes are divided into four classes (Degtyarenko, 1995; Peterson, 1998):

- Class I: These mainly occur in mitochondrial systems and most bacteria. The electron systems consist of a FAD-domain as reductase and a further iron-sulfur protein.
- Class II: These are often located in the endoplasmic reticulum and require only a single protein for the electron transfer, a FAD/FMN-reductase.
- Class III: These do not require reduction equivalents. The P450 enzymes directly convert peroxygenated substrates which have already "incorporated" the oxygen.
- Class IV: Only one enzyme is known, which receives its electrons directly from NADH.

Postulated Reaction Cycle

The postulated reaction cycle of P450 enzymes is shown in Figure 3.8 (see also Fig. 3.11). This reaction typifies how enzymatic oxygenation is carried out under physiological conditions using molecular oxygen as oxidant (Urlacher et al., 2004). Activation of oxygen takes place at the iron-protoporphyrin IX (heme), which is six-fold coordinated. It has a conserved thiolate residue as the fifth ligand and, in the inactive ferric form, a water molecule as its sixth ligand (stage 1 in Fig. 3.8). First, the substrate is bound and the water molecule is displaced (stage 2), after which the ferric

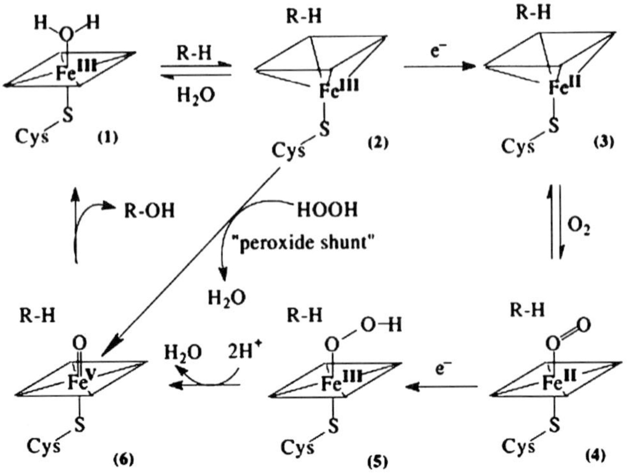

Fig. 3.8 Reaction cycle of a hydroxylation catalyzed by a P450-monooxygenase.

enzyme is reduced to a ferrous state by an one-electron transfer (stage 3). Molecular oxygen is bound, resulting in a ferrous dioxy species (stage 4). A second reduction followed by a proton transfer leads to an iron-hydroperoxo intermediate (stage 5) which, upon cleavage of the O–O bond releases water and an activated iron-oxo ferryl species (stage 6). This iron-oxo ferryl oxidizes the substrate, and the product is subsequently released. During this catalytic cycle several highly reactive oxygen species are formed, which are believed to cause the rather rapid inactivation of P450 enzymes. Those P450s which are able to perform hydroxylation using the "peroxide shunt" pathway, utilizing hydrogen peroxide as electron and oxygen donor, are very useful in biocatalysis.

Currently, eight enzyme structures of CYP have been resolved which, despite their low sequence homology, share a close structural similarity. As the activated intermediates are not covalently bound, only three structures in complex with their natural substrates (for CYP101, $P450_{cam}$, and CYP102) have been determined.

Several concepts to bypass the cofactor regeneration of NADPH have been suggested, such as electrochemical methods or the use of cobalt(III)sepulchrate, which can be reduced by cheap zinc dust (Schwaneberg et al., 2000). The most convenient idea is the shunt pathway (stages **2** $\rightarrow$ **6** in Fig. 3.8), using hydrogen peroxide instead of NADH or NADPH, which has been described for $P450_{cam}$ (Joo et al., 1999) and P450 BM-3 (Farinas et al., 2001).

Further information on P450s can be found in a number of books and reviews (Cirino and Arnold, 2002; Li et al., 2002; Urlacher and Schmid, 2002; Urlacher et al., 2004).

Synthesis of Hydroxylated Carboxylic Acids

Hydroxylated carboxylic acids have various (potential) applications as polymer building blocks or as intermediates in antibiotic synthesis. In lactonized form they can serve as perfume ingredients.

A considerable number of P450s are known to catalyze the hydroxylation of fatty acids, though in most cases only medium- to long-chain fatty acids (C_{12}–C_{18}) are accepted as substrates. In addition, hydroxylation occurs at several positions, namely terminal (ω-position) and subterminal (ω–1, ω–2-positions) to yield a product mixture (Schwaneberg and Bornscheuer, 2000). Recently, it was shown that the substrate specificity of a P450 enzyme from *Bacillus megaterium* could be efficiently altered by means of rational protein design, and especially by directed evolution (see also Chapter 2, Section 2.11.2). The P450 produced by the strain *Bacillus megaterium* (P450-BM3) is able to catalyze the hydroxylation of fatty acids with the highest turnover numbers yet reported for P450 monooxygenases (in the range of >1000 Eq min^{-1}). The encoding gene was cloned and functionally expressed in *E. coli* (Narhi and Fulco, 1986, 1987; Ruettinger, 1989). P450 BM-3 (CYP102) is especially suitable for biocatalysis, as it is a water-soluble natural fusion protein containing the P450 and reductase part on one polypeptide chain. In addition, its crystal structure has been determined. The replacement of Arg_{47} with Glu resulted in an ability of this P450 BM-3 mutant to hydroxylate *N*-alkyltrimethylammonium compounds, which was explained by an inversion of the substrate binding conditions (Oliver et al., 1997). P450 BM-3, heterologously expressed in *E. coli*, has been used *in vivo* to produce mixtures of chiral 12-, 13- and 14-hydroxypentadecanoic acid on a preparative scale at high optical purity (Schneider et al., 1998). Furthermore, P450 BM-3 and its mutant Phe87Ala can be expressed in gram scale and efficiently purified in a single step for further enzyme-based biotransformation reactions on a preparative scale (Schwaneberg et al., 1999b).

The key to the successful engineering of P450 BM-3 was the development of an elegant chromophoric assay (Fig. 3.9), which allows determination of the P450 BM-3 wild-type and mutant activity without background reaction (Schwaneberg et al., 1999a). As shown in Figure 3.9, hydroxylation of the terminal position of the substrate, yields an unstable hemiacetal which dissociates spontaneously into the ω-oxo-

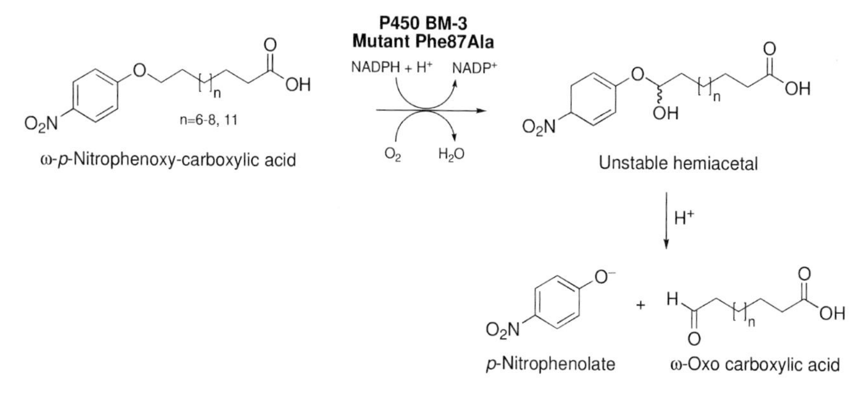

Fig. 3.9 Principle of the colorimetric pNCA assay allowing the determination of the fatty acid hydroxylating activity of P450 BM-3 from *Bacillus megaterium* mutant F87A (Schwaneberg et al., 1999a).

carboxylic acid and the chromophore *p*-nitrophenolate. The latter can be easily quantified at 410 nm using a spectrophotometer. This pNCA assay created the basis for the directed evolution of P450-BM3 as it allows screening of variants in a high-throughput format (Schwaneberg et al., 2001).

The development of P450 BM-3 mutants with improved hydroxylation activity began with the identification of eight mutation sites from the X-ray crystallographic structure of P450 BM-3, which were then subjected to random mutagenesis. After screening with the pNCA assay, the best variants were combined, yielding a biocatalyst with five mutations, which efficiently hydroxylated C_8-pNCA while maintaining its activity for C_{10} and C_{12} fatty acids (Li et al., 2001). Arnold and coworkers discovered that P450 BM-3 also hydroxylates short-chain alkanes such as octane, yielding a mixture of 4-octanol, 3-octanol, 2-octanol, 4-octanone, and 3-octanone. More recently, the hydroxylation of propane to yield 2-propanol was reported (Glieder et al., 2002). Although the reaction rates and yields are not yet satisfying, this can be regarded as a major breakthrough, as upon further improvement commodity chemicals might become available by biotransformation routes. Directed evolution using an assay similar to the pNCA-test described above enabled identification of mutants with five-fold increased specific activity over wild-type P450 BM-3 (Farinas et al., 2001). Later, other variants were created by directed evolution, which also exhibited enantioselectivity resulting in the formation of (*S*)-2-octanol (40 % ee) or (*R*)-2-hexanol (40–55 % ee) (Peters et al., 2003).

A triple mutant of P450 BM-3 (Phe87Val, Leu188Gln, Ala74Gly) was found to efficiently hydroxylate indole to indigo and indirubin (Li et al., 2000). In addition, this variant also hydroxylates alkanes, cycloalkanes, arenes and heteroarenes (Appel et al., 2001) for which the wild-type shows no or only very little activity.

Directed evolution was also applied to alter the substrate specificity of $P450_{cam}$ from *Pseudomonas putida* to convert naphthalene more efficiently into hydroxylation products. Mutants were identified by coexpressing them with horseradish peroxidase (HRP), which converts the products of the P450 reaction into fluorescent compounds amenable to digital imaging screening (Joo et al., 1999) (Fig. 3.10).

The examples mentioned above are summarized in Figure 3.11. Further details and examples can be found in recent reviews (Burton, 2003; Cirino and Arnold, 2002; Li et al., 2002; Urlacher and Schmid, 2002; Wong, 2002).

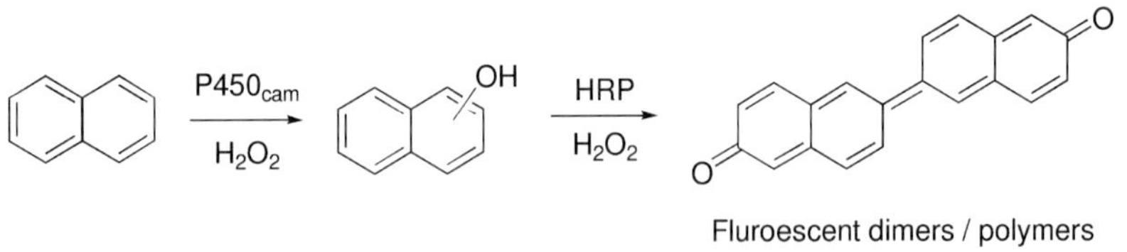

Fig. 3.10 Principle of the assay used to identify $P450_{cam}$ variants hydroxylating naphthalene. The resulting naphthols are oxidatively coupled with horseradish peroxidase (HRP) to fluorescent polymers. All reactions occur intracellularly in the recombinant *E. coli* cell.

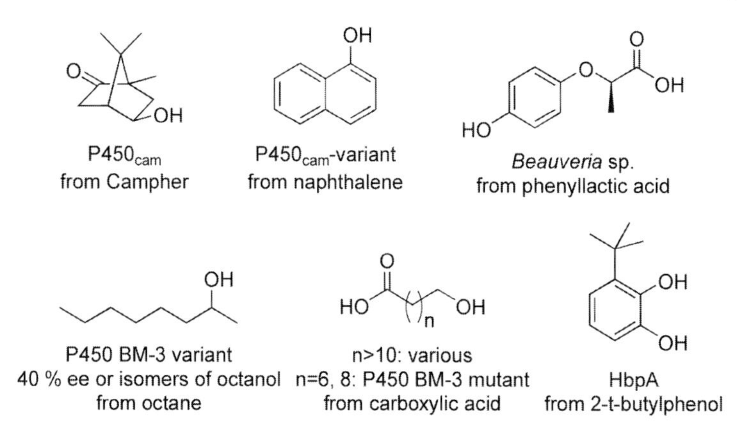

Fig. 3.11 Selected examples of products obtained from monooxygenase-catalyzed hydroxylations. Note that most examples represent the use of engineered P450s.

Non-heme Monooxygenases

Two non-heme monooxygenases have recently been engineered by directed evolution. The substrate specificity and activity of a 2-hydroxybiphenyl-3-monooxygenase (HbpA) from *Pseudomonas azelaica* HBP1 in the regioselective *o*-hydroxylation of 2-substituted phenols to yield catechols was altered (Meyer et al., 2002, 2003). One mutant (expressed in *E. coli*) hydroxylated 2-*t*-butylphenol to the catechol. The whole-cell process was thoroughly optimized and allowed a productivity of 64 mg L^{-1} h^{-1}, which is equal to 1.5 g per day.

Baeyer–Villiger Monooxygenases (EC 1.14.13.16, 1.14.13.22)

Baeyer–Villiger monooxygenases (BVMO) catalyze the biocatalytic counterpart of the chemical Baeyer–Villiger oxidation (Walsh and Chen, 1988), which was first described by Adolf Baeyer and Victor Villiger more than a century ago (Baeyer and Villiger, 1899). The chemical oxidation usually uses peracids. The mechanism is generally accepted to proceed by a two-step process, which was initially proposed by Criegee (Criegee, 1948). However, the chemical Baeyer–Villiger oxidation is not stereoselective unless organometallic reagents are employed. Even then, the optical purity of the products is unsatisfactory in most cases (Bolm, 1997; Bolm et al., 1994; Frisone et al., 1993; Strukul, 1998).

The first examples of enzymatic Baeyer–Villiger oxidation (BVO) date back to 1948 for a fermentation of cholestanone with *Proactinomyces erythropolis* (Turfitt, 1948). A few years later, a double BVO was suggested for the side-chain degradation of progesterone (Fig. 3.12) (Fried et al., 1953).

During the early 1980s, a range of publications about biocatalysis with newly discovered BVMOs appeared. One enzyme originates from *Pseudomonas putida* (Grogan et al., 1992, 1993; Jones et al., 1993; Taylor and Trudgill, 1986), the other is produced by the S2-organism *Acinetobacter calcoaceticus* (Gagnon et al., 1994). For both enzymes, it could be demonstrated, that they accept a relatively broad range of

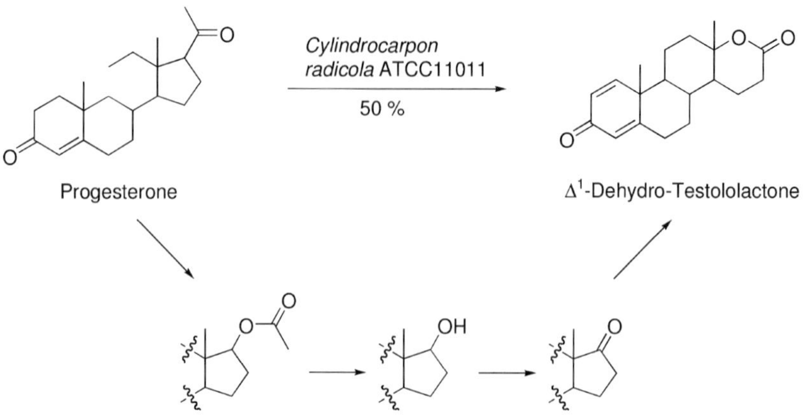

Cylindrocarpon radicola ATCC11011
50 %
Progesterone
Δ1-Dehydro-Testololactone
OH

Fig. 3.12 Early example of an enzyme-catalyzed Baeyer–Villiger oxidation (Fried et al., 1953).

Top
Side
Front
M
R
OH
Front
Top
Side
HO
1.54Å

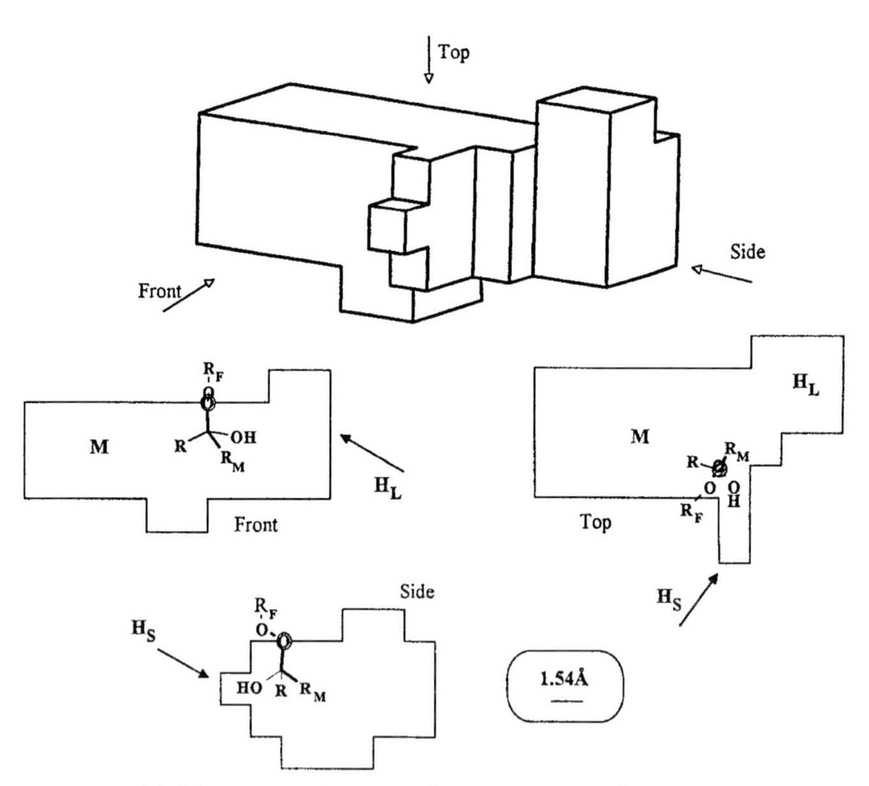

Fig. 3.13 Model of the active-site of a Baeyer–Villiger monooxygenase from *Acinetobacter* sp.

mono- and bicyclic ketones. In the case of racemic substrates, the oxidation often proceeded with good to high enantioselectivity. During the 1990s, the BVMO from *Acinetobacter* sp. was cloned and functionally expressed in yeast (Kayser et al., 1999; Stewart et al., 1996) and later also in *E. coli* (Mihovilovic et al., 2001). Moreover, it could be shown, that a whole-cell biotransformation is possible, although side-reactions – such as a reduction of the ketone substrate to an alcohol by an endogeneous alcohol dehydrogenase present in the whole-cell system – can occur. In addition, other BVMO-producing strains were identified and the application of these enzymes in biocatalysis was described (see also below).

Until recently, no three-dimensional structure of a BVMO has been reported. In its absence a simplified cartoon-like model based on the substrate specificity of the enzyme from *Acinetobacter calcoaceticus* has been developed (Ottolina et al., 1996), which somehow rationalized experimental observations. The application of this model (Fig. 3.13) to predict substrate specificity and especially to design mutants for rational protein design is very difficult.

However, the newly resolved 3D-structure of a BVMO (Malito et al., 2004) will be very helpful to verify this model.

Mechanism

During the enzymatic Baeyer–Villiger oxidation, one oxygen atom is introduced into a ketone precursor molecule, yielding either the corresponding lactone or an ester (depending on whether a cyclic or a non-cyclic ketone is used). The second oxygen atom is converted into water.

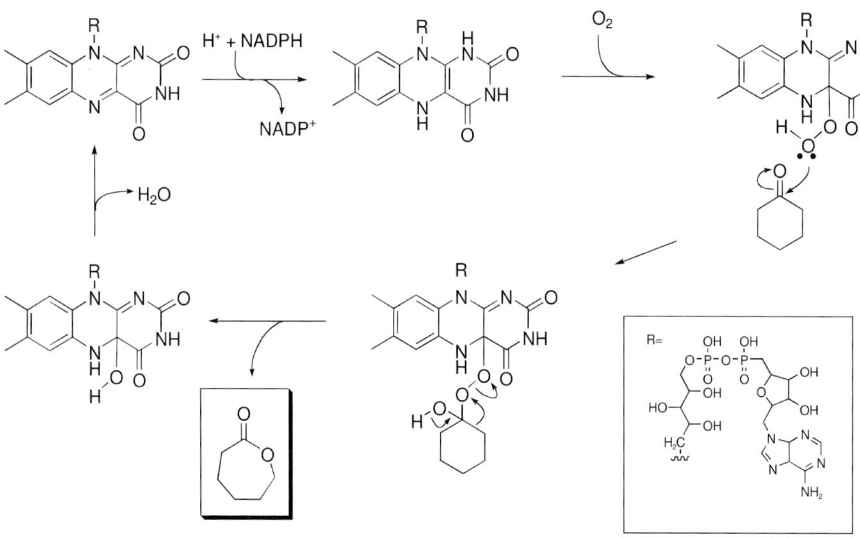

Fig. 3.14 Proposed mechanism of an enzyme-catalyzed Baeyer–Villiger oxidation (Donoghue et al., 1976).

For their catalytic action, BVMOs require a cofactor (NADH or NADPH) and they are usually flavin-dependent (FAD or FMN). Figure 3.14 shows the mechanism for the oxidation of cyclohexanone to ε-caprolactone by BVMO from *Acinetobacter calcoaceticus* NCIMB 9871 (Donoghue et al., 1976).

By far the best-studied BVMO is the enzyme produced by *Acinetobacter calcoaceticus* NCIMB 9871, which is also available as recombinant enzyme, as mentioned above. Some examples of the application of this biocatalyst to the conversion of mono- and bicyclic ketones are shown in Figure 3.15; more details can be found in a recent excellent review (Mihovilovic et al., 2002). A quite similar substrate spectrum (Fig. 3.15) was reported for BVMO from campher-induced *Pseudomonas putida*, which produces three different BMVOs; two of these almost identical enzymes require NADH (each type is generated by induction with either (–) or (+)-camphor), while the third BVMO utilizes NADPH.

Acetophenone-converting BVMOs were also described originating from, for example, *Arthrobacter* (Cripps, 1975), *Nocardia* (Cripps et al., 1978), and *Pseudomonas* species (Kamerbeek et al., 2001; Tanner and Hopper, 2000). These enzymes accept a broad range of ring-substituted acetophenones, which were converted into the corresponding achiral acetate esters.

The conversion of aliphatic straight-chain ketones was described for monooxygenases from *Mycobacterium* sp. (Hartmans and deBont, 1986), *Pseudomonas* (Forney

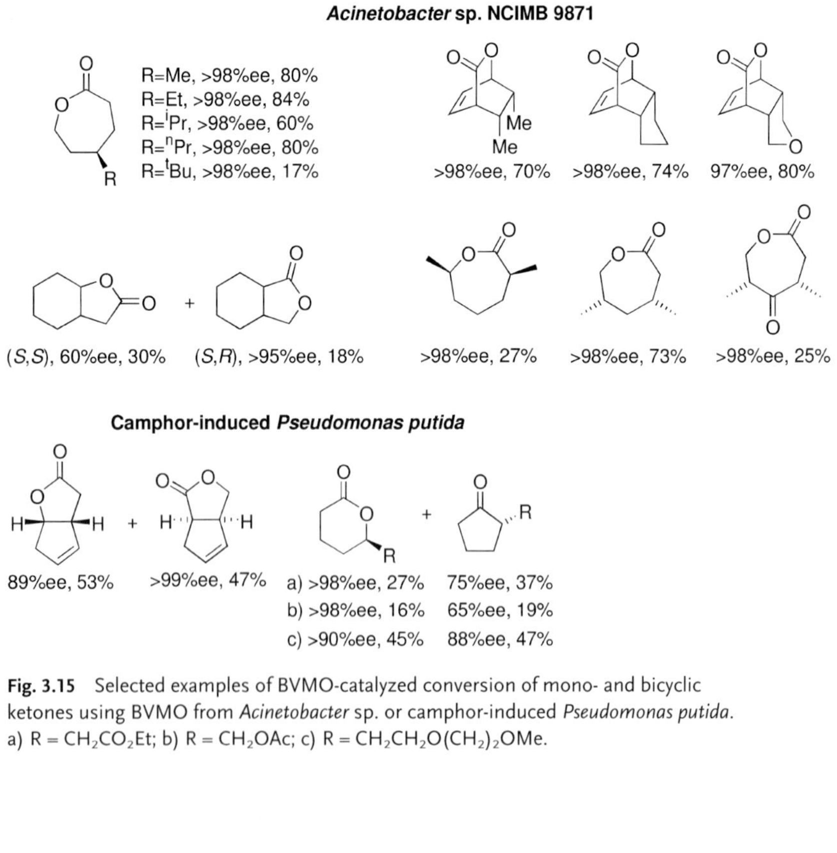

Fig. 3.15 Selected examples of BVMO-catalyzed conversion of mono- and bicyclic ketones using BVMO from *Acinetobacter* sp. or camphor-induced *Pseudomonas putida*. a) R = CH_2CO_2Et; b) R = CH_2OAc; c) R = $CH_2CH_2O(CH_2)_2OMe$.

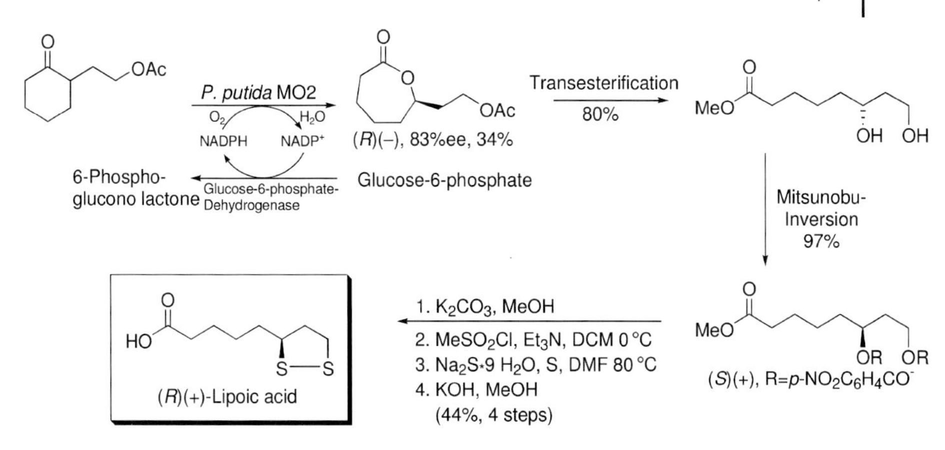

Fig. 3.16 Chemoenzymatic synthesis of (*R*)-(+)-lipoic acid using a BVMO from *Pseudomonas putida* (Adger et al., 1995).

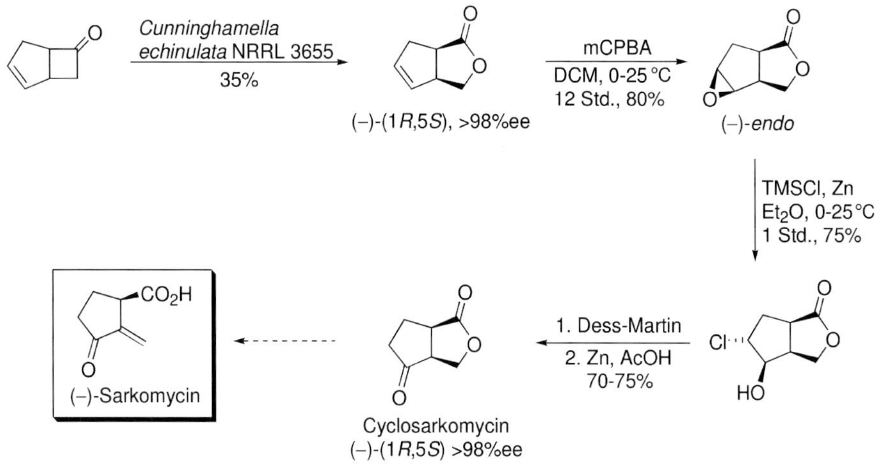

Fig. 3.17 Chemoenzymatic synthesis of (–)-Sarkomycin using a BVMO from *Cunninghamella echinulata* (Königsberger and Griengl, 1994).

et al., 1967), and *Nocardia* sp. (Britton et al., 1974). However, simple ketones were studied and no conversion of racemic starting materials to produce optically active compounds was investigated. It is believed, that these enzymes play a role in the subterminal oxidation of long-chain alkanes to carboxylic acids.

It was also demonstrated, that BMVOs can be used in the chemoenzymatic synthesis of natural compounds, such as lipoic acid (Fig. 3.16; Adger et al., 1995) and Sarkomycin (Fig. 3.17; Königsberger and Griengl, 1994).

In addition, it was shown that BMVOs also catalyze the synthesis of chiral sulfoxides (Fig. 3.18; Secundo et al., 1993).

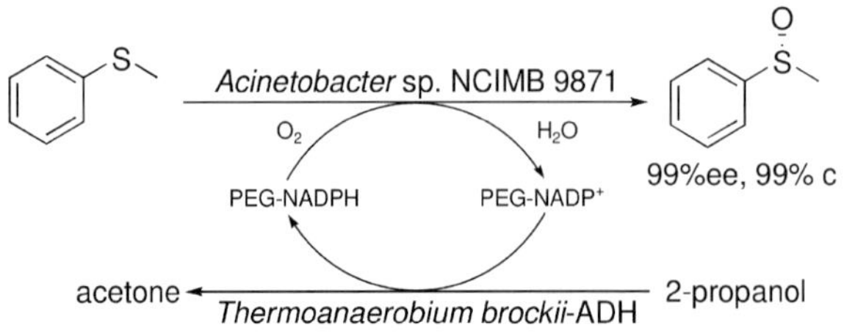

Fig. 3.18 Synthesis of optically active sulfoxides using a Baeyer–Villiger monooxygenase (BVMO) with integrated cofactor recycling (Secundo et al., 1993).

3.2.1.3 Peroxidases (e.g., EC 1.11.1.10)

Heme iron-containing peroxidases are capable of oxidizing organic substrates using a peroxide, usually hydrogen peroxide, as oxidant. The only important enzyme for organic synthesis is a chloroperoxidase from *Caldariomyces fumago*, which has been shown to catalyze stereoselective oxidation of aromatic compounds, yielding the corresponding alcohols, and the epoxidized alkenes (Fig. 3.19). Also, (*Z*)-3-heptene was converted into its isomeric alcohols (Zaks and Dodds, 1995).

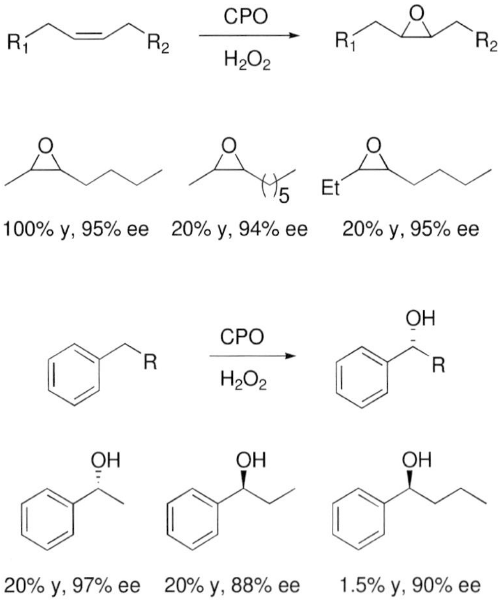

Fig. 3.19 Examples of epoxidations and benzylic hydroxylations catalyzed by chloroperoxidase (CPO) from *Caldariomyces fumago* (Zaks and Dodds, 1995).

3.2.2 Hydrolases (EC 3.1)

3.2.2.1 Lipases (EC 3.1.1.3)

Lipases (triacylglycerol hydrolases) are probably the most frequently used hydrolases in organic synthesis (Kazlauskas and Bornscheuer, 1998; Schmid and Verger, 1998). They are widely found in nature (animals, man, bacteria, yeast, fungi, plants), and a considerable number of enzymes are commercially available. Their natural function is the hydrolysis and re-esterification of triglycerides – that is, natural fats and oils. The reaction is catalyzed by a catalytic triad composed of Ser, His, and Asp (sometimes Glu) similar to serine peptidases (see Figure 2.5a, p. 37) and carboxyl esterases (EC 3.1.1.1). The mechanism for ester hydrolysis or formation is essentially the same for lipases and esterases, and is composed of four steps:

1. The substrate reacts with the active-site serine, yielding a tetrahedral intermediate stabilized by the catalytic His- and Asp-residues.
2. The alcohol is released and a covalent acyl-enzyme complex is formed.
3. Attack of a nucleophile (water in hydrolysis, alcohol in (trans-)esterification) forms again a tetrahedral intermediate.
4. The intermediate collapses to yield the product (an acid or an ester) and free enzyme (Fig. 3.20).

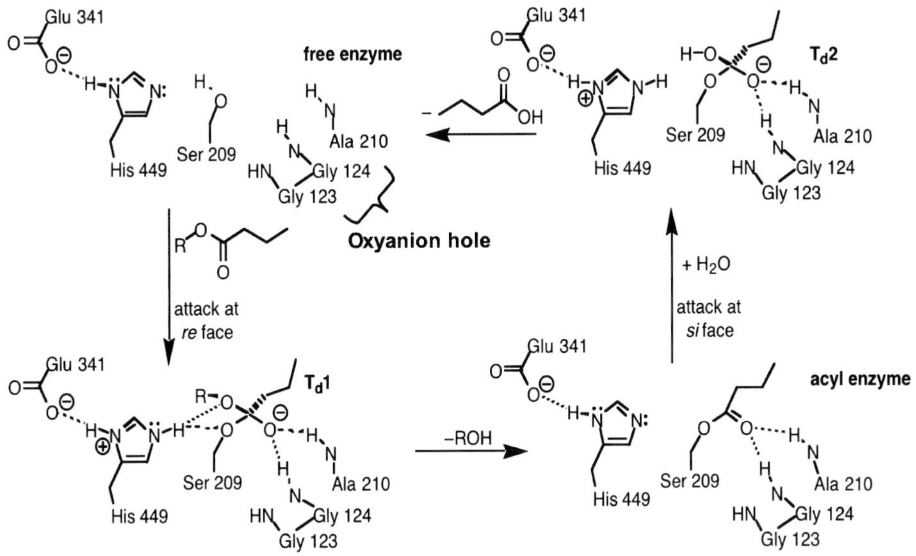

Fig. 3.20 Mechanism of lipase-catalyzed ester hydrolysis of a butyrate ester. Numbering of amino acid residues is for lipase from *Candida rugosa* (CRL).

Beside their use in organic synthesis for the production of optically active compounds (which are covered below in detail), lipases are used in a variety of other areas. Applications include laundry detergents, cheese-making, modification of natural fats and oils (e.g., the synthesis of cocoa-butter equivalents, monoglycerides, incorporation of polyunsaturated fatty acids for human nutrition), synthesis of sugar esters and simple ester used in personal care (e.g., myristyl myristate, decyl cocoate).

Lipases are distinguished from esterases (see below) by their substrate specificity: lipases accept long-chain fatty acids (in triglycerides) as substrates, whereas esterases prefer short-chain fatty acids. More generally, it can be stated that lipases readily accept water-insoluble substrates, while esterases prefer water-soluble compounds. A further difference was found in the 3-D structures of these enzymes: lipases contain a hydrophobic oligopeptide (often called a *lid* or *flap*) – which is not present in esterases – covering the entrance to the active site. Lipases preferentially act at a water–organic solvent (or oil) interface, which presumably accounts for a movement of the *lid* making the active site accessible for the substrate. This phenomenon is referred to as "interfacial activation". Further characteristic structural features of lipases are the α,β-hydrolase fold (Ollis et al., 1992) and a consensus sequence around the active-site serine (Gly-X-Ser-X-Gly, where X denotes any amino acid). It could be shown that after removal of the lid by genetic engineering, the activity of a lipase was improved in solution, mainly for applications in the laundry/detergent area.

The further classification of lipases can be based on substrate specificity towards triglycerides: non-specific lipases hydrolyze all three fatty acids of a triglyceride, while *sn*1,3-regiospecific enzymes hydrolyze only the primary ester bonds of triglycerides, yielding *sn*2-monoglycerides. Some lipases also show distinct fatty acid chain length and saturation selectivity. For instance, lipase B from *Geotrichum candidum* shows high selectivity towards *cis*Δ9 fatty acids, such as oleic acid (Baillargeon and McCarthy, 1991). Other classifications are based on their molecular weight, amino acid sequences, or sequence motifs (Pleiss et al., 2000).

The main interest in the application of lipases in organic chemistry is due to the following reasons:

- Lipases are highly active in a broad range of non-aqueous solvents.
- They often exhibit excellent stereoselectivity.
- They accept a broad range of esters other than triglycerides.
- They accept nucleophiles other than water (i.e., alcohols, amines).

For the synthesis of optically pure compounds, a kinetic resolution is usually the method of choice yielding (under optimum conditions) 50 % of either enantiomer in optically pure form. A kinetic resolution can be performed by hydrolysis of a racemic substrate (i.e., an acetate; Fig. 3.21) or by acylation in an organic solvent (Fig. 3.22). Note that if the enzyme is (*S*)-selective, the (*S*)-alcohol is formed in the hydrolysis reaction, but the (*R*)-alcohol is left unreacted in the acylation reaction.

As lipases are very stable in a range of organic solvents, acylation is often the preferred reaction. In order to drive the kinetic resolution to completion, either one substrate is used in excess or activated acyl donors are used. Using enol esters such as vinyl acetate (Fig. 3.22) or isopropenyl acetate, the reaction is practically irreversible

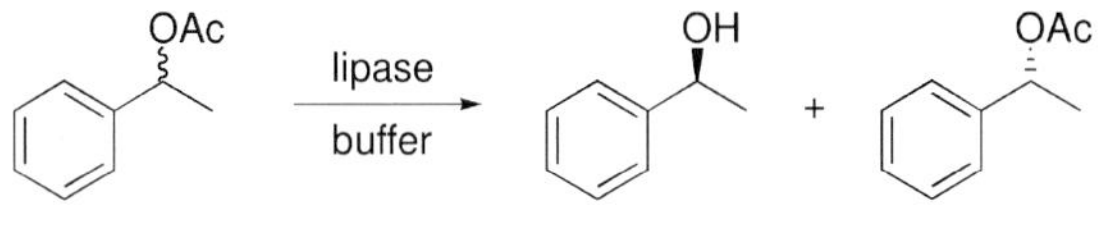

Fig. 3.21 Example of a lipase-catalyzed kinetic resolution by hydrolysis. A preparative separation of an ester/alcohol mixture is straight forward by distillation or chromatography.

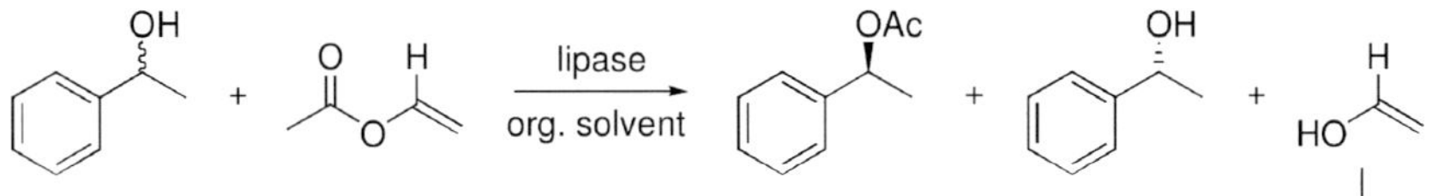

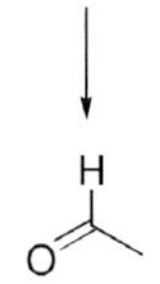

Fig. 3.22 Example of a lipase-catalyzed kinetic resolution by acylation. The use of the enol ester vinyl acetate ensures an irreversible reaction as the by-product acetaldehyde is generated via a keto-enol tautomerization from the vinyl alcohol.

as the vinyl alcohol generated undergoes keto–enol tautomerization to carbonyl compounds (i.e., acetaldehyde from vinyl esters, acetone from isopropenyl esters) (Fig. 3.23). However, acetaldehyde might inactivate the enzyme via formation of a Schiff base with lysine residues. Trifluoroethyl esters are more expensive, and the thiols generated from *S*-ethyl thioesters have an undesirable flavor. Anhydrides such as acetic acid anhydride are cheap, but the free acid generated causes a drop in pH and might also participate in a slower background acylation.

Currently, more than 1000 examples (!) for the resolution with lipases can be found in the literature (Bornscheuer and Kazlauskas, 1999). In general, reactions proceed with good to excellent enantioselectivity for secondary alcohols, but only moderate selectivity is found for primary alcohols. Researchers have tried to predict the outcome of a lipase-catalyzed resolution based on the substrate structure and the

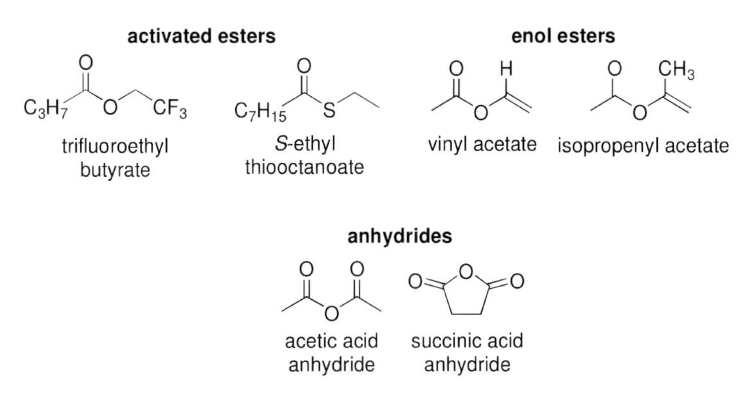

Fig. 3.23 Some acyl donors used in lipase-catalyzed kinetic resolutions.

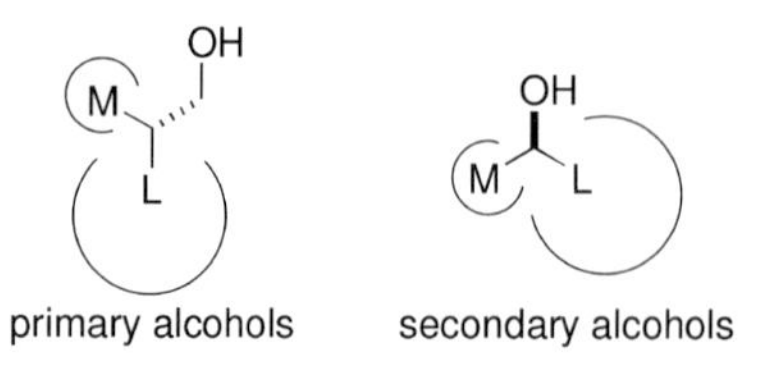

Fig. 3.24 An empirical rule ('Kazlauskas rule') developed for lipase from *Burkholderia cepacia* (BCL) summarizing the enantiopreference for primary and secondary alcohols. The scheme shows the favored enantiomer. For primary alcohols, this rule is only reliable if no oxygen is attached to the stereocenter. Note that BCL shows an opposite enantiopreference for primary alcohols.

type of lipase. Kazlauskas and coworkers developed empirical rules which allow the prediction which enantiomer of a primary or a secondary alcohol will react faster than the other (Kazlauskas et al., 1991; Weissfloch and Kazlauskas, 1995). According to this rule, high enantioselectivities can be achieved for substrates bearing a medium size (e.g., methyl-) and large size (e.g., phenyl-) substituent (Fig. 3.24).

From the considerable number of lipases available, only a few have been shown to be broadly applicable, to exhibit high enantioselectivity, and to have sufficient stability. These are porcine pancreatic lipase (PPL), lipase B from *Candida antarctica* (CAL-B, tradenames: Novozyme®, Chirazyme® L2), lipases from *Burkholderia cepacia* (BCL, former names: *Pseudomonas cepacia, Ps. fluorescens*), *Candida rugosa* (CRL, former name: *Candida cylindracea*) and *Rhizomucor miehei* (RML, tradename Lipozyme®).

Figures 3.25 and 3.26 show a few selected examples of substrates resolved using these lipases.

In contrast to secondary and primary alcohols, fewer examples for the application of lipases can be found for the kinetic resolution of carboxylic acids (Fig. 3.27). Interestingly, lipases which show high selectivity for alcohols (e.g., PCL, CAL-B), are much less selective in the resolution of carboxylic acids, and *vice versa* (e.g., CRL).

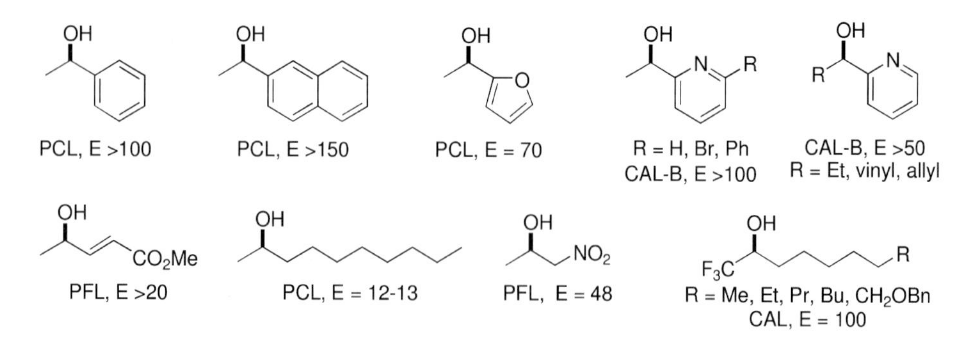

Fig. 3.25 Some examples of secondary alcohols resolved using lipases. In all cases vinyl acetate served as acyl donor. Note the changes in enantioselectivity with varying substrate structure.

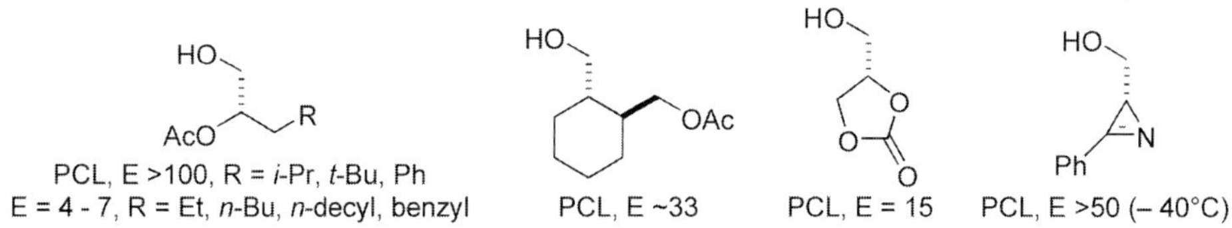

HO AcO R
PCL, E >100, R = *i*-Pr, *t*-Bu, Ph
E = 4 - 7, R = Et, *n*-Bu, *n*-decyl, benzyl

HO OAc
PCL, E ~33

HO O O O
PCL, E = 15

HO Ph N
PCL, E >50 (– 40°C)

Fig. 3.26 Some examples of primary alcohols resolved using lipase from *Pseudomonas cepacia* (PCL). In all cases vinyl acetate served as acyl donor. Note the changes in enantioselectivity with varying substrate structure and the lower E-values compared to secondary alcohols.

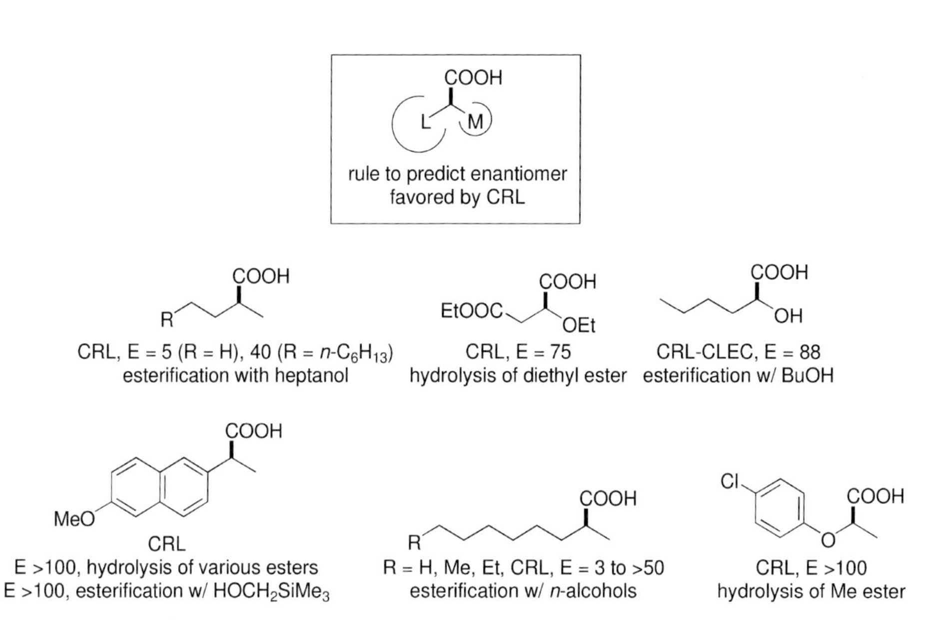

Fig. 3.27 Some examples of carboxylic acid resolved using lipase from *Candida rugosa* (CRL). CRL-CLEC is a crosslinked enzyme crystal preparation of CRL. Higher E-values might be due to the removal of interfering isoenzymes during crystallization.

Conversion and Resolution of Tertiary Alcohols

Until recently, only a few lipases (e.g., CRL and lipase A from *Candida antarctica*) were known to accept tertiary alcohols. Due to their bulky structure, it was believed that these compounds do not fit into the active site of the enzyme. It was later discovered, that a certain amino acid motif (GlyGlyGlyX-motif, where X denotes any amino acid) located in the oxyanion binding pocket of lipases and esterases determines activity towards tertiary alcohols (Henke et al., 2002). All enzymes bearing this motif (e.g., pig liver esterase, several acetyl choline esterases, an esterase from *Bacillus subtilis*) were active towards several acetates of tertiary alcohols, while enzymes bearing the more common GlyX-motif did not hydrolyze the model com-

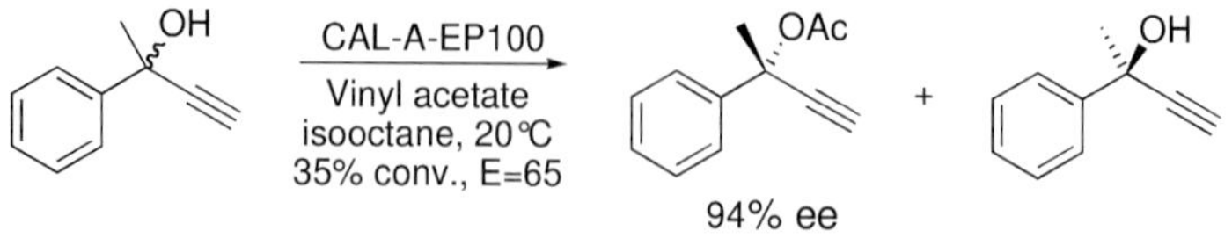

Fig. 3.28 Kinetic resolution of a tertiary alcohol using lipase A from *Candida antarctica* immobilized on a polypropylene carrier (CAL-A-EP100).

pounds. However, the enantioselectivity of these enzymes is usually rather low. This could be overcome by rational protein design and site-directed mutagenesis (Henke et al., 2003) yielding a variant of an esterase from *Bacillus subtilis* (Gly105Ala) with an E-value of 19 (wild-type: E = 3). Alternatively, careful optimization of the reaction conditions (solvent, carrier for immobilization, temperature) allowed a good kinetic resolution by transesterification (Krishna et al., 2002) (Fig. 3.28).

Industrial Applications

Optically pure amines are versatile synthons for a wide variety of products. Interestingly, lipases were found most efficient for the synthesis of these compounds by kinetic resolution. An example is the highly enantioselective acylation of (*R*,*S*)-1-phenethyl amine in a process established at BASF (Fig. 3.29). The (*R*)-amide is separated from the (*S*)-amine by distillation or extraction, and the free (*R*)-amine is released through basic hydrolysis. As the lipase from *Burkholderia* sp. shows broad substrate specificity, a wide variety of aryl alkyl amines, alkyl amines and amino alcohols could be resolved, in some cases on the multi-tonne scale. Undesired enantiomers can be racemized and the acylating agent (such as methoxyacetic acid ethylester) can be recovered. BASF produces more than 2500 tonnes annually with this process (Balkenhohl et al., 1997).

Researchers at DSM developed a lipase-catalyzed process for the production of a Captopril intermediate, (*R*)-3-chloro-2-methyl propionate. All lipases preferentially hydrolyzed the (*S*)-enantiomer, and up to 98 % ee at 64 % conversion was observed with a lipase from *Candida cylindracea* (Fig. 3.30).

In addition, further industrial processes based on lipase catalysis have been established in the past few years (Breuer et al., 2004; Liese et al., 2000).

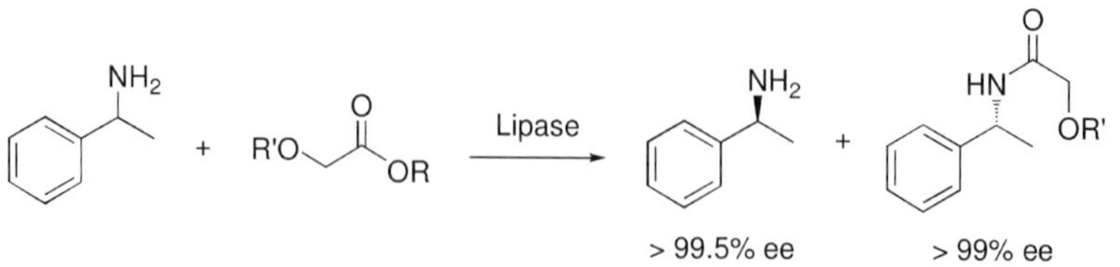

Fig. 3.29 Lipase-catalyzed kinetic resolution of amines in the BASF-process.

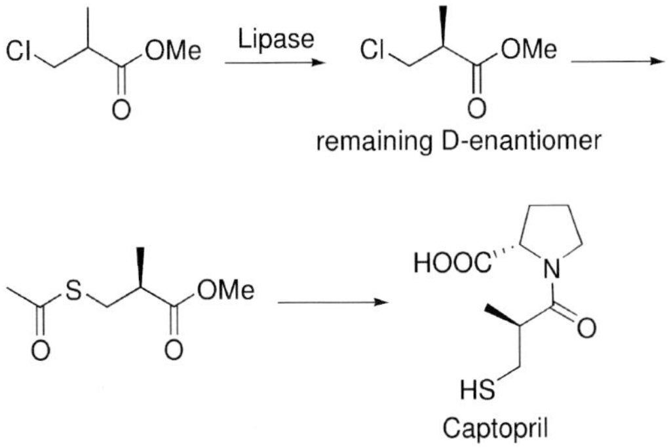

Fig. 3.30 Lipase-catalyzed kinetic resolution of a building block for the synthesis of Captopril.

Dynamic Kinetic Resolutions (DKR)

As already outlined in Section 3.1.1, a kinetic resolution of a racemate can only yield at maximum 50 % product. In order to achieve a complete conversion of both enantiomers, a dynamic kinetic resolution can be used. Such a strategy can also make the synthesis of an optically pure compounds more competitive to an asymmetric synthesis using, for example, alcohol dehydrogenases and a prochiral substrate.

The requirements for a DKR are: (1) the substrate must racemize faster than the subsequent enzymatic reaction proceeds; (2) the product must not racemize; and (3) as in any asymmetric synthesis, the enzymatic reaction must be highly stereoselective. The principle for this is exemplified in Figure 3.31, and many examples have been described in recent reviews (El-Gihani and Williams, 1999; Kim et al., 2002; Pàmies and Bäckvall, 2003; Pàmies and Bäckvall, 2004).

The earliest example of DKR was the synthesis of optically pure α-amino acids from hydantoins. Racemization of the hydantoin occurs at alkaline pH or with the aid of a racemase (see Section 3.2.2.6). Later, dynamic kinetic resolutions have been described for desymmetrizations of chemically labile secondary alcohols, thiols and amines (i.e., cyanohydrins, hemiacetals, hemithioacetals). More recently, *in-situ* deracemization via nucleophilic displacement has been demonstrated for 2-chloropropionate (92 % yield, 86 % ee) using lipase from *Candida cylindracea* in an aminolysis supported by triphenylphosphonium chloride (Bdjìc et al., 2001).

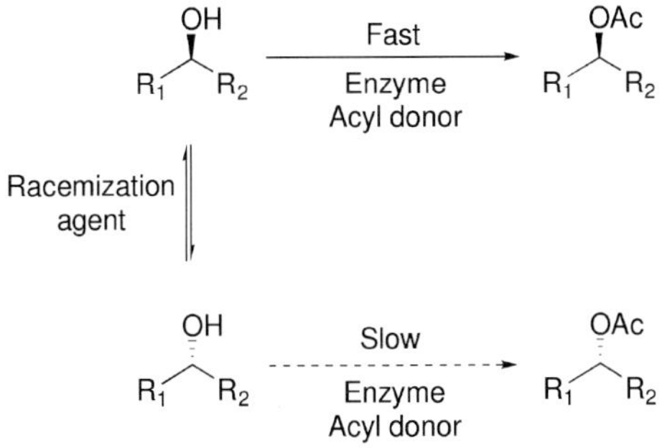

Fig. 3.31 Principle of dynamic kinetic resolutions.

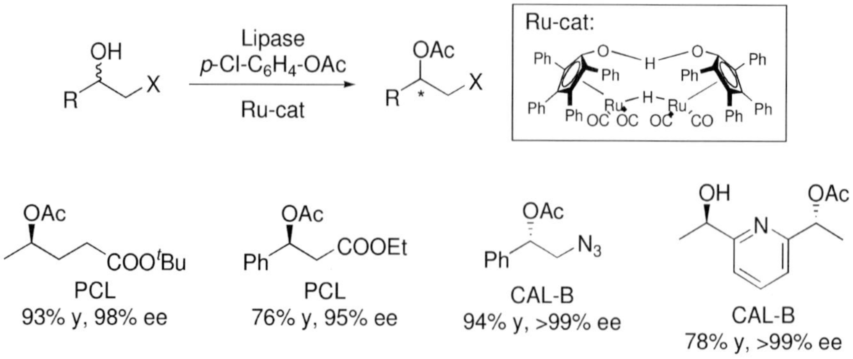

Fig. 3.32 Examples of the dynamic kinetic resolution of secondary alcohols using a ruthenium catalyst.

Other approaches are combinations of enzymatic resolution with metal-catalyzed racemization. These reactions usually proceed either via hydrogen transfer or via π-allyl-complex formation. Bäckvall and coworkers developed a hydrogen transfer system based on a ruthenium catalyst with *p*-chloroethyl acetate as acyl donor. (It should be noted that the use of metals, and especially ruthenium, can lead to problems in recycling/regeneration, and there may be environmental problems associated with the overall process.) Enolesters – with the exception of isopropenyl acetate – cannot be used due to side reactions. On the other hand, no additions of ketones or external bases are required, which often affect the reaction performance. Selected examples are shown in Figure 3.32.

Kim and coworkers improved the DKR of allylic acetates using Pd(0) catalysts in tetrahydrofuran. 2-Propanol serves as acyl acceptor, and the unreactive enantiomer is racemized by $Pd(PPh)_3$ with added diphosphine at room temperature (Fig. 3.33). A series of linear allylic acetates were deracemized in high enantiomeric excess (97–99 % ee) and with moderate to good yields (61–78 %).

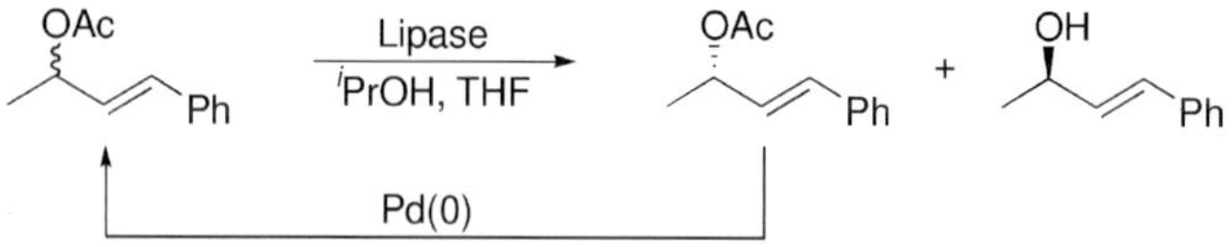

Fig. 3.33 Examples of the dynamic kinetic resolution of an allylic alcohol using Pd(0).

3.2.2.2 Esterases (EC 3.1.1.1)

Although a considerable number of carboxyl esterases are known and have been overexpressed in suitable hosts (Table 3.4), only a few of them have been used for the synthesis of optically pure compounds. The main reasons for this are the limited commercial availability and the often-observed moderate enantioselectivity. For a number

Table 3.4 Comparison of (recombinant) microbial esterases.

Origin[a]	*Biochemical properties*	*Specific substrates*	*Remarks*
Burkholderia gladioli ATCC10248 (EstB)	392 aa, 42 kDa	Triglycerides ($\leq C_6$), deacylates cephalosporins	S-x-x-K-motif, β-lactamase-like
Burkholderia gladioli ATCC10248 (*EstC*)	298 aa, 32 kDa	–	G-x-S-x-G-motif, homology to plant hydroxynitrile lyases
Pseudomonas fluorescens DSM50106	36 kDa, $T_{opt.}$ 43 °C	Lactones, ethyl caprylate, moderate enantioselectivity	G-x-S-x-G-motif, homology to a haloperoxidase
Pseudomonas fluorescens SIKW1	27 kDa, homodimer	High E-value for α-phenyl ethanol	Altered substrate specificity and improved enantioselectivity by directed evolution
Pseudomonas putida MR2068	29 kDa, homodimer, $T_{opt.}$ 70 °C	Alkyl-dicarboxylic acid methyl esters, high selectivity ($E > 100$)	–
Bacillus acidocaldarius	34 kDa, $T_{opt.}$ 70 °C	Moderate selectivity ($E \sim 18$)	Homology to hormone-sensitive lipase
Bacillus subtilis NRRL B8079	489 aa, 54 kDa, $T_{opt.}$ 52 °C (66.5 °C for best mutant)	*p*-Nitrobenzyl ester of Loracarbef	Evolved by directed evolution for increased stability in DMF and thermostability
Bacillus stearothermophilus	–	Moderate enantioselectivity	Thermostable mutants
Bacillus subtilis (Thai18)[e]	32 kDa, $T_{opt.}$ 35–55 °C	High E for 2-arylpropionic acids	Structure known, stable mutants by SDM[b]
Pyrococcus furiosus DSM3638	$T_{opt.}$ 100 °C, $t_{1/2}$ 50 min at 126 °C	MU-Ace[b]	–
Lactococcus lactis[d]	258 aa, 30 kDa	Tributyrin, C_6-phospholipids	G-x-S-x-G-motif, function unclear
Rhodococcus ruber DSM 43338[c]	–	Linalool-acetate ($E > 100$)	Two esterases with opposite enantiopreference
Rhodococcus sp. H1	34 kDa, tetramer	Heroin	G-x-S-x-G-motif, conserved His_{86}
Rhodococcus sp. MB1	574 aa, 65 kDa, monomer	Cocaine	Homology aminopeptidases
Streptomyces diastatochromogenes	326 aa, 31 kDa	Moderate enantioselectivity	–
Saccharomyces cerevisiae IFO2347	28 kDa, homodimer, $T_{opt.}$ 25 °C	Isoamyl acetate, isobutyl acetate	–

[a] Overexpressed in *E. coli*, if not stated otherwise. [b] MU-Ace, 4-methylumbelliferyl-acetate; SDM, site-directed mutagenesis. [c] Non-recombinant purified enzymes. [d] Overexpressed in *L. lactis*. [e] Overexpressed in *B. subtilis*.

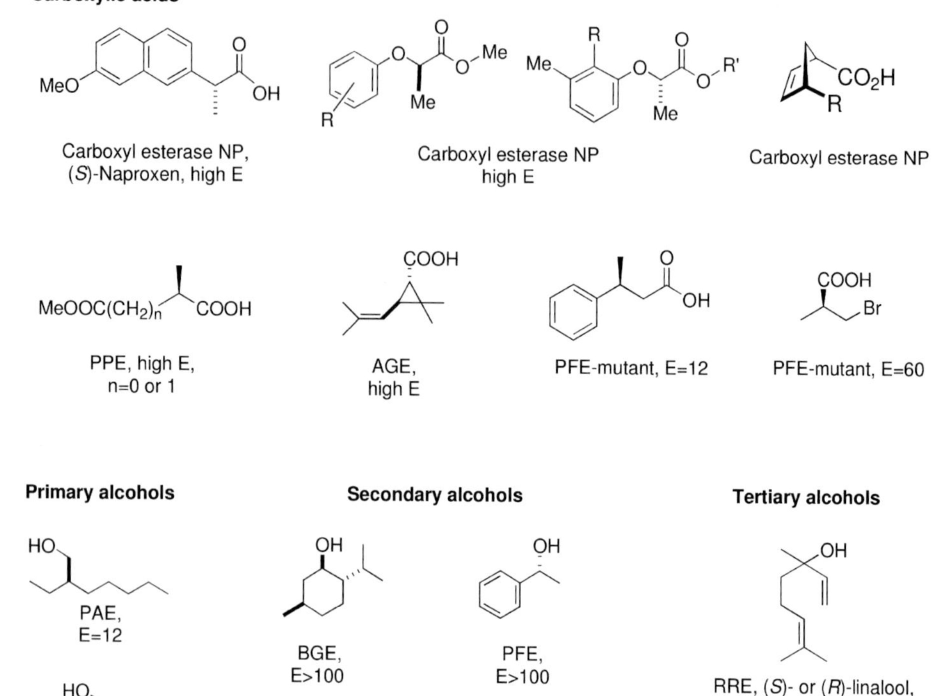

Fig. 3.34 Examples of esterase-catalyzed resolutions (adapted from Bornscheuer, 2002). AGE, *Arthrobacter globiformis* esterase; BCE, *Bacillus coagulans* esterase; BGE, *Burkholderia gladioli* esterase; BSE, *Bacillus stearothermophilus* esterase; PAE, *Pseudomonas aeruginosa* esterase; PFE, *Pseudomonas fluorescens* esterase; PPE, *Pseudomonas putida* esterase; RRE, *Rhodococcus ruber* esterase; SDE, *Streptomyces diastatochromogenes* esterase

of years now, several esterases have been available commercially from a variety of suppliers (e.g., Fluka, Amano, Jülich Fine Chemicals, Diversa, Roche Diagnostics).

Probably the best-studied enzyme is the carboxyl esterase NP (NP from Naproxen, a non-steroidal anti-inflammatory drug) originating from *Bacillus subtilis* (Quax and Broekhuizen, 1994). Beside Naproxen, various other 2-arylpropionic acids are resolved with high enantioselectivity (Fig. 3.34) (Azzolina et al., 1995). Carboxyl esterase NP has a molecular weight of 32 kDa, a pH optimum between pH 8.5 and 10.5, and a temperature optimum between 35 and 55 °C. Carboxyl esterase NP is produced as an intracellular protein, the structure of which is unknown. In a pilot-scale process, (*R*,*S*)-naproxen methylester was hydrolyzed in the presence of Tween 80 to increase substrate solubility at pH 9.0. The (*S*)-acid was separated from the remaining (*R*)-methylester and the latter racemized using an organic base. This reaction yields (*S*)-Naproxen with excellent optical purity (99 % ee) at an overall yield of 95 % (Quax and Broekhuizen, 1994).

Another efficient kinetic resolution was achieved in the synthesis of (+)-*trans*-(1*R*,3*R*) chrysanthemic acid, which is an important precursor of pyrethrin insecticides (Fig. 3.34). Here, an esterase from *Arthrobacter globiformis* catalyzed the sole formation of the desired enantiomer (>99 % ee, at 38 % conversion). The enzyme was purified and the gene cloned in *E. coli* (Nishizawa et al., 1995). In a 160-g scale process, hydrolysis was performed at pH 9.5 at 50 °C. The acid produced was separated through a hollow-fiber membrane module and the esterase proved to be very stable over four cycles, each of 48 h.

Further selected examples of the application of microbial carboxyl esterases in the synthesis of optically pure compounds are summarized in Figure 3.34.

Pig Liver Esterase

Pig liver esterase (PLE) is probably the most useful carboxyl esterase for organic synthesis, as it accepts a broad range of substrates, which are often converted with excellent stereoselectivity. Beside the kinetic resolution of various racemates, PLE was shown to be efficient in the desymmetrization of prostereogenic and *meso*-compounds (Jones, 1990; Jones et al., 1985; Lam et al., 1986).

PLE is isolated from pig liver by extraction, and consists of several isoenzymes with the α-, β-, and γ-subunits as the most dominant ones. In the literature, a debate continues as to whether these isoenzymes differ (Öhrner et al., 1990) or not (Lam et al., 1988) in their enantioselectivity. Recently, the first functional overexpression of active PLE (the γ-isoenzyme) in the yeast *Pichia pastoris* was reported (Lange et al., 2001). This allowed the production of recombinant PLE (rPLE) as a stable product without the interfering influence of other isoenzymes and hydrolases. More importantly, rPLE shows substantially higher E-values in the resolution of acetates of secondary alcohols. For instance, the resolution of (*R*,*S*)-1-phenyl-2-butyl acetate proceeded with E=1–4 using commercial PLE, but with E>100 using the recombinant enzyme, which contains only a single isoenzyme (Fig. 3.35) (Musidlowska et al., 2001).

The gene encoding rPLE shares 97 % identity with the published nucleotide sequence of porcine intestinal carboxyl esterase (PICE). By using site-directed mutagenesis, 22 nucleotides encoding 17 amino acids were exchanged step-wise from the PLE gene, yielding the recombinant PICE sequence and eight intermediate mutants. All esterases were successfully produced in *Pichia pastoris* as extracellular proteins (Musidlowska-Persson and Bornscheuer, 2003a). Again, significant differenc-

OAc
rPLE, E>>100
PLE (Fluka, Sigma, Roche), E=1.4-4.0
OH
+
OAc

Fig. 3.35 Recombinant pig liver esterase (rPLE) shows significantly higher enantioselectivity towards 1-phenyl-2-butyl acetate than crude preparation from commercial suppliers.

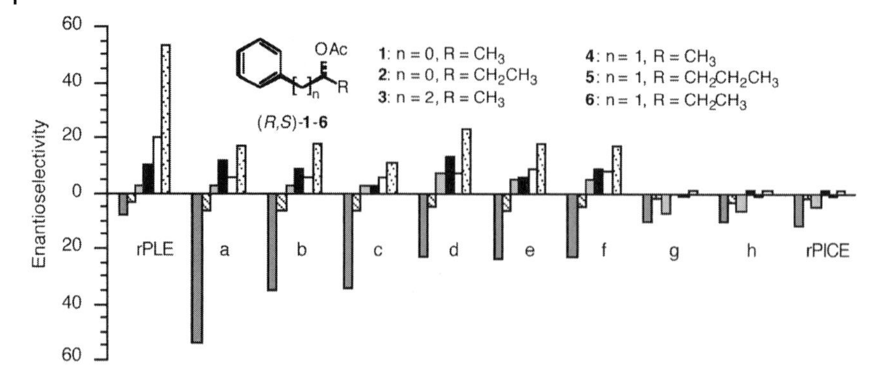

Fig. 3.36 Changes in enantioselectivity observed for recombinant pig liver esterase (rPLE), mutants (a–h) and recombinant porcine intestinal carboxyl esterase (rPICE). Enantioselectivity was studied towards six acetates of secondary alcohols (1–6) and is shown in the corresponding columns (acetates 1–6 from left to right bars). Note that not only enantioselectivity values change, but also enantiopreference is altered by amino acid substitutions.

es were found for the enantioselectivity of these new esterases in the hydrolysis of a range of acetates of secondary alcohols (Musidlowska-Persson and Bornscheuer, 2003b) (Fig. 3.36). An up to six-fold increase in enantioselectivity (E = 46) compared to rPLE (E = 8) was observed in the hydrolysis of (*R*,*S*)-1-phenylethyl acetate using a variant containing only two mutations (a in Fig. 3.36). For other substrates, a switch in enantiopreference was observed with the introduction of certain mutations.

3.2.2.3 **Peptidases, Acylases, and Amidases**

Peptidases (EC 3.4.-.-) and amidases (EC 3.5.-.-) catalyze both the formation and hydrolysis of amide links. Although their natural role is hydrolysis, they are also used to form amide bonds. Two different strategies have been applied for this, namely a thermodynamic or a kinetic control (Fig. 3.37; see also Chapter 2, Sections 2.6 and 2.8).

In thermodynamically controlled syntheses, the reaction conditions are changed to shift the equilibrium toward synthesis instead of hydrolysis. The hydrolysis of peptides is favored by –5 $kJmol^{-1}$, and is driven mainly by the favorable solvation of the carboxylate and ammonium ions; it also depends on the pH (see Fig. 2.20, p. 69). One common way to shift the equilibrium toward synthesis is to replace the water with an organic solvent. The organic solvent suppresses the ionization of the starting materials, and also reduces the concentration of water. Other common ways of shifting the equilibrium are to increase the concentrations of the starting materials, or to choose protective groups that promote precipitation of the product.

In kinetically controlled syntheses, an activated carboxyl component (usually an ester or amide) is used; this reacts with the enzyme to form an acyl enzyme intermediate, which then reacts either with an amine to form the desired amide, or with water

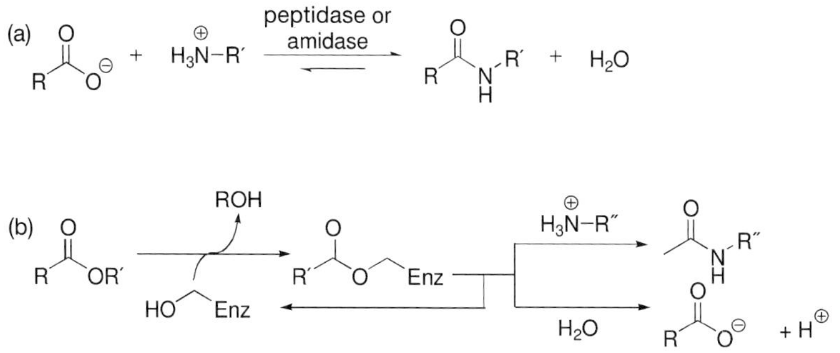

Fig. 3.37 Synthesis of amide bonds using proteases and amidases. (a) Thermodynamic control shifts the equilibrium toward synthesis by changing the reaction conditions. For example, organic solvents are added to reduce the concentration of water and suppress ionization of the starting materials. (b) Kinetic control starts with an activated carboxyl component (e.g., an ester or an amide) and forms an acyl enzyme intermediate. The acyl enzyme intermediate then reacts with an amine to form the amide. In a competing side reaction, water may react with the acyl enzyme intermediate.

to form a carboxylic acid. Because the starting material is an activated carboxyl component, reactions are faster in the kinetically controlled approach than in the thermodynamically controlled approach (see Fig. 2.9). Because the kinetically controlled approach requires an acyl enzyme intermediate, only serine hydrolases (e.g., the peptidase subtilisin, the amidase penicillin G amidase) are suitable. Metallo peptidases such as thermolysin function only in thermodynamically controlled syntheses.

Although peptidase-catalyzed peptide synthesis was first reported in 1901 (Savjalov, 1901), peptidases have been used much more frequently in peptide synthesis since the late 1970s (for reviews, see Bordusa, 2002; Drauz and Waldmann, 2002; Kasche, 2001; Schellenberger and Jakubke, 1991). The advantages of an enzyme-catalyzed peptide synthesis are mild conditions, no racemization, a minimal need for protective groups, and high regio- and enantioselectivity.

The largest scale application (producing hundreds to thousands of tonnes) of peptidase-catalyzed peptide synthesis is the thermolysin-catalyzed synthesis of aspartame, a low-calorie sweetener (Fig. 3.38) (Oyama, 1992). Precipitation of the product drives this thermodynamically controlled synthesis, and the high regioselectivity of thermolysin ensures that only the α-carboxyl group in aspartate reacts. Thus, there is no need to protect the β-carboxylate. The high enantioselectivity allows the use of racemic amino acids as only the L-enantiomer reacts.

An example of a kinetically controlled peptide synthesis is the α-chymotrypsin-catalyzed production of kyotorphin (Tyr-Arg), an analgesic dipeptide (Fig. 3.39; see also Fig. 2.23) (Fischer et al., 1994). In order to minimize the hydrolysis of the acyl enzyme intermediate, a high concentration of the nucleophile was used, which was possible only because the charged maleyl protective group increased the solubility of the carboxyl component.

Subtilisin accepts a broader range of substrates than other peptidases, so it is also used for amide couplings involving unnatural substrates (Moree et al., 1997). When

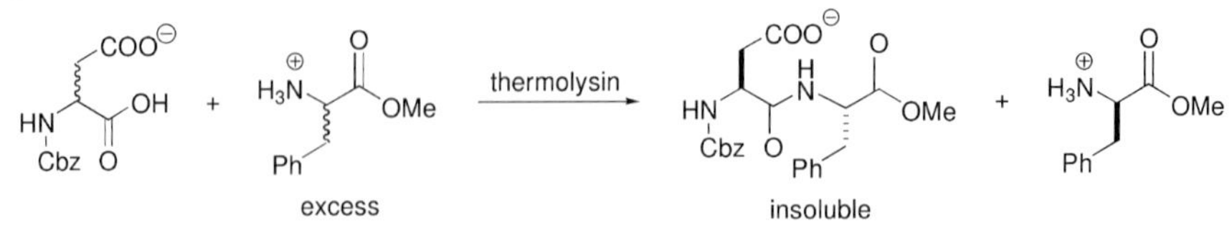

Fig. 3.38 Commercial process for the production of aspartame (α-L-aspartyl-L-phenylalanine methyl ester). Thermolysin catalyzes the coupling of an *N*-Cbz-protected aspartic acid with phenylalanine methyl ester. The product forms an insoluble salt with excess phenylalanine methyl ester. This precipitation drives this thermodynamically controlled peptide synthesis. The high regioselectivity of thermolysin for the α-carboxylate allows the β-carboxylate in aspartate to be left unprotected. The enantioselectivity of thermolysin allows the use of racemic starting materials.

coupling a D-amino acid, it is best to use it as the nucleophile rather than as the carboxyl donor because subtilisin is more tolerant of changes in the nucleophile than in the carboxyl group.

Acylases and amidases also catalyze the formation of peptide bonds,and an important application here is in the hydrolysis and synthesis of the β-lactam peptide antibiotics (penicillins and cephalosporins). About 40 000 tonnes penicillin are hydrolyzed each year in an equilibrium-controlled process catalyzed by penicillin amidases to produce 6-aminopenicillanic acid (6-APA) (see Fig. 1.7; see Chapter 2, Exercise 14; see also Chapter 9, Section 9.2). New processes for the synthesis of a range of semi-synthetic β-lactam antibiotics starting from 6-APA or 7-aminocephalosporanic acid (7-ACA) have been developed (see Chapter 9, Section 9.2; Liese et al., 2000). For the synthesis, penicillin amidase is used in a kinetically controlled approach, and the antibiotic can be obtained in yields >95 % (of 6-APA or 7-ACA) in aqueous solution when the synthesis is performed with a more than 2-fold excess of activated substrate (Fig. 3.40). The production of 7-ACA from cephalosporin C is described in detail in Chapter 9, Section 9.2.

Biochemical properties and the mechanism of peptidases and amidases are covered in more detail in Figure 2.5 (see also Chapter 2, Sections 2.7 and 2.10). More in-

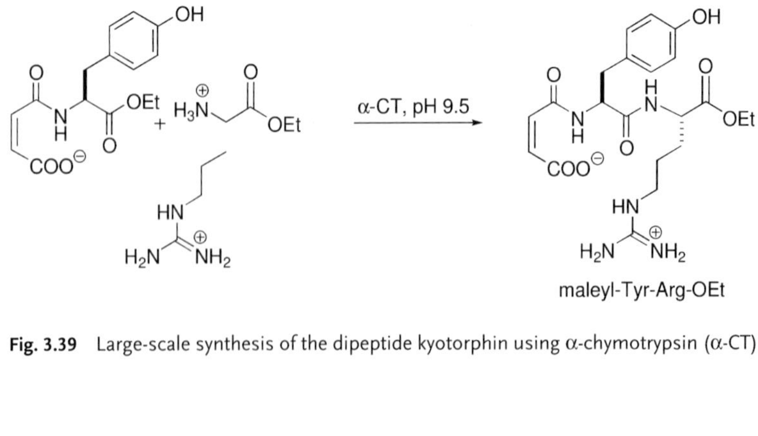

Fig. 3.39 Large-scale synthesis of the dipeptide kyotorphin using α-chymotrypsin (α-CT).

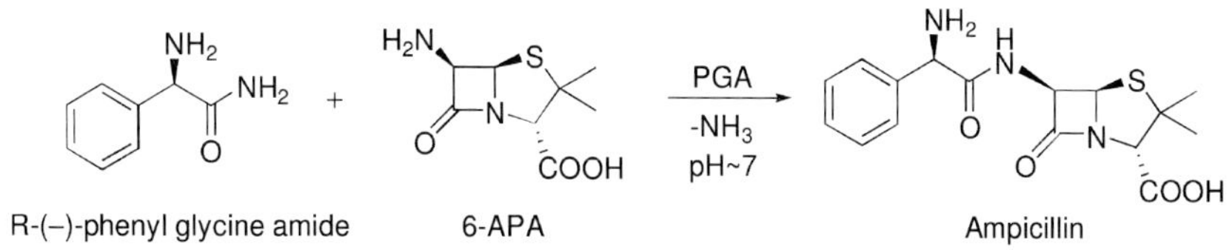

R-(–)-phenyl glycine amide + 6-APA → (PGA, $-NH_3$, pH~7) Ampicillin

Fig. 3.40 The penicillin acylase (PGA)-catalyzed coupling of 6-amino penicillanic acid (6-APA) with (*R*)-(–)phenylglycine to produce ampicillin, a β-lactam antibiotic.

formation can be also found in a variety of books (e.g., Bornscheuer and Kazlauskas, 1999; Drauz and Waldmann, 2002).

3.2.2.4 **Epoxide Hydrolases (EC 3.3.2.3)**

Optically pure epoxides are versatile building blocks in organic synthesis. Several chemical methods for the preparation of optically pure epoxides have been developed, as described by Sharpless or Jacobsen-Katsuki. However, the Sharpless epoxidation (Katsuki and Martin, 1996) is limited to allylic alcohols, while the Jacobsen–Katsuki method (Hosoya et al., 1994; Linker, 1997) often gives poor enantiomeric excess (≤60 % ee) with *gem*- or *trans*-disubstituted olefins. These and related chemical methods are used on an industrial scale (Breuer et al., 2004).

Alternatively, epoxides can be resolved by using epoxide hydrolases, which catalyze the hydrolysis of an epoxide to furnish the corresponding vicinal diol. The reaction proceeds via a S_N2-specific opening of the epoxide, leading to the formation of the corresponding *trans*-configurated 1,2-diols.

Epoxide hydrolases (EHs) do not require cofactors and have been found in a variety of sources, including mammals, plants, insects, yeasts, filamentous fungi, and bacteria. These enzymes are catalytically active in the presence of organic solvents, and often show high regio- and enantioselectivity. Although they have been intensively studied with respect to their metabolic function, mammalian EHs are not applied in biocatalysis, mainly due to their limited availability.

The first reports on microbial EHs were published in 1991 and 1993, but since then a broad range of EHs from various microorganisms have been identified and extensively characterized. The structures of three EHs from mammalian, bacterial and fungi origin have been determined, and have shown a typical α/β-hydrolase fold.

The proposed mechanism of epoxide ring-opening involves the attack of a nucleophilic carboxylate residue at one end of the epoxide which has been activated by protonation. This leads to an α-hydroxyester intermediate covalently bound to the active site of the enzyme. This intermediate is hydrolyzed by the nucleophilic attack of a water molecule which is activated by a histidine, followed by the release of the diol product and regeneration of the enzyme. From the crystal structure it was concluded that the activation of the epoxide moiety is guided by two tyrosine residues (Fig. 3.41) (Nardini et al., 1999).

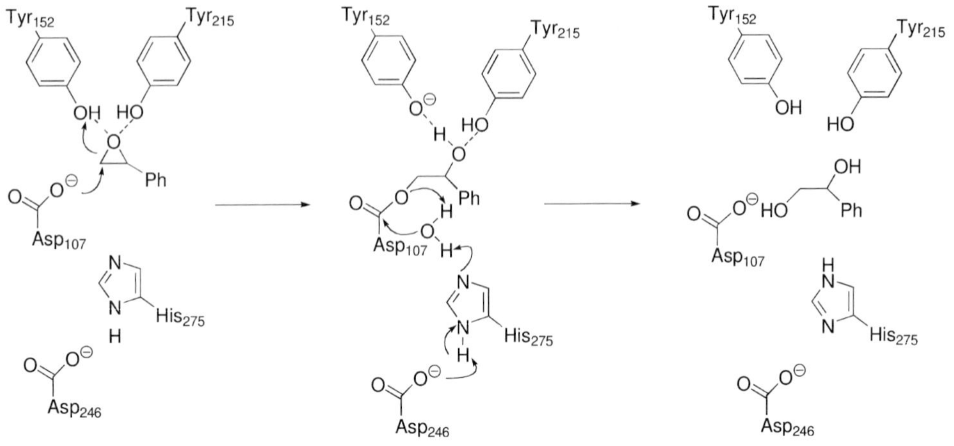

Fig. 3.41 Mechanism proposed for epoxide hydrolase from *Agrobacterium radiobacter.*

The enzymatic hydrolysis of terminal epoxides may proceed via attacking either the less-hindered oxirane carbon, leading to retention of configuration (most common), or at the stereogenic center, which results in an inversion of configuration (Faber et al., 1996) (Fig. 3.42).

Initially, Faber's group discovered that the biocatalyst preparation SP409 (from *Rhodococcus* sp.) produced by Novozymes also exhibited EH activity, and converted a wide variety of mono- and 1,1-disubstituted epoxides with low to moderate enantioselectivity. In addition, nucleophiles other than water (e.g., azide) were accepted (Fig. 3.43) (Mischitz and Faber, 1994). In general, EHs are used as washed whole cells or as lyophilized cell-free extracts, because isolated enzymes show less stability.

During the past decade, a wide range of microbial EHs have been identified, and examples of the resolution of epoxides using enzymes of bacterial origin are shown in Figure 3.44. Aliphatic and arylaliphatic substrates were converted with high enantioselectivity, though in general higher selectivities are observed for monosubstituted arylaliphatic epoxides.

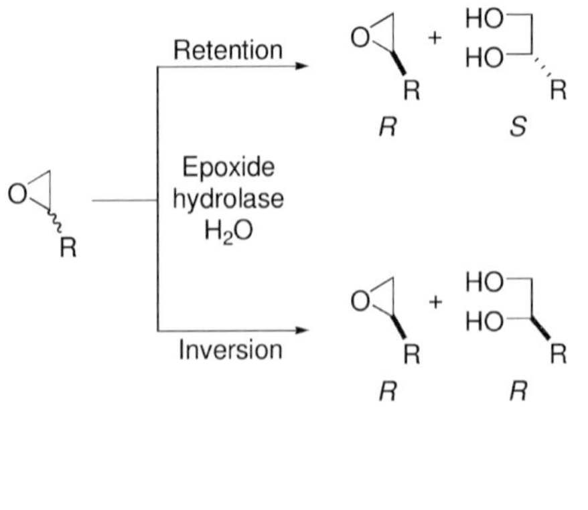

Fig. 3.42 Hydrolysis of epoxides can proceed with retention or inversion of configuration.

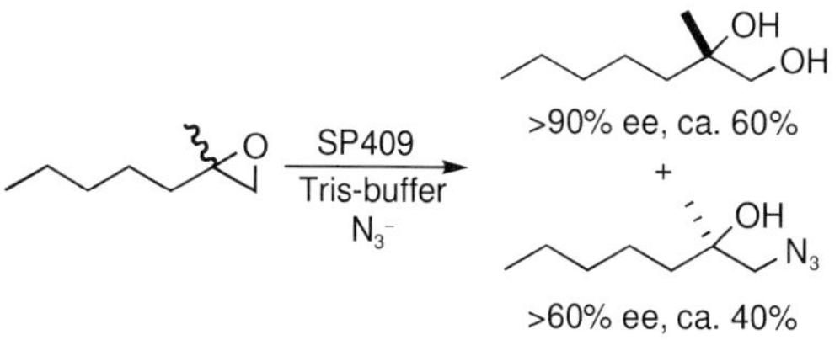

OH
OH
>90% ee, ca. 60%
+
O
SP409
Tris-buffer
N_3^-
OH
N_3
>60% ee, ca. 40%

Fig. 3.43 *Rhodococcus* sp. SP409 catalyzes epoxide ring-opening and accepts an azide as nucleophile (Mischitz and Faber, 1994).

Several strains from *Nocardia* sp. (H8, EH1, TB1, later designated as *Rhodococcus ruber*) have been identified as EH producers. In particular *Nocardia* sp. EH1 shows high enantioselectivity at 50 % conversion in the resolution of 2-methyl-1,2-epoxyheptane (E > 100). However, the introduction of a phenyl group into the side chain decreases enantioselectivity (E = 5.6). The enzyme was purified to homogeneity via a four-step procedure. It is a monomer with a molecular weight of 34 kDa, a pH optimum of 8–9, and a temperature optimum of 35–40 °C. The pure enzyme is much less stable than a whole-cell preparation, but it is stabilized by the addition of Tween 80 or Triton X-100. Immobilization on DEAE-cellulose doubled the specific activity and allowed five repeated batch reactions, though enantioselectivity was slightly lowered (Kroutil et al., 1998a,b). Using *Nocardia* sp. EH1, the synthesis of naturally occuring (*R*)-(–)-mevalonolactone was achieved by deracemization of 10 g 2-benzyl-2-methyloxirane. The enzymatic reaction gave the corresponding (*S*)-diol, while the addition of catalytic amounts of sulfuric acid hydrolyzed the remaining (*R*)-epoxide with inversion of configuration, thus allowing the isolation of (*S*)-diol in an overall yield of 94 % at 94 % ee. Subsequent chemical steps afforded (*R*)-(–)-mevalonolactone in a total yield of 55 % (Orru et al., 1997, 1998).

Hydrolysis of (±)-*cis*-2,3-epoxyheptane with rehydrated lyophilized cells of *Nocardia* sp. EH1 proceeded in an enantioconvergent fashion, and only (2*R*,3*R*)-heptane-2,3-diol was obtained as the sole product (Kroutil et al., 1996).

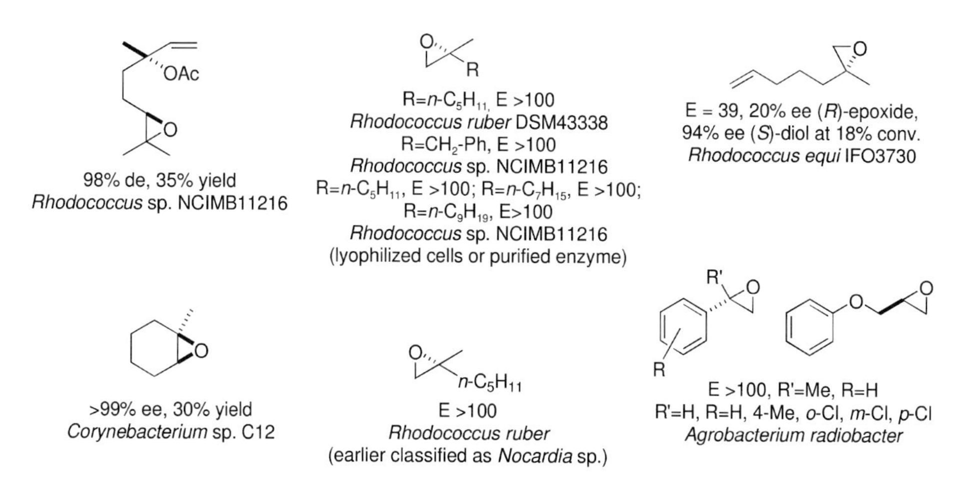

Fig. 3.44 Examples of the resolution of epoxides using bacterial epoxide hydrolases.

The recombinant microbial EH cloned from *Agrobacterium radiobacter* AD1 and overexpressed in *E. coli* (Rink et al., 1997) accepts a broad range of styrene oxide derivatives and phenyl glycidyl ether, which are converted with excellent enantioselectivity (Spelberg et al., 1998). The enzyme has a molecular weight of 34 kDa, and the catalytic triad was proposed to consist of Asp_{107}, His_{275} and Asp_{246}.

The first preparative-scale epoxide hydrolysis was reported by the group of Furstoss, who discovered that the fungus *Aspergillus niger* enantioselectively converts epoxy geraniol-*N*-phenylcarbamate to yield the (*S*)-epoxide in high optical purity (96 % ee), (Zhang et al., 1991). The resolution of *p*-nitrostyrene oxide proceeded with acceptable enantioselectivity (E = 41), and up to 20 % DMSO could be added without any significant loss of activity. EH activity was also discovered in *Beauveria sulfurescens* ATCC7159, which converts styrene oxide as well as indene and tetrahydronaphthalene oxides with high enantioselectivity (Pedragosa-Moreau et al., 1996). These and further examples are summarized in Figure 3.45.

Interestingly, *Aspergillus niger* and *Beauveria sulfurescens* produced the (*R*)-diol in the hydrolysis of styrene oxide. Thus, the reaction catalyzed by *A. niger* proceeded with retention of configuration (via attack at C-2), whereas the hydrolysis with *B. sul-*

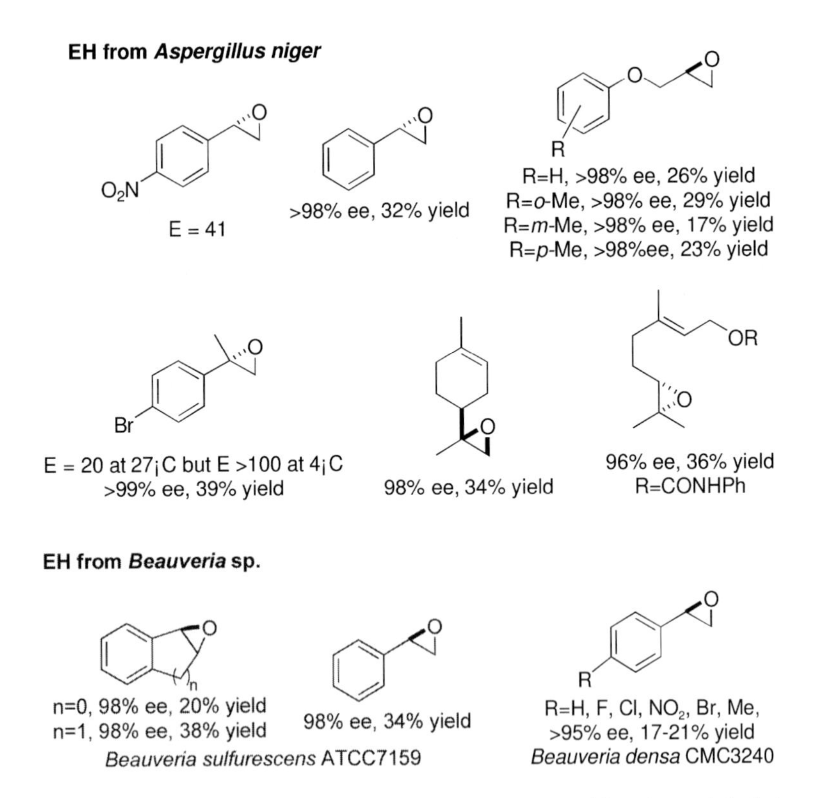

Fig. 3.45 Examples of the resolution of epoxides using yeast and fungal epoxide hydrolases.

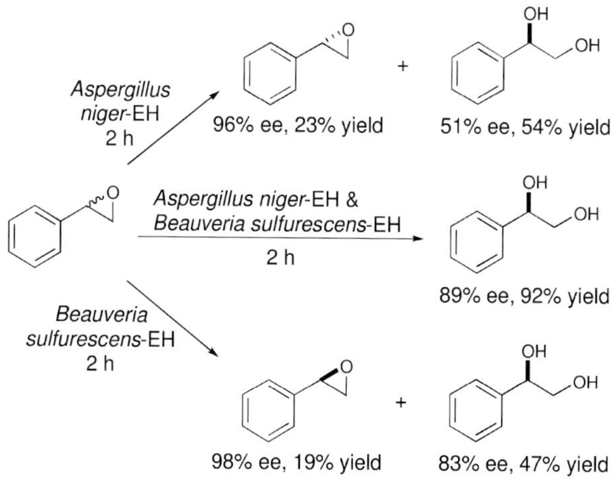

Fig. 3.46 Resolution of styrene oxide using fungal epoxide hydrolases from *Aspergillus niger* or *Beauveria sulfurescens* or a mixture of both for an enantio-convergent synthesis.

furescens occurred with inversion of configuration (via attack at C-1, benzylic position). Employing a mixture of both organisms permitted the enantioconvergent synthesis of (*R*)-1-phenyl-1,2-dihydroxyethane in 92 % yield and 89 % ee (Fig. 3.46).

In the search for an enantioselective EH capable of resolving indene oxide, a precursor to the side chain of HIV peptidase inhibitor MK639, researchers at Merck found that out of 80 fungal strains investigated, *Diploida gossipina* ATCC16391 and *Lasiodiploida theobromae* MF5215 showed excellent enantioselectivity, yielding exclusively the desired (1*S*,2*R*)-enantiomer. Two other strains from *Gilmaniella humicola* MF5363 and from *Altenaria enius* MF4352 showed opposite enantiopreference. Preparative biotransformation using whole cells of *Diploida gossipina* ATCC16391 allowed isolation of optically pure (1*S*,2*R*)-indene oxide in 14 % yield after a 4 h reaction time (Fig. 3.47) (Zhang et al., 1995).

More examples of the characterization and application of EHs can be found in recent reviews (Archelas and Furstoss, 2001; Smit, 2004; Steinreiber and Faber, 2001).

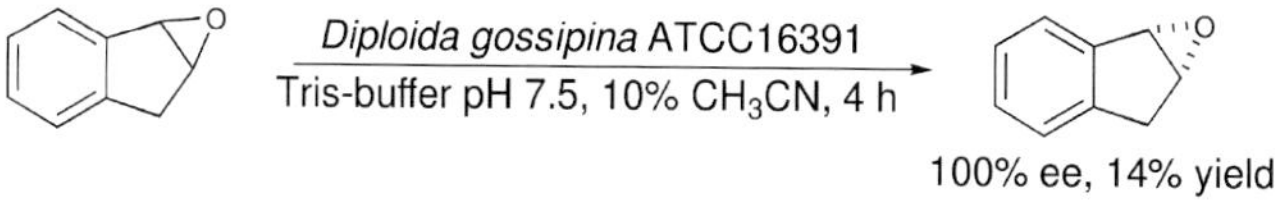

Fig. 3.47 Resolution of indene oxide catalyzed by epoxide hydrolase from *Diploida gossipina* yields a HIV protease inhibitor precursor.

3.2.2.5 Nitrilases (EC 3.5.5.1)

Nitriles are important precursors for the synthesis of carboxylic acid amides and carboxylic acids. The chemical hydrolysis of nitriles requires strong acid or base at high temperatures. Nitrile-hydrolyzing enzymes have the advantage that they react under mild conditions and do not produce large amounts of by-products. In addition, the enzyme can be regio- and stereoselective. Two different enzymatic pathways can be used to hydrolyze nitriles (Fig. 3.48). Nitrilases (EC 3.5.5.1) directly catalyze the conversion of a nitrile into the corresponding acid plus ammonia. In the other pathway, a nitrile hydratase (NHase; EC 4.2.1.84; a lyase) catalyzes the hydration of a nitrile to the amide, which may be converted to the carboxylic acid and ammonia by an amidase (EC 3.5.1.4).

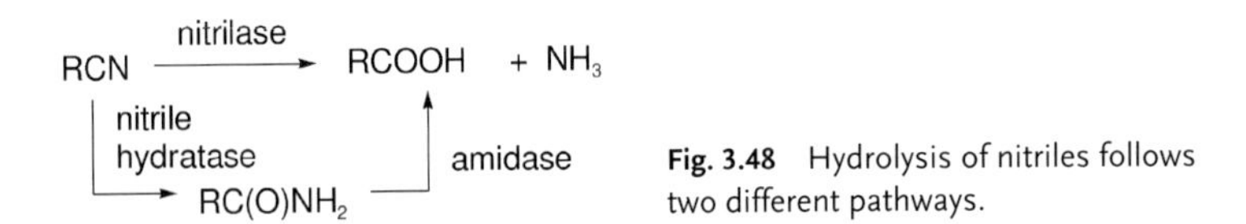

Fig. 3.48 Hydrolysis of nitriles follows two different pathways.

Pure nitrilases and nitrile hydratases are usually unstable, and thus the biocatalyst is often used as a whole-cell preparation. The nitrile-hydrolyzing activity must be induced first,and common inducers are benzonitrile, isovaleronitrile, crotononitrile, and acetonitrile, though the inexpensive inducer urea also works. In addition, inducing with ibuprofen or ketoprofen nitriles can yield enantioselective enzymes (Layh et al., 1997). After induction, preparative conversion is usually performed by adding the nitriles either during cultivation or by employing resting cells. The most commonly used strains are from *Rhodococcus* sp., and the most important ones are subspecies of *Rh. rhodochrous.*

Nitrilases are cysteine hydrolases which act via an enzyme-bound imine intermediate using a Glu-Lys-Cys catalytic triad. All nitrilases are inactivated by thiol reagents (e.g., 5,5′-dithiobis(2-nitrobenzoic acid)), indicating that they are sulfhydryl enzymes.

Two different groups of nitrile hydratases have been described, which require Fe(III) or Co(III) ions (Nagasawa et al., 1991). The *Rh. rhodochrous* J1 strain produces two kinds of NHases that differ in their molecular weight (520 kDa and 130 kDa). The high molecular-weight NHase acts preferentially on aliphatic nitriles, whereas the smaller enzyme also has high affinity towards aromatic nitriles. Most nitrile hydratases accept aliphatic nitriles only (e.g., from *Arthrobacter* sp. J-1, *Brevibacterium* R312, *Pseudomonas chlororaphis* B23); however, strains have also been described which can act on arylalkylnitriles, arylacetonitriles, and heterocyclic nitriles.

Yamada's group showed, by the use of electron spin resonance (ESR) studies, that nitrile hydratases are non-heme ferric iron-containing enzymes. Further spectroscopic studies suggested that the enzyme also binds the cofactor pyrroloquinoline quinone (PQQ), leading to the proposed mechanism shown in Figure 3.49 (Sugiura et al., 1987).

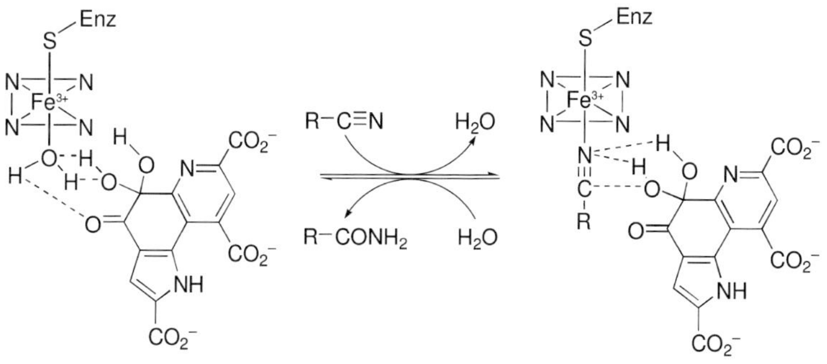

Fig. 3.49 Proposed mechanism of nitrile hydration catalyzed by a nitrile hydratase involving Fe(III) and PQQ (Sugiura et al., 1987).

The crystal structure of nitrile hydratase from *Rhodococcus* sp. R312 suggested that the enzyme is composed of two subunits (α,β), which contain one iron atom per α,β unit. The α-subunit is composed of a long N-terminal and a C-terminal domain that forms a novel fold, which can be described as a four-layered α-β-β-α structure. The two subunits form a tight heterodimer that is the functional unit of the enzyme. The active site is located in a cavity at a subunit–subunit interface. The iron center is formed by residues from the α-subunit only – three cysteine thiolates and two main chain amide nitrogen atoms are ligands – although the iron center is located between the α- and β-subunits. Three possible catalytic roles for the metal ion in NHases were proposed, in which the metal ion always acts as a Lewis acid activating the nitrile for hydration. Further information on structure, regulation, and application of metallo nitrile hydratases can be found in a review (Kobayashi and Shimizu, 1998).

Various NHase genes have been cloned and characterized, and the amidase gene was found to be closely located to the NHase gene; this supported the theory that both enzymes are involved in the two-step degradation of nitriles to carboxylic acids.

Two applications based on nitrile hydratase from *Rhodococcus rhodochrous* (Nagasawa and Yamada, 1995) have been commercialized (Fig. 3.50). The large-scale pro-

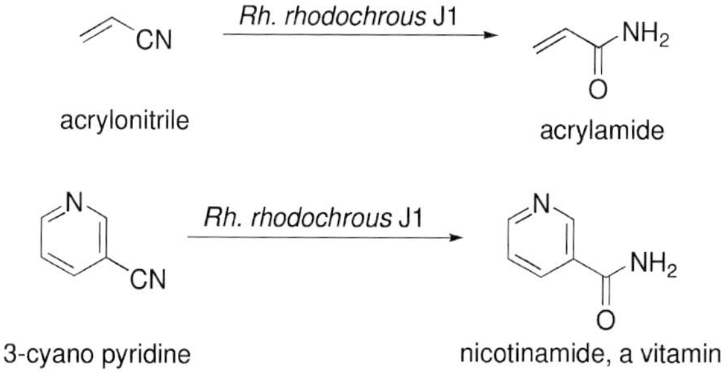

Fig. 3.50 Commercial production of acrylamide and nicotinamide using resting cells of *Rhodococcus rhodochrous* J1.

duction of the commodity chemical acrylamide is performed by Nitto Chemical (Yokohama, Japan) on a >40 000 tonnes per year scale. Initially, strains from *Rhodococcus* sp. N-774 or *Pseudomonas chlororaphis* B23 were used, but the current process uses the 10-fold more productive strain *Rhodococcus rhodochrous* J1. The productivity is >7000 g acrylamide per gram of cells at a conversion of acrylonitrile of 99.97 %. The formation of acrylic acid is barely detectable at the reaction temperature of 2–4 °C. In laboratory-scale experiments with resting cells, up to 656 g acrylamide per liter reaction mixture was achieved. Besides acrylamide, a wide range of other amides can also be produced, including acetamide (150 g L^{-1}), isobutyramide (100 g L^{-1}), methacrylamide (200 g L^{-1}), propionamide (560 g L^{-1}), and crotonamide (200 g L^{-1}).

Furthermore, *Rh. rhodochrous* J1 also accepts aromatic and arylaliphatic nitriles as substrates. For example, the conversion of 3-cyanopyridine to nicotinamide (a vitamin in animal feed supplementation) is catalyzed, and the process was industrialized by Lonza (Switzerland) on a >3000 tonne per year scale.

A wide variety of dinitriles can be hydrolyzed with moderate to excellent regioselectivity to the corresponding monocarboxylic acids (Fig. 3.51).

In the NHase/amidase system, it is usually only the amidase that shows stereoselectivity, while the nitrile hydratase is non-selective. Examples of kinetic resolutions include precursors of non-steroidal anti-inflammatory drugs (e.g., ketoprofen, ibuprofen, naproxen) (Fig. 3.52).

The strain *Alcaligenes faecalis* ATCC8750 can be used for the production of mandelic acid (Yamamoto et al., 1991). *A. faecalis* contains a nitrilase and an amidase, but no NHase. Using resting cells, (*R*)-(–)-mandelic acid was formed in excellent optical purity (100 % ee). Moreover, the remaining (*S*)-mandelonitrile racemized, resulting in an overall yield of 91 % mandelonitrile after a 6 h reaction time. It was suggested that the rapid racemization was due to an equilibrium between mandelonitrile and benzaldehyde/HCN, because mandelic acid could also be obtained when only benzaldehyde and HCN are used as substrates (Fig. 3.53).

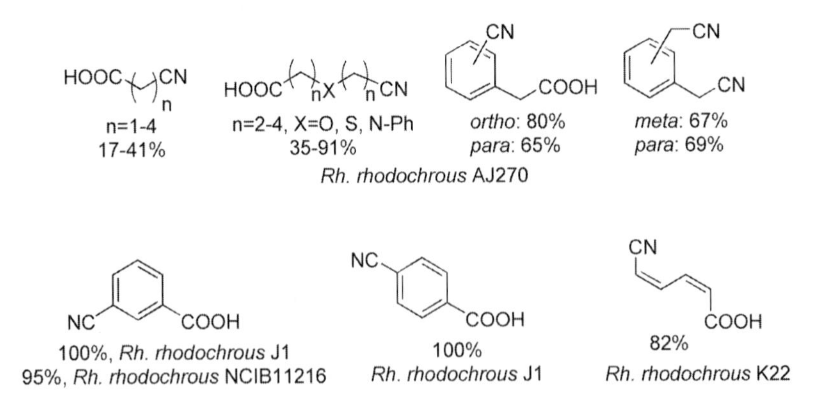

Fig. 3.51 Examples of the regioselective hydrolyses using nitrile hydratases and nitrilases.

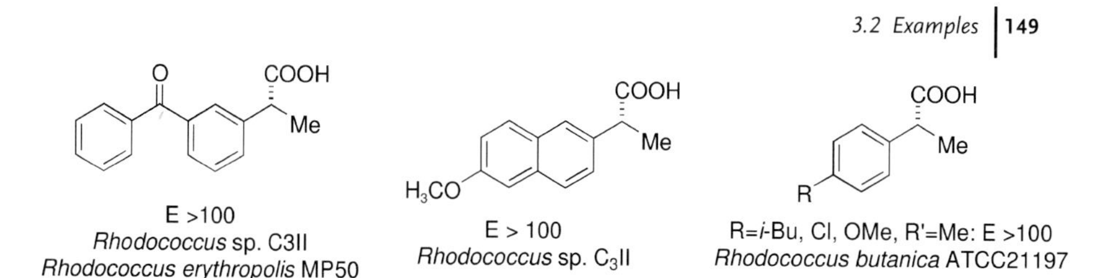

Fig. 3.52 Examples of the synthesis of non-steroidal anti-inflammatory drugs by hydrolysis of nitriles.

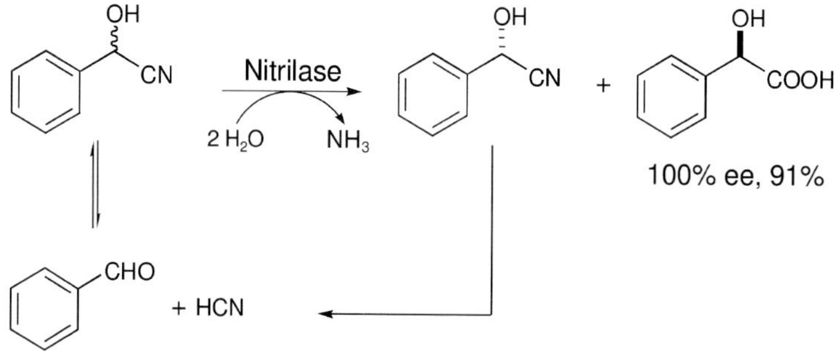

Fig. 3.53 Synthesis of (*R*)-(–)-mandelic acid by a nitrilase present in *A. faecalis* resting cells involves in-situ recycling of (*S*)-mandelonitrile by disproportion in benzaldehyde and HCN followed by formation of (*R*,*S*)-mandelonitrile.

Until recently, only approximately 15 nitrilases had been described. Researchers at Diversa (San Diego, USA) have now discovered more than 200 unique nitrilases from genomic libraries obtained from DNA extracted from environmental samples (DeSantis et al., 2002). Twenty-seven enzymes afforded mandelic acid in >90 % ee under conditions of dynamic kinetic resolution (Fig. 3.53). Out of these biocatalysts, one nitrilase afforded (*R*)-mandelic acid in 86 % yield and 98 % ee. Further studies showed that this nitrilase also converted a broad range of mandelic acid derivatives and analogues with high activity and similar stereoselectivity. Another nitrilase exhibited high activity and broad substrate tolerance towards aryllactic acid derivatives, which were also converted in a dynamic kinetic resolution (Table 3.5).

A few groups have already investigated the hydrolysis of prochiral dinitriles. The hydrolysis of 3-hydroxyglutaronitriles revealed that Bn- or Bz-protecting groups were required to achieve acceptable enantiomeric excess for the monocarboxylic acid (Crosby et al., 1992). For the nitrilases discovered by Diversa, it could be shown that a range of enzymes show high conversion (>95 %) and selectivity (>90 % ee). The best enzyme gave 98 % yield and 95 % ee for the (*R*)-product (DeSantis et al., 2002). In addition, 22 enzymes that afford the opposite enantiomer with 90–98 % ee were discovered (Fig. 3.54). In a later study, the most effective (*R*)-nitrilase was optimized by directed evolution (see Chapter 2, Section 2.11.2) to withstand high sub-

Table 3.5 Preparation of optically active aryllactic acid derivatives using a nitrilase under dynamic kinetic resolution conditions (modified from DeSantis et al., 2002).

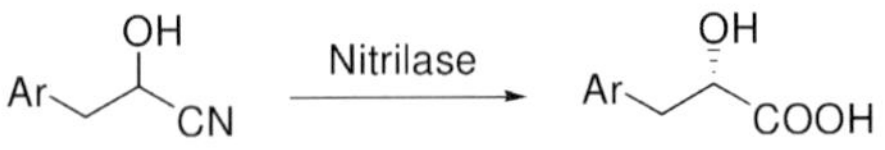

Ar	*Relative activity [%]*	*ee [%]*
C_6H_5	25	96
2-Me-C_6H_5	160	95
2-Br-C_6H_5	121	95
3-F-C_6H_5	22	99
1-naphthyl	64	96
2-pyridyl	10.5	99

strate concentrations while maintaining high enantioselectivity. The best variant obtained by a "gene-site saturation mutagenesis" technique contained a single mutation (Ala 190 His) and allowed the production of the (*R*)-acid at 3 M substrate concentration with 96 % yield at 98.5 % ee (DeSantis et al., 2003).

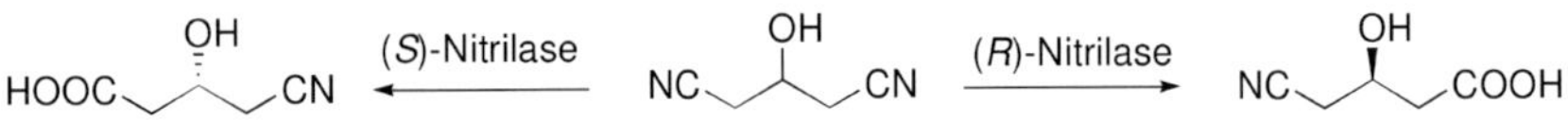

Fig. 3.54 Desymmetrization of prochiral-hydroxyglutaronitrile using engineered nitrilases.

More examples on the characterization and application of nitrilases and nitrile hydratases/amidases can be found in recent reviews (Bornscheuer and Kazlauskas, 1999; Kobayashi and Shimizu, 1994; Wieser and Nagasawa, 1999).

3.2.2.6 **Hydantoinases (EC 3.5.2.-)**

Hydantoinases are valuable enzymes for the production of optically pure D- and L-amino acids. (Hydantoinase is the commonly used name, more specifically they are cyclic amidases. An alternative name is dihydropyrimidinase.) These enzymes catalyze the reversible hydrolytic cleavage of hydantoins and 5'-monosubstituted hydantoins. In combination with carbamoylases (EC 3.5.1.-), the reaction yields L- or D-amino acids, depending on the stereoselectivity of the enzymes (Fig. 3.55). To shift the yield above 50 %, a dynamic kinetic resolution starting from racemic hydantoins is also possible by either working at slightly alkaline pH values (pH > 8) or using hydantoin-specific racemases. Since racemic hydantoins are readily available by chemical synthesis (e.g., Strecker synthesis), the synthesis of non-natural amino acids is also feasible, if the enzyme has an appropriate substrate specificity.

It has been known since the 1940s that some microorganisms can grow on D,L-5-monosubstituted hydantoins, if these were used as sole carbon and nitrogen source.

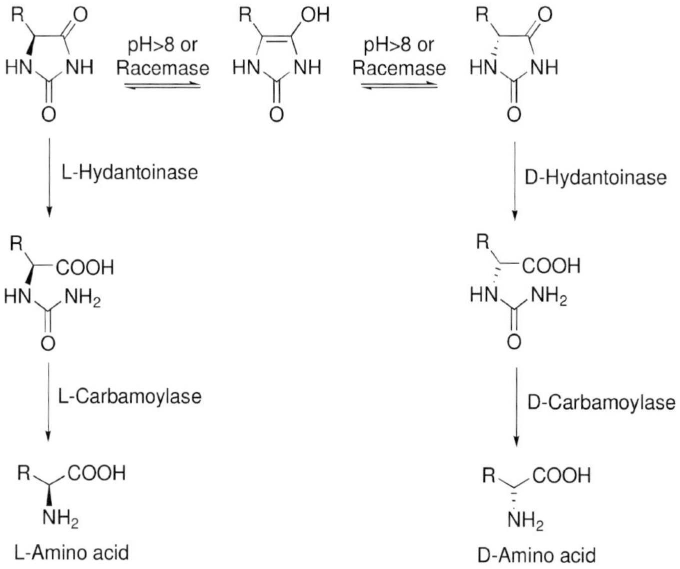

Fig. 3.55 Synthesis of L- or D-amino acids using a combination of hydantoinase and carbamoylase. Complete conversion of racemic D,L-hydantoin can be achieved by racemization at alkaline pH or specific racemases.

Especially since the 1970s, a broad range of hydantoinases and carbamoylases have been discovered by screening, owing to their importance in the synthesis of optically pure amino acids. Nowadays hydantoinases from, for example, *Arthrobacter* sp., *Nocardia* sp., *Bacillus* sp., and *Pseudomonas* sp., have been described and a range of D-amino acids are accessible by using them. For a hydantoinase from *Arthrobacter aurescens* DSM3745 it was shown, that about 2.5 mol Zn^{2+} is required per mol subunit, and that these metal ions have both a catalytic and a structural function. Crystal structures have been elucidated for hydantoinases from *Arthrobacter aurescens* and *Thermus* sp. and the D-carbamoylase from *Agrobacterium*. For this D-carbamoylase, the catalytic center was identified to consist of a glutamine, a lysine and a cysteine residue (Altenbuchner et al., 2001; Syldatk et al., 1999).

Two different strategies have been described for the hydantoinase-based synthesis of D- or L-amino acids. In one approach, the carbamoyl formed is chemically hydrolyzed (with chemical racemization of the non-wanted enantiomer). Alternatively, a combined one-pot process ("hydantoinase process") using a hydantoinase, a carbamoylase and a racemase can be employed (Fig. 3.55). This approach has the major advantage that an insufficient enantioselectivity of a hydantoinase can be overcome by the use of a more selective carbamoylase, racemization during the process enables directly up to 100 % yield, and the use of chemicals and solvents can be avoided or considerably reduced.

More recently, the hydantoinase process for the production of L-methionine was improved by directed evolution (see Chapter 2, Section 2.11.2) for the hydantoinase from *Arthrobacter* sp. DSM9771 expressed in *E. coli*. A combination of error-prone PCR and saturation mutagenesis led to the identification of a variant with higher L-selectivity and five-fold increased specific activity. Recombinant technology was also used to clone and express together a D-hydantoinase from *Bacillus stearothermophilus* SD1 and a D-N-carbamoylase from *Agrobacter tumefaciens* in *E. coli*. Both enzymes form approximately 20 % of the total cell protein, and both proteins were expressed at a comparable level (ratio 1:1.2). Thus, D,L-*p*-hydroxyphenylhydantoin (30 g L^{-1}) was efficiently converted by this recombinant whole-cell catalyst to D-*p*-hydroxyphenylglycine – an important precursor in the synthesis of semisynthetic β-lactam antibiotics – within 15 h at 96 % yield (Park et al., 2000). A hydantoinase process using a genetically engineered *E. coli* expressing all three enzymes under the control of a rhamnose-inducible promotor was also recently described. With this system, L-tryptophan was produced with six-fold more efficiency compared to the wild-type strain *Arthrobacter aurescens*.

3.2.3
Lyases (EC 4)

3.2.3.1 Hydroxynitrile Lyases (EC 4.1.2.-)

Hydroxynitrile lyases (HNL, often also named oxynitrilases) catalyze the reversible stereoselective addition of hydrogen cyanide (HCN) to aldehydes and ketones (Fig. 3.56). Thus, HNLs allow the synthesis of optically pure compounds from prostereogenic substrates at (theoretically) quantitative yield. In addition, many HNLs show excellent stereoselectivity (for reviews, see Effenberger, 2000; Griengl et al., 2000; Effenberger et al., 2000; Schmidt and Griengl, 1999). The resulting α-hydroxynitriles are versatile intermediates for a broad variety of chiral synthons, which can be obtained by subsequent chemical synthesis (Fig. 3.57).

One of the earliest reports on biocatalysis involved the use of a HNL, when Rosenthaler used a hydroxynitrile lyase-containing extract (emulsin) for the preparation of (*R*)-mandelonitrile from benzaldehyde and HCN (Rosenthaler, 1908). However, little attention was paid to this discovery until the 1960s, when the corresponding enzyme was isolated, characterized and used for the production of enantiomerically enriched (*R*)-cyanohydrins.

More than 3000 plant species are known to release HCN from their tissues, and for approximately 300 plants the HCN is released from a cyanogenic glycoside or lipid. These cyanide donors can be cleaved either spontaneously or by the action of an enzyme such as hydroxynitrile lyase.

Until recently, all HNLs were isolated from different plant sources, and about 11 enzymes with either (*S*)- or (*R*)-stereoselectivity were described. The oxynitrilase

$$\mathrm{RCHO} + \mathrm{HCN} \xrightarrow{\mathrm{HNL}} \mathrm{R{-}CH(OH){-}CN}$$

Fig. 3.56 Example of the synthesis of optically active cyanohydrins from aldehydes and hydrogen cyanide catalyzed by a hydroxynitrile lyase.

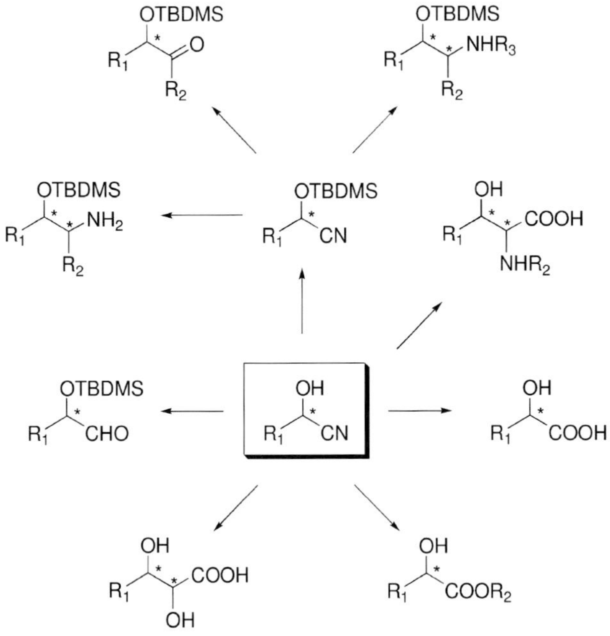

Fig. 3.57 Examples of building blocks, which can be obtained by chemical synthesis from chiral cyanohydrins. Products with two stereocenters are accessible by these routes (adapted from Schmidt and Griengl, 1999).

from *Prunus* sp. contains the cofactor FAD, but this is not involved in redox reactions. Instead, it seems to have a structure-stabilizing effect. Some of the HNLs are glycosylated and most of them are composed of subunits. The most thoroughly studied enzymes are listed in Table 3.6. As only small amounts of HNLs were available by extraction from plant sources, the enzymes from *Hevea brasiliensis* (rubber tree), *Manihot esculenta* (cassava) and *Linum usitatissimum* (flax) have been cloned and

Table 3.6 Hydroxynitrile lyases for organic synthesis.

Enzyme	*Specificity*	*Molecular weight [kDa]*	*pH-optimum*	*Substrate spectra*
Sorghum bicolor	*S*	105	n.d.	Aromatic aldehydes and ketones
Manihot esculenta	*S*	92–124	5.4	Broad
Hevea brasiliensis	*S*	58	5.5–6.0	Broad
Linum usitatissimum	*R*	82	5.5	Aliphatic aldehydes and ketones
Prunus sp.	R	55–80	5.5	broad

overexpressed in *E. coli* or *P. pastoris*. For instance, the gene encoding the enzyme from *Hevea brasiliensis* was expressed at high levels in the yeast *Pichia pastoris* under the control of the AOX1 promoter. On laboratory scale, a production of 23 g L^{-1} of pure HNL was achieved in a high-cell density fermentation, and the protein could be recovered by an one-step ion-exchange chromatography (Griengl et al., 2000).

The crystal structures have been elucidated for enzymes from *Hevea brasiliensis, Sorghum bicolor, Manihot esculenta* and *Prunus amygdalus* (almond). Interestingly, the enzymes share an α/β-hydrolase fold and also contain an active site serine which is usually embedded in a GXSXG-motif, similar to lipases and esterases. This might indicate an evolutionary relationship between these two enzyme classes, hydrolases and lyases, although recent studies have suggested that HNLs obey a different mechanism.

The following mechanism was proposed for the HNL from *Sorghum bicolor*. In contrast to a histidine residue serving as a general base in serine peptidases and lipases, the carboxylate group of a C-terminal tryptophan (Trp_{270} in *Sorghum bicolor*) abstracts a proton from the cyanohydrin hydroxyl group. A water molecule bound to the active site appears to be involved in proton transfer. Thus, the entering cyanohy-

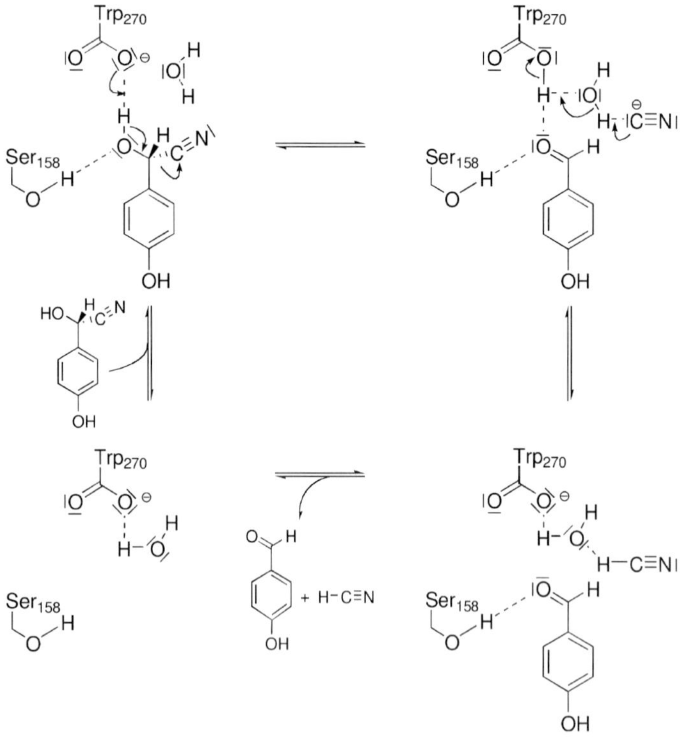

Fig. 3.58 Suggested reaction mechanism for hydroxynitrile lyase (HNL) from *Sorghum bicolor* (Lauble et al., 2002).

drin is hydrogen-bonded to Ser_{158} and Trp_{270}, which abstracts a proton from the OH-group of the substrate. Next, a proton is transferred from the tryptophan via the active site water to the nitrile leaving group. Protonation of the cyanide ion results in the products, 4-hydroxybenzaldehyde and HCN (Fig. 3.58). This model can also explain the (*S*)-stereoselectivity observed for the reverse reaction (Lauble et al., 2002).

Beside the availability of HNLs, two further problems had to be solved to allow an efficient application of HNLs in organic synthesis: (1) performing the reactions in water-immiscible organic solvents; and (2) reaction at low pH, typically in the range of pH 4–5. Both methods suppress the competing chemical reaction, which results in lower optical purities of the product. This is shown in Table 3.7 for the synthesis of several aliphatic α-hydroxynitriles using HNL from *Prunus amygdalus.*

In contrast to the enzymes from *H. brasiliensis* and *M. esculenta,* where aliphatic and aromatic aldehydes function as substrates, the HNL from *Sorghum bicolor* only catalyzes the formation and cleavage of aromatic (*S*)-cyanohydrins. The other enzymes listed in Table 3.6 show (*R*)-selectivity.

A broad range of aldehydes and ketones were converted into the corresponding α-hydroxynitriles using HNLs. Figure 3.59 shows some products obtained at high yields and with good to excellent optical purities using the enzymes listed in Table 3.6. As mentioned above, the use of water-immiscible solvents is recommended for organic synthesis with HNLs in order to avoid a non-stereoselective chemical reaction. As seen in Table 3.7, higher yields and excellent optical purities were only possible using PaHNL immobilized on Avicel (a cellulose membrane) in diisopropylether. In contrast, reactions in water/ethanol mixtures gave products with inferior optical purity, for example, only 11 % ee for the *m*-substituted phenyl derivative. This effect was less pronounced for reactions using ketones, however (Table 3.8).

An alternative to using the highly toxic HCN is to perform the reaction as a transcyanation (or transhydrocyanation), in which aromatic or aliphatic aldehydes are reacted with acetone cyanohydrin as shown for the (*R*)-HNL from almond (*Prunus*

Table 3.7 Preparation of (*R*)-cyanohydrins from aldehydes using HNL from *Prunus amygdalus* (PaHNL) in organic solvent or buffer (modified from Fessner, 2000).

R_1CHO + HCN → ((*R*)-PaHNL; *i*Pr_2O/Avicel or H_2O/EtOH) → $R_1CH(OH)CN$

R_1	H_2O/EtOH		*i*Pr_2O/Avicel	
	Yield [%]	*ee [%]*	*Yield [%]*	*ee [%]*
Ph	99	86	96	99
3-PhO-C_6H_4	99	11	99	98
C_3H_7	75	69	99	98
$(CH_3)_3C$	56	45	84	83

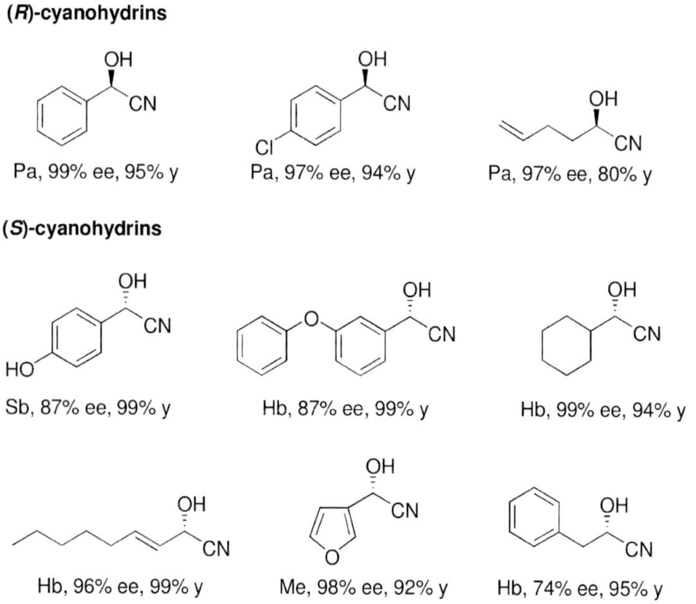

(*R*)-cyanohydrins

Pa, 99% ee, 95% y — Pa, 97% ee, 94% y — Pa, 97% ee, 80% y

(*S*)-cyanohydrins

Sb, 87% ee, 99% y — Hb, 87% ee, 99% y — Hb, 99% ee, 94% y

Hb, 96% ee, 99% y — Me, 98% ee, 92% y — Hb, 74% ee, 95% y

Fig. 3.59 Selected examples of the hydroxynitrile lyase (HNL)-catalyzed synthesis of chiral cyanohydrins. Pa, HNL from *Prunus amygdalus*; Sb, HNL from *Sorghum bicolor*; Hb, HNL from *Hevea brasiliensis*; Me, HNL from *Manihot esculenta* (Griengl et al., 2000).

Table 3.8 Preparation of (*R*)-cyanohydrins from ketones using HNL from *Prunus amygdalus* (PaHNL) in organic solvent or buffer (modified from Fessner, 2000).

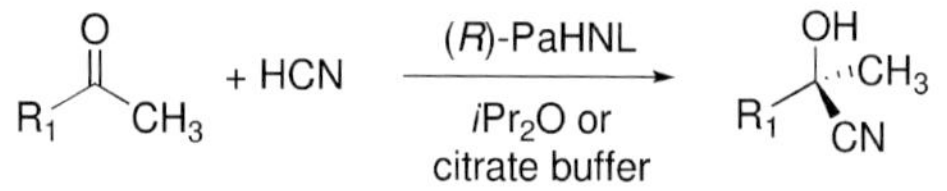

$R_1C(O)CH_3$ + HCN → ((*R*)-PaHNL, *i*Pr_2O or citrate buffer) → $R_1C(OH)(CH_3)CN$

R_1	*iPr₂O/Avicel*		*Citrate buffer*	
	Yield [%]	*ee [%]*	*Yield [%]*	*ee [%]*
C_3H_7	70	97	78	95
C_4H_9	90	98	94	98
C_5H_{11}	88	57	56	96
$(CH_3)_2CHCH_2$	57	98	40	98

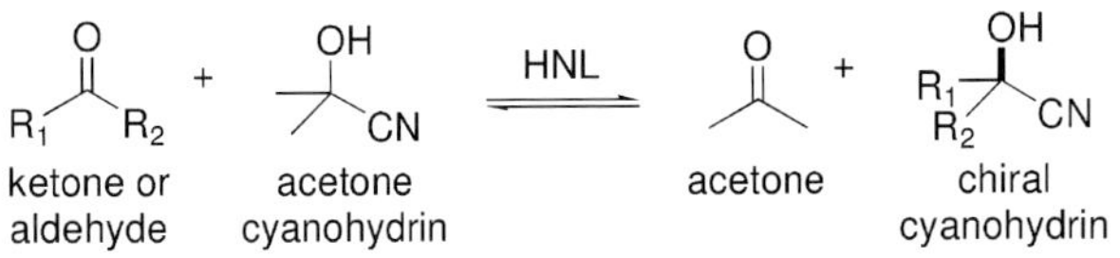

ketone or aldehyde + acetone cyanohydrin ⇌(HNL) acetone + chiral cyanohydrin

Fig. 3.60 The principle of transhydrocyanation.

amygdalus) (Fig. 3.60). However, the optical purity of the resulting products was slightly lower compared to reactions involving free HCN.

Very recently, processes for the production of (*S*)-cyanohydrins catalyzed by HNLs from *Hevea brasiliensis* and *Manihot esculenta* and the subsequent chemical hydrolysis to (*S*)-hydroxycarboxylic acids have been developed. The synthesis of (*S*)-*m*-phenoxybenzaldehyde cyanohydrin – an intermediate in the synthesis of pyrethroids – in a biphasic system has been commercialized (Fig. 3.61) by DSM Chemie Linz and Nippon Shokubai.

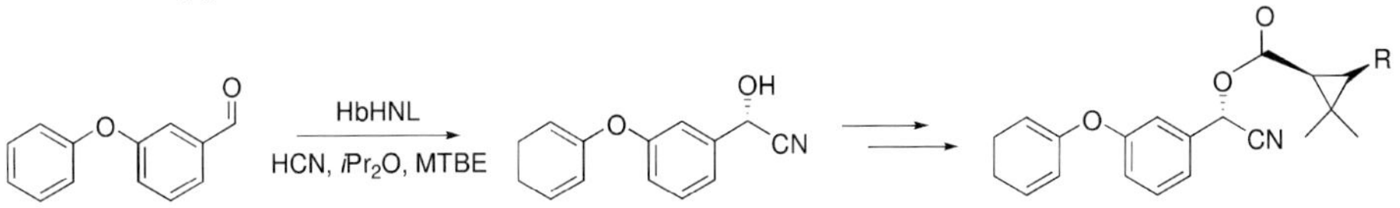

HbHNL

HCN, *i*Pr$_2$O, MTBE

Fig. 3.61 Synthesis of (*S*)-*m*-phenoxybenzaldehyde cyanohydrin as an intermediate of pyrethroids using a recombinant HNL from *Hevea brasiliensis*.

3.2.3.2 **Aldolases (EC 4.1.2.-; 4.1.3.-)**

Aldolases catalyze the biological equivalent to the chemical Aldol-reaction, the formation of carbon–carbon bonds by (reversible) stereocontrolled addition of a nucleophilic ketone to an electrophilic aldehyde acceptor. Aldolases are usually classified according to the nature of the nucleophilic component into: (1) pyruvate (and phosphoenolpyruvate)-; (2) dihydroxyacetone phosphate-; (3) acetaldehyde-; and (4) glycine-dependent enzymes. The most important ones in organic synthesis use dihydroxyacetone phosphate (DHAP) as they allow the formation of two new stereocenters in a single reaction. A range of corresponding enzymes was also identified. The ability of aldolases to accept a variety of unnatural acceptor substrates and to generate new stereocenters of known absolute and relative stereochemistry makes them powerful tools for asymmetric synthesis.

Depending on their mechanism, aldolases are classified into Type I and Type II enzymes. Type I enzymes are predominantly found in higher plants and animals and are metal-cofactor independent. The free amino group of a lysine residue in the active site reacts with DHAP under the formation of a Schiff's base intermediate. An enamine is formed after deprotonation, which then attacks the aldehyde. Finally, the Schiff's base intermediate product decomposes after reaction with water and the aldol and enzyme are released (Fig. 3.62). Type II aldolases occur mostly in bacteria and fungi, and are Zn^{2+}-dependent. The zinc ion acts as a Lewis acid, which polarizes the carbonyl group in DHAP (Fig. 3.62).

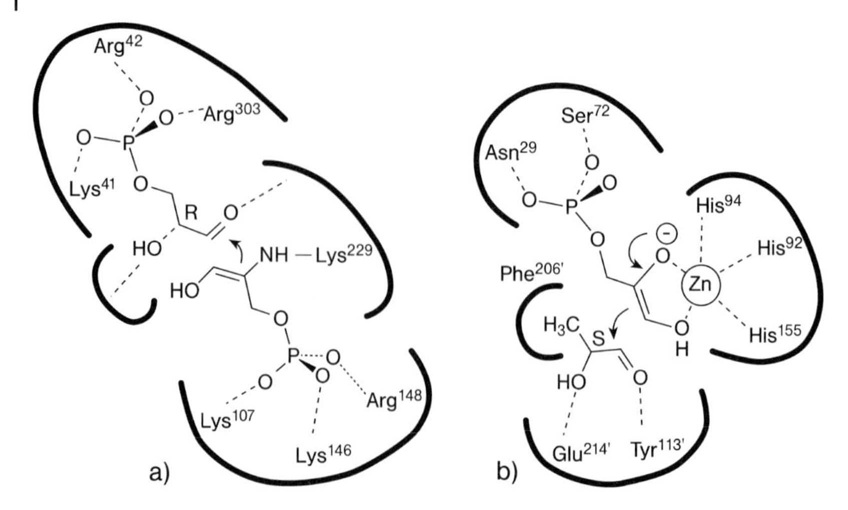

Fig. 3.62 Mechanism of Type I and Type II aldolases. The example shows dihydroxyacetone phosphate (DHAP) as nucleophilic ketone.

As pointed out above, aldolases are highly specific and the stereochemical outcome of an aldol reaction can be usually predicted independently of the substrate structure. Consequently, the synthesis of all four diastereomers accessible from DHAP and an aldehyde is possible by using four different aldolases (Fig. 3.63).

Numerous examples for the application of aldolases can be found in a number of excellent reviews and book chapters (Fessner, 2000; Fessner and Helaine, 2001; Wong, 2002). Only two types of aldolases are described in the following paragraphs, although it should be noted that catalytic antibodies with aldolase activity have also been described (Gildersleeve et al., 2003).

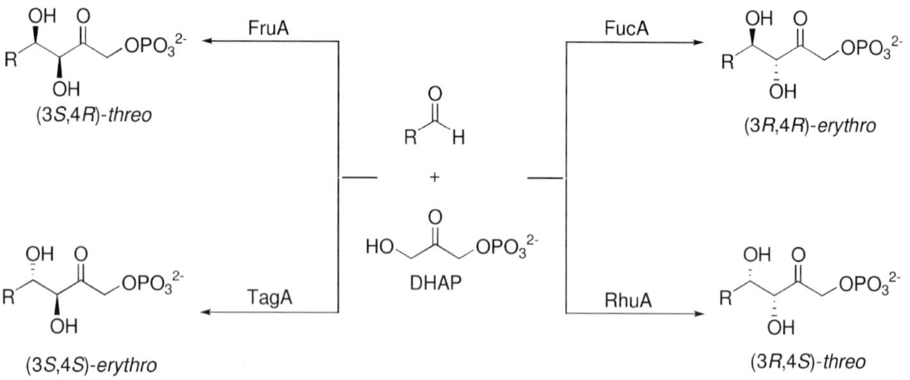

FruA — (3S,4R)-*threo*
FucA — (3R,4R)-*erythro*
TagA — (3S,4S)-*erythro*
RhuA — (3R,4S)-*threo*
RCHO + DHAP

Fig. 3.63 Aldolase reactions catalyzed by the four stereocomplementary aldolases FruA (fructose-1,6-diphosphate aldolase), FucA (Fuculose-1-phosphate aldolase), TagA (Tagatose-1,6-diphosphate aldolase) and RhuA (Rhamnulose-1-phosphate aldolase).

DHAP-dependent Aldolases

The most often-used DHAP-dependent enzymes are Fru-aldolases (FruA, often also abbreviated to FDP aldolases; EC 4.1.2.13), which catalyze the reaction between DHAP and D-glyceraldehyde-3-phosphate to form D-fructose-diphosphate (FDP). The equilibrium constant for this reaction is ~10^4 M^{-1} in favor of FDP formation. The enzyme has been isolated from various eucaryotic and procaryotic sources, and both type I and type II biocatalysts have been described. Type I enzymes are usually tetramers of 160 kDa molecular weight, while type II FDP-aldolases are dimers (~80 kDa). Sequence homologies are usually very low between type I and II aldolases, and especially for in-between type II enzymes. The crystal structures of the FDP-aldolase from rabbit muscle (RAMA) and others have been determined. RAMA and several type II aldolases from microbial sources have been cloned and overexpressed, which substantially facilitates access to the enzymes. Aldolases have been widely employed in carbohydrate synthesis, and a few examples of products are shown in Figure 3.64.

The phosphorylated ketone DHAP must be available for these aldolase-catalyzed reactions. Although DHAP is, in principle, available via the reverse reaction – that is, from a retro-aldol reaction with FDP aldolase from FDP – this approach requires triosephosphate isomerase and can hardly be used, if a different aldolase is used in the forward aldol reaction, which then leads to a complex product mixture. The kinase-catalyzed phosphorylation of dihydroxyacetone is hampered by the requirement for ATP regeneration. A chemical alternative is a multi-step synthesis, which usually provides a DHAP dimer that is stable and can easily be converted into DHAP by acid hydrolysis. Alternatively, arsenate derivatives of DHAP can be used, albeit at lowered aldolase activity.

Recently, an efficient approach based on phosphorylation of glycerol using a phytase – a cheap and readily available enzyme – and inorganic pyrophosphate was described. The feasibility of this reaction was demonstrated for the synthesis of 5-deoxy-5-ethyl-D-xylulose in a one-pot reaction combining four enzymatic steps with three different enzymes starting from glycerol (Fig. 3.65). First, phosphorylation of glycerol by reaction with pyrophosphate in the presence of phytase at pH 4.0 in 95 % glycerol afforded racemic glycerol-3-phosphate quantitatively. The L-enantiomer is then oxidized with glycerol phosphate oxidase (GPO) to DHAP under aerobic conditions at pH 7.5. Hydrogen peroxide is removed with the aid of catalase. In-situ-gen-

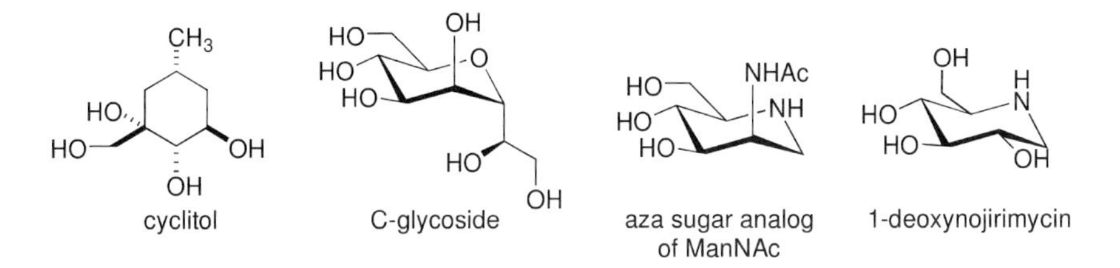

Fig. 3.64 Selected examples of molecules synthesized with FDP aldolase (ManNAc, *N*-acetyl-D-mannosamine).

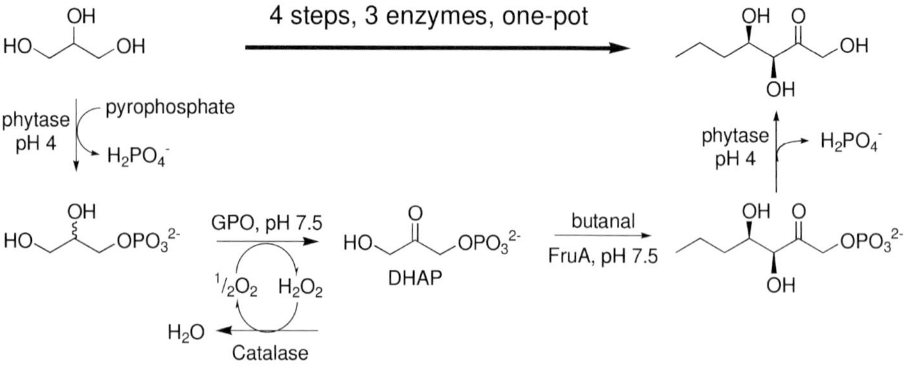

Fig. 3.65 Enzymatic total synthesis of 5-deoxy-5-ethyl-xylulose with in-situ formation of dihydroxyacetone phosphate (DHAP). The key to success was the switch in pH from 4.0 (phytase reaction) to 7.5 (glycerolphosphate oxidase (GPO) and aldolase (FruA) reaction).

erated DHAP reacts with butanal using FruA, followed by dephosphorylation of the aldol adduct, again using phytase at pH 4. Overall, 5-deoxy-5-ethyl-D-xylulose was obtained in 57 % yield from L-glycerol-3-phosphate. The phytase "on/off-switch", by changing the pH value, was the key to controlling phosphorylation and dephosphorylation (Schoevaart et al., 2000).

Pyruvate/phosphoenolpyruvate-dependent Aldolases

The best-studied enzyme utilizing pyruvate is *N*-acetylneuraminate (NeuAc) aldolase (EC 4.1.3.3). This catalyzes the reversible condensation of pyruvate with *N*-acetylmannosamine (ManNAc) to form sialic acid. The initial products of aldol cleavage are α-ManNAc and pyruvate. Although *in vivo* the equilibrium favors the retro-aldol reaction (K_{eq} ~12.7 M^{-1}), the aldol reaction for organic synthesis can be achieved

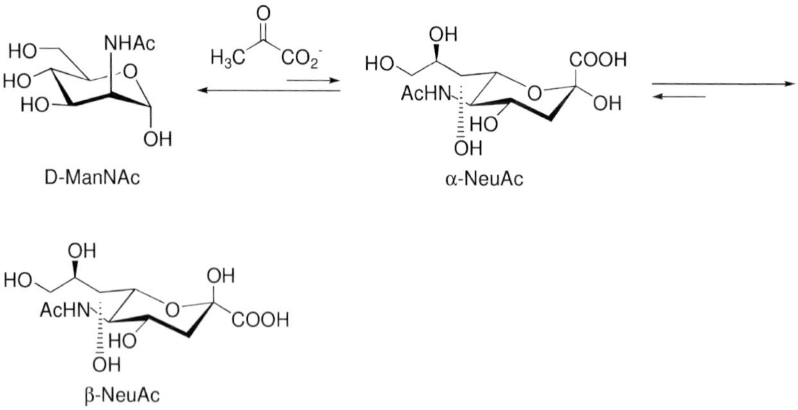

Fig. 3.66 In-vivo aldol reaction catalyzed by NeuAc-aldolase.

using excess pyruvate, which yields the β-anomer of NeuAc (Fig. 3.66). NeuAc aldolase is a Schiff's base type I aldolase, and has been isolated from both bacteria and animals. The enzymes from *Clostridium* and *E. coli* are commercially available, and the *E. coli* enzyme has been cloned and overexpressed. The optimum pH is 7.5, but activity is retained at between pH 6 and 9. NeuAc aldolase can be used in solution, in immobilized form, or enclosed in a membrane system. With this aldolase, glycoconjugates – which play important roles in cell–cell interactions and cell adhesion – have been synthesized on the multigram scale.

3.2.4
Isomerases (EC 5)

The most important enzymes for organic synthesis in this class are racemases and epimerases (both EC 5.1.-.-), both of which catalyze the inversion of stereocenters. Racemases convert an enantiomer in a racemate, while epimerases convert one diastereomer selectively to another diastereomer. Other well-studied enzymes in this class are glucose-fructose isomerase (EC 5.3), which is used in the synthesis of high-fructose corn syrup (see Sections 6.4.1 and 9.4.2) and *cis-trans* isomerases.

Although racemases "destroy" the optical purity of a given chiral compound, they are very important for biocatalysis. If in a kinetic resolution one enantiomer is not needed, it can be racemized and again subjected to a resolution which eventually yields to 100 % product after several iterative cycles (recycling), or directly in a dynamic kinetic resolution. This has already been exemplified for the hydantoinase process in Section 3.2.2.6. In addition, racemases have also been used to increase access to the D-enantiomers of amino acids.

Most racemases described in the literature act on α-amino carboxylic acids, with the prominent exception of mandelate racemase (Kenyon et al., 1995; St. Maurice and Bearne, 2004). The majority of racemases and epimerases act at a stereocenter adjacent to a carbonyl functionality, and reversibly cleave a C–H bond by lowering the pK_a of the hydrogen through stabilization of the resulting anion. As this anionic intermediate is planar and reprotonation can occur from both sides at identical probability, a racemic (or epimeric) product is formed.

Racemases are classified into two different types, depending on whether they require the cofactor pyridoxal-5′-phosphate (PLP), or not. The reaction mechanism for PLP-dependent enzymes is based on the formation of an imine linkage between the PLP-cofactor and the substrate. This greatly acidifies the α-proton, resulting in a lower pK_a and consequently an easier abstraction of the proton. This proton is then returned to the imine, the result being a 1:1 mixture of both enantiomers (Fig. 3.67).

For PLP-independent racemases (e.g., glutamate racemase) a two-base mechanism (Fig. 3.68) was suggested based on mechanistical studies with deuterated substrates. Mutational studies on glutamate racemase from *Lactobacillus fermenti* identified two cysteine residues (Cys_{73} and Cys_{184}) as catalytic acid/base residues.

Alanine racemase (EC 5.1.1.1) is a bacterial enzyme that catalyzes the racemization of alanine and requires PLP. It plays an important role in the bacterial growth by providing D-alanine used in peptidoglycan assembly. Several enzymes have been

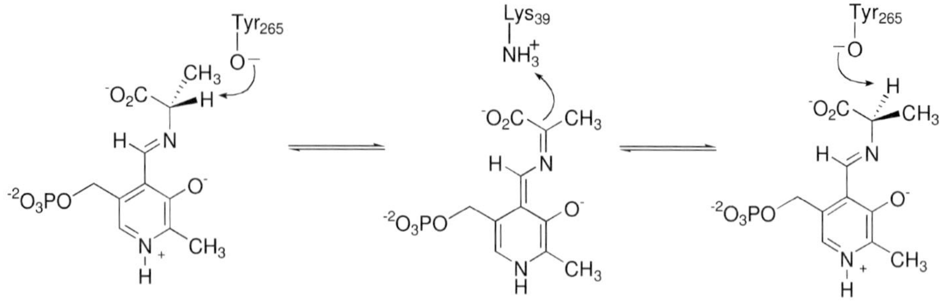

Fig. 3.67 Mechanism of the pyridoxal-5'-phosphate (PLP)-dependent alanine racemase.

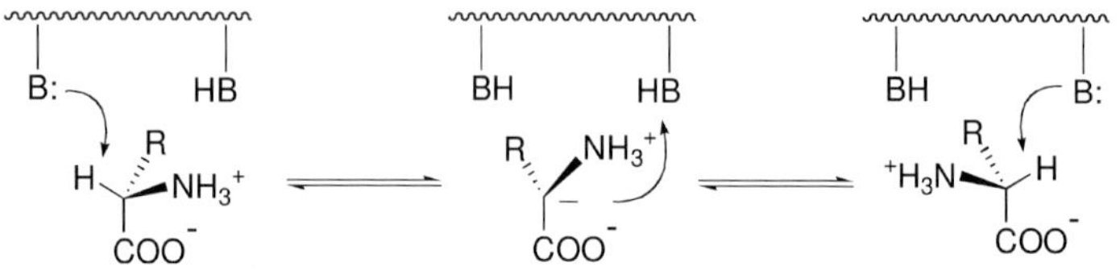

Fig. 3.68 Mechanism of a pyridoxal-5'-phosphate (PLP)-independent racemase, i.e., glutamate racemase.

identified from different bacterial sources, and these have also been cloned. The enzyme has been used in biocatalysis for the production of various D-amino acids by combination of L-alanine-dehydrogenase, D-amino acid aminotransferase and formate dehydrogenase (Fig. 3.69). The simultaneous expression of all four genes has also been described.

Another PLP-dependent racemase used in biocatalysis is an α-amino-ε-caprolactam racemase (Fig. 3.70), which allows the synthesis of L-lysine. However, this process is not competitive for the fermentative production of this amino acid.

Glutamate racemases have been identified in various microorganisms, and the genes from *Lactobacillus* sp. and *Bacillus* sp. have been cloned and overexpressed. The structure of the enzyme from *Aquifex pyrophilus* has been determined using

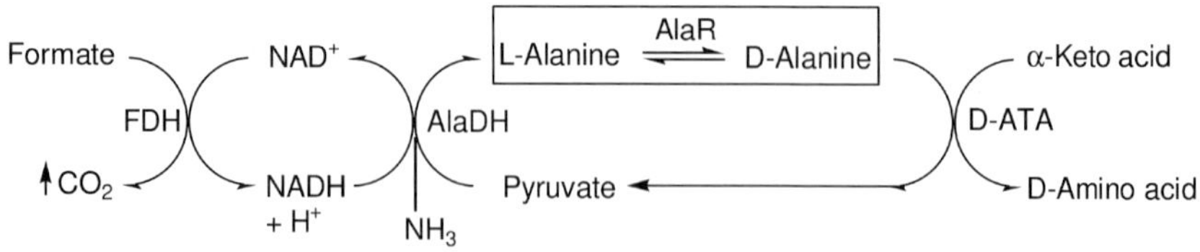

Fig. 3.69 Synthesis of a D-amino acid from a α-keto acid, formate, alanine and ammonia using a set of four enzymes (FDH, formiate dehydrogenase; AlaDH, alanine dehydrogenase; AlaR, alanine racemase; D-ATA, D-amino acid transferase). The reaction catalyzed by the racemase is boxed.

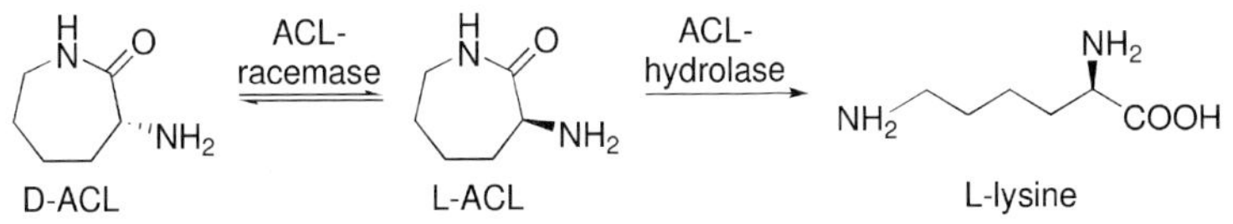

D-ACL ⇌ (ACL-racemase) L-ACL → (ACL-hydrolase) L-lysine

Fig. 3.70 Enzymatic synthesis of L-lysine from racemic D,L-α-amino-ε-caprolactam (ACL) using an ACL-racemase and a hydrolase.

X-ray crystallography, and revealed that the enzyme is composed of two α/β-fold domains. The application in the synthesis of L-glutamate has been described similarly to the examples shown in Figure 3.69, and allowed the production of D-valine, D-alanine, D-aspartate, and other D-amino acids at high yields.

N-Acylamino Acid Racemases

N-Acylamino acids racemize more easily than the corresponding amino acids. Therefore, the combination of chemical racemization and enantioselective hydrolysis of *N*-acylamino acids by aminoacylase (EC 3.5.1.14) has been used for the synthesis of L-tryptophan. Unfortunately, the reaction conditions for the chemical and the enzymatic step are very different, and therefore the resolution step and the racemization must be performed separately. Researchers at Takeda Chemical Industries (Japan) succeeded in identifying a racemase from *Streptomyces atratus* Y-53 that acts only on the *N*-acylamino acid but not on the free amino acid; this facilitates a dynamic kinetic resolution starting from D,L-*N*-acylamino acids. This racemase acts on *N*-acetyl D,L-methionine, -L-valine, -L-tyrosine and *N*-chloroacetyl L-valine, but not on methyl or ethyl esters of *N*-acyl-amino acids.

3.3 Exercises

1. Explain the concept of a dynamic kinetic resolution.
2. Discuss the pros and cons of a process for the synthesis of an optically pure alcohol using either a dynamic kinetic resolution with a hydrolase as key enzyme, or an asymmetric synthesis using the reduction of a ketone by an alcohol dehydrogenase.
3. Discuss the pros and cons of a P450-catalyzed hydroxylation in a whole-cell system or with isolated enzymes. What determines which process is the method of choice?
4. Which enzymes can be used for the synthesis of optically pure α-amino acids?
5. It is known that acetaldehyde released in the lipase-catalyzed resolution using vinyl esters as acyl donors can inactivate the biocatalyst. How can you circumvent this problem?
6. How would you screen a culture collection for the presence of racemase activity?

Literature

Textbooks

These provide general information relating to the use of enzymes in organic synthesis.

Bommarius, A.S., Riebel, B.R., *Biocatalysis*, Wiley-VCH, Weinheim, **2004**

Bornscheuer, U.T., Kazlauskas, R.J., *Hydrolases in Organic Synthesis – Regio- and Stereoselective Biotransformations*, Wiley-VCH, Weinheim, **1999**

Drauz, K., Waldmann, H., (Eds), *Enzyme Catalysis in Organic Synthesis*, 2nd edition, Vols. 1–3, Wiley-VCH, Weinheim, **2002**

Faber, K., *Biotransformations in Organic Chemistry*, 5th edition, Springer, Berlin, **2004**

Patel, R.N., (Ed.), *Stereoselective Biocatalysis*, Marcel Dekker, New York, **2000**

Roberts, S.M., *Biocatalysts for Fine Chemical Synthesis*, Wiley-VCH, Weinheim, **1999**

References

Adger, B., Bes, M.T., Grogan, G., McCague, R., Pedragosa-Moreau, S., Roberts, S.M., Villa, R., Wan, P.W.H., Willetts, A.J., Application of enzymic Baeyer-Villiger oxidations of 2-substituted cycloalkanones to the total synthesis of (R)-(+)-lipoic acid, *J. Chem. Soc., Chem. Commun.* **1995**, 1563–1564

Altenbuchner, J., Siemann-Herzberg, M., Syldatk, C., Hydantoinases and related enzymes as biocatalysts for the synthesis of unnatural chiral amino acids, *Curr. Opin. Biotechnol.*, **2001**, *12*, 559–563.

Appel, D., Lutz-Wahl, S., Fischer, P., Schwaneberg, U., Schmid, R.D., A P450 BM-3 mutant hydroxylates alkanes, cycloalkanes, arenes and heteroarenes, *J. Biotechnol.*, **2001**, *88*, 167–171

Archelas, A., Furstoss, R., Synthetic applications of epoxide hydrolases, *Curr. Opin. Chem. Biol.*, **2001**, *5*, 112–119

Azzolina, O., Vercesi, D., Collina, S., Ghislandi, V., Chiral resolution of methyl 2-aryloxypropionates by biocatalytic stereospecific hydrolysis, *Il Farmaco*, **1995**, *50*, 221–226

Baeyer, A., Villiger, V., Einwirkung des Caro'schen Reagens auf Ketone, *Chem. Ber.*, **1899**, *32*, 3625–3633

Baillargeon, M.W., McCarthy, S.G., *Geotrichum candidum* NRRL Y-553 lipase: purification, characterization and fatty acid specificity, *Lipids* **1991**, *26*, 831–836

Balkenhohl, F., Ditrich, K., Hauer, B., Ladner, W., Optically active amines via lipase-catalyzed methoxyacetylation, *J. Prakt. Chem.*, **1997**, *339*, 381–384

Bdjìc, J.D., Kadnikova, E.N., Kostic, N.M., Enantioselective aminolysis of an α-chloroester catalyzed by *Candida cylindracea* lipase encapsulated in sol-gel silica glass, *Org. Lett.*, **2001**, *3*, 2025–2028

Bolm, C., Catalyzed Baeyer-Villiger reactions, *Adv. Catal. Processes*, **1997**, *2*, 43–68

Bolm, C., Schlingloff, G., Weickhardt, K., Optically active lactones from a Baeyer-Villiger-type metal-catalyzed oxidation with molecular oxygen, *Angew. Chem. Int. Ed. Engl.*, **1994**, *33*, 1848–1849

Bommarius, A.S., Reduction of C=N bonds, in: Drauz, K., Waldmann, H., (Eds), *Enzyme catalysis in organic synthesis*, Vol. 3, pp. 1047–1063. VCH, Weinheim, **2002**

Bommarius, A.S., Drauz, K., Groeger, U., Wandrey, C., Membrane bioreactors for the production of enantiomerically pure a-amino acids, in: Collins, A.N., Sheldrake, G.N., Crosby, J., (Eds), *Chirality in Industry*, pp. 371–397. Wiley, London, **1992**

Bordusa, F., Proteases in organic synthesis, *Chem. Rev.*, **2002**, *102*, 4817–4867

Breuer, M., Ditrich, K., Habicher, T., Hauer, B., Keßeler, M., Stürmer, R., Zelinski, T., Industrial methods for the production of optically active intermediates, *Angew. Chem. Int. Ed.*, **2004**, *43*, 788–824

Britton, L.N., Brand, J.M., Markovetz, A.J., Source of oxygen in the conversion of 2-tridecanone to undecyl acetate by *Pseudomonas cepacia* and *Nocardia* species, *Biochim. Biophys. Acta*, **1974**, *369*, 45–49

Burton, S.G., Oxidizing enzymes as biocatalysts, *Trends Biotechnol.*, **2003**, *21*, 543–549

Chen, C.S., Fujimoto, Y., Girdaukas, G., Sih, C.J., Quantitative analyses of biochemical kinetic resolutions of enantiomers, *J. Am. Chem. Soc.*, **1982**, *104*, 7294–7299

Chen, C.S., Wu, S.H., Girdaukas, G., Sih, C.J., Quantitative analyses of biochemical kinetic resolution of enantiomers. 2. Enzyme-catalyzed esterifications in water-organic solvent biphasic systems, *J. Am. Chem. Soc.*, **1987**, *109*, 2812–2817

Cirino, P.C., Arnold, F.H., Protein engineering of oxygenases for biocatalysis, *Curr. Opin. Chem. Biol.*, **2002**, *6*, 130–135

Criegee, R., The rearrangement of decahydronaphthalene peroxide esters resulting from cationic oxygen. *Justus Liebigs Ann. Chem.*, **1948**, *560*, 127–135

Cripps, R.E., The Microbial Metabolism of Acetophenone, *Biochem. J.*, **1975**, *152*, 233–241

Cripps, R.E., Trudgill, P.W., Whateley, J.G., The metabolism of 1-phenylethanol and acetophenone by *Nocardia* T5 and an *Arthrobacter* species, *Eur. J. Biochem.*, **1978**, *86*, 175–186

Crosby, J.A., Parratt, J.S., Turner, N.J., Enzymic hydrolysis of prochiral dinitriles, *Tetrahedron: Asymmetry*, **1992**, *3*, 1547–1550

Degtyarenko, K.N., Structural domains of P450-containing monooxygenase systems, *Prot. Eng.*, **1995**, *8*, 737–747

DeSantis, G., Wong, K., Farwell, B., Chatman, K., Zhu, Z., Tomlinson, G., Huang, H., Tan, X., Bibbs, L., Chen, P., Kretz, K., Burk, M.J., Creation of a productive, highly enantioselective nitrilase through gene site saturation mutagenesis (GSSM), *J. Am. Chem. Soc.*, **2003**, *125*, 11476–11477

DeSantis, G., Zhu, Z., Greenberg, W.A., Wong, K., Chaplin, J., Hanson, S.R., Farwell, B., Nicholson, L.W., Rand, C.L., Weiner, D.P., Robertson, D.E., Burk, M.J., An Enzyme library approach to biocatalysis: Development of nitrilases for enantioselective production of carboxylic acid derivatives, *J. Am. Chem. Soc.*, **2002**, *124*, 9024–9025

Donoghue, N.A., Norris, D.B., Trudgill, P.W., The purification and properties of cyclohexanone oxygenase from *Nocardia globerula* CL1 and *Acinetobacter* NCIB 9871, *Eur. J. Biochem.*, **1976**, *63*, 175–192

Effenberger, F., Hydroxynitrile lyases in stereoselective synthesis, in: Patel, R.N., (Ed.), *Stereoselective Biocatalysis*, pp. 321–342. Marcel Dekker, New York, **2000**

Effenberger, F., Förster, S., Wajant, H., Hydroxynitrile lyases in stereoselective catalysis, *Curr. Opin. Biotechnol.*, **2000**, *11*, 532–539

El-Gihani, M.T., Williams, J.M.J., Dynamic kinetic resolutions, *Curr. Opin. Biotechnol.*, **1999**, *3*, 11–15

Faber, K., Mischitz, M., Kroutil, W., Microbial epoxide hydrolases, *Acta Chem. Scand.*, **1996**, *50*, 249–258

Farinas, E.T., Schwaneberg, U., Glieder, A., Arnold, F.H., Directed evolution of a cytochrome P450 monooxygenase for alkane oxidation, *Adv. Synth. Catal.*, **2001**, *343*, 601–606

Fessner, W.-D., Enzymatic synthesis using aldolases, in: Patel, R.N., (Ed.), *Stereoselective Biocatalysis*, pp. 239–265. Marcel Dekker, New York, **2000**

Fessner, W.-D., Helaine, V., Biocatalytic synthesis of hydroxylated natural products using aldolases and related enzymes, *Curr. Opin. Biotechnol.*, **2001**, *12*, 574–586

Fischer, A., Bommarius, A.S., Drauz, K., Wandrey, C., A novel approach to enzymic peptide synthesis using highly solubilizing N^{α}-protecting groups of amino acids, *Biocatalysis*, **1994**, *8*, 289–307

Forney, F.W., Markovetz, A.J., Kallio, R.E., *J. Bacteriol.*, **1967**, *36*, 155–158

Fried, J., Thoma, R.W., Klingsberg, A., Oxidation of steroids by microorganims. III. Side chain degradation, ring D-cleavage and dehydrogenation in ring A, *J. Am. Chem. Soc.* **1953**, *75*, 5764–5765

Frisone, M.d.T., Pinna, F., Strukul, G., Baeyer-Villiger oxidation of cyclic ketones with hydrogen peroxide catalyzed by cationic complexes of Platinum(II): Selectivity properties and mechanistic studies, *Organometallics* **1993**, *12*, 148–156

Gagnon, R., Grogan, G., Levitt, M.S., Robets, S.M., Wan, P.W.H., Willetts, A.J., Biological Baeyer-Villiger oxidation of some monocyclic and bicyclic ketones using monooxygenases from *Acinetobacter calcoaceticus* NCIMB 9871 and *Pseudomonas putida* NCIMB 10007, *J. Chem. Soc., Perkin Trans. 1*, **1994**, 2537–2543

Gildersleeve, J., Varvak, A., Atwell, S., Evans, D., Schultz, P.G., Development of a high-throughput screen for protein catalysts: Application to the directed evolution of antibody aldolases, *Angew. Chem. Int. Ed.*, **2003**, *42*, 5971–5973

Glieder, A., Farinas, E.T., Arnold, F.H., Laboratory evolution of a soluble, self-sufficient, highly active alkane hydroxylase, *Nat. Biotechnol.*, **2002**, *20*, 1135–1139

Goldstein, J.A., Faletto, M.B., Advances in mechanisms of activation and deactivation of environmental chemicals., *Environ Health Perspect.*, **1993**, *100*, 169–176

Griengl, H., Schwab, H., Fechter, M., The synthesis of chiral cyanohydrins by oxynitrilases, *Trends Biotechnol.*, **2000**, *18*, 252–256

Grogan, G., Roberts, S., Wan, P., Willetts, A.J., Camphor-grown *Pseudomonas putida*, a multifunctional biocatalyst for undertaking Baeyer-Villiger monooxygenase-dependent biotransformations, *Biotechnol. Lett.*, **1993**, *15*, 913–918

Grogan, G., Roberts, S., Willetts, A., Biotransformations by microbial Baeyer-Villiger monooxygenases stereoselective lactone formation *in vitro* by coupled enzyme systems, *Biotechnol. Lett.*, **1992**, *14*, 1125–1130

Hartmans, S., deBont, J.A.M., Acetol monooxygenase from *Mycobacterium* Py1 cleaves acetol into acetate and formaldehyde, *FEMS Microbiol. Lett.*, **1986**, *36*, 155–158

Henke, E., Bornscheuer, U.T., Schmid, R.D., Pleiss, J., A molecular mechanism of enantiorecognition of tertiary alcohols by carboxylesterases, *ChemBioChem.*, **2003**, *4*, 485–493

Henke, E., Pleiss, J., Bornscheuer, U.T., activity of lipases and esterases towards tertiary alcohols: insights into structure-function relationships, *Angew. Chem. Int. Ed.*, **2002**, *41*, 3211–3213

Hosoya, N., Hatayama, A., Irie, R., Sasaki, H., Katsuki, T., Rational design of Mn-Salen epoxidation catalysts: Preliminary results, *Tetrahedron*, **1994**, *50*, 4311–4322

Hummel, W., Schütte, H., Schmidt, E., Wandrey, C., Kula, M.-R., Isolation of L-phenylalanine dehydrogenase from *Rhodococcus* sp. M4 and its application for the production of L-phenylalanine, *Appl. Microbiol. Biotechnol.*, **1987**, *26*, 409–416

Jones, J.B., Esterases in organic synthesis: present and future, *Pure Appl. Chem.*, **1990**, *62*, 1445–1448

Jones, J.B., Hinks, R.S., Hultin, P.G., Enzymes in organic synthesis. 33. Stereoselective pig liver esterase-catalyzed hydrolyses of meso cyclopentyl-, tetrahydrofuranyl-, and tetrahydrothiophenyl-1,3-diesters, *Can. J. Chem.*, **1985**, *63*, 452–456

Jones, K.H., Roy, T.S., Trudgill, P.W., Diketocamphane enantiomer-specific 'Baeyer-Villiger' monooxygenases from camphor-grown *Pseudomonas putida* ATCC 17453, *J. Gen. Microbiol.*, **1993**, *139*, 797–805

Joo, H., Lin, Z., Arnold, F.H., Laboratory evolution of peroxide-mediated cytochrome P450 hydroxylation, *Nature*, **1999**, *399*, 670–673

Kamerbeek, N.M., Mooen, M.J.H., van der Ven, J.G.M., van Berkel, W.J.H., Fraaije, M.W., Janssen, D.B., 4-Hydroxyacetophenone monooxygenase from *Pseudomonas fluorescens* ACB, *Eur. J. Biochem.*, **2001**, *268*, 2547–2557

Kasche, V., Proteases in peptide synthesis, in: Beynon, M., Bond., J., (Eds), *Proteolytic enzymes. A practical approach*, pp. 265–292.: Oxford University Press, Oxford, **2001**

Katsuki, T., Martin, V.S., Asymmetric epoxidation of allylic alcohols: The Katsuki-Sharpless epoxidation reaction, *Org. React. (N.Y.)*, **1996**, *48*, 1–299

Kayser, M., Chen, G., Stewart, J., Designer Yeast: an enantioselective oxidizing reagent for organic synthesis, *Synlett.*, **1999**, *1*, 153–158

Kazlauskas, R.J., Bornscheuer, U.T., Biotransformations with Lipases, in: Rehm, H.J., Reed, G., Pühler, A., Stadler, P.J.W., Kelly, D.R., (Eds), *Biotechnology*, Vol. 8a, pp. 37–191. Wiley-VCH, Weinheim, **1998**

Kazlauskas, R.J., Weissfloch, A.N.E., Rappaport, A.T., Cuccia, L.A., A rule to predict which enantiomer of a secondary alcohol reacts faster in reactions catalyzed by cholesterol esterase, lipase from *Pseudomonas cepacia*, and lipase from *Candida rugosa*, *J. Org. Chem.*, **1991**, *56*, 2656–2665

Kenyon, G.L., Gerlt, J.A., Petsko, G.A., Kozarich, J.W., Mandelate racemase: structure–function studies of a pseudosymmetric enzyme, *Acc. Chem. Res.*, **1995**, *28*, 178–186

Kim, J.-M., Ahn, Y., Park, J., Dynamic kinetic resolutions and asymmetric transformations by enzymes coupled with metal catalysis, *Curr. Opin. Biotechnol.*, **2002**, *13*, 578–587

Kobayashi, M., Shimizu, S., Versatile nitrilases: Nitrile-hydrolysing enzymes, *FEMS Microbiol. Lett.*, **1994**, *120*, 217–224

Kobayashi, M., Shimizu, S., Metalloenzyme nitrile hydratase: structure, regulation, and application to biotechnology, *Nature Biotechnol.*, **1998**, *16*, 733–736

Königsberger, K., Griengl, H., Microbial Baeyer-Villiger reaction of bicyclo[3.2.0]heptan-6-ones – A novel approach to Sarkomycin A, *Bioorg. Med. Chem.*, **1994**, *2*, 595–564

Kragl, U., Gödde, A., Wandrey, C., Kinzy, W., Cappon, J.J., Lugtenburg, J., Repetitive batch as an efficient method for preparative scale enzymic synthesis of 5-azido-neuraminic acid and ^{15}N-L-glutamic acid, *Tetrahedron: Asymmetry*, **1993**, *4*, 1193–1202

Krishna, S.H., Persson, M., Bornscheuer, U.T., Enantioselective transesterification of a tertiary alcohol by lipase A from *Candida antarctica*, *Tetrahydron: Asymmetry*, **2002**, *13*, 2693–2696

Kroutil, W., Genzel, Y., Pietzsch, M., Syldatk, C., Faber, K., Purification and characterization of a highly selective epoxide hydrolase from *Nocardia* sp. EH1, *J. Biotechnol.*, **1998a**, *61*, 143–150

Kroutil, W., Mischitz, M., Plachota, P., Faber, K., Deracemization of (±)-*cis*-2,3-epoxyheptane *via* enantioconvergent biocatalytic hydrolysis using *Nocardia* EH1-epoxide hydrolase, *Tetrahedron Lett.*, **1996**, *37*, 8379–8382

Kroutil, W., Orru, R.V.A., Faber, K., Stabilization of *Nocardia* EH1 epoxide hydrolase by immobilization, *Biotechnol. Lett.*, **1998b**, *20*, 373–377

Lam, L.K.P., Brown, C.M., Jeso, B.d., Lym, L., Toone, E.J., Jones, J.B., Enzymes in organic synthesis. 42. Investigation of the effects of the isozymal composition of pig liver esterase on its stereoselectivity in preparative-scale ester hydrolyses of asymmetric synthetic value, *J. Am. Chem. Soc.*, **1988**, *110*, 4409–4411

Lam, L.K.P., Hui, R.A.H.F., Jones, J.B., Enzymes in organic synthesis. 35. Stereoselective pig liver esterase catalyzed hydrolyses of 3-substituted glutarate diesters. Optimization of enantiomeric excess via reaction conditions control, *J. Org. Chem.*, **1986**, *51*, 2047–2050

Lange, S., Musidlowska, A., Schmidt-Dannert, C., Schmitt, J., Bornscheuer, U.T., Cloning, functional expression, and characterization of recombinant pig liver esterase, *ChemBioChem*, **2001**, *2*, 576–582

Lauble, H., Miehlich, B., Förster, S., Wajant, H., Effenberger, F., Crystal structure of hydroyxnitrile lyase from *Sorghum bicolor* in complex with the inhibitor benzoic acid: a novel cyanogenic enzyme, *Biochemistry*, **2002**, *41*, 12043–12050

Layh, N., Hirrlinger, B., Stolz, A., Knackmuss, H.-J., Enrichment strategies for nitrile-hydrolysing bacteria, *Appl. Microbiol. Biotechnol.*, **1997**, *47*, 668–674

Lewis, D.F.V., *Cytochromes P450: Structure, Function and Mechanism*, Wiley-VCH, Weinheim, **1996**

Li, Q.-S., Schwaneberg, U., Fischer, F., Schmitt, J., Pleiss, J., Lutz-Wahl, S., Schmid, R.D., Rational evolution of a medium chain-specific cytochrome P-450 BM-3 variant, *Biochim. Biophys. Acta*, **2001**, *1545*, 114–121

Li, Q.-S., Schwaneberg, U., Fischer, P., Schmid, R.D., Directed evolution of the fatty-acid hydroxylase P450 BM-3 into an indole-hydroxylating catalyst, *Chem. Eur. J.*, **2000**, *6*, 1531–1536

Li, Z., van Beilen, J.B., Duetz, W.A., Schmid, A., deRaadt, A., Griengl., H., Witholt, B., Oxidative biotransformations using oxygenases, *Curr. Opin. Chem. Biol.*, **2002**, *6*, 136–144

Liese, A., Seelbach, K., Wandrey, C., *Industrial Biotransformations*, Wiley-VCH, Weinheim, **2000**

Linker, T., The Jacobsen-Katsuki epoxidation and its controversial mechanism, *Angew. Chem. Int. Ed. Engl.*, **1997**, *36*, 2060–2062

Malito, E., Alfari, A., Fraaije, M. W., Mattevi, A., Crystal structure of a Baeyer–Villiger-monooxygenase, *Proc. Natl. Acad. Sci. USA*, **2004**, *101*, 13157–13162

Meyer, A., Held, M., Schmid, A., Kohler, H.-P.E., Witholt, B., Synthesis of 3-tert-butylcatechol by an engineered monooxygenase, *Biotechnol. Bioeng.*, **2003**, *81*, 518–524

Meyer, A., Schmid, A., Held, M., Westphal, A.H., Röthlisberger, M., Kohler, H.-P.E., van Berkel, W.J.H., Witholt, B., Changing the substrate specificity of 2-hydroxybiphenyl-3-monooxygenase from *Pseudomonas azelaica* HBP1 by directed evolution, *J. Biol. Chem.*, **2002**, *277*, 5575–5582

Mihovilovic, M.D., Müller, B., Kayser, M.M., Stewart, J.D., Fröhlich, J., Stanetty, P., Spreitzer, H., Baeyer-Villiger oxidations of representative heterocyclic ketones by whole cells of engineered *Escherichia coli* expressing cyclohexanone monooxygenase, *J. Mol. Catal. B: Enzym.*, **2001**, *11*, 349–353

Mihovilovic, M.D., Müller, B., Stanetty, P., Monooxygenase-Mediated Baeyer-Villiger Oxidations, *Eur. J. Org. Chem.*, **2002**, 3711–3730

Mischitz, M., Faber, K., Asymmetric opening of an epoxide by azide catalyzed by an immobilized enzyme preparation from *Rhodococcus* sp., *Tetrahedron Lett.*, **1994**, *35*, 81–84

Moree, W.J., Sears, P., Kawashiro, K., Witte, K., Wong, C.H., Exploitation of subtilisin BPN′ as catalyst for the synthesis of peptides containing noncoded amino acids, peptide mimetics and peptide conjugates, *J. Am. Chem. Soc.*, **1997**, *119*, 3942–3947

Musidlowska, A., Lange, S., Bornscheuer, U.T., By overexpression in the yeast *Pichia pastoris* to enhanced enantioselectivity: New aspects in the application of pig liver esterase, *Angew. Chem. Int. Ed.*, **2001**, *40*, 2851–2853

Musidlowska-Persson, A., Bornscheuer, U.T., Recombinant porcine intestinal carboxyl esterase: cloning from the pig liver esterase gene by site-directed mutagenesis, functional expression and characterization, *Prot. Eng.*, **2003**a, *16*, 1139–1145

Musidlowska-Persson, A., Bornscheuer, U.T., Site directed mutagenesis of recombinant pig liver esterase yields in mutant with altered enantioselectivity, *Tetrahedron: Asymmetry*, **2003**b, *14*, 1341–1344

Nagasawa, T., Takeuchi, K., Yamada, H., Characterization of a new cobalt-containg nitrile hydratase purified from urea-induced cells of *Rhodococcus rhodochrous* J1, *Eur. J. Biochem.*, **1991**, *196*, 581–589

Nagasawa, T., Yamada, H., Microbial production of commodity chemicals, *Pure Appl. Chem.*, **1995**, *67*, 1241–1256

Nardini, M., Ridder, S. I., Rozeboom, H.J., Kalk, K.H., Rink, R., Janssen, D.B., Dijkstra, B.W., The X-ray structure of epoxide hydrolase from *Agrobacterium radiobacter* AD1, *J. Biol. Chem.*, **1999**, *274*, 14579–14596

Narhi, L.O., Fulco, A.J., Characterization of a catalytically self-sufficient 119,000 Dalton cytochrome P-450 monooxygenase induced by barbiturates in *Bacillus megaterium*, *J. Biol. Chem.*, **1986**, *261*, 7160–7169

Narhi, L.O., Fulco, A.J., Identification and characterization of two functional domains in cytochrome P-450 BM-3, a catalytically self-sufficient monooxygenase induced by barbiturates in *Bacillus megaterium*, *J. Biol. Chem.*, **1987**, *262*, 6683–6690

Nelson, D.R., Cytochrome P450 and the individuality of species, *Arch. Biochem. Biophys.*, **1999**, *369*, 1–10

Nelson, D.R., Kamataki, T., Waxman, D.J., Guengerich, F.P., Estabrook, R.W., Feyereisen, R., Gonzalez, F.J., Coon, M.J., Gunsalus, I.C., Gotoh, O., et al., The P450 superfamily: update on new sequences, gene mapping, accession numbers, early trivial names of enzymes, and nomenclature, *DNA Cell. Biol.*, **1993**, *12*, 1–51

Nishizawa, M., Shimizu, M., Ohkawa, H., Kanaoka, M., Stereoselective production of (+)-*trans*-chrysanthemic acid by a microbial esterase: cloning, nucleotide sequence, and overexpression of the esterase gene of *Arthrobacter globiformis* in *Escherichia coli*, *Appl. Environ. Microbiol.*, **1995**, *61*, 3208–3215

Öhrner, K., Mattson, A., Norin, T., Hult, K., Enantiotopic selectivity of pig liver esterase isoenzymes, *Biocatalysis*, **1990**, *4*, 81–88

Ohshima, T., Soda, K., Stereoselective biocatalysis: Amino acid dehydrogenases and their applications, in: Patel, R.N., (Ed.), *Stereoselective Biocatalysis*, p. 877. Marcel Dekker, New York, **2000**

Oliver, C.F., Modi, S., Pimrose, W.U., Lian, L.Y., Roberts, G.C., Engineering the substrate specificity of *Bacillus megaterium* cytochrome P-450 BM3: hydroxylation of trimethylammonium compounds, *Biochem. J.*, **1997**, *327*, 537–544

Ollis, D.L., Cheah, E., Cygler, M., Dijkstra, B., Frolow, F., Franken, S., Harel, M., Remington, S.J., Silman, I., The α/β hydrolase fold, *Protein Eng.*, **1992**, *5*, 197–211

Omura, T., Sato, R.J., The carbon monooxide-binding pigment of liver microsomes. I. Evidence for its hemoprotein nature, *J. Biol. Chem.*, **1964**, *239*, 2370–2378

Orru, R.V.A., Kroutil, W., Faber, K., Deracemization of (±)-2,2-disubstituted epoxides *via* enantioconvergent chemoenzymic hydrolysis using *Nocardia* EH1 epoxide hydrolase and sulfuric acid, *Tetrahedron Lett.*, **1997**, *38*, 1753–1754

Orru, R.V.A., Osprian, I., Kroutil, W., Faber, K., An efficient large-scale synthesis of (*R*)-(−)-mevalonolactone using simple biological and chemical catalysts, *Synthesis*, **1998**, 1259–1263

Ottolina, G., Carrea, G., Colonna, S., Ruckemann, A., A predictive active site model for cyclohexanone monooxygenase catalyzed Baeyer-Villiger oxidations, *Tetrahedron: Asymmetry*, **1996**, *7*, 1123–1136

Oyama, K., Industrial production of aspartame, in: Collins, A.N., Sheldrake, G.N., Crosby, J., (Eds), *Chirality in Industry*, pp. 237–247. Wiley, Chichester, **1992**

Pàmies, O., Bäckvall, J.-E., Combination of enzymes and metal catalysts. A powerful approach in asymmetric catalysis, *Chem. Rev.*, **2003**, *103*, 3247–3262

Pàmies, O., Bäckvall, J.-E., Chemoenzymatic dynamic kinetic resolution, *Trends Biotechnol.*, **2004**, *22*, 130–135

Park, J.H., Kim, G.J., Kim, H.S., Production of D-amino acid using whole cells of recombinant *Escherichia coli* with separately and coexpressed D-hydantoinase and N-carbamoylase, *Biotechnol. Prog.*, **2000**, *16*, 564–570

Pedragosa-Moreau, S., Archelas, A., Furstoss, R., Microbiological transformations. 32. Use of epoxide hydrolase mediated biohydrolysis as a way to enantiopure epoxides and vicinal diols: Application to substituted styrene oxides, *Tetrahedron*, **1996**, *52*, 4593–4606

Peters, M.W., Meinhold, P., Glieder, A., Arnold, F.H., Regio- and enantioselective alkane hydroxylation with engineered cytochromes P450 BM-3, *J. Am. Chem. Soc.*, **2003**, *125*, 13442–13450

Peterson, J.A., Graham, S.E., A close family resemblance: the importance of structure in understanding cytochromes P450, *Structure*, **1998**, *6*, 1079–1085

Pleiss, J., Fischer, M., Peiker, M., Thiele, C., Schmid, R.D., Lipase engineering database understanding and exploiting sequence-structure-function relationships, *J. Mol. Catal B: Enzymatic*, **2000**, *10*, 491–508

Quax, W.J., Broekhuizen, C.P., Development of a new *Bacillus* carboxyl esterase for use in the resolution of chiral drugs, *Appl. Microbiol. Biotechnol.*, **1994**, *41*, 425–431

Rink, R., Fennema, M., Smids, M., Dehmel, U., Janssen, D.B., Primary structure and catalytic mechanism of the epoxide hydrolase from *Agrobacterium radiobacter* AD1, *J. Biol. Chem.*, **1997**, *272*, 14650–14657

Rosenthaler, L., Durch Enzyme bewirkte asymmetrische Synthese, *Biochem. Z.*, **1908**, *14*, 238–253

Ruettinger, R.T., Wen, L.-P., Fulco, A.J., Coding nucleotide, 5'-regulatory, and deduced amino acid sequences of P-450 BM-3, a single peptide cytochrome P-450:NADPH-P-450 reductase from *Bacillus megaterium*, *J. Biol. Chem.*, **1989**, *264*, 10987–10995

Savjalov, W.W., Zur Theorie der Eiweissverdauung, *Pflügers Arch. Ges. Physiol.*, **1901**, *85*, 171

Schellenberger, V., Jakubke, H.D., Protease-catalyzed kinetically controlled peptide synthesis, *Angew. Chem. Int. Ed. Engl.*, **1991**, *30*, 1437–1449

Schmid, R.D., Verger, R., Lipases – interfacial enzymes with attractive applications, *Angew. Chem. Int. Ed. Engl.*, **1998**, *37*, 1608–1633

Schmidt, M., Griengl, H., Oxynitrilases: From Cyanogenesis to asymmetric synthesis, in: Fessner, W.-D., (Ed.), *Topics Curr. Chem.*, Vol. 200, pp. 193–226. Springer, Berlin, **1999**

Schneider, S., Wubbolts, M.G., Sanglard, D., Witholt, B., Biocatalyst engineering by assembly of fatty acid transport and oxidation activities for in vivo application of cytochrome P-450BM-3 monooxygenase., *Appl. Environ. Microbiol.*, **1998**, *64*, 3784–3790

Schoevaart, R., Rantwijk, F.v., Sheldon, R.A., A four-step cascade for the one-pot synthesis of non-natural carbohydrates from glycerol, *J. Org. Chem.*, **2000**, *65*, 6940–6943

Schwaneberg, U., Appel, D., Schmitt, J., Schmid, R.D., P450 in biotechnology: zinc driven ω-hydroxylation of *p*-nitrophenoxydodecanoic acid using P450 BM-3 F87A as a catalyst, *J. Biotechnol.*, **2000**, *84*, 249–257

Schwaneberg, U., Bornscheuer, U.T., in: Bornscheuer, U.T., (Ed.), *Enzymes in Lipid Modification.*: Wiley-VCH, Weinheim, **2000**

Schwaneberg, U., Otey, C., Cirino, P.C., Farinas, E., Arnold, F.H., Cost-effective whole-cell assay for laboratory evolution of hydroxylases in *Escherichia coli*, *J. Biomol. Screen.*, **2001**, *6*, 111–117

Schwaneberg, U., Schmidt-Dannert, C., Schmitt, J., Schmid, R.D., A continuous spectrophotometric assay for P450 BM-3, a fatty acid hydroxylating enzyme, and its mutant F87A, *Anal. Biochem.*, **1999**a, *269*, 359–366

Schwaneberg, U., Sprauer, A., Schmidt-Dannert, C., Schmid, R.D., P450 monooxygenase in biotechnology I. Single-step, large-scale purification method for cytochrome P450 BM-3 by anion-exchange chromatography, *J. Chromatogr. A*, **1999**b, *848*, 149–159

Secundo, F., Carrea, G., Riva, S., Battistel, E., Bianchi, D., Cyclohexanone monooxygenase catalyzed oxidation of methyl phenyl sulfide and cyclohexanone with macromolecular NADP in a membrane reactor, *Biotechn. Lett.*, **1993**, *155*, 865–870

Smit, M.S., Fungal epoxide hydrolases: new landmarks in sequence-activity space, *Trends Biotechnol.*, **2004**, *22*, 123–129

Spelberg, J.H.L., Rink, R., Kellogg, R.M., Janssen, D.B., Enantioselectivity of a recombinant epoxide hydrolase from *Agrobacterium radiobacter, Tetrahedron: Asymmetry*, **1998**, *9*, 459–466

St. Maurice, M., Bearne, S.L., Hydrophobic nature of the active site of mandelate racemase, *Biochemistry*, **2004**, *43*, 2524–2532

Steinreiber, A., Faber, K., Microbial epoxide hydrolases for preparative biotransformations, *Curr. Opin. Biotechnol.*, **2001**, *12*, 552–558

Stewart, J.D., Reed, K.W., Zhu, J., Chen, G., Kayser, M.M., A 'designer yeast' that catalyzes the kinetic resolutions of 2-alkyl-substituted cyclohexanones by enantioselective Baeyer-Villiger oxidations, *J. Org. Chem.*, **1996**, *61*, 7652–7653

Strukul, G., Transition metal catalysis in the Baeyer-Villiger oxidation of ketones, *Angew. Chem. Int. Ed.*, **1998**, *37*, 1198–1209

Sugiura, Y., Kuwahara, J., Nagasawa, T., Yamada, H., Nitrile hydratase: the first non-heme iron enzyme with a typical low-spin Fe(III)-active center, *J. Am. Chem. Soc.*, **1987**, *109*, 5848–5850

Syldatk, C., May, O., Altenbuchner, J., Mattes, R., Siemann, M., Microbial hydantoinases – industrial enzymes from the origin of life?, *Appl. Microbiol. Biotechnol.*, **1999**, *51*, 293–309

Tanner, A., Hopper, D.J., Conversion of 4-hydroxyacetophenone into 4-phenyl acetate by a flavin adenine dinucleotide-containing Baeyer-Villiger-type monooxygenase, *J. Bacteriol.*, **2000**, *182*, 6565–6569

Taylor, D.G., Trudgill, P.W., Camphor revisited: studies of 2,5-diketocamphane 1,2-monooxygenase from *Pseudomonas putida* ATCC 17453, *J. Bacteriol.*, **1986**, *165*, 489–497

Turfitt, G.E., Microbiological degradation of steroids. IV. Fission of the steroid molecule. *Biochem. J.*, **1948**, *42*, 376–383

Urlacher, V., Schmid, R.D., Biotransformations using prokaryotic P450 monooxygenases, *Curr. Opin. Biotechnol.*, **2002**, *13*, 557–564

Urlacher, V.P., Lutz-Wahl, S., Schmid, R.D., Microbial P450 enzymes in biotechnology, *Appl. Microbiol. Biotechnol.*, **2004**, *64*, 317–325

Walsh, C.T., Chen, Y.C.J., Enzymatische Baeyer-Villiger-Oxidation durch flavinabhängige Monooxygenasen, *Angew. Chem. Int. Ed.*, **1988**, *27*, 333–343

Weissfloch, A.N.E., Kazlauskas, R.J., Enantiopreference of lipase from *Pseudomonas cepacia* toward primary alcohols, *J. Org. Chem.*, **1995**, *60*, 6959–6969

Wieser, M., Nagasawa, T., Stereoselective nitrile-converting enzymes, in: Patel, R., (Ed.), *Stereoselective Biocatalysis*, pp. 461–486. Marcel Dekker, New York, **1999**

Wong, C.H., Formation of C-C bonds, in: Drauz, K., Waldmann, H., (Eds), *Enzyme catalysis in organic synthesis*, Vol. 2, pp. 931–974. VCH, Weinheim, **2002**

Yamamoto, K., Oishi, K., Fujimatsu, I., Komatsu, K.-I., Production of *R*-(–)-mandelic acid from mandelonitrile by *Alcaligenes faecalis* ATCC 8750, *Appl. Environm. Microbiol.*, **1991**, *57*, 3028–3032

Zaks, A., Dodds, D.R., Chloroperoxidase-catalysed asymmetric oxidations: substrate specificity and mechanistic study, *J. Am. Chem. Soc.*, **1995**, *117*, 10419–10424

Zhang, J., Reddy, J., Roberge, C., Senanayake, C., Greasham, R., Chartrain, M., Chiral bioresolution of racemic indene oxide by fungal epoxide hydrolase, *J. Ferment. Bioeng.*, **1995**, *80*, 244–246

Zhang, X.M., Archelas, A., Furstoss, R., Microbiological transformations. 19. Asymmetric dihydroxylation of the remote double bond of geraniol: A unique stereochemical control allowing easy access to both enantiomers of geraniol-6,7-diol, *J. Org. Chem.*, **1991**, *56*, 3814–3817

4
Enzyme Production and Purification

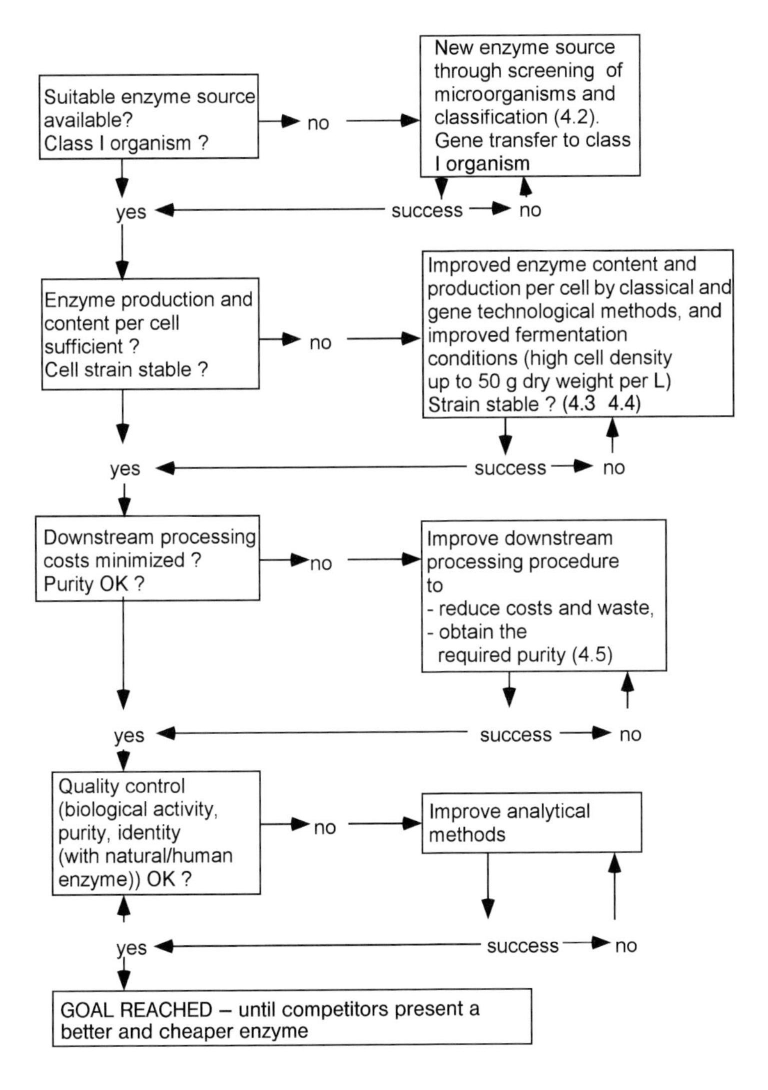

Biocatalysts and Enzyme Technology. K. Buchholz, V. Kasche, U.T. Bornscheuer

ISBN: 3-527-30497-5

For enzyme processes and for the use of enzymes in therapy and for analytical purposes, a steady and safe enzyme supply is required. In technical applications the enzyme costs must be less than 1–10 % of the product costs, with the low and high values applying to low- and high-value products, respectively. In order to achieve this, the production and purification of enzymes can be improved as shown below.

4.1 Introduction

The use of isolated enzymes as biocatalysts in enzyme processes is limited by their unavoidable inactivation. The enzyme cost in a process can be reduced by using more stable enzymes with the same or better activity or *vice versa*, provided that they can be produced with the same yield as the original enzyme. From this it follows that enzyme processes must be coupled with the production and purification of enzymes. The latter must be designed to minimize the enzyme costs, and to provide a stable supply for the enzyme with the required purity. The first thing to do to reach this aim is to find a suitable enzyme source. Generally, the natural sources (animal and plant tissues, wild-type microorganisms) produce insufficient quantities of the desired enzyme, as natural systems have not been optimized for high enzyme contents during evolution. Therefore, the next step in the production of an enzyme with wild-type organisms that can be used for the safe production of enzyme (GRAS or class I organisms) is the optimization of the enzyme production.[1)]

When the enzyme-producing organism is not a GRAS or class I organism, it should not be used for the production of the enzyme, based on safety criteria. In this case the gene for the desired enzyme can be transferred to and expressed in a GRAS or class I organism. The optimization of enzyme production in wild-type or recombinant microorganisms involves manipulation of the transcriptional, translational and post-translational processes that influence the yield of active enzyme. This can be achieved with and without gene technological methods. Finally, the enzyme produced must be purified to the desired purity for its application as a biocatalyst. The purification is much easier for extracellular enzymes as they can be isolated directly from the medium without the disruption of cells, as is required for intracellular enzymes. The purification costs decrease with the enzyme content in the medium or homogenate, and increase with the purity of the enzyme that is required for its final

1) GRAS (Generally recognized as safe) organisms are those that have been used without health problems for a long time in food processing. Class I organisms have been classified as unlikely to cause human or animal disease. The list of class I organisms that can be used for production of enzymes is smaller than the list of such organisms that can be used in laboratories. This classification is in most countries regulated by law and continuously reviewed by international and national health authorities (e.g., World Health Organization (WHO), Food and Agricultural Organization (FAO) (United Nations); Food and Drug Authority (FDA) (USA); Zentrale Kommission für Biologische Sicherheit (ZKBS) (Germany)). See also the literature list and Internet resources in Appendix I (p. 419).

use. Estimated costs of technical enzymes used as biocatalysts are between ≈ 10 € (amylases, pectinases) and up to about 100 000 € (penicillin amidase) per kg. These costs are lower than for enzymes used for analytical and therapeutical purposes, where a much higher purity is required than for technical enzymes.

This chapter will provide an overview of how enzyme costs can be reduced by increasing the yield in the production and downstream processing to reach the desired purity. For more detailed information, the interested reader can consult the textbooks and reviews given in the literature list at the end of the chapter.

4.2
Enzyme Sources

4.2.1
Animal and Plant Tissues

Classical enzyme sources are animal and plant tissues from the processing of animals and plants, especially those that are residues not used for food. To these belong the pancreas (used earlier in the treatment of hides in the production of leather; see Section 1.3), calf stomach (used in cheese-making), kidney, liver, parts of plants (papaya, ananas) used for the tenderization of meat, or sodom apple leaves used for milk clotting in the production of cheese in central Africa. These processes, except the use of pancreas, are up to 5000 years old. They were developed empirically, long before it was known that enzymes were essential as biocatalysts in these processes. The enzymes involved in such processes have been identified during the development of biochemistry and enzymology over the past 150 years. Some of the enzyme sources used here have a large enzyme content and can produce up to 1 % enzyme of the wet tissue weight per day (Table 4.1). To these belong tissues producing digestive enzymes (pancreas, seeds), or expressing large quantities of enzyme for tissue-specific metabolism (liver, or muscle such as heart). Due to problems caused by difficulties in the controlled recovery of enzymes from these tissues or their alternative uses, and in order to avoid shortages in the enzyme supply, alternative enzyme sources had to be found. This is illustrated by the use of pancreas as an enzyme source. Until it was discovered (in 1921) that the pancreas also produces insulin, the tissue was mainly used as a source of technical enzymes (see Section 1.3) and digestive enzymes for substitution therapy. However, after this time the tissue became too expensive as a source of technical enzymes. The enzymes still produced from the pancreatic tissue (chymotrypsin and trypsin) were by-products in the production of insulin. Today, insulin is mainly produced in recombinant organisms (*E. coli* or yeast cells), and therefore the price of those enzymes that have established applications in therapy and animal cell culture technology has increased. For other technical enzymes – for which the pancreas was once a source – alternative supplies, mainly from microorganisms, have been found in the meantime.

Today, animal and plant tissues can be used for the production of heterologeous proteins by the transfer of genes (*pharming*) (Miele, 1997; Rudolph, 1999). Until

Table 4.1 Biologically active enzyme content and enzyme productivity in animal and plant tissues and microorganisms (Albee et al., 1990; Amneus, 1977; Deshpande et al., 1994; Rua et al., 1998; Pilone and Pollegioni, 2002; Christiansen et al., 2003; Kasche et al., 2004).

Enzyme source	*Enzyme*	*Enzyme content [g g^{-1} cell dry weight]*	*Enzyme productivity [g L^{-1} tissue or fermenter per day]*
Animal tissues			
Pancreas	Digestive enzymes such as chymotrypsin, lipase, nuclease, trypsin (e)[a]	These enzymes are activated outside the pancreas cells; proenzyme content ≈0.04	Up to 10
Liver	Aldolase (i)[a]	0.001	
Muscle	Aldolase (i)	0.03	
Porcine kidney	D-Aminoacid oxidase (i)	0.001	
Plant tissues			
Papaya	Papain (e)	<0.01	
Microorganisms			
S. cerevisiae	Aldolase (i)	0.016	
E. coli			
– wild-type	Penicillin amidase (p)[a]	0.004	0.25
– recombinant	Penicillin amidase (p)	0.04-0.1	1–4
– recombinant	Glutaryl amidase (p)		7
– recombinant	Lipase (e)[b]	0.04	
– recombinant	D-Amino acid oxidase (i)	0.04	
B. clausii			
– recombinant	Savinase (a subtilisin)	0.016	0.4

a) (i) intracellular; (p) periplasmic (outside cell membrane inside cell wall); (e) extracellular (outside cell).
membrane and cell wall;

b) *B. thermocatenulatus* lipase is extracellular but was mainly produced intracellularly in *E. coli*.

now, this approach has mainly been considered for therapeutical proteins, but the potential advantage lies in the reduced cost of protein production compared to that by microorganisms in fermenters. Pharming might be used to produce therapeutical proteins, though its use to supply technical enzymes that are not normally produced in these tissues remains questionable. In animals, this would lead to the production of antibodies, and in both animal and plant tissues these enzymes might catalyze unwanted reactions that would have negative influences on animal and plant growth. Another problem with enzymes derived from animal tissues is their possible contamination with prions (e.g., bovine spongiform encephalitis, BSE) or viruses (human immunodeficiency virus, HIV) that are harmful to humans.

4.2.2 Wild-type Microorganisms

Microorganisms that have long been used safely in food production, mainly as preserving agents – that is, GRAS organisms and class I organisms – are ideal sources for the production of enzymes for enzyme processes. Enzymes from these sources are used mainly in food processing, where the demand to use non-recombinant enzymes is high, and regulations are strict.

Wild-type microorganisms are also used to screen for "new" enzymes that would catalyze the desired conversion or degradation of a compound, A. When this conversion involves only one enzyme it can also be used to catalyze the synthesis of A from the products. Different methods exist to screen for these enzymes:

- from a sample with different microorganisms (Fig. 4.1),
- from a sample with the same microorganisms but with a different enzyme content per cell (Figs. 4.2 and 4.3).

These methods can also be used to screen for enzymes that are active under desired process conditions (pH and temperature). For processes to be carried out under alkaline or acidic pH conditions, the screening must be performed with organisms or metagenomes gathered from alkaline or acidic environments, respectively (see Chapter 2, Section 2.11). "Metagenomes" are DNA from entire microbial consortia that cannot be cultivated directly (see Section 2.11, p. 82). Genes from the metagenomes are transferred to microorganisms, which in turn express the enzymes coded by the metagenome genes. For processes which are are optimal at higher temperatures (up to 100 °C), it is necessary to screen organisms from hot springs (extremophiles) (Niehaus et al., 1999), after which the organisms producing the desired enzyme must be classified.

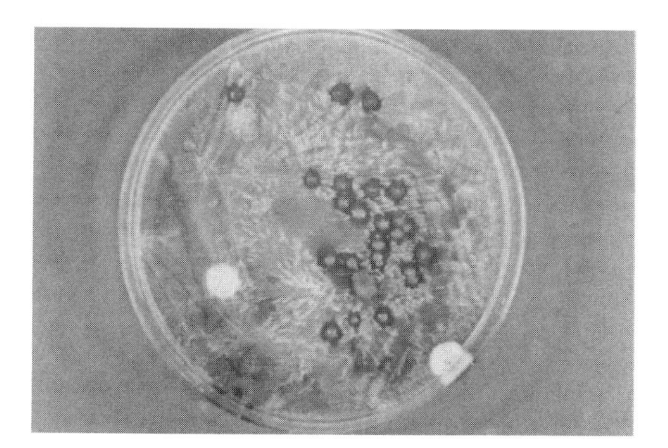

Fig. 4.1. Screening for microorganisms from a soil sample that can degrade pyrene. The cells are grown isolated on the Petri dish with pyrene as the only C-source. The clones that can degrade pyrene and divide are directly observed (Kästner, 1994).

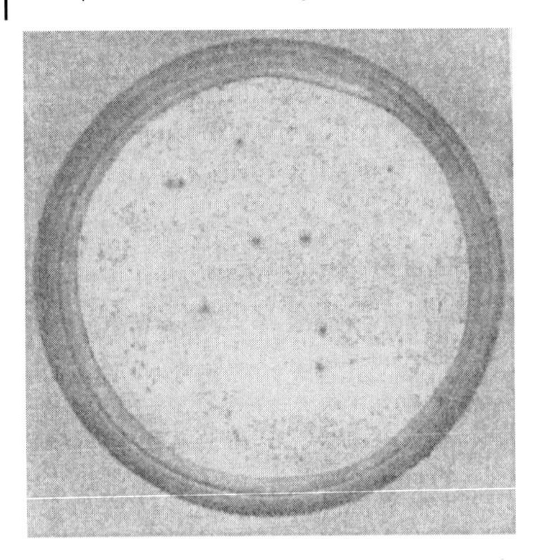

Fig. 4.2. Screening for *E. coli* cells that produce penicillin amidase. A filter paper with a substrate selectively hydrolyzed by this enzyme that yields a yellow product is placed on the Petri dish with isolated cell clones. The clones producing penicillin amidase color the filter paper yellow (here seen as dark spots) above the clone. The cells that produce more enzyme develop the yellow color more rapidly than those producing less of the enzyme.

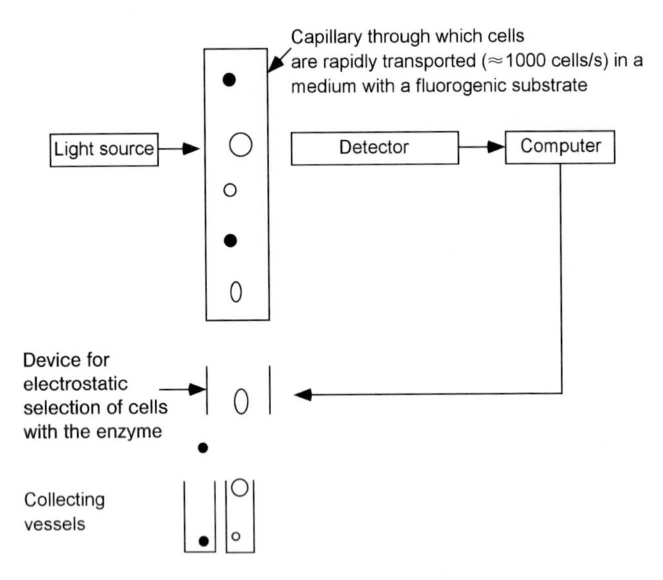

Fig. 4.3. A fluorescence-activated cell sorter (FACS) for high-throughput screening (HTS) of cells with respect to different properties such as size, DNA content, or content of an intracellular or periplasmic enzyme, E. In the latter case, single cells are transported through a capillary in a solution with a fluorogenic substrate that can only be converted to a fluorescent product by the enzyme E. The cells with the enzyme (marked black) become fluorescent and are detected by the measuring device (detector and computer); this gives a signal to a device for electrostatic selection of these cells, and they are collected into a separate vessel.

4.2.2
Recombinant Microorganisms

When the enzyme yield is low in wild-type class I organisms, or the desired enzyme is not found in class I organisms, then alternative enzyme sources must be found. This can easily be achieved by transferring the enzyme gene to, and expressing it as a recombinant enzyme in, a suitable class I organism. For this purpose, frequently used microorganisms include bacteria and yeasts such as *Aspergillus* sp., *Bacillus* sp., *Escherichia coli* (only Class I strains), *Saccharomyces* sp., and *Streptomyces* sp. Nowadays, most enzymes used in enzyme processes are produced using recombinant technology.

4.3
Improving Enzyme Yield

4.3.1
Processes that Influence Enzyme Yield

The transcriptional, translational and post-translational processes that influence the yield of active enzyme are shown schematically in Figure 4.4. These processes include (where stages VI to VIII apply only to extracellular and periplasmic enzymes that are transported through the cell membrane):

I. For transcription to occur, the gene must first be activated. This requires an external inducer that in wild-type organisms is normally a molecule, the conversion of which is catalyzed by the enzyme encoded by the activated gene. The repressor protein that inhibits the binding of RNA polymerase to the gene in the double-stranded DNA, required for transcription, loses this property when it binds to an inducer molecule. These activation mechanisms are also used in the production of recombinant enzymes.

II. mRNA is rapidly hydrolyzed *in vivo*. Recently it has been found that there exists a natural mechanism for gene-silencing. In this, antisense nucleotides (short interference RNA, siRNA) that can inhibit the translation or accelerate the degradation of the complementary mRNA are formed by transcription (Scherer and Rossi, 2003). This reduces the number of pre-pro-enzyme polypeptides that can be translated by mRNA.

III. Intracellular proteolysis of unfolded polypeptides by peptidases that are required for protein turnover (Enfors, 1992). This protein turnover is a normal process *in vivo* to recycle amino acids from folded proteins no longer in use (Wickner et al., 1999; Henge and Buckau, 2003).

IV. When the concentration of (pre-pro-)enzyme polypeptides that are synthetized is high, neighboring polypeptides may interact with each other during folding. This leads to the formation of insoluble protein aggregates (inclusion bodies) that must be unfolded and renatured to become biologically active (Mitraki et

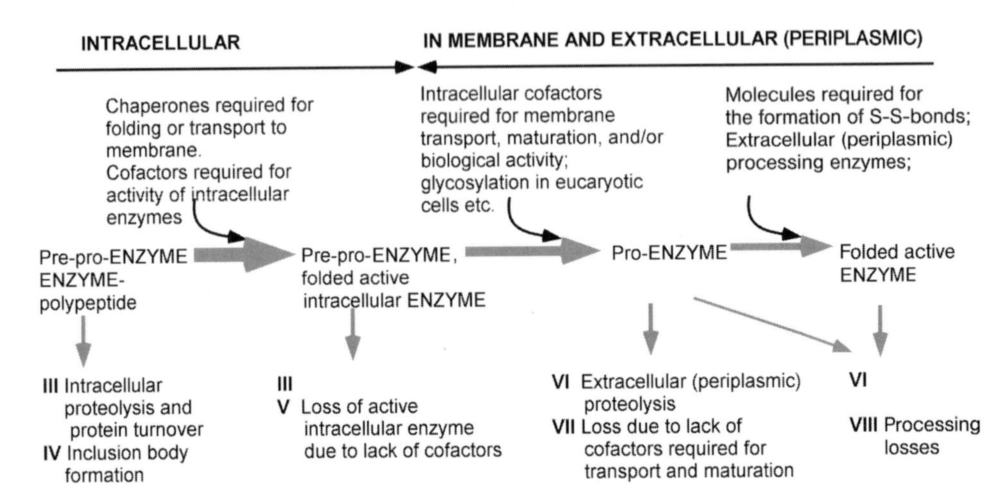

Fig. 4.4. Transcriptional, translational and post-translational processes that influence the yield of intracellular, extracellular, and periplasmic enzymes. The latter two are synthesized as pre-pro-enzymes. The signal peptide (pre-) is required for the transport through the cell membrane, and is hydrolyzed off by signal peptidases located on the outside of the cell membrane. The biologically active enzyme is written in bold-face letters.

al., 1991). This process is prevented by chaperones that act as folding catalysts of intracellular enzymes or extracellular pre-pro-enzymes that must be folded before they are transported through the cell membrane. For extracellular pre-pro-enzymes that must be unfolded during the membrane transport, other proteins that assist their transport to the membrane prevent the folding in the cytoplasm, intracellular proteolysis and inclusion body formation.

V. Additions of cofactors to intracellular enzymes required for their biological activity. They are bound covalently (such as phosporylation) or non-covalently (such as Zn^{2+} to hydantoinases (Altenbuchner et al., 2001) or FAD to D-amino acid oxidase (Pilone and Pollegioni, 2002). An insufficient supply of these cofactors reduces the yield of active enzyme. The intracellular concentrations of ions such as Ca^{2+} and Zn^{2+} is < 1 μM; the binding constants for these ions to the enzymes that require them for their activity must therefore be much lower than 1 μM in order that they are saturated with the required metal ion (Jones et al., 2002).

VI. Extracellular (periplasmic) enzymes can be degraded by extracellular (periplasmic) peptidases.

VII. For pre-pro-enzymes that require cofactors (such as metal ions; see stage V) for the transport through the membrane, maturation and/or biological activity, a lack of these leads to the accumulation of inactive pre-pro-enzyme or pro-enzyme in the cytoplasm.

VIII. Pro-enzymes that are matured by proteolytically processing generally exist as several different processed forms that differ in activity.

Table 4.2 How the yield of biologically active enzyme can be increased on the transcriptional, translational, and post-translational level by influencing processes I–VIII in Figure 4.4.

Process influencing the yield per cell (Fig. 4.4)	*Yield can be increased without gene technology by*	*Yield can be increased with gene technology by*
I. Gene activation by induction and rate of mRNA synthesis	Select optimal inducer concentration	Increase the number of genes per cell and the rate of mRNA synthesis
II. mRNA hydrolysis	Select cells with reduced rate of mRNA hydrolysis	Reducing the content of different RNases
III. Intracellular proteolysis	Select cells with lower intracellular peptidase content	Increase chaperone content to reduce proteolysis losses
IV. Inclusion body formation	Reduce rate of polypeptide synthesis by reducing the temperature	Increase chaperone content to reduce inclusion body formation.
V. Cofactor binding to intracellular enzymes	Increase extracellular cofactor concentration (metal ions)	Stimulate intracellular cofactor synthesis and increase extracellular cofactor concentration (metal ions)
VI. Extracellular (periplasmic) proteolysis	Select cells with a low content in extracellular (periplasmic) peptidases	Reducing the content of periplasmic peptidases
VII. Cofactor binding to extracellular and periplasmic enzymes and membrane transport	Increase cofactor content, select cells with a high membrane transport rate	Increase cofactor concentration and number of protein translocation systems in the membrane; change signal peptide
VIII. Processing losses		Site-directed mutagenesis of peptide bond in pro-enzyme to prevent further processing

[a] In all gene transfers with plasmids to microorganisms used for large-scale production of enzymes, the use of antibiotic resistance selection markers must be avoided to minimize the risk to select antibiotic-resistant microorganisms.

How these processes can be influenced to increase the enzyme yield with and without gene technology is shown in Table 4.2. When the yield has been optimized in the cells in shake-flask cultures, as indicated in Table 4.2, it can only be improved by increasing the cell density during the fermentation. When the cells grow exponentially, medium with necessary energy, C-, N-, S- and P- sources and different ions required as cofactors must be added exponentially in order to achieve this. In practice, cell densities up to 50 g cell dry weight per liter can be obtained in such high cell density fermentations. The upper limit for the concentration of an active enzyme synthesized under these conditions is about 10 g L^{-1}, or about 40 % of the total cell protein. This is about the same order of magnitude as in the best enzyme-producing tissues (e.g., pancreas; see Table 4.1). How this has been achieved will be illustrated mainly for the post-translational processes III–VIII, based on recent studies, for extracellular (lipases) and periplasmic (penicillin amidases) enzymes.

4.4 Increasing the Yield of Periplasmic and Extracellular Enzymes

Most technical enzymes are currently produced in microorganisms (bacteria, yeasts, fungi), and this applies to both non-recombinant and recombinant enzymes. For the latter type of enzymes, *E. coli*, *Bacillus* sp., *Aspergillus* sp., and yeast cells (only Class I strains) are frequently used hosts. The major hydrolases used in technical processes are either extracellular or periplasmic – that is, they have been transported through the cell membrane. How their yields can be improved in wild-type and recombinant *E. coli* cells (as outlined in Table 4.2), will be illustrated for the hydrolases penicillin amidase and lipase, both of which are used as biocatalysts in large-scale enzyme processes (Makrides, 1996). This illustration is based on the yield of active enzyme, and not for inclusion bodies of the (pre-)pro-enzyme. Larger yields (in terms of protein content) can be obtained for inclusion bodies, and these could be activated after purification. However, this requires additional processing steps, such as refolding to the native pro-enzyme and proteolytic processing *in vitro* that not only increase the processing costs but also reduce the yield of active enzyme Middelberg (2002).

4.4.1 Penicillin Amidase

The penicillin amidases (EC 3.5.1.11) from *E. coli*, *A. faecalis*, *P. rettgeri*, *K. citrophila*, *B. megaterium*, and *A. viscosus* are used for the hydrolysis of about 30 000 tonnes of penicillin G (see Fig. 1.7) each year. Recently, a new application of this enzyme as a biocatalyst for the kinetically controlled synthesis of semisynthetic penicillins and cephalosporins has been industrialized (see Chapter 2, Section 2.6 and Chapter 9, Section 9.2; see also Bruggink, 2001). The penicillin amidases can be produced as periplasmic enzymes in *E. coli* cells. The active enzyme contains a tightly bound Ca^{2+}-ion as cofactor that is not directly involved in the biological function (McDonough et al., 1999; Kasche et al., 2004). When produced in *E. coli*, the inactive pro-

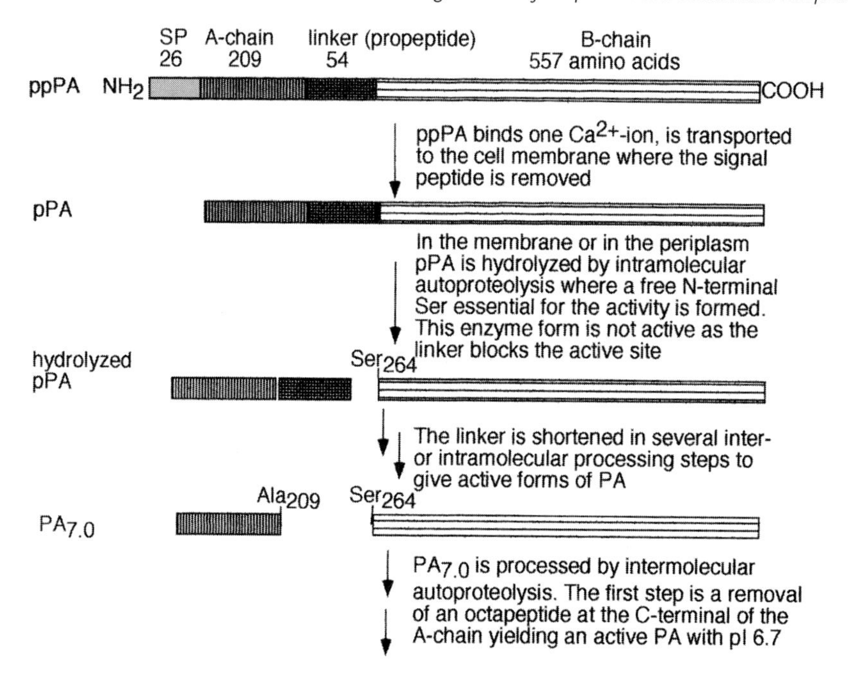

Fig. 4.5. The processing of penicillin amidase (PA_{EC}, EC 3.5.1.11) from *E. coli*. The pre-pro-enzyme (ppPA) is one polypeptide chain that from the N-terminal consists of the signal peptide (SP) that directs the periplasmic enzyme to the membrane translocation system, the A-chain, the linker or propeptide that must be removed to obtain an active enzyme and the B-chain. The number of amino acids in these are given in the second line from above. The dominating active PA-form in samples of technical enzyme ($PA_{7.0}$) from wild-type cells consists of the A- and the B-chain and has an isoelectric point (pI) of 7.0.

enzyme is first activated by intramolecular autoproteolysis, and then by intermolecular (auto)proteolysis, as shown in Figure 4.5 for the *E. coli* enzyme (PA_{EC}) (Kasche et al., 1999). This activation occurs after or during translocation into the periplasm. Penicillin amidases produced from other microorganisms have similar processing schemes (Ignatova et al., 1998).

On the transcriptional and translational level, higher enzyme yields can be obtained by stabilizing the mRNA and increasing the rate of synthesis of pre-pro-protein (ppPA). This can be achieved by increasing the number of genes by transforming the cells with high-copy plasmids that contain the gene of interest, or by increasing the temperature. There is, however, an upper limit in the rate of synthesis of ppPA. When, during the ribosomal polypeptide synthesis, adjacent peptides approach each other they can be folded together,and this leads to the formation of inclusion bodies (Mitraki et al., 1991). To avoid this, the rate of polypeptide synthesis can be reduced by lowering the cultivation temperature. Partly due to this the yield of active PA_{EC} in wild-type *E. coli* could be increased by a factor of more than 20 by reducing the temperature from 37 to 28 °C, as shown in Table 4.3. In these cells, however, some of the pre-pro-enzyme is lost by intracellular proteolysis.

The intracellular proteolysis is an energy-requiring process that can be influenced by the content of intracellular proteases and the medium. For wild-type *E. coli* it has been found essential to keep the glucose content in the medium below 50 μM (Table 4.3) in order to increase the yield of PA_{EC}. To reduce the loss due to intracellular proteolysis in the recombinant production of PA_{EC}, peptidase-deficient cells should be used for the synthesis, but even here the composition of the medium is of major importance (Table 4.3).

Table 4.3 Effect of various factors that influence the transcriptional, translational and post-translational processes in Table 4.2 and Figure 4.4, on the active PA_{EC} content in wild-type and recombinant *E. coli* cells cultivated in shake-flasks and a fermenter (Ignatova et al., 2000, 2003; Kasche et al., 2004).

Factor	*Process in Fig. 4.4 mainly influenced*	*Cell type/cultivation conditions*	*Active PA_{EC} content [mg g^{-1} CDW]*
Temperature [°C]	III, IV	Wild-type/	
24	Inclusion body	shake-flask	1.7
28	formation;	LB medium	1.9
32	Intracellular		0.5
36	proteolysis		<0.1
Glucose content [μM]	III	Wild-type/	
0	Intracellular	shake-flask	1.9
28	proteolysis	28 °C LB medium	1.6
56			1.5
140			1.0
280			0.2
560			<0.1
Cell strain	All processes	Shake-flask/28 °C	
Wild-type		Minimal medium	3.3
DH5			3.1
K5			7.0
JM109			10.8
TOP10			5.7
BL21DE			52.0
Ca^{2+}-content [μM]	III, VII	Recombinant	
3	Cofactor addition in	BL21DE/shake-flask	1.9
10	cytoplasm required	28 °C	7.9
100	for membrane	Minimal medium	14.0
1000	translocation		20.2
High cell density fermentation with exponential glucose feed with different Ca^{2+}-content [mM]	III, VII Cofactor addition in cytoplasm required for membrane translocation	Recombinant BL21DE/2-L fermenter 28 °C	
0			1
0.48			16
2.4			40
4.8			35

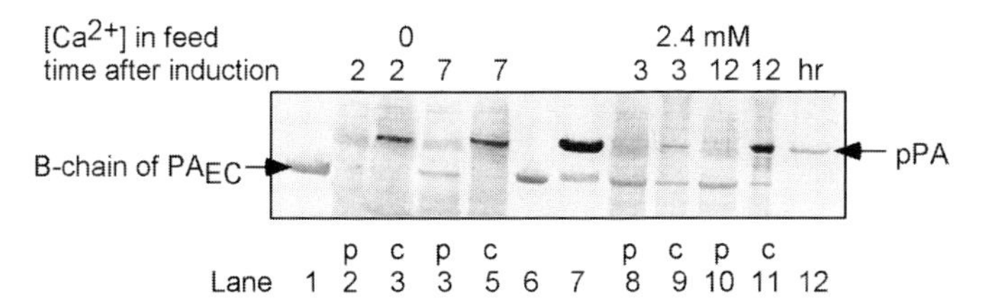

Fig. 4.6. Western blots of SDS gels of the periplasmic (p) and cytoplasmic (c) fraction of recombinant BL21(D3) *E. coli* cells producing *E. coli* penicillin amidase (PA_{EC}) after fed-batch cultivations with different Ca^{2+} concentrations in the exponential feed (see Fig. 4.7). Lanes 1 and 6: PA_{EC}; lanes 7 and 12: a slowly processing pPA from *E. coli* that has bound a Ca^{2+}-ion. The proteins are separated with respect to molecular weight in the SDS-gel electrophoresis. Only pPA, the B-chain of PA_{EC} and intracellular proteolysis products of pPA, can react with the monoclonal antibody used to visualize the bands of these proteins in the Western blot. That the band marked with pPA was the pro-enzyme was confirmed by N-terminal sequencing. Note that pPA accumulates in the cells grown without Ca^{2+} in the feed or/and when the active PA content levels off in the presence of Ca^{2+} in the feed (see also Fig. 4.7).

The penicillin amidases from the above-mentioned organisms contain one tightly bound Ca^{2+} per molecule as cofactor. The intracellular concentration of this ion is very low, and the question arises as to where this ion is added – before or after membrane translocation? In both wild-type and recombinant cells producing PA_{EC} it has been found that only $ppPA_{EC}$ or pPA_{EC} having a bound Ca^{2+} ion can be translocated through the membrane (Fig. 4.6) (Kasche et al., 2004). Thus, as expected, the concentration of the cofactor influences the yield of PA_{EC}. It should be pointed out however that the above results were obtained in shake-flask cultures where only a low cell density (up to 1–2 g dry cell weight) can be achieved.

In order to increase the PA_{EC} further, the cells must be cultivated to a high cell density in a fermenter. For the recombinant BL21(D3) cells this can be achieved with an exponential feed starting from a medium that was optimal in the shake-flask cultures. The feed contains a C-source (glucose), N-sources and ions required for growth, especially Ca^{2+} that is required as cofactor. The glucose content in the medium must be low to avoid formation of acetic acid that reduces the PA_{EC}-yield.

The results of such fermentations with and without Ca^{2+} in the feed are shown in Figure 4.7. At optimal Ca^{2+}-concentrations, up to ≈10 % of the protein in the fermenter was active PA_{EC}, while for the PA from *A. faecalis* the corresponding figure was 20 %. These results demonstrate the necessity of supplying sufficient amounts of Ca^{2+} to the pre-pro-enzymes or pro-enzymes that require this ion for membrane transport and the proteolytic processing to yield the active enzyme.[2)]

2) Other periplasmic or extracellular enzymes important in enzyme technology that have tightly bound Ca^{2+} which is not directly involved in enzyme function include α-amylases (Nielsen and Borchert, 2000), subtilisins (Bryan, 2000), and some lipases (see Section 4.4.2).

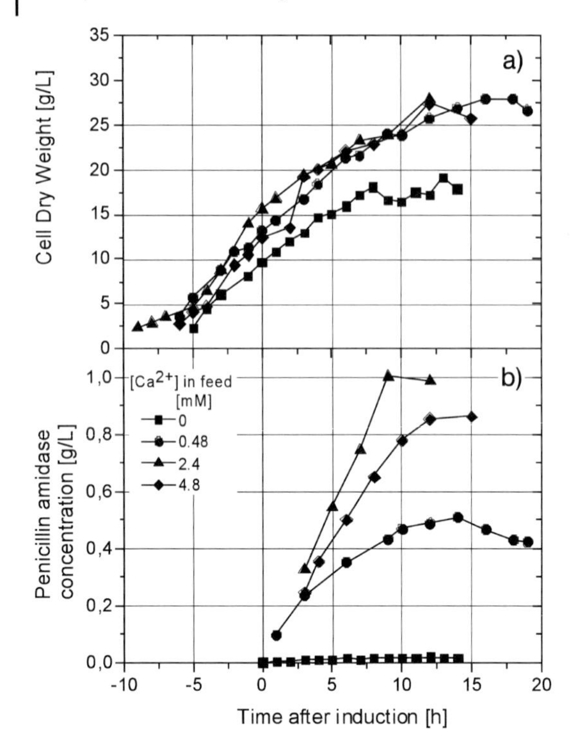

Fig. 4.7. Production of penicillin amidase from *E. coli* in recombinant *E. coli* BL21(DE) cells grown to high cell density (CDW, g L^{-1}). The cells were fed exponentially with glucose and different concentrations of Ca^{2+} at 10 h before induction with ITPG. The cells were induced to produce the pre-pro-enzyme at time 0 (Kasche et al., 2004).

4.4.2
Lipase

Lipases (EC 3.1.1.3) from different microorganisms are used as biocatalysts for a large number of different enzyme processes (see Chapter 3, Section 3.2.2.1). These lipases are all extracellular, and can, in contrast to penicillin amidase, also be transported through the cell wall of their wild-type hosts into the medium. Different mechanisms for the post-translational processing and maturation of lipases are shown in Figure 4.8. Some procaryotic lipases (from *Pseudomonas aeruginosa; B. glumae; C. visosum*) that are produced as pre-enzymes require simultaneously synthetized proteins such as lipase-specific chaperones (lipase-specific foldases; Rosenau et al., 2004). These protect the enzymes against intracellular proteolysis and the formation of inclusion bodies, and are also required for membrane transport and folding in the periplasm (Fig. 4.8b). The regulation of their gene expression, folding and secretion has been qualitatively reviewed (Rosenau et al., 2004). Some of these lipas-

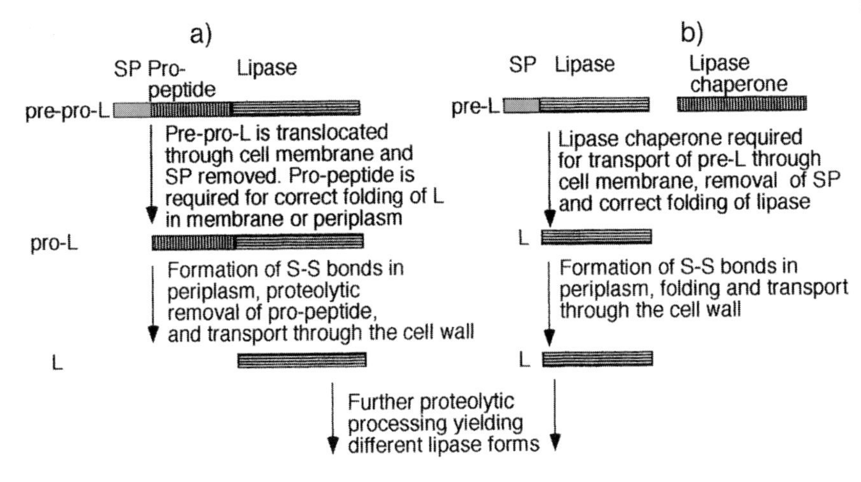

Fig. 4.8. Mechanism for the post-translational processing of extracellular lipases (L, EC 3.1.1.3). from different microorganisms. SP = signal peptide. The schemes apply for: (a) Eucaryotic lipases (such as *Rh. miehei, Rh. oryzae*); (b) Procaryotic lipases (such as *Pseudomonas* sp.; *Burkholderia* sp.; *C. viscosum*) (Jaeger et al., 1999; Ueda et al., 2002).

es also contain a tightly bound Ca^{2+}-ion (Nardini et al., 2000), though at what stage this ion is added and whether it is required for the activity has not yet been elucidated.

Eucaryotic lipases from filamentous fungi (*Rh. miehei, Rh. oryzae*) are, as in the case of penicillin amidase, produced as pre-pro-enzymes (Fig. 4.8a). The pro-enzyme of these lipases has been shown to be more active than the mature enzyme (Beer et al., 1998). For these enzymes the pro-peptide acts as an internal folding catalyst.

The influence of certain post-translational factors that influence enzyme yield (see Table 4.2) have been studied for some lipases. The procaryotic lipase from *P. cepacia* expressed in *E. coli* was mainly produced as inclusion bodies. The expression of the active enzyme could be increased when the wild-type signal peptide was exchanged to an *E. coli* signal peptide, and the pro-lipase inclusion bodies could be renatured in the presence of the lipase-specific foldase that had been separately expressed as inclusion bodies in *E. coli*. The final yield of active lipase was ≈ 0.2 g g^{-1} cell dry weight – that is, about 40 % of the cellular protein (Quyen et al., 1999). This was somewhat larger than the yields for recombinant active enzymes given in Table 4.1, although the production required more processing steps than did the direct production of active enzymes.

The eucaryotic lipase from *Rh. oryzae* has been expressed in *E. coli* and *S. cerevisiae* (Beer et al., 1998; Ueda et al., 2002). At 42 °C, mainly inclusion bodies were produced in *E. coli*, but by reducing the temperature to 39.4 °C some active enzyme was formed. The amount formed was larger when the wild-type signal peptide was exchanged to an *E. coli* signal peptide.

4.5
Downstream Processing of Enzymes

The costs for the downstream processing of enzymes – that is, purification to a desired degree of purity and conditioning for final use – generally account for more than 50 % of the total enzyme production costs. The downstream costs are higher for enzymes used for therapy and diagnostics than for technical enzymes; hence, in order to reduce these costs it is important to improve the downstream processing.

Following their production, the enzymes are located either inside the cell membrane (intracellular), outside the cell membrane and inside the cell wall (periplasmic), or outside the cell wall (extracellular). The first downstream processes in the isolation of the different enzymes are shown in Figure 4.9. Following the clarification step, the solutions with intracellular, periplasmic or extracellular enzymes should be concentrated using either ultra- or nano-filtration.[3)]

For this step it is important to know the molecular weight of the enzyme. In the concentration step, low molecular-weight impurities and ions that can interfere in the further purification steps, are removed. This is also favorable for any further chromatographic purification, as less adsorbent is required to adsorb the enzymes in the sample when it is concentrated. In addition to the subject enzyme, the con-

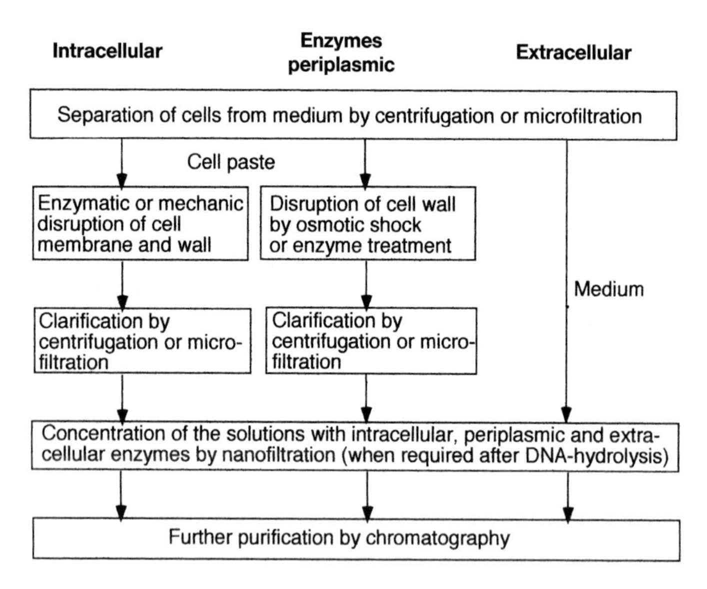

Fig. 4.9. The first downstream processing steps in the isolation of intracellular, periplasmic, and extracellular enzymes after their production in the fermenter.

3) Previously, solutions containing up to 50 % $(NH_4)_2SO_4$ were used to precipitate and concentrate enzymes in homogenates, or in the solution before crystallization (see Fig. 4.11 a and d). This should be avoided as this salt must be removed in the wastewater treatment. The better alternative here to concentrate enzymes now is to use ultra- or nanofiltration.

centrated solutions contain other enzymes, proteins, nucleic acids and polysaccharides. Although the content of these impurities is larger for intracellular enzymes, it is not negligible in solutions containing extracellular enzymes. Indeed, after the high cell density production of periplasmic or extracellular proteins, DNA concentrations of up to 100 mg L^{-1} have been observed in the medium (Castan and Enfors, 2000). This is also illustrated by the fact that concentrated solutions of technical extracellular enzymes exhibit an ultra-violet absorption spectrum that more closely resembles that for DNA/RNA than for protein (Scheidat, 1999).

Whether these impurities must be separated from the concentrated enzyme solutions depends on the final use of the enzyme, and whether the impurities disturb the process in which the enzyme is to be used as a catalyst. For recombinant enzymes it may also be necessary to remove or hydrolyze the contaminating DNA in order to minimize the risk of unwanted gene transfer in the final use of the enzyme. For these separations it is essential to know the following properties of the enzyme:

1. Molecular weight (MW): this can be determined by sodium dodecyl sulfate (SDS)-gel electrophoresis under denaturing conditions (see Fig. 4.6). Before separation, the proteins are unfolded in a SDS solution and the S–S bonds reduced. They are then separated with respect to molecular weight in the gel and the MW scale is obtained with MW standards. For oligomeric enzymes, the MW of the monomer is determined. For enzymes consisting of several polypeptide chains (e.g., penicillin amidase) the MW of the different peptides are determined. After separation, the proteins are visualized either by protein staining or with immunological methods (Western blot) that visualizes only the protein and its processing products that interact with the antibody against the protein (see Fig. 4.6).

2. The isoelectric point (pI) – that is, the pH at which the net charge of the protein is 0. This can be determined by isolectric focusing (IEF), and is essential for selecting the separation conditions for ion-exchange or bifunctional (multimode) chromatography. This separation is, in contrast to SDS-gel electrophoresis, carried out under non-denaturing conditions. The pI can be determined as shown in Figure 4.10. In IEF, the proteins are focused where their pI equals the pH on the gel. The pH scale is determined by standards with different pI. After separation, the pI of the enzyme can be determined either with an enzyme-specific chromogenic substrate that is converted to a colored product on the gel, or immunologically with an enzyme-specific antibody, as in the Western blot in Figure 4.6.

3. Cofactors, especially ions required for optimal enzyme activity and stability, such as Ca^{2+} for penicillin amidase and lipase discussed in Section 4.4 that may be lost during the purification procedures.

4. The pH-range where the enzyme is stable, within which the purification must occur.

Enzymes are generally purified by chromatography. The enzyme properties used for their separation with different chromatographic adsorbents are listed in Table 4.4. Previously, the purification of an enzyme from a concentrated clarified homogenate involved up to more than five separation steps involving different types of chroma-

Table 4.4 Adsorbents that can be used for the chromatographic separation based on different properties of enzymes.

Chromatographic adsorbents for	*Enzyme properties*
Size-exclusion chromatography	Molecular weight
Anion-exchange chromatography	Surface charge (at pH > pI)
Cation-exchange chromatography	Surface charge (at pH < pI)[a]
Hydrophobic chromatography	Hydrophobicity of enzyme surface
Metal-affinity chromatography	Histidines at the surface; or His-tags[b]
Bifunctional or multimode chromatography	Surface charge and hydrophobicity of the surface (hydrophobic adsorption at pH > pI and desorption by electrostatic repulsion at pH < pI
Biospecific chromatography (Enzyme-inhibitor; enzyme-antibody; biotin-streptavidin; peptide-peptide antibody)	Specific hydrophobic and electrostatic binding interactions with inhibitor or antibody or biospecific tags (streptavidin or peptide)[b]

a) In cation- *but not in anion-* exchange chromatography metal ions bound to the enzyme that are essential for the activity may be removed by the adsorbent.

b) A His-tag with up to 8 His is added to the N- or C-terminal of the enzyme. This tag binds to the metal ions (Co^{2+}, Ni^{2+}) adsorbed on the surface of the adsorbent. It is only successful when the C- or N-terminal is located on the enzyme surface, and when the tag allows a correct folding of the enzyme. This and other tags fused to the enzyme allow an one-step isolation of enzymes from homogenates of enzymes that can be 'tagged' without loss of activity. They can also be used to make an intracellular enzyme extracellular. These tags are to be removed, when the enzyme must be identical with the wild-type enzyme in the final use (Nygren et al., 1994; Terpe, 2003). The removal of the tag requires at least two additional purification steps. One- or two-step chromatographic isolations of enzymes can be obtained without tags using biospecific, ion-exchange or bifunctional adsorbents (Figs. 4.10 and 4.11).

tography. In each step, more than 10 % of the enzyme could be lost, and this led to a low recovery of the enzyme. Hence, to reduce the enzyme costs it was necessary to lower the number of processing steps in order to increase the recovery. The process engineering of chromatography has improved considerably during the past 10–20 years, and this includes the development of continuous chromatographic processes such as simulated moving bed (SMB), continuous separation (CSEP) and continuous annular chromatography (CSEP), all of which have recently been introduced into the downstream processing of proteins (Gottschlich et al., 1996; Imamoglu, 2002; Uretschlager and Jungbauer, 2002). The manufacturers of chromatographic adsorbents have also developed a wide variety of adsorbents suitable for the chromatographic purification of enzymes. This – combined with knowledge of the pI, MW and biospecific interactions of the enzyme – allows a more rational design of the enzyme purification, with a higher recovery. Examples of older and newer enzyme purification schemes are shown in Figures 4.10 and 4.11. The newer processes allow a one-step isolation and purification of the enzyme from a concentrated homogenate with bifunctional or multimode (Fig. 4.11B and C) and biospecific adsor-

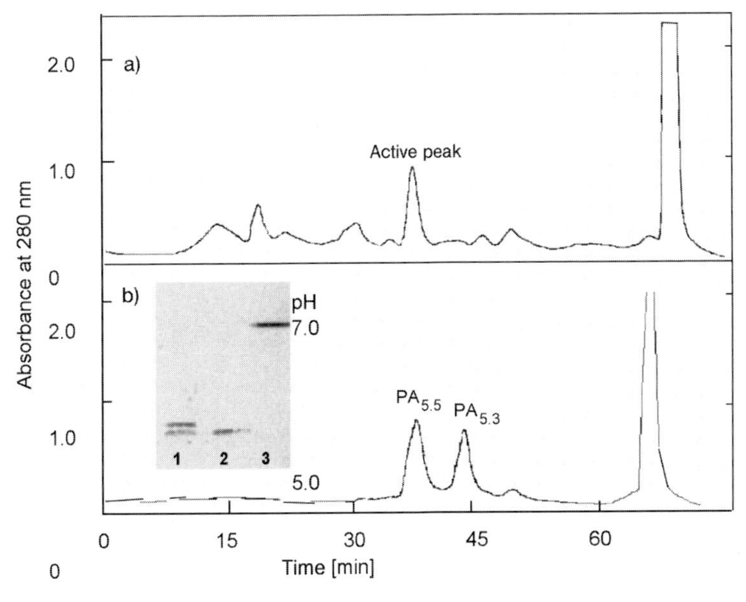

Fig. 4.10. A two-step chromatographic purification of *A. faecalis* penicillin amidase (PA) produced in *E. coli* from the clarified periplasmic fraction. Elution profile of the anion-exchange chromatography (a) and re-chromatography (a) of the active fraction from (b) with a Mono Q 10/10 column. After equilibration with 30 mM Tris-HCl, pH 7.5 at 2 ml min^{-1}, and application of the sample in the same buffer, the column was eluted as follows: 0–60 min with a linear salt gradient 0–20 % 1 M NaCl in 30 mM Tris-HCl; 60–80 min with 30 mM Tris-HCl containing 1 M NaCl. The insert shows Coomassie-stained IEF of the active bands from the rechromatography (lane 1: $PA_{5.5}$; lane 2: $PA_{5.3}$ that is rapidly formed from $PA_{5.5}$ by proteolytic processing; lane 3: purified PA from *E. coli* with pI 7.0). The last peak in (a) and (b) contains mainly DNA. Cation-exchange chromatography to avoid adsorption of DNA could not be performed due to the low stability of the PA from *A. faecalis* below pH 4.5. The subscripts indicate the pI of the different enzyme forms (Ignatova et al., 1998).

bents (Gottschlich et al., 1996), or two-step processes using the same ion-exchange adsorbent (Fig. 4.10). With these methods, the enzyme recovery can be increased to more than 80 %.

4.5.1 Technical Enzymes

When the clarified homogenate containing a technical enzyme does not also contain contaminants that disturb the enzyme's use as a biocatalyst, it must not be purified further. This applies especially to enzymes that are immobilized. However, when the final process requires a DNA-free enzyme, or an enzyme preparation with a higher specific activity (in U g^{-1} protein), the clarified homogenate must undergo further purification. DNA can be hydrolyzed in the clarified homogenate by adding a DNase, after which the hydrolysis products and contaminating proteins with mo-

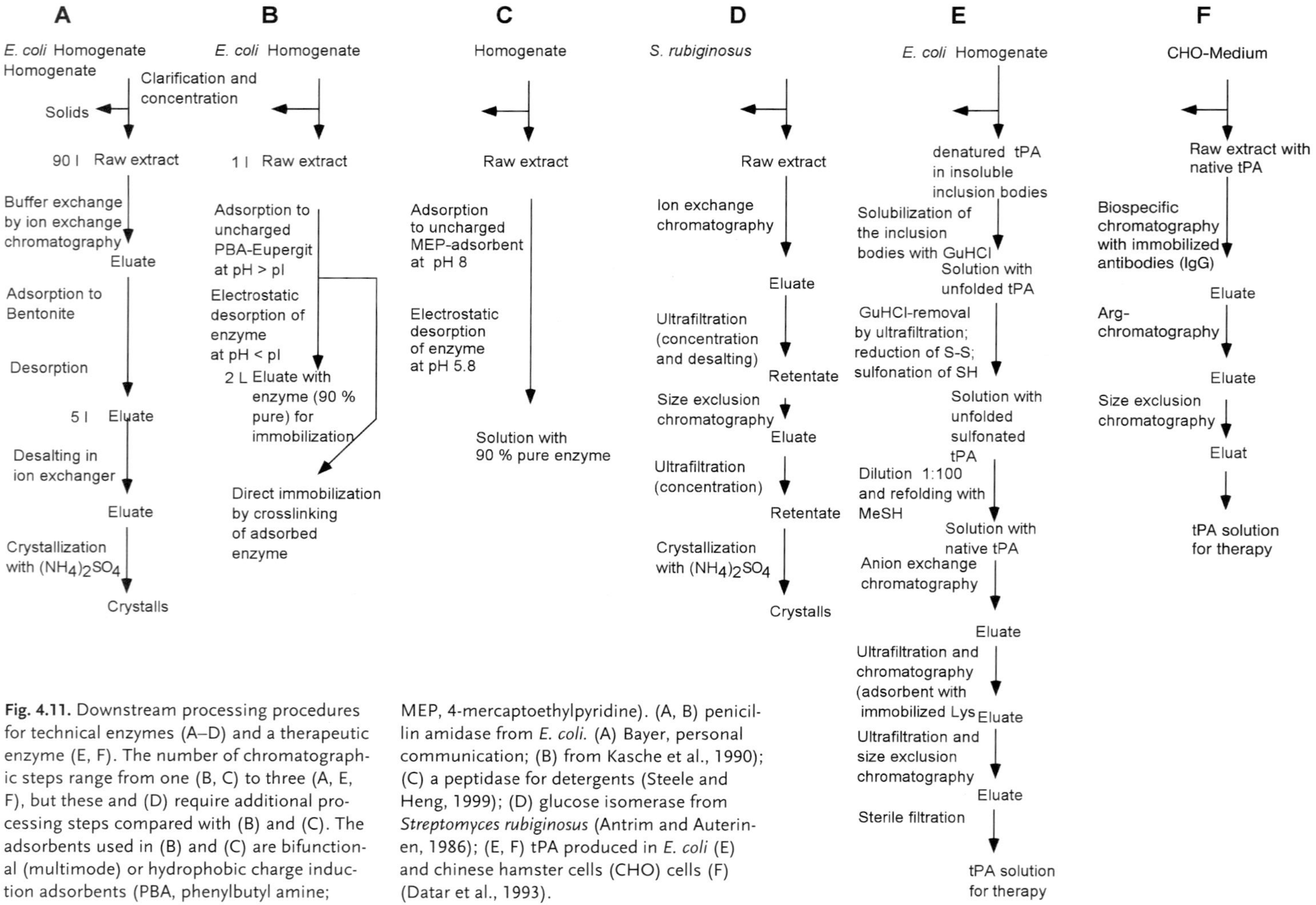

A

E. coli Homogenate
Homogenate
Clarification and concentration
Solids
90 l Raw extract
Buffer exchange by ion exchange chromatography
Eluate
Adsorption to Bentonite
Desorption
5 l Eluate
Desalting in ion exchanger
Eluate
Crystallization with $(NH_4)_2SO_4$
Crystalls

B

E. coli Homogenate
1 l Raw extract
Adsorption to uncharged PBA-Eupergit at pH > pI
Electrostatic desorption of enzyme at pH < pI
2 L Eluate with enzyme (90 % pure) for immobilization
Direct immobilization by crosslinking of adsorbed enzyme

C

Homogenate
Raw extract
Adsorption to uncharged MEP-adsorbent at pH 8
Electrostatic desorption of enzyme at pH 5.8
Solution with 90 % pure enzyme

D

S. rubiginosus
Raw extract
Ion exchange chromatography
Eluate
Ultrafiltration (concentration and desalting)
Retentate
Size exclusion chromatography
Eluate
Ultrafiltration (concentration)
Retentate
Crystallization with $(NH_4)_2SO_4$
Crystalls

E

E. coli Homogenate
denatured tPA in insoluble inclusion bodies
Solubilization of the inclusion bodies with GuHCl
Solution with unfolded tPA
GuHCl-removal by ultrafiltration; reduction of S-S; sulfonation of SH
Solution with unfolded sulfonated tPA
Dilution 1:100 and refolding with MeSH
Solution with native tPA
Anion exchange chromatography
Eluate
Ultrafiltration and chromatography (adsorbent with immobilized Lys
Eluate
Ultrafiltration and size exclusion chromatography
Eluate
Sterile filtration
tPA solution for therapy

F

CHO-Medium
Raw extract with native tPA
Biospecific chromatography with immobilized antibodies (IgG)
Eluate
Arg-chromatography
Eluate
Size exclusion chromatography
Eluat
tPA solution for therapy

Fig. 4.11. Downstream processing procedures for technical enzymes (A–D) and a therapeutic enzyme (E, F). The number of chromatographic steps range from one (B, C) to three (A, E, F), but these and (D) require additional processing steps compared with (B) and (C). The adsorbents used in (B) and (C) are bifunctional (multimode) or hydrophobic charge induction adsorbents (PBA, phenylbutyl amine; MEP, 4-mercaptoethylpyridine). (A, B) penicillin amidase from *E. coli.* (A) Bayer, personal communication; (B) from Kasche et al., 1990); (C) a peptidase for detergents (Steele and Heng, 1999); (D) glucose isomerase from *Streptomyces rubiginosus* (Antrim and Auterinen, 1986); (E, F) tPA produced in *E. coli* (E) and chinese hamster cells (CHO) cells (F) (Datar et al., 1993).

lecular weights lower than that of the enzyme can be removed by ultrafiltration using a membrane that has a cut-off about 10 kDa below the MW of the enzyme.

When DNA cannot be removed using such a procedure, bifunctional or multimode chromatography (Fig. 4.11B and C), or one- or two-step ion-exchange chromatography (see Fig. 4.10) can be used. Bifunctional or multimode (also called charge induction) chromatography is used to separate enzymes on the basis of two properties (see Table 4.4). In this technique, a hydrophobic molecule containing phenyl or other groups and an amino group is immobilized in the chromatographic adsorbent particle. At a pH above the p*K*-value of the amino group, the adsorbent is uncharged and practically all enzymes in the clarified and concentrated homogenate are adsorbed hydrophobically (Kasche et al., 1990; Burton and Harding, 1998). At high ionic strength, the DNA in the homogenate is not adsorbed and is thus separated from the enzyme. When the pH is decreased, the positive charges on the adsorbent and the bound enzyme increase; this causes a repulsion that eventually will be stronger than the hydrophobic binding, and leads to desorption of the enzyme by electrostatic repulsion. By using these adsorbents, DNA-free technical enzymes of high purity can be obtained in either one-step discontinuous or continuous chromatographic separation procedures.

Technical enzymes are distributed to the end users as either solutions or powders, or immobilized to different porous supports (see Chapter 6). The solutions must have a defined active enzyme content that can be obtained by concentration (ultrafiltration) or dilution. To avoid bacterial growth and to maintain a desired activity over a given time, preservation and stabilization agents may be added to these solutions. The powder can be either freeze-dried (producing small particles that can be inhaled) or dried in a spray-dryer and covered with a protecting layer in particle sizes that cannot be inhaled. All of these conditioning steps must be carried out in closed production units in order to avoid exposure of employees or the environment to small enzyme particles that might cause an allergic response.

4.5.2 Enzymes for Therapy and Diagnostics

These enzymes must be very pure. In particular, therapeutical enzymes must be free from pyrogens and have a very low DNA content to guarantee that they are free from viruses and unwanted genes. To achieve this, more purification steps are required than for technical enzymes (see Fig. 4.11E and F). These two separation procedures for recombinant tissue plasminogen activator (tPA), an enzyme used to dissolve blood clots in the treatment of stroke, illustrate different strategies in the production of recombinant enzymes or proteins for therapy. The enzyme can be produced as biologically inactive inclusion bodies in bacteria, or as active enzyme in eucaryotic cells. In the latter case (but not in bacteria), the enzyme can also be produced in a glycosylated form, which is frequently required for full biological activity. The enzyme yield (in g L^{-1} fermenter) is much higher for the inclusion bodies ($\approx$5 g L^{-1}) than in Chinese hamster ovary (CHO) cells (0.05 g L^{-1}). The tPA in the inclusion bodies can be renatured in a multi-step process involving chaotropic solvents (urea or guanadinium hydrochloride, GuHCl), in addition to oxidizing and reducing (mercaptoethanol, MeSH) agents. The final refolding to the active enzyme must be performed in dilute solutions to avoid the formation of inactive tPA complexes. The refolding yield is also much less than 100%. All of these steps increase the costs of producing and purifying tPA from bacteria however, and consequently the low yield production in CHO cells may in fact be more economical than production via inclusion bodies in bacteria (Datar et al., 1993). The production of biologically active tPA in bacteria with a high yield should be possible under conditions discussed in Section 4.3 and Table 4.2.

Finally, the purified enzyme solutions are sterilized by filtration, and generally distributed to the final users as freeze-dried powders that are either ingested (digestive enzymes for substitution therapy) or dissolved in sterile buffers before use.

The quality of these enzymes must be checked with respect to three criteria: purity, biological activity, and identity (with the human enzyme). The methods available for performing this validation are detailed in Table 4.5.

Table 4.5 Methods to determine the purity, biological activity and identity of therapeutic enzymes (Geisow, 1991).

Quality criteria	*Method*
Purity (this includes the determination of impurities such as DNA and other compounds)	PCR (DNA) Electrophoresis (IEF and SDS) Chromatography Mass spectrometry
Biological activity	Determination of enzyme activity Immunological methods
Identity (with the human enzyme)	Chromatography N-terminal sequencing Mass spectrometry Peptide mapping after hydrolysis with a peptidase

4.6 Exercises

1. Is *E. coli* generally a class I organism?
2. *S. cerevisiae* is a class I and GRAS organism that is used by some companies as a host for technical enzymes such as chymosin (used in the production of cheese) and savinase (used in washing powder; see Table 4.1). In order to select the cells that have been transformed, the enzyme gene is transferred with a selection marker. Previously (and for laboratory use) the main selection marker

was and is antibiotic resistance (in bacteria a gene for β-lactamase, an enzyme that hydrolyses β-lactam antibiotics). Why should this be avoided in the recombinant production of technical enzymes and enzymes for therapy? What policy – and alternatives – do the enzyme-producing companies have here?

3. *E. coli* cells contain ≈ 70 % H_2O and 15 % protein. The cells can be considered to be cylindrical. What is the maximum protein amount that can be produced per liter in a fermentation? Is this smaller or larger than for the real cells that are spherical and not planar at the end of the cylinder?

4. How much penicillin amidase could be produced when the cells are grown to the maximum density in Exercise 3? Discuss whether this is possible to achieve.

5. What processes that influence the enzyme yield shown in Figure 4.4 and Table 4.2 have been studied for penicillin amidase and lipases in Section 4.4?

6. Why must enzymes that bind cofactors such as Ca^{2+} or Zn^{2+} in the cell, have very low equilibrium constants (dissociation constants) for the binding of these ions?

7. Is it sustainable to produce inclusion bodies?

8. Technical enzymes from extreme thermophilic organisms can be produced in *E. coli*. Discuss how they can be easily separated from a homogenate, based on information provided in Table 2.9. Heat-stable proteins tend to be more slowly hydrolyzed by peptidases than less-stable proteins. This frequently correlates with allergenic potential, caused by the uptake of the whole unhydrolyzed protein in the intestine. What must therefore be studied for heat-stable enzymes before they are used in the food industry?

9. Explain the separation procedures in Figure 4.11B and C.

10. What is the difference in the purification of tPA in Figure 4.11E and F?

11. Why does the medium with extracellular enzymes contain DNA? How can it be separated from the enzyme? When the medium is separated from the cells by filtration, must the cells be washed to increase the enzyme yield? Why should the solution with the enzyme be concentrated by ultrafiltration before chromatographic purification steps?

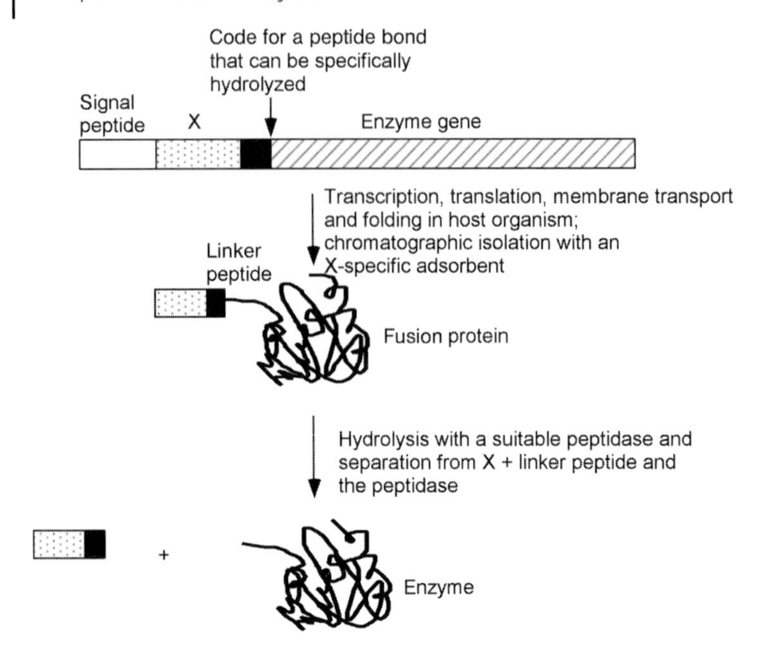

12. The above scheme illustrates the use of so-called tags X or extracellular proteins in the production of extracellular fusion proteins that can be easily separated using X-specific chromatographic adsorbents. What are basic requirements for the success of this? (Hint: are the N- or C-ends of the desired protein on the surface (see Fig. 2.6)? The enzyme must be correctly folded.) Discuss advantages and disadvantages of this method to produce and isolate an enzyme compared with other methods described in this chapter (Nygren et al., 1994).

Literature

Textbooks

The following are general sources on enzyme biosynthesis and purification, classification of microorganisms and on safety aspects of enzyme production and use.

Subramanian, G. (Ed.), *Bioseparation and Bioprocessing*, Vols. I and II, Wiley-VCH, Weinheim, 1998

AMFEP (Association of Manufactures of Fermentation Enzyme Products) the homepage (www.amfep.org) contains important information on safety aspects on enzyme production and use, and has links the main producers of enzymes and to international and national organizations that formulate the regulations on the use of enzymes, especially in the food sector

FDA (U.S. Food and Drug Administration) (2001), Partial list of microorganisms and microbial-derived ingredients that are used in foods. www.cfsan.fad.gov/~dms/opa-micr.-html

JECFA (Joint FAO/WHO expert committee on food additives) (2001), Compendium of food additive specifications (http://apps.3.fao.org/ jecfa) contains an extensive list of references on safety aspects of enzyme production and use

RKI (Robert Koch Institut), The German classification of microorganisms (www.rki.de/Gentec/ZKBS/ZKBS.HTM

References

Albee, K.R., Butler, M.H., Wrigtht, B.E., Cellular concentration of enzymes and their substrates, *J. Theor. Biol.* **1990**, *143*, 163–195

Altenbuchner, J., Siemann, M., Syldatk, C., Hydantoinases and related enzymes as biocatalysts for the synthesis of unnnatural chiral amino acids, *Curr. Opin. Biotechnol.*, **2001**, *12*, 559–563

Amneus, H., *On the use of pancreatic proteases for indication of genetic mutation*, Dissertation, Univ. Uppsala, Sweden, 1977

Antrim, R.L., Auterinen, A.-L., A new regenerable immobilized glucose isomerase, *Starch/Stärke*, **1986**, *38*, 132–127

Beer, H.D., McCarthy, J.E.G., Bornscheuer, U.T., Schmid, R.D., Cloning, expression, characterization and role of the leader sequence of a lipase from *Rhizopus oryzae*, *Biochim. Biophys. Acta*, **1998**, *1399*, 173–180

Bruggink, A. (Ed.), *Synthesis of ß-lactam antibiotics*, Kluwer Academic Publishers, Dordrecht, 2001

Bryan, P.N., Protein engineering of subtilisin, *Biochim. Biophys. Acta*, **2000**, *1543*, 203–222,

Burton, S.C., Harding, D.R.K., Hydrophobic charge induction chromatography: salt independent protein adsorption and facile elution with aqueous buffers, *J. Chromatogr. A*, **1998**, *814*, 71–81

Castan, A., Enfors, S.O. Characterization of a DO-controlled fed-batch culture of *Escherichia coli*. *Bioproc. Eng.* **2000**, *22*, 509–515

Christiansen, T., Michaelsen, S. Wümpelmann, M., Production of savinase and population viability of *Bacillus clausii* during high cell density fermentations, *Biotechnol. Bioeng.*, **2003**, *83*, 344–352

Datar, R.V., Carwright, T., Rosen, C.-G., Process economics of animal cell and bacterial fermentations: a case study analysis of tissue culture plasminogen activator, *Bio/Technology*, **1993**, *11*, 349–357

Deshpande, S.S., Ambedkar, V.K., Sudhakaran, V.K., Shewale, J.G., Molecular biology of β-lactam acylase, *World J. Microbiol. Biotechnol.*, **1994**, *10*, 129–138

Enfors, S.O., Control of proteolysis in fermentation of recombinant proteins, *Trends Biotechnol.*, **1992**, *10*, 310–315

Geisow, M.J., Characterizing recombinant proteins, *Bio/Technology*, **1991**, *9*, 921–924

Gottschlich, N., Weidgen, S., Kasche, V., Continuous biospecific affinity purification of enzymes by simulated moving bed chromatography: Theoretical description and experimental results, *J. Chromatogr.*, **1996**, *719*, 267–274

Henge, R., Bukau, B., Proteolysis in prokaryotes: protein quality control and regulatory principles, *Mol. Microbiol.*, **2003**, *49*, 1451–1462

Ignatova, Z., Stoeva, S., Galunsky, B., Hörnle, C., Nurk, A.,Piotraschke, E., Voelter, W., Kasche, V., Proteolytic processing of penicillin amidase from *A. faecalis* in *E. coli* yields several active forms, *Biotechnol. Lett.* **1998**, *20*, 977–982

Ignatova, Z., Taruttis, S., Kasche, V., Role of the intracellular proteolysis in the production of the periplasmatic penicillin amidase in *Escherichia coli*, *Biotechnol. Letters*, **2000**, *22*, 1727–1732

Ignatova, Z., Mahsunah, A., Geogieva, M., Kasche, V., Improvement of Posttranslational Bottlenecks in the Production of Penicillin Amidase in Recombinant *Escherichia coli* strains, *Appl. Environ. Microbiol.*, **2003**, *69*, 1237–1245

Imamoglu, S., Simulated moving bed chromatography (SMB) for application in bioseparation, *Adv. Biochem. Eng. Biotechnol.*, **2002**, *76*, 211–231

Jaeger, K-E., Dijkstra, B.W., Reetz, M.T., Bacterial biocatalysts: molecular biology, three-dimensional structures, and biotechnological applications of lipases, *Annu. Rev. Microbiol.*, **1999**, *53*, 315–351

Jones, H.E., Holland, I.B., Campbell, A.K., Direct measurement of free Ca^{2+} shows different regulation of Ca^{2+} between the periplasm and the cytosol of *Escherichia coli*, *Cell Calcium*, **2002**, *32*, 183–192

Kästner, M., personal communication, **1994**

Kasche, V., Scholzen, T. Boller, Th., Krämer, D.M., Löffler, F., Rapid protein purification using PBA-Eupergit™ a novel method for large-scale procedures, *J. Chromatogr.*, **1990**, *510*, 149–154

Kasche, V., Lummer, K., Nurk, A., Piotraschke, E., Rieck s, A., Stoeva, S., Voelter, W., Intramolecular autoproteolysis initiates the maturation of penicillin amidase from *E. coli*, *Biochim. Biophys. Acta*, **1999**, *1433*, 76–86

Kasche, V., Galunsky, B., Ignatova, Z., Fragments of pro-peptide activate mature penicillin amidase of *Alcaligenes faecalis*. *Eur. J. Biochem.* **2003**, *270*, 4721–4728

Kasche, V., Ignatova, Z., Märkl, H., Plate W., Punckt N., Schmidt D., Wiegandt K., Ernst B., Cofactor (Calcium ion) content as a yield determining factor in penicillin amidase production in high cell density fermentations, submitted, **2004**

Makrides, S.C., Strategy for achieving high-level expression of genes in *Escherichia coli*. *Microbiol. Rev.*, **1996**, *60*, 512–538

McDonough, M.A., Klei, H.E., Kelly, J.A., Crystal structure of penicillin G acylase from the Bro 1 mutant strain of *Providencia rettgeri*, *Prot. Sci.*, **1999**, *8*, 1971–1981

Middelberg, A.P.J., Preparative protein refolding, *Trends Biotechnol.*, **2002**, *20*, 437–443,

Miele, L., Plants as bioreactors for biopharmaceuticals; regulatory considerations, *Trends Biotechnol.*, **1997**, *15*, 45–50

Mitraki, A., Fane, B., Haase-Pettigell, C., Sturtevant, J., King, J., Global suppression of protein folding defects and inclusion body formation, *Science*, **1991**, *290*, 54–58

Nardini, M., Lang, D.A., Liebeton, K., Jaeger, K.-E., Dijkstra, B.W., Crystal structure of *Pseudomonas aeruginosa* Lipase in the Open Conformation. *J. Biol. Chem.*, **2000**, *275*, 31219–31225

Niehaus, F., Bertoldo, C., Kähler, M., Antranikian, G., Extremophiles as a source of novel enzymes for industrial application, *Appl. Microbiol. Biotechnol.*, **1999**, *51*, 711–729

Nielsen, J.E., Borchert, T.V., Protein engineering of bacterial α-amylases. *Biochim. Biophys. Acta*, **2000**, *1543*, 253–274

Nygren, P.Å., Ståhl, S., Uhlén, M., Engineering proteins to facilitate bioprocessing, *Trends Biotechnol.*, **1994**, *12*, 184–188

Pilone, M.S., Pollegioni, L., D-amino acid oxidase as an industrial biocatalyst. *Biocatal. Biotrans.*, **2002**, *20*, 145–159

Quyen, D.T., Schmidt-Dannert, C., Schmid, R.D., High-level formation of active *Pseudomonas cepacia* lipase after heterologeous expression of the encoding gene and its modified chaperone in *Escherichia coli* and rapid in vitro refolding, *Appl. Env, Microbiol.*, **1999**, *65*, 787–794

Rosenaua, F., Tommasen, J., Jaeger, K-E., Lipase-specific foldases, *ChemBiochem.*, **2004**, *5*, 152–161

Rudolph, N.S., Biopharmaceutical production in transgenic livestock, *Trends Biotechnol.*, **1999**, *17*, 367–374

Scheidat, B., *Bioverfahrenstechnische Aspekte zum Einsatz von technischen Enzymen am Beispiel der kommunalen Abwasserreinigung*, Thesis, TU Hamburg-Harburg, **1999**

Scherer, L.J., Rossi, J.J., Approaches for the sequence-specific knockdown of mRNA, *Nature Biotechnol.*, **2003**, *21*, 1457–1465

Steele, L., Heng, M., *Rapid protease variant purification and scale-up using novel hydrophobic charge induction chromatography*, Recovery of biological products IX, Whistler, BC, May 23–28, **1999**

Terpe, K., Overview of tag protein fusions: from molecular and biochemical fundamentals to commercial systems, *Appl. Microbiol. Biotechnol.*, **2003**, *60*, 523–533

Ueda, M., Takahashi, S., Washida, M., Shiraga, S., Tanaka, A., Expression of *Rhizopus oryzae* lipase gene in *Saccharomyces cerevisiae*, *J. Mol. Cat. B: Enzymatic*, **2002**, *17*, 113–124

Uretschlager, A., Jungbauer, A., Preparative continuous annular chromatography (P-CAC), a review, *Bioprocess. Biosyst. Eng.*, **2002**, *25*, 120–140

Wickner, S., Maurizi, M.R., Gottesman, S., Posttranslational quality control: folding, refolding, and degrading proteins, *Science*, **1999**, *286*, 1888–1893

5 Application of Enzymes in Solution: Soluble Enzymes and Enzyme Systems

Topics/Purposes/Problems	Solutions to problems
Reaction/transformation • technical application • preparative purpose	Enzymes in solution vs. immobilized biocatalyst (criteria see below)
Unsoluble high molecular substrates	Enzymes in solution
Enzyme amount required at • given substrate concentration • time • temperature	Calculations based on kinetics, productivity, inactivation kinetics
Cost of enzymes relating to cost of product	Calculation based on productivity • low cost: – application of enzymes in solution • high cost: – optimization of enzyme production (Chapter 4) – membrane reactor (Chapter 5, Section 5.4) – immobilization (Chapters 6 and 7)
Further criteria • residual enzyme in product • reaction dependent on cosubstrates • substrate insoluble in water	

Biocatalysts and Enzyme Technology. K. Buchholz, V. Kasche, U. T. Bornscheuer

ISBN: 3-527-30497-5

5.1 Introduction and Areas of Application

Soluble enzymes continue to dominate technical applications, most notable in the areas of

- detergent formulations with proteases, lipases and cellulases; and
- food manufacturing, with starch hydrolysis and further processing, bread, cheese and fruit juice manufacture.

These applications represent the most important with respect to volume and turnover, and a general survey of the situtation is presented in Table 5.1. In recent years, applications have extended into the areas of paper and pulp processing, textile manufacture, and the degumming of oil, with one-step reactions and hydrolytic enzymes continuing to dominate the scene.

However, in recent years newer areas of application and perspectives of synthesis in organic and pharmaceutical chemistry have gained much interest, and these are described in more detail in Chapters 3 and 9.

The impact of genetic engineering, notably with regard to recombinant techniques, has played a major role both with regard to significantly increased enzyme yields with consequent reductions on cost, together with improved performance in terms of stability and productivity. In addition, the range of applications for enzymes has also been significantly extended, and their selectivity greatly modified. In this respect, one important factor is that of safety aspects, notably in food manufacture, where the GRAS (generally recognized as safe, see p. 203) status plays a key role.

Several conditions govern the potential for the application of soluble enzymes, and their advantages and/or shortcomings as compared to chemical reactions (c.f. in starch hydrolysis) or to the application of immobilized enzymes (Table 5.2).

Soluble enzymes in production processes are limited to singular application, with the condition of low price (5–20 € kg^{-1} or € L^{-1} of concentrate, respectively), or their application in small amounts, as in analytical testing or medical/pharmaceutical use.

The hydrolysis of high molecular-weight, or insoluble, or adsorbed substrates proceeds efficiently only with the use of soluble enzymes. With immobilized biocatalysts, a slow diffusion and poor accessibility of macromolecules, or adsorbed substances would imply an extremely low efficiency. Thus, the application areas of proteases, cellulases and lipases for detergents in laundry, as well as glycosidases in starch hydrolysis, fruit juice and vegetable processing, remain the major areas for the utilization of soluble enzymes (see Section 5.3).

Membrane processes allow for the continuous use of soluble enzymes in the conversion of low molecular-weight substrates, with these being notably favorable for systems requiring coenzymes and/or multi-enzyme systems (see Section 5.4).

The total turnover in industrial enzyme sales was estimated as being in the range of € 1 billion in 1995 (Godfrey and West, 1996), and € 2 billion in 2000 (Poulson,

Table 5.1 Industrial applications of enzymes: major selected areas.

Market share[a]	*Enzyme*	*Purpose, application in solution*	*Membrane systems*	*Immobilized systems*
	Hydrolases			
Detergents:	Proteases	Detergents		
35–40 %	Lipases			
	...	Textile, oil		
Food:	Rennin	Cheese		
40–45 %	Chymosin	manufacture		
	Glycosidases			
(Starch:	Amylases	Starch hydrolysis		
11–15 %)[b]	Amyloglucosidases			(Dextrin hydrolysis)
	Pectinases[c]	Juice manufacture		
	Lactase			Lactose hydrolysis
	Esterases			
	Acylases		Amino acid synthesis	Amino acid synthesis
	Penicillin amidase			Penicillin hydrolysis/ synthesis
	Oxidoreductases			
	Glucose	Drinks manufacture		Glucose analyzer
	Oxidase	Analytics		
	Lipooxygenase	Baking		
	Transferases			
	Aminotransferase			Amino acid manufacture
	Cyclodextrin transferase			Cyclodextrin manufacture
	Isomerases			
(12 %)[b]	Glucose isomerase			Manufacture of High Fructose Corn Sirup (HFCS)
2000 Mio €[d]	**Total turnover** Worldwide			

a) Technical enzymes, worldwide, estimates.
b) Included in food.
c) Including other activities.
d) 2000.

Table 5.2 Advantages and disadvantages of the application of enzymes in solution and immobilized enzymes, respectively.

Advantage for soluble enzymes	*Disadvantage for soluble enzymes*	*Advantage for immobilized enzymes*	*Disadvantage for immobilized enzymes*
Low enzyme price	High enzyme price Allergy by enzymes in product	High enzyme price requires immobilization	High cost of immobilization
Singular application (analytical, therapeutic)	Limited productivity	Continuous processes, high productivity	
Low amounts (analytical, therapeutic application)			
Efficiency with insoluble, or adsorbed, or high molecular-weight substrates[a]		Fewer by-products[b]	More by-products[c]
Complex systems with coupled reaction path			Low efficiency with complex systems
Coenzyme-dependent reactions			No coenzyme retention
Application of membrane reactor systems (retention or recycling of enzymes)		Easy re-use	

[a] Enzyme remains in the product.
[b] Example: less trisaccharides due to decreasing lactose gradient in the carrier in lactose hydrolysis.
[c] Example: more reversion products due to increasing product gradient in the carrier with amyloglucosidase (dextrin hydrolysis).

2001). The most important areas of this market relate to the food sector, which was estimated at 40–45 %, including 11–15 % for starch processing and 12 % for glucose isomerase, while enzymes in detergents accounted for 35–40 %.

5.1.1 The Impact of Genetic Engineering

The impact of genetic engineering, with recombinant techniques, has played a major role in extension of the fields of enzyme applications, with over 50 % market share for recombinant enzymes being reached in 1992. Moreover, yields for enzymes in fermentation processes have also increased tremendously, by factors up to

100, and this in turn has led to much reduced prices, typically lower by a factor of 10 as compared to enzymes from conventional sources (Poulson, 2001).

Thermal stability could be improved by site-directed and random mutagenesis, as well as by directed evolution. Although these are now established techniques, they must be worked out for each enzyme individually (for details, see Chapter 2, Section 2.11). Site-directed mutagenesis may also be used to exchange amino acids that are unstable at elevated temperature. Examples of this include asparagine and glutamine which undergo deamidation, and lysine in amylases and glucose isomerase which tends to form Schiff's bases with reducing sugars (Sicard et al., 1990; Quax et al., 1991). The modification of regions involved in unfolding is also a target. The exchange of between 4 and 12 amino acids may be required for technically relevant effects, such as shifting the application range by >5 °C towards a higher operation temperature (see Chapter 2, Section 2.11). Thermostable amylases obtained from hyperthermophilic microorganisms are active up to 130 °C, and are used industrially in the range of 105–110 °C for about 5 minutes during starch processing. Other important improvements concern ion requirements (e.g., amylases which no longer need Ca^{2+} ions to be added to substrate solutions) or shifts in the pH application range (e.g., lowering the pH of amylases and approaching the operational range of amyloglucosidases) so that ion-exchange operations to adjust the pH of substrate solutions may be omitted.

A range of successful modifications has been summarized in Chapter 2 (Section 2.11) and Chapter 3. Despite these successes however, native enzymes are generally required for food processing in several European countries, mainly due to customer preferences.

5.1.2
Medium Design

Medium design is essential in order to improve enzyme stability. The protective effect of salts and polyols is well known, and applied to improve the storage stability of commercial enzyme preparations. Under operating conditions, the stabilizing effects may also be effective at both elevated temperatures and pressures. A stabilizing effect was seen to be optimal for salts, including ammonium or potassium and sulfuric acid ions, and is outlined in the following lyotropic series (Foster et al., 1995; Curtis et al., 2002; Morgan and Clark, 2004):

- Cations: $NH_4^+ > K^+ > Na^+ > Mg^{2+} > Ca^{2+}$ (in some cases Ca^{2+} is essential for stability).
- Anions: $SO_4^{2-} > Cl^- > NO_3^-$.

The protective effect of polyols has also been investigated by several authors (Monsan and Combes, 1984; Ye et al., 1988). Prominent among these is the example of sorbitol stabilizing β-galactosidase by more than two orders of magnitude at elevated concentrations (Athes and Combes, 1998; see also data below). These effects are most significant in industrial processes for hydrolyzing starch, and for the isomerization, synthesis and transformation of sugars, where high concentrations of substrates and products are favorable for product recovery (Cheetham, 1987). Polymers have also been used as enzyme stabilizers; thus, polysaccharides such as dextran, and synthetic polymers such as polyvinylalcohol and polyvinylpyrolidone, and polyvinylsaccharides were investigated with respect to their protecting and stabilizing effects, by the addition of up to 1 % to the solution (Gibson et al., 1993; Bryiac and Novorita, 1994; Kühlmeyer, 2000).

Enzyme modification – for example at the external protein surface – has been applied to both stabilization and specific applications. Certain reagents – for example, acyl derivatives or epoxides – may react with amino acid functional groups, thus modifying the surface charge and/or the hydrophobic character of the protein. Crosslinking with bifunctional reagents can improve the thermal stability by reducing the protein flexibility and its tendency to unfold (for reviews, see Laane et al., 1987; Ballesteros et al., 1998). Recombinant strategies however have proven to be more successful in this respect.

The formulation of detergent enzymes for improved stability (notably against high pH) and bleaches such as perborates or percarbonates may include several stabilizing agents. One successful approach used different coating layers of salt, antioxidant and polymers such as polyvinylalcohol. The encapsulation effected protection of the proteases as well as a slow release of the enzyme from the granule, and this resulted in a much improved storage stability, with 80 %, as compared to 40 %, residual activity in a comparative test (Gaertner et al., 2000). In many applications, granulation by agglomeration of the enzyme concentrate in the presence of different salts and polysaccharides is a well-established method for making particles of appropriate size in the range of 0.5 to 1 mm, mainly in order to avoid allergic reactions (Eriksen 1996).

5.1.3 Safety Aspects

Safety aspects are very important, not only for food manufacture but also for products where contact with the protein is possible (e.g., detergents). A joint FAO/WHO Expert Committee on Food Additives (JECFA; see Chapter 4, p. 194) considered the problems of the safety of enzyme preparations in food processing, and presented recommendations and laid down criteria for evaluating enzymes according to the source material (Denner and Gillanders, 1996). Enzymes obtained from the edible tissues of animals or plants commonly used as foods, and from microorganisms traditionally accepted as constituents of, or for the preparation of foods, were all regarded as foods. For enzymes obtained from non-pathogenic microorganisms commonly found as contaminants of food, the committee recommended short-term toxicity tests. For enzymes from other microorganisms used as sources, more extensive toxicological studies were required. For enzymes from genetically modified organisms, the safety evaluation must include the donor, the host and the resulting cloned organism. For enzyme purity, specifications were published in 1972 by FAO (Denner and Gillanders, 1996). For food additives and allergic reactions, the Acceptable Daily Intake (ADI) represents the amount of food additive that can be taken daily in the diet without risk (see WHO homepage: *www.WHO.org; www.elc-eu.org/addit3.htm*). For the production of recombinant enzymes, microorganisms with GRAS status[1] (see also Chapter 4, Section 4.1) should be used preferentially as host, including for example *Bacillus* sp., *A. niger* or yeast used in food manufacture.

5.2
Space-Time-Yield and Productivity

The kinetics of enzyme processing were outlined in Chapter 2, and form the basis for calculating or estimating the amount of enzyme and/or the reaction time (and thus the reactor volume) required for a given transformation in the technical scale. These aspects are summarized by the space-time-yield (*STY*), giving the substrate amount reacted, or product produced per unit of time and volume (see Chapters 2.7 and 8.6).

Catalyst productivity is the second economic key parameter, and indicates the amount of product that can be obtained per unit amount of biocatalyst (this will be dealt with in more detail at the end of this paragraph). It is important to bear in mind that transformations on the technical scale normally must be performed at non-optimal conditions with respect to the reaction rate and stability, such as high conversion, elevated temperature and non-ideal (technical) reactors (thus far from natural conditions). The important process conditions are summarized in Table 5.3.

The selection of reaction conditions aims at maximizing the yield and minimizing the overall cost of the process: the cost for raw material, investment, energy, enzyme, disposal of residual material and environmental protection (waste water, ex-

Table 5.3 Selection of reaction conditions in technical processes.

Reaction conditions	*Reasons*	*Consequences*
High concentrations	Simplified, efficient product recovery; avoiding infections; high STY	Substrate inhibition
High conversion	As before; minimizing substrate cost	Product inhibition
Temperature	Optimization of reaction rate and yield; avoiding infections	Enzyme inactivation
pH	Shift of equilibrium towards optimal yield; adaption to side and subsequent reactions; corrosion	Non-optimal conditions for reaction rate and stability

1) GRAS (FDA Terminus): General recognized as safe: For general recognition, there must be an expert consensus that the substance is safe for use as a component of food, and this expert consensus of safety must be based on either: (a) generally available data and information to show common use of the substance in animal feed prior to 1958; or (b) scientific procedures, which require the same quantity and quality of scientific data needed for FDA approval of the substance as a food additive. In addition, this information must be published in the scientific literature.

haust gas, etc.). Enzyme costs form a minor part of the total in general (1–10 %), whereas raw material expenditure dominates in processes at the large scale with bulk products (e.g., 30–50 % for products prepared from starch). In these cases, as well as in detergent applications, the enzyme costs typically comprise about 1 %. *Expense and investment costs* for upstream and downstream operations – for example, for product isolation and purification – are in general significantly higher than those for the transformation step; thus, the latter must be optimized with respect to these steps (high product yield and concentration, minimal side products).

The basic kinetic equations which allow for the calculation of process data and parameters discussed previously were indicated in Chapter 2, and are summarized in Box 5.1. It must be remembered that a transformation can proceed towards an equilibrium (e.g., in glucose isomerization) or under kinetic control (e.g., antibiotics- or peptide synthesis). For consistent and precise calculations, the corresponding equations must be applied (see Section 2.7). The kinetics of synthetic reactions with more than one substrate and different reaction pathways – as well as the hydrolysis of high molecular-weight substrates – may be much more complex. An example of this is the oligosaccharide synthesis with glucansucrases from sucrose and another sugar as acceptor (Böker et al., 1994; Demuth et al., 1999). It may be appropriate then to use simlipified kinetics for a limited range of fixed conditions, such as Michaelis–Menten kinetics for constant acceptor concentration, or first-order kinetics for protein hydrolysis (Prenosil et al., 1987).

For the application of enzymes in solution, batch (discontinuous) processes are standard, and the concentration of educts and products is then a function of time (Fig. 5.1). Hence, for the calculation of conversion rates the integrated equations given in Box 5.1 must be applied.

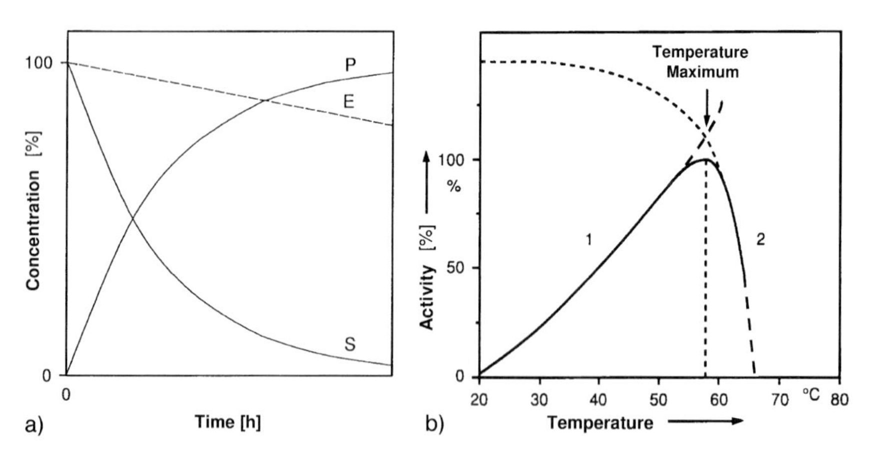

Fig. 5.1 (a) Concentration of substrate [S] and product [P] as a function of time during a discontinuous batch process. Also shown is the enzyme activity [E] as function of time due to inactivation. (b) Enzyme activity as a function of temperature with the different effects of increasing reaction rate (graph 1) and increasing inactivation rate (graph 2).

Box 5.1

Equations for the calculation or estimation, respectively, of the biocatalyst amount expressed as an activity V_{max} per unit reactor volume required for a defined substrate conversion in a given time[1]. From this, the biocatalyst productivity (amount of substrate converted to product)/V_{max} can be calculated (for derivations, see Chapter 2).

Without inhibition and inactivation of the biocatalyst:

The rate equation is (Eq. 2.8):

$$v = -\frac{d[S]}{dt} = \frac{k_{cat}[E][S]}{K_m + [S]} = \frac{V_{max}[S]}{K_m + [S]} \tag{5.1}$$

the dependence of V_{max} and K_m on pH, T, and inhibitors I is shown in Chapter 2, Section 2.7.2. Integration of Eq. (5.1) gives

$$V_{max} = \frac{([S]_0 - [S]) + K_m \ln([S]_0/[S])}{t} \tag{5.2}$$

where the subscript 0 denotes the initial substrate concentration or biocatalyst activity.

The biocatalyst productivity is then:

$$([S]_0 - [S])/V_{max} \tag{5.3}$$

Inhibition but no inactivation of the biocatalyst:

For **non-competitive substrate inhibition** (Eq. 2.25), Eq. (5.1) is modified to

$$v = \frac{V_{max}[S]}{K_m + [S]\left(1 + \frac{K_m}{K_i}\right) + \frac{[S]^2}{K_i}} \tag{5.4}$$

integration gives

$$V_{max} = \frac{([S]_0 - [S])\left(1 + \frac{K_m}{K_i}\right) + K_m \ln([S]_0/[S]) + \frac{[S]_0^2 - [S]^2}{2K_i}}{t} \tag{5.5}$$

1) The biocatalyst amount for free enzymes is usually expressed in U (see Chapter 2, Section 2.7.1), i.e., the biocatalyst amount that converts 1 μmol substrate per minute at a defined temperature. The calculated V_{max}-values can be easily converted to U.

For **competitive** product inhibition (Eq. 2.36), Eq. (5.1) is modified to

$$v = \frac{V_{max}[S]}{K_m + \frac{K_m}{K_i}[S]_o + [S]\left(1 - \frac{K_m}{K_i}\right)} \tag{5.6}$$

integration gives

$$V_{max} = \frac{([S]_o - [S])\left(1 - \frac{K_m}{K_i}\right) + \left(K_m + \frac{K_m}{K_i}[S]_o\right)\ln([S]_o/[S])}{t} \tag{5.7}$$

These equations apply to penicillin hydrolysis by penicillin amidase (see Chapter 2, Section 2.7.2.3).

Inhibition and inactivation of the biocatalyst:

The inactivation of the biocatalyst can be either a first-order process

$$V_{max} = V_{max,0}\, e^{-k_i t} \tag{5.8}$$

or a second-order process. The latter occurs in the autoproteolysis of peptidases, or when peptidases are present in the biocatalyst used. With first-order biocatalyst inactivation the denominator t in Eqs. (5.2), (5.5) and (5.7) is replaced by

$$\frac{1 - e^{-k_i t}}{k_i} \geq t \tag{5.9}$$

i.e., the biocatalyst productivity decreases when the biocatalyst is inactivated.

Enzyme inactivation is of major importance. The enzyme is lost after the transformation (this is different for membrane systems; see Section 5.4). Thus, the temperature must be selected for a rapid transformation, but taking into account that enzyme inactivation should not lead to low reaction rates prior to achievement of the desired conversion. In general, activation energies for enzyme catalysis are in the range of 40–80 kJ mol^{-1}, whilst those for enzyme inactivation are distinctly higher, in the range of 60–400 kJ mol^{-1}. Rather simple estimations with simplified kinetics allow for the identification of an appropriate compromise (see Exercise 2).

In order to calculate an enzymatic conversion, the procedure is as follows: the kinetic constants are taken from the literature (see Table 2.4 for important technical data; see also Bergmeyer (1983), Atkinson and Mavituna (1983), Schomburg and Stephan (1996), Schomburg and Schomburg (2000); and internet resources in Appendix I). If no data are available, they must be determined experimentally. The conversion required is fixed, thus for example 98 %, which corresponds to $([S_o] - [S]/[S_o]) = 0.98$, then the reaction time (and thus the *STY*) is also fixed (e.g.,

24 h). These figures are then introduced into the appropriate integrated equation and the required amount of enzyme $V_{max,0}$ is calculated. Note: It is important to take both substrate and/or product inhibition into account, or the results may be totally incorrect.

Finally, the loss of enzyme activity by inactivation must be calculated. Few data are available in the literature, so that this parameter – inactivation kinetics, a key for economic application – often must be determined experimentally (see Prenosil et al., 1987). It must be taken into account that enzyme stability under process conditions often differs from that in model solutions. In many cases, stabilizing conditions can be identified, notably in food manufacture, where carbohydrates are processed in concentrated solutions.

Thus, at high sucrose concentration (>50 %) a considerable stabilization was observed for (immobilized) invertase in the temperature range of 65–70 °C with a productivity of 10 tonnes (product) per kg (biocatalyst) (Monsan and Combes, 1984). An analogous effect was found for a protease developed for detergents (Adler-Nissen, 1986). For the application at elevated temperatures, enzymes from thermophiles ex-

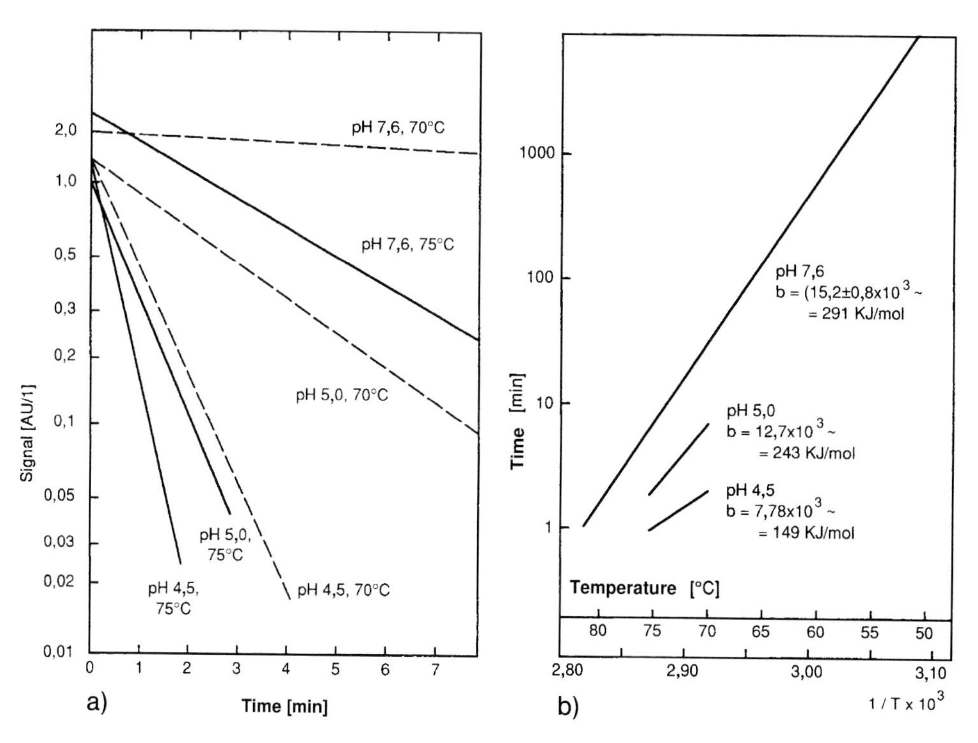

Fig. 5.2 (a) Inactivation of a protease (Alcalase) which is stable under alkaline conditions at different pH and temperatures (activity given in logarithmic scale as a function of time in min) (Adler-Nissen, 1986). (b) Correlation according to Arrhenius for t_D (t_D in min corresponds to the time during which the activity decreases by a factor of 10, thus the *practical maximal time of operation*) (Adler-Nissen, 1986).

hibiting high thermostability are available, and these can be cloned into appropriate enzyme-producing organisms. Experimental results for a protease are shown in Figures 5.2a and b. The influence of pH on stability is clear. For a thermostable α-amylase from *Bacillus licheniformis*, the following kinetic data for inactivation rates were obtained (de Cordt et al., 1992) (all subsequent data for k_i reported refer to a reaction following first-order with the dimension of min^{-1}): $k_i = 0.02 - 0.06$ (when the commercial enzyme preparation was diluted by factors of 1: 100 – 1: 500, 95 °C, pH 8,5, Ca^{2+}: 70 ppm); activation energy EA = 426 kJ mol^{-1} (for further data on kinetics of inactivation, including temperature and pressure as parameters, see Weemaes et al., 1996; see also Section 5.5, Task 2). Further data published by Feng et al. (2002) for inactivation kinetics of two thermostable amyloglucosidases, including stabilization by sorbitol are provided in Exercise 2.

Additional data on the kinetics of inactivation were summarized by Fullbrock (1996), and referred to proteases, β-galactosidase, α-amylases, and glucose isomerase. Among the highest stabilities reported so far was that of an esterase from *Pyrococcus furiosus* which was cloned in *E. coli* and exhibited a half-life of 50 minutes at 126 °C (Ikeda and Clark, 1998). A two-stage series-type mechanism of inactivation was observed for immobilized β-galactosidase (Illanes et al., 1998).

The biocatalyst (or enzyme) productivity can be calculated using Eq. (5.3) (see p. 205). Based on this calculation the enzyme costs as part of the product costs can be deduced. Examples of the enzyme amount required, the productivity and half-life (together with simple calculations) are provided in Section 5.5.

5.3
Examples for the Application of Enzymes in Solution

5.3.1
Survey

Since the first technical enzyme preparations were introduced during the 19th century, a broad range of diverse industrial applications has been developed. With respect to economic relevance however, two areas have dominated: (1) enzymes in detergents, where proteases (and more recently cellulases), lipases and amylases are important activities; and (2) amylases and amyloglucosidases in starch processing (see Table 5.1). Some major companies which supply enzymes are listed in Table 5.4; for a more extensive list, see Godfrey and West 1996, Data Index 2, pp. 545–554).

More extensive details of technical enzymes and their commercial applications are listed in Table 5.5 (for more detailed data on commercial enzymes, see Godfrey and West 1996, Data Index 3, pp. 555–581). One major field is that of detergents, and notably of washing powders and *liquid preparations*, where proteases, cellulases, lipases, and amylases are all applied (see Section 5.3.3). The success of lipase application evolved with genetically modified preparations that were active in aqueous solution (native lipases are primarily active at interfaces, such as oil/water). Most appli-

Table 5.4 Selected major enzyme supplying/selling companies (see Appendix I).

Company	***Country***
AB Tech (former Röhm)	England, Germany
Amano Pharmaceutical Co.	Japan
Biocatalysts Ltd.	Great Britain
Christian Hansen AS	Denmark
Danisco (also with Genencor enzymes)	Denmark
DSM (with former Gist-Brocades enzymes)	Netherlands
Enzyme Devt. Corp. (with Röhm enzymes and other)	USA
Fluka Chemie GmbH	Switzerland
Genencor International	Finland, USA
Meiji Seika Kaisha Ltd	Japan
Nagase Biochemicals	Japan
Novozymes	Denmark
Primalco Co. (associated with AB Tech/Röhm)	Finland
Roche (with former Boehringer Mannheim enzymes)	Switzerland
Sankyo Co.	Japan
Sigma Chemical Co.	USA

cations have been – and continue to be – within the food sector, mainly in areas of food processing, modification, and upgrading (see Table 5.5).

Improved economics of processing, with enhanced yield, fewer by-products and improved environmental protection (e.g., less waste water) are also of major importance, and in this respect one or more of the following aspects may play a role:

- modified or new products (e.g., sweeteners, bakery ingredients, confectionery);
- nutritive value, functional food;
- taste, texture;
- physical properties (water activity, viscosity); and
- economics.

Starch hydrolysis and processing is one of the most important processes, and incorporates a wide range of different products (see Section 5.3.2). Furthermore, microbial amylases and other glucanases are used in brewing and alcohol production, in addition to malt enzymes.

Table 5.5 Applications of technical enzymes (selected according to economic relevance).

Enzyme	*Source*	*Reaction type*	*Application*
Proteases			
Alkaline protease	*Bacillus, Aspergillus*	Protein degradation	Detergents, leather manufacturing
Rennin	Calf	Specific casein hydrolysis	Cheese manufacturing
Pepsin and other	*Mucor*; pancreas of pig, cattle	Protein degradation	Meat production
Papain	*Carica papaya*	Protein hydrolysis	Bread manufacture; brewing; meat production
Glycosidases[a]			
α-Amylases	*Bacillus, Aspergillus*	Endo-hydrolysis of starch to dextrins	Starch hydrolysis, dextrin production; alcohol production; brewing; bread manufacture
Amyloglucosidases	*Aspergillus*	Exo-hydrolysis of dextrins to glucose	Glucose syrup production
Pullulanase Isoamylase		Hydrolysis of α-1,6-bonds in amylopectin	Glucose syrup production; brewing
Cellulases	*Aspergillus, Trichoderma*	Hydrolysis of cellulose	Detergents; processing of plant material
Pectinases[b], including galactanases, xylanases, arabanases	*Aspergillus*	Hydrolysis of corresponding polysaccharides in plant cell walls	Processing of fruit, vegetables, production of wine, juices, pureés, etc.; brewing
Pentosanases (similar as before except pectinases)	*Humicola, Trichoderma*	Hydrolysis of pentosan fractions	Bread manufacture; bakery products
Lactase	*Aspergillus*	Hydrolysis of lactose	Milk products; whey processing
Invertase	*Saccharomyces*	Hydrolysis of sucrose	Production of invert sugar syrup; bakery, confectionery
Esterases, Amidases			
Lipases	*Pseudomonas, Aspergillus, Candida*	Fat, triglyceride hydrolysis, transesterification, organic synthesis	Surfactants; fat processing, modification, fatty acid production; milk products
Acetylesterases			Amino acid-, speciality production
Hydantoinases/carbamoylases		Hydrolysis of hydantions	Amino acid production
Transferases			
Cyclodextrin transferases	*E. coli*	Glycosyl transfer/cyclization	Cyclodextrin production
Transglutaminase			Meat processing
Oxidoreductases			
Glucose oxidase	*Aspergillus, Penicillium*	Oxidation of glucose	Glucose analysis; elimination of oxygen in juices, beer, wine
Catalase	*Aspergillus, Micrococcus*	Degradation of hydrogen peroxide	(As before, together with glucose oxidase)
Lipoxygenase	*Soy beans*		Bread manufacturing

a) In many cases mixtures with different enzymes or activities, respectively.
b) Mixtures, in general offered as pectinases.

5.3.1.1 Food Applications

Bread manufacture has become an important field for enzymes such as amylases, pentosanases (e.g., galactanases, xylanases, arabanases), glucose oxidase, lipases and lipoxygenase. The different activities act synergistically, as is the case with most complex substrates (c.f. pectinases; Section 5.3.3.2). The motive for this is the improved processing of doughs, a better or more pleasant structure and volume of the bread, and also improved storage quality (anti-staling effect). The amounts applied are low, in the range of 0.1–10 g enzyme per 100 kg of flour. Furthermore, proteases are applied when there is a need to partially hydrolyze the gluten present – this makes the dough less viscous, and its processing more easy. Lipases can replace emulsifiers by promoting emulsion formation. Glucose oxidase (including catalase) is used for strengthening doughs by oxidation of the free sulfhydryl units and the formation of disulfide linkages in gluten protein. Granulation of the enzyme preparations prevents dust formation during application. All of these enzymes are inactivated at elevated temperatures during the baking process. Similar enzymes mixtures, including amylases, glucoamylases, cellulases, xylanases, and proteases, are applied in brewing processes (O'Rourke, 1996). In addition, a rapidly growing market with a considerable volume (>100 million €) has recently developed for enzymes applied to feed production, especially β-glucanases, endoxylanases, and phytases.

In cheese manufacturing, rennet or chymosin (or related proteases) are used to coagulate the milk protein, while further proteases and lipases may be added in order to improve the taste. Lactase (β-galactosidase) provides the hydrolysis of lactose in milk and soft ice products for lactose-intolerant consumers.

For fruit and vegetable processing, a broad range of pectinases including hemicellulases (glycanases) are used in order to improve yields, and/or to produce pureés by maceration and accelerating the filtration and pressing stages (see Section 5.3.1.2). It should be noted that the GRAS status of the microorganism is a major factor in enzyme production for all fields of application within the food sector.

5.3.1.2 Textile Applications

A remarkable amount of enzymes – some 10 % of the market share – is purchased by the textile industries. For example, adhesive sizes are used in order to protect fabrics during weaving, and these are starch-based in most applications. Desizing after machining requires that the starch is solubilized, and this is done using amylases of different origin. Surfactants must also be added during desizing, and these must be largely non-ionic with small amounts of cationic types in order not to damage the enzymes; $CaCl_2$ must also be added to provide an appropriate enzyme performance. Generally, both convential and thermostable bacterial amylases are applied in amounts of 0.025 up to 0.25 % (w/v), during 2–4 hours, in the temperature range of 75–80 °C (Godfrey, 1996).

Another major application of enzymes is in paper and pulp processing. The motives here are to produce a cost-effective and environmentally benign technology, one precondition being an improved understanding of the interactions between enzymes, materials (substrate) and the process. Mixtures of cellulases and hemicellulases are used to increase the rate of pulp dewatering by the partial solubilization of fines and the cleaning of small fibrils on the cellulose fiber surface, thus improving the process economics. Furthermore, xylanases are used to aid bleaching of Kraft pulp, where lignin is removed from the cellulose fiber, and to reduce the use of chlorine-based bleaching compounds (Tolan, 1996; Bajpai, 1999).

A much larger number of enzymes is applied in analytical procedures, but these are not included in the statistics of technical enzymes. Although the total amount produced is perhaps rather low, their purity and high added value means that their economic relevance is rather high.

Following this survey, a few selected examples will be treated in more detail. Aspects concerning reaction mechanisms are referred to in order to highlight their role with respect to selectivity and by-product formation. The reaction conditions (concentrations, pH, temperature, reaction or residence time, and effectors such as metal ions) are described for some examples as these play a key role in process and product stability, and also in the process economics. These parameters reflect the complexity of the process, while their treatment and optimal design require insight into basic scientific aspects as well as a broad empirical knowledge and experimental experience.

5.3.2
Starch Processing

Starch is the most important carbon and energy source among plant carbohydrates, and it is second following cellulose, at 10^{10} tonnes per annum, in total biosynthesis. Its industrial production is about 45×10^6 t a^{-1} (EU ca. 7.5×10^6 t a^{-1}). The most important starch sources are corn (maize, 3.6×10^6 t a^{-1} starch in the EU), wheat and potato. Starch is composed of two polysaccharides: 15–30 % is amylose, a linear polyglucan with α-D-(1→4)-linked glucopyranosyl units (up to 6000) and a molar mass ranging from 3×10^5 up to 10^6 Da (Fig. 5.3a); special corn types ("high-amylose") have been developed which contain up to 80 % amylose. Conventional starch is composed of 70–85 % amylopectin, a polyglucan with main chains as in amylose, with α-D-(1→4)-linked glucopyranosyl units, but with side chains linked by α-D-(1→6)-glycosidic bonds to the main chain, which has about 5 % branch points (Fig. 5.3b). The molar mass of amylopectin ranges from 16 to 160×10^6 Da, corresponding to about 10^6 glucose units. Seeds or tubers contain starch as compact granules of several μm diameter, with partially crystalline amylose and amylopectin, which are essentially inaccessible to enzymes (Robyt, 1998). However, when heated in water to about 60 °C, these granules take up large amounts of water and begin to swell, being transformed into a gel, and strongly increasing the viscosity of the suspension. Eventually, under shear forces the granules disintegrate, and it is only after this transformation that starch can be efficiently hydrolyzed by enzymes. The degree of hydrolysis is expressed as the DE (dextrose equivalent) value, which is determined by the reducing groups (equivalent to the amount of hydrolyzed bonds) taken as glucose equivalents per dry mass. Hence, DE 100 is pure glucose, and DE 0 is unhydrolyzed long-chain starch.

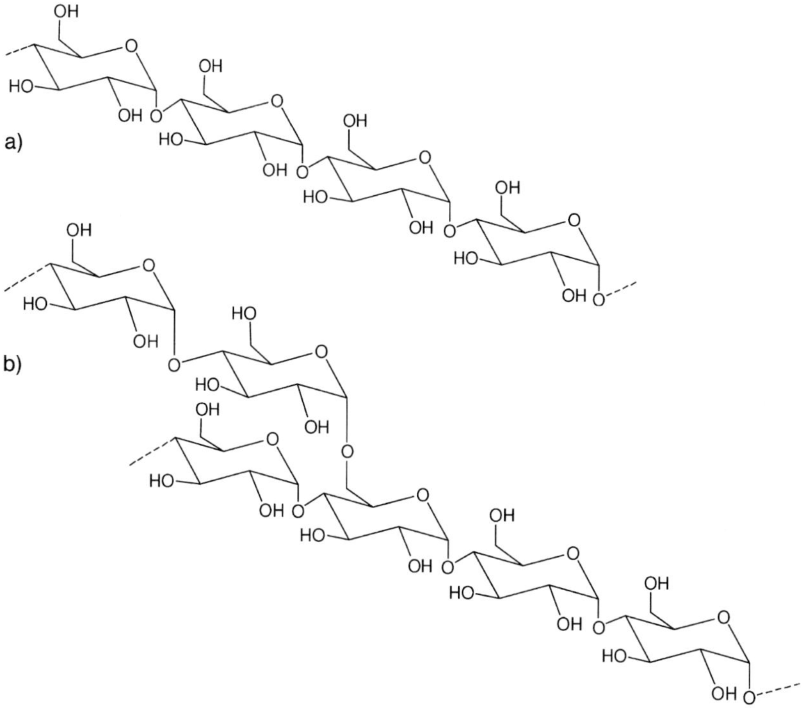

Fig. 5.3 Structure of starch molecules: (a) amylose; (b) amylopectin.

A low degree of hydrolysis (DE 10–15) results in so-called "soluble starch", with a considerably lower viscosity than that of the original solution. Oligosaccharide fractions with 4 to 20 glucosyl units are called dextrins. Such products are obtained by acid or α-amylase hydrolysis in one step from starch. Further hydrolysis requires either combined acid and enzymic hydrolysis, or two enzymatic steps involving α-amylase and glucoamylase. Syrups with higher degree of DE up to those with essentially pure glucose, which can be crystallized to yield crystalline glucose, can thus be produced. However, these require additional downstream processing, since besides low amounts of oligosaccharides from starch (i.e., maltose), other oligosaccharides – so-called reversion products with α-(1→6) glycosidic bonds (i.e., isomaltose) – are also formed by condensation. This is due to the high concentration of the reaction solutions (>30 %) required for economic reasons.

Enzymes for starch processing, encompassing α-amylase, glucoamylase, β-amylase, pullulanase, isoamylase and glucose isomerase, comprise about 30 % of the world's industrial enzyme production. Only the first two enzymes will be described extensively here, although a wide range of different products such as starch hydrolysates, glucose, high-fructose corn syrups (HFCS), and cyclodextrins are available commercially.

5.3.2.1 Mechanism

There are basically four groups of starch-converting enzymes: (1) endo-amylases; (2) exo-amylases; (3) debranching enzymes; and (4) glucosyltransferases.

The third of these groups, which comprises isoamylases and pullulanases, exclusively hydrolyses α-(1→6) bonds. The last group is remarkable insofar as it exhibits only very minor hydrolase activity – that is, it excludes water from the active site, allowing only sugars as nucleophiles for attack of the covalently bound glucosyl group. The cyclomaltodextrin glucanotransferases are among the most efficient enzymes known, increasing the reaction rate 10^{15}-fold over that when no enzyme is present (van der Maarel et al., 2002).

A number of starch-converting enzymes belong to a single family: the α-amylases or Family 13 glycosyl hydrolases (Coutinho and Henrissat, 1999). This group of enzymes shares a number of common characteristics such as a $(\beta/\alpha)_8$ barrel structures, the hydrolysis or formation of α-glycosidic bonds, and a number of conserved amino acid residues in the active site. The three-dimensional structures, mechanistic principles deduced from structure–function relationships, and properties such as kinetics, selectivity and stability of these enzymes have been investigated, reported and summarized in several reviews (Gottschalk et al., 1998; van der Maarel et al., 2002).

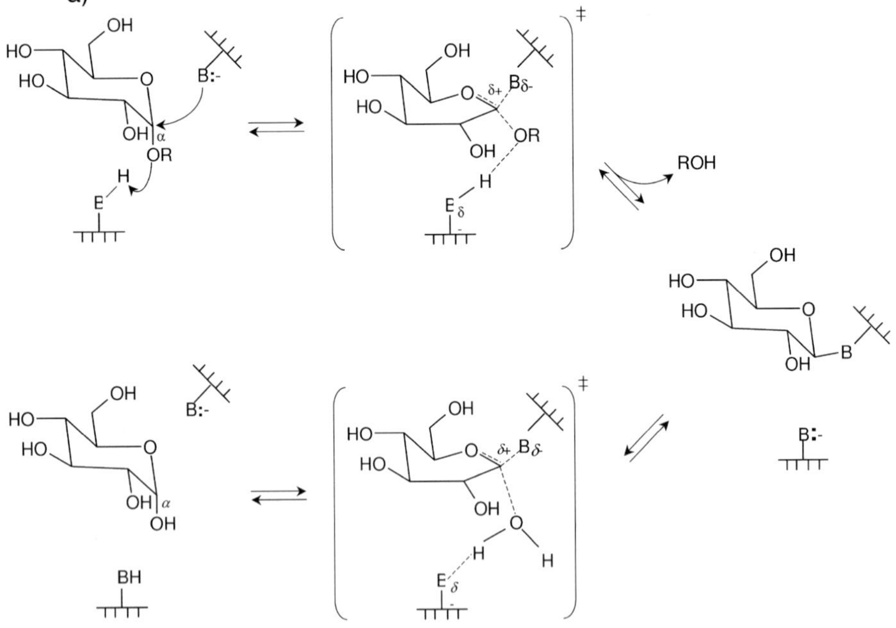

Fig. 5.4 a Active center and subsites in the amylase family. (a) Simplified scheme of the mechanism with nucleophilic attack of the glycosidic bond by the nucleophile B, Glu or Asp, respectively, followed by formation of an intermediate with a covalently bound glycosyl residue and its release by nucleophilic displacement by a water molecule (van der Maarel et al., 2002).

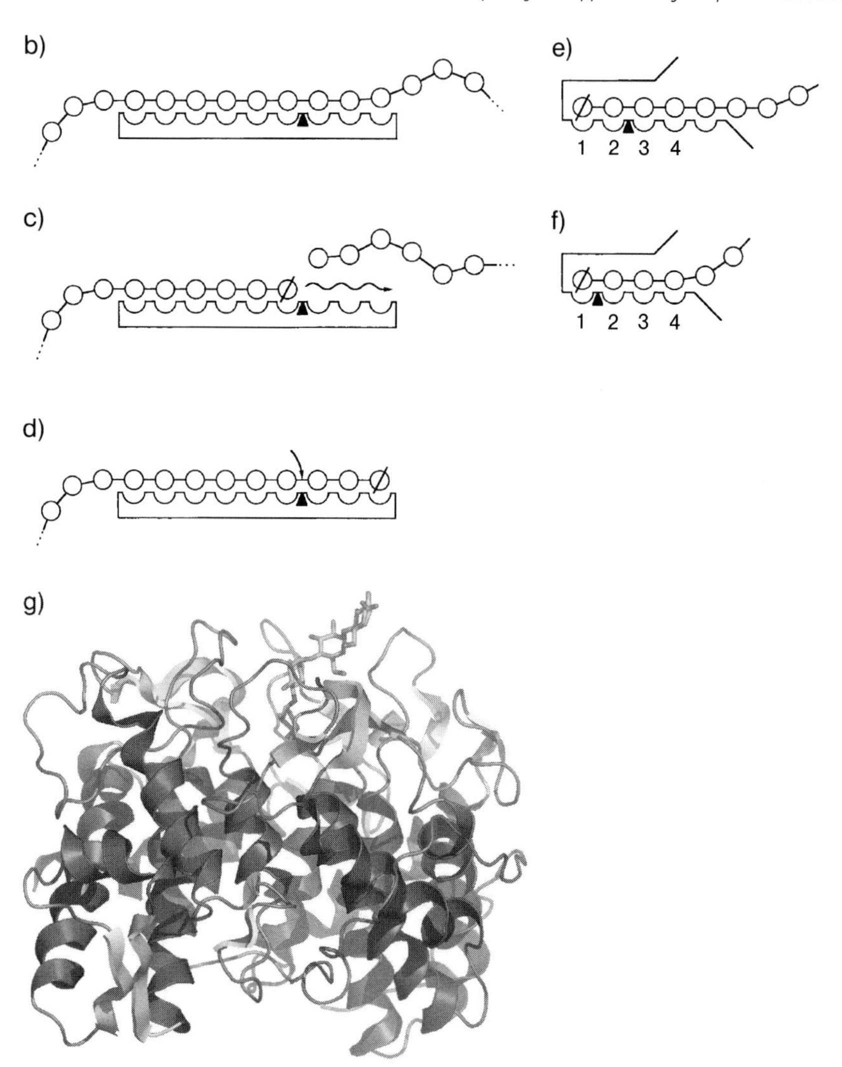

Fig. 5.4 b–g Active center and subsites in the amylase family. (b) Scheme of subsite in α-amylase. (c) Random hydrolytic reaction. (d) Reaction from the substrate end. (e) Scheme of subsite in β-amylase. (f) Scheme of subsite in glucoamylase. (g) Side view of the three-dimensional structure of glucoamylase with a $(\beta/\alpha)_8$ barrel structure (Coutinho and Reilly, 1994; Reilly 2003); the α-helices (dark) forming the typical barrel may be identified, the active site amino acids are located in the loops on top.

The established catalytic mechanism of the α-amylase family, with α-retaining activity, involves two catalytic residues in the active site: a glutamate residue as general acid catalyst and an aspartate residue as nucleophile (Glu_{230} and Asp_{206} in *Aspergillus niger* numbering) (Fig. 5.4a). The glutamate donates a proton to the glycosidic bond between two substrate glucosyl units bound in the active site, at subsites –1 and +1; the nucleophile aspartate attacks the C1 of the glucosyl unit at subsite –1; a covalent intermediate is formed via an (assumed) oxocarbonium-like transition state; the protonated glucosyl unit at subsite +1 leaves the active site while a water molecule moves in; the glutamate accepts a hydrogen from water, the oxygen atom of which attacks the covalent bond between the glucosyl unit at subsite –1 and the aspartate, going in turn via an oxycarbonium-like transition state and forming a new hydroxyl group at the C1 position of the glycosyl unit at subsite –1. When another glucose molecule, instead of water, enters the active site, a new glycosidic bond, preferentially with an α-(1→6) bond, is formed. This occurs more frequently with increasing glucose concentrations, leading to reversion products such as isomaltose (α-D-glucopyranosyl-(1→6)-D-glucose). A third conserved residue, a second aspartate (Asp_{297}) binds to the OH-2 and OH-3 groups of the substrate through hydrogen bonds, and plays an important role in substrate distortion (Fig. 5.5) (van der Maarel et al., 2002).

Enzymes of the α-amylase family vary widely in their substrate and product specificities, which can be attributed to the attachment of different domains to the catalytic core, or to extra sugar-binding subsites around the catalytic site (van der Maarel et al., 2002).

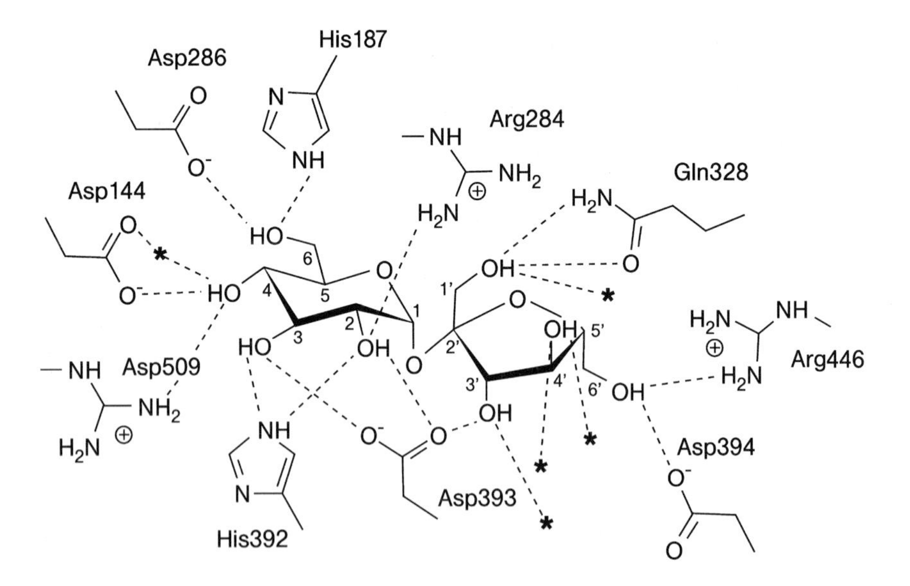

Fig. 5.5 Binding of sucrose to amino acids in the active site of amylosucrase (Mirca et al., 2001).

The tight binding of a substrate in the active site is illustrated by sucrose bound to a mutant of amylosucrase, a glucosyltransferase-forming amylose with high regio- and stereoselectivity (α-1,4-bound glucose units). It is clear that every OH-group of sucrose is bound to one or two amino acids in the active site (Fig. 5.5) (Mirza et al., 2001).

Amylolytic enzymes bind substrate glucosyl residues at an array of consecutive subsites that extends throughout the active-site cleft. In typical endo-acting enzymes such as α-amylases, the substrate-binding region comprises from 5 to 11 subsites (Svensson et al., 1999). The structure of the binding region determines substrate and product selectivity.

For α-amylases the formation of products with a preferential distribution of molecular weights can in principle be predicted, as can the limits of hydrolysis of products from amylopectin with side chains ("α-limit dextrins"), which cannot be accommodated in the active center. Low molecular weight products form nonproductive complexes, which may inhibit enzyme activity. Enzymes from different sources exhibit considerable differences in binding-site structure, so they produce product mixtures with different molecular weight distributions.

Major efforts have been devoted to researching thermostable α-amylases with much success, revealing structural determinants responsible for the high thermostability of *Bacillus* enzymes (Declerck et al., 2000). The results of stability studies on barley α-amylase isoenzymes suggested that stabilization by electrostatic interactions might play a role (Svensson et al. 1999).

β-Amylase also is an exo-hydrolase with inverting activity. Its active site is a deep flap-covered cleft, into which the non-reducing ends of starch and maltooligosaccharides are bound (Fig. 5.4e). This enzyme differs from glucoamylase in producing maltose, which is expelled when the flap opens. The same substrate chain is attacked repeatedly, this being an example of single-chain attack.

Glucoamylase (known industrially as amyloglucosidase) is an exo-acting inverting glycoside hydrolase (α-(1→4)-D-glucan glucohydrolase) that catalyses the release of β-D-glucose. It is exo-acting since it hydrolyses bonds only from the non-reducing end of starch, and it is inverting since β-glucose is formed from an α-glycosidic bond. However, in solution the initial product undergoes spontaneous mutarotation to the equilibrium mixture of α- and β-D-glucose.

Unlike β-amylase, which has an active-site cleft which is freely accessible to starch and oligosaccharide chains, the active site of glucoamylase is in the shape of a well (Fig. 5.4f, g), forcing exo-activity and multichain attack, in which different chains have essentially equal probability of being bound. Glu_{179} and Glu_{400} in *A. niger* glucoamylase are the general acid and base catalysts, respectively (Sierks et al., 1989; Harris et al., 1993). A C-terminal starch-binding domain is bound to the catalytic domain by a glycosylated linker. Detailed studies on substrate, oligosaccharide and inhibitor binding yielded structural and mechanistic insights as well as kinetic and thermodynamic data, including free enthalpies of binding of single glucosyl units and individual OH groups (Olsen et al., 1993; Sigurskjold et al., 1998; Sierks and Svensson, 2000).

The formation of reversion side-products by glucoamylase can be explained in a straightforward manner by the reaction mechanism. The covalently bound glucosyl residue can undergo nucleophilic attack by another glucose or oligosaccharide molecule instead of by water, notably at high sugar concentrations, with preferential formation of an α-(1→6)-glycosidic bond, leading to isomaltooligosaccharides. This reaction pathway allows glycosidases to form oligosaccharides at reduced or low water activities and high saccharide concentrations. Glycosyltransferases are far more selective at this, however (see Chapter 6, Section 6.4).

Glucoamylase is produced by filamentous fungi, mainly *Aspergillus* sp., in higher tonnage than almost any other industrial enzyme, and thus it is available at low price. This was a motive for major efforts to resolve its structure and mechanism (Aleshin et al., 1994, 1996; Sauer et al., 2000). A review summarizing recent data on protein properties, enzyme mechanism and kinetic data is available (Reilly, 2003).

Two major problems were the motives for extensive research to be conducted into the genetic modification of glucoamylase. First, at high dissolved solid concentrations it condenses some of the glucose formed to di-, tri- and tetrasaccharides, the most important being isomaltose. Second, glucoamylase is not as stable as α-amylase and glucose isomerase, being used at 60 °C in starch processing. Combining favorable mutations was successful in decreasing the kinetics of the enzyme towards the formation of isomaltose, increasing its glucose yield from 96 % to 97.5 % (equivalent to 200000 extra tons at actual production scale) (Reilly, 1999). One condition for the successful design of mutations of single amino acids in glucoamylase is advanced knowledge of its structure and function. Thus, stiffening α-helices by Gly→Ala and Ser→Pro mutations, as well as creating disulfide bonds across two loops, contributed to a four-fold increase in the thermostability of *A. niger* glucoamylase (Reilly, 1999; Suvd et al., 2001).

Genetic engineering and recombinant technologies thus had significant impact on enzyme technology by:

- creating a tremendous increase in the productivity of enzyme fermentation with recombinant organisms (as mentioned above);
- creating a significant improvement in thermostability (α-amylases, glucoamylases), by the screening of thermophilic organisms, directed evolution and rational design via site-directed mutagenesis; and
- providing process optimization by improved selectivity (fewer side products) and reduced or no Ca^{2+} dependence (α-amylase).

5.3.2.2 **Process Steps**

For the first process step of starch hydrolysis, starch liquefaction, thermostable α-amylases are available which are active in the range of 85–120 °C. Thus, thermal treatment at temperatures beyond 105 °C (gelatinization) and hydrolysis (liquefaction) can be partially integrated. An overview of the enzymes used in starch processing is provided in Table 5.6.

The technological conditions for hydrolysis by α-amylase are pH 5.8–7 and 30–40 % (dry solids) starch concentration with 0.4–1.2 L $tonne^{-1}$ (dry matter) en-

Table 5.6 Enzymes used in industrial starch hydrolysis (Olsen, 1995; Bentley and Williams, 1996).

Enzyme	*Source*	*Type*[a]	*Min DP substrate*	*Min DP*[b] *product*	*Amount applied*[c]	pH_{opt} *(Ca)*[e]	T_{opt} *[°C]*[d]
α-Amylase	*Aspergillus oryzae*	endo-α-(1→4)	PS[f]	OS[f]		4–5 (50)	55–70
	A. niger					55–65	
	Bacillus subtilis				0.8–1.2	6–7 (150)	60–85
	B. licheniformis				0.4–1	5.8–7 (20)	95–105
β-Amylase		exo-α-(1→4)	3	2	0.3–0.7		55–60
Isoamylase[g]	*Pseudomonas sp.*, *Bacillus sp.*	endo-α-(1→6)				40–50	
Pullulanase[g]	*Pullularia*, *Streptococus*	endo-α-(1→6)	>3	3		3.5–5	55–65
Glucoamylase	*A. oryzae*, *A. niger*, *Rhizopus oryzae*	exo-α-(1→4) and α-(1→6) (slow)	2	1	0.5–1.1 (300 U mL^{-1})	4–5	55–65

[a] Type of glycosidic bond hydrolyzed.
[b] Minimal degree of polymerization (DP).
[c] Preparation of mean activity (kg enzyme per tonne starch dry substance).
[d] Optimal application temperature.
[e] Ca^{2+} required, ppm.
[f] PS: polysaccharide, OS: oligosaccharides.
[g] Applied in blends with glucoamylase.

Table 5.7 Process data for starch hydrolysis (Olsen, 1995; Bentley and Williams, 1996).

Substrate	*Enzyme applied*	*Process step*	*Conditions*
Starch suspension 30–40 % d.s.[a]	α-Amylase (thermostable)	Heating, gelatinization[b]	ca. 105 °C, pH 6.5, 5–10 min
(DE 8–15) 25–40 % d.s.	α-Amylase	Partial hydrolysis (solubilization), cooling	85–95 °C, 1–2 h Ca-ions (3–50 ppm)
Starch/dextrins (DE 10–15)	β-Amylase	Hydrolysis to maltose syrup[c]	60 °C, pH 4.5–5, 24–40 h
Partial hydrolysate (DE 8–15)	Glucoamylase	Hydrolysis to glucose syrup	55–60 °C, pH 4.5, 24–48 h
Hydrolysate (DE 42) or glucose syrup (DE 95)		Decoloring (active carbon), deionization (ion exchange), concentration	
Product: crystalline glucose		Crystallization, centrifugation, drying	

[a] Dry substance.
[b] Jet cooker.
[c] Maltose content of 60 up to 80 %, with higher maltooligosaccharides.

zyme present. The initial step (gelatinization) is conducted at 105–110 °C over a 5–10 minute period with steam injected through a jet, and with the steam–enzyme–starch slurry mixture proceeding to a tubular reactor. This is followed by liquefaction at 85–95 °C for 1–2 hours (Tables 5.6 and 5.7) (for different process variants, see Bentley and Williams, 1996). Ca^{2+} in the range of 20–150 ppm is required for α-amylase stabilization. This gives maltodextrin of DE 10–15 as a trade product (93 % with DP > 4; 4 % maltotriose, 3 % maltose). The degree of hydrolysis may be controlled by rapid enzyme inactivation via acid and/or thermal treatment. New recombinant α-amylases do not require Ca^{2+} and are active at pH-values down to about 5, so that little shift in pH is required for further hydrolysis by glucoamylase (Pedersen, 2002). This substantially reduces upstream processing by an ion-exchange operation.

Commercial products from extensive α-amylase hydrolysis typically are near DE 42, with about 19 % glucose, 14 % maltose, 12 % maltotriose, and 55 % higher oligosaccharides. The production of crystalline glucose or of FCS requires a product with a DE of 95–96 %. For crystallizing glucose the syrup must be concentrated to 70–78 % d.s., followed by crystallization, centrifugation, and drying. The production of glucose-fructose syrup is dealt with in Chapter 6, Section 6.4.1 and Chapter 9, Section 9.4.2.

Further hydrolysis to produce a glucose syrup (saccharification) requires another process step with glucoamylase at different conditions (Table 5.7). The fungal enzyme (Table 5.6) releases α-(1→4)-bound glucose from the non-reducing end of amylose or amylopectin, respectively. It also hydrolyzes α-(1→6)-bonds at a much lower rate. This step proceeds considerably faster when pullulanase or isoamylase is added, and this occurs in industrial processes. The reaction is carried out with substrate solutions of 25–40 % (d.s.) at pH 4.3–4.5, 55–60 °C and reaction times of 24–48 hours, depending on the degree of hydrolysis required. After completion of the reaction the enzymes are inactivated at 120 °C and pH >4.5, or removed by ion exchange, in order to minimize the formation of reversion products. Decoloring and ion exchange (elimination of metal ions) are essential subsequent steps. Elevated temperatures must be applied to avoid crystal formation while storing the product syrup.

5.3.3
Further Examples and Perspectives

5.3.3.1 Detergents

The economically most important application of technical enzymes – besides starch hydrolysis – is that in washing powder (often a liquid) by the detergent industries (see Table 5.1). Although this is a traditional application, it has gained dramatic progress due to recombinant technology for all enzyme components, proteases, lipases, cellulases, and amylases. The enzymes are applied, with optimal activity under alkaline conditions at moderate temperature, corresponding to the requirements of washing procedures (Fig. 5.6). Enzyme manufacturers succeeded in screening bacteria that produced enzymes with the rather unconventional properties required (good stability and optimal activity at pH ~11 and temperatures of 40 °C or higher). These hydrolyze proteins from milk, fat from butter and other compo-

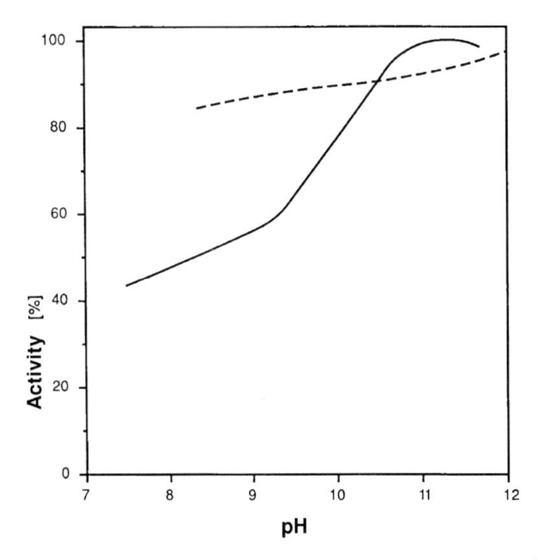

Fig. 5.6 Activity (%) of an alkaline protease as a function of pH (substrates: casein, reaction at 40 °C, —; hemoglobin (denatured), reaction at 25 °C, broken line) (after International Bio-Synthetics, 1988).

nents of food. The cellulases hydrolyze the ends of broken fibers which scatter light, giving the fabric a grayish appearance, and tend to protect adsorbed or entrapped impurities from being eliminated by surfactants.

Peptidase (subtilisin) variants stabilized towards the inactivation caused by peroxidases, and applied with a hydrogen peroxide precursor for improved bleaching in detergents, were obtained by site-directed mutagenesis (substitution of methionine). For details of further modifications and improvements made, see Chapter 2, Section 2.11.

A major success relates to lipases which hydrolyze fat (triglycerides) into a mixture of free fatty acids, di- and monoglycerides, and glycerol – all of which can be removed much more easily than fat. Lipases normally are not active during washing in the aqueous phase, due to a "lid" which must be opened to give access to a substrate; opening occurs at non-polar interfaces, and not in the aqueous phase (Eriksen, 1996). Removal of this "lid" by genetic engineering of lipases, significantly improved their efficiency during the washing procedure (Svendsen, 2000). Further improvements were introduced by modified electrostatic surface charge, including a N-terminal extension with positive charges, in order to have activity in the presence of anionic detergents. The problem was solved by "classical" protein engineering. For improved thermostability, a rather broad spectrum of methods was applied, comprising computational chemistry with substrate docking, molecular dynamics and localized random mutagenesis (error-prone PCR of relevant gene fragments, combined with doped DNA-oligonucleotides) and family shuffling (see Chapter 2, Section 2.11). The approach was successful, and led to the production of a variant which was active up to and over 80 °C (Vind, 2002).

The different forms of enzymes used in washing powders are granulated, dust-free preparations, with an inclusion formulation of the protein inside a water-soluble inert coating. This technique is one of the most important unit operations of enzyme confectioning, where the formation of protein dust and contact by inhalation or with the skin, and thus allergic reactions are avoided.

5.3.3.2 Food Processing

Another important field of application of enzymes to be discussed in more detail at this point is fruit and vegetable processing (c.f. Grassin and Fauquembergue, 1996; Benen et al., 2003). Pectic enzymes in fruit processing provide for higher juice yields, increased filtration rates, and improved color (e.g., in red wine). Further-

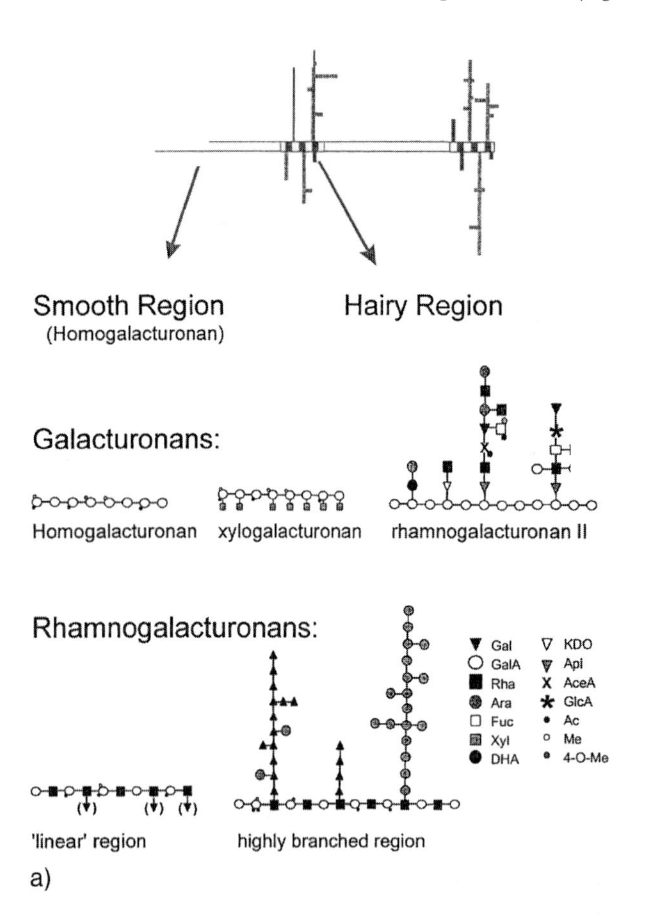

Fig. 5.7 (a) Structural elements of pectin. Gal: galactose; GalA: galacturonic acid; Rha: rhamnose; Ara: Arabinose; Fuc: fucose; Xyl: xylose; DHA: 3-deoxy-D-lyxo-2-heptulosaric acid; KDO: 2-keto-3-deoxy-D-manno-octulosonic acid; Api: apiose; AceA: acetic acid; GlcA: glucuronic acid; Ac: acetyl group; Me: Methyl ester; 4-O-Me: 4-O-methyl ether (Schols and Voragen, 2003).

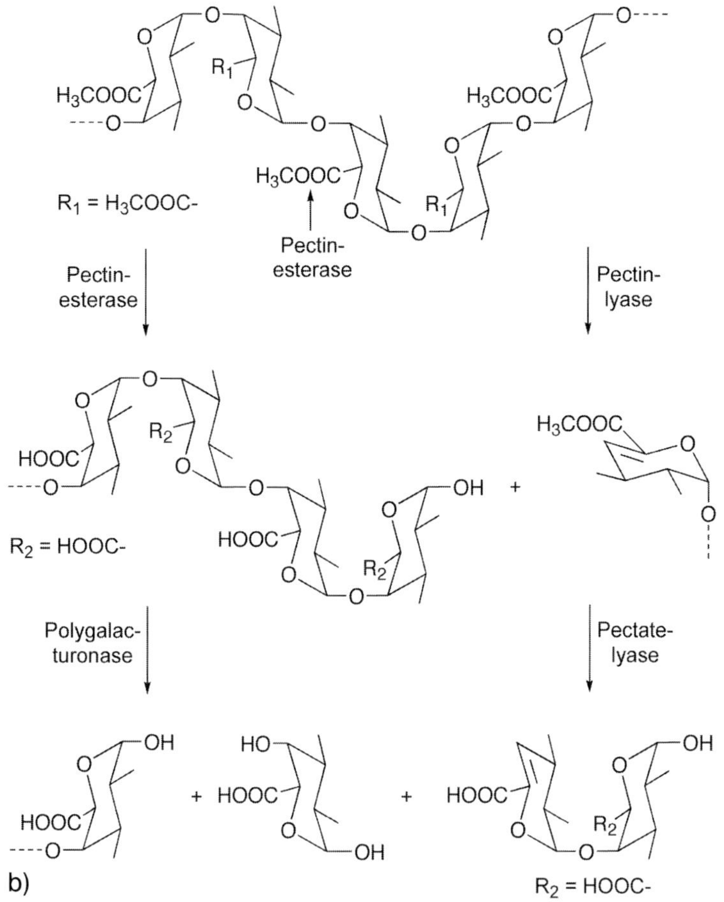

Fig. 5.7 (b) Simplified scheme of the action of some key pectic enzymes.

more, they assist in the clarification of juices or the cloud stabilization of pulpy drinks. Other applications are in processing of vegetables for maceration products, pastes or pureés of tomatoes, carots or potatoes, and the preparation of pectins for specific applications and further fields of the food industries (Benen et al., 2003).

For application, it is advisable to have an insight into the substrate's composition and structure. The structure of pectic material is highly complex, but has been elucidated in detail for a range of plant materials used in industrial food processes (Fig. 5.7a) (Schols and Voragen, 2003). The strength of the plant cell wall is provided by cellulose fibrils, which also exhibit major stability against acid and enzyme hydrolysis. The fibrils are linked to other polysaccharides, such as xylan, xyloglucan, araban, arabinogalactan and galactan (often termed hemicelluloses). These in turn are cross-linked to pectic substances, highly heterogeneous polymers which may be considered to be a binder material, as well as providing the elasticity of the cell wall. Pectin roughly comprises linear homogalacturonan, linearly branched xylogalacturonan carrying xylose residues, highly branched galacturonan, and rhamnogalacturonan, where the backbone is made up of α-(1→2)-linked L-rhamnosyl residues and α-(1→4)-linked D-galacturonic acid residues ramnified at the rhamnosyl residue with varying oligosaccharide side chains, including L-arabinan, D-galactan and mixed oligosaccharides (Fig. 5.7a) (Schols and Voragen, 2003).

The most important activities of pectinase preparations, rhamnogalacturonan hydrolases, and pectin acetyl esterases are summarized in Table 5.8. Partial hydrolysis of the pectin results in disintegration of the cell wall material, and the release of fiber material including cellulosic and hemicellulosic components, as well as soluble pectic and hemicellulosic material. Endo-acting enzymes quickly reduce the molar mass of the polymeric substrate, and this is easily detected by viscosimetry. Methyl- and acetyl esterases release methanol and acetic acid, respectively, shifting the substrates characteristics and the susceptibility to enzymatic attack from pectin-hydrolyzing towards pectate-hydrolyzing lyases. In this regard, the most relevant activities are:

- Pectin methyl-esterases, releasing methanol and carboxylic acid functions in the polymer backbone.
- Endo-pectinases, hydrolyzing the polymer backbone statistically (at methylated positions).
- Endo-pectin lyases, degrading the polymer backbone statistically (in methylated positions) by a transelimination reaction, forming a double bond at the non-reducing end of the product (see Fig. 5.7b).
- Endo-polygalacturonase, hydrolyzing the polymer backbone statistically.
- Endo-pectate lyases, acting similarly to endo-pectin lyases, but adjacent to carboxyl groups of the polygalacturanate.

Table 5.8 Enzyme activities in pectinase preparations (Uhlig, 1998; Benen et al., 2003).[a]

Enzyme	*Activity*[b] *[U g^{-1}]*	*pH of application, or optimum*
Pectin methylesterase	50–100	3–5
Endo-pectinase (polymethyl galacturonase)		
Endo-polygalacturonase	1000–2000	4–4.5
Exo-polygalacturonase		
Endo-pectin lyase (-transeliminase)	100–300	
Endo-pectate lyase (-polygalacturonan–transeliminase)		ca. 5
Further activities		
Xylanases		
L-Arabanases		
Galactanases		
Cellulases		

a) Source: *A. niger* in general.
b) Commercial preparations, liquid or solid (granulated), with activities as given above, stabilized by salts, sugars, dextrins and/or glycerol.

Analytical determination of the enzymes is difficult when individual activities are to be quantified. Simple methods include viscosimetry of polymer degradation, which is sensible, as the splitting of 2–3 % of the glycosidic bonds may result in a 50 % reduction in viscosity. Colorimetric methods determine the reducing end groups, and photometry the formation of unsaturated products by the action of lyases. By using high-performance liquid chromatography (HPLC) or high-performance anion-exchange chromatography (HPAEC) the formation of individual mono- and oligosaccharides allows for a detailed analysis of the different activities (Lieker, 1993; Schols and Voragen, 2003).

As is typical for many plant substrates processed in the food industries, a range of different enzyme activities is required to act synergistically – that is, to function much more efficiently in complex than in singular activity. The optimal mix of activities depends on the purpose of application, and process details have been provided by Uhlig (1991, 1998). Kinetic data for polygalacturonases, which may assist in process development, have been summarized by Benen et al. (2003). Moreover, a wide range of different commercial preparations are available.

When examining details of the dominant areas of application of industrial enzymes discussed above, it is clear that soluble enzymes are superior to immobilized systems, with the following principles (or difficulties) playing a variety of roles:

- the conversion of high molecular-weight substrates;
- the conversion of insoluble and/or adsorbed substrates, for example in washing procedures (proteins, starch, pectin, hemicelluloses); and
- the conversion of highly complex substrates requiring the synergistic action of different activities (such as glycosidases and pectinases in fruit processing).

In addition to the applications addressed above, a considerable number of further developments are currently being investigated, and consequently economic questions relating to the success of these remain unanswered. Some examples of these applications will be examined subsequently (see also Literature: general, p. 238).

The modification and synthesis of sugars, oligosaccharides and derivatives has gained much interest since recognition of their role in biological signal transfer, recognition and cell–cell interactions. New and diverse applications have been derived in the food and pharmaceutical industries and markets, as food additives, prebiotics, and so-called "neutraceuticals" with protective and immune-stimulating functions (Pierre et al., 1997; Reddy, 1998; Rastall, 2003; Buchholz and Seibel, 2003). Saccharide and oligosaccharide synthesis by aldolases is described in Chapter 3, Section 3.2.3.2. For Leloir glycosyltransferases see Hendrix and Wong (1998). Industrially established processes with immobilized enzymes such as glucose isomerization, oligosaccharide and and cyclodextrin production with glucosyltransferases will be discussed in Chapter 6, Section 6.4.

The modification of mono- and oligosaccharides has also gained much interest, as chemical routes in general require difficult and many-fold steps, such as the introduction of protecting groups, chemical activation for glycosidation, organic solvents, deprotection steps, especially as they generally give low yields (the exceptions are straight catalytic reductions of sugars, such as sorbitol production). In contrast, en-

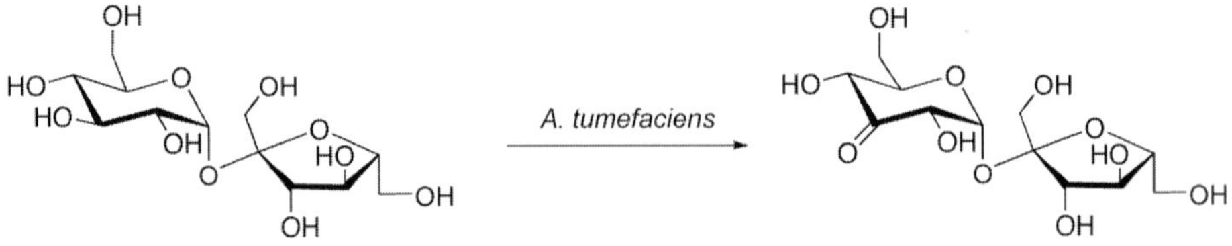

Fig. 5.8 Selective oxidation of sucrose by *Agrobacterium tumefaciens.*

zymatic routes operate directly in aqueous solution, with good or high regio- and stereoselectivity and good yields in general, notably when appropriate reaction engineering is applied. The modification of sucrose is of major interest as it is available in large quantities, at high purity, and at a low price. However, chemical routes have proven most difficult as the molecule has eight hydroxyl groups, each of similar reactivity. Sucrose can be oxidized by *Agrobacterium tumefaciens* with high regioselectivity to 3-keto-sucrose (α-D-ribo-hexopyranosyl-3-ulose-β-D-fructofuranoside) with up to 75 % yield (Fig. 5.8) (Stoppok and Buchholz, 1999). Likewise, other disaccharides such as maltose, lactose, and glucosyl-sorbitol can be oxidized via this route, though resting cells must be used in order to provide for cofactor regeneration of the dehydrogenase. 3-Keto-sucrose can be used as a building block for a range of subsequent chemical steps, including catalytic reduction to allosucrose and, when hydrolyzed, to yield allose (Timme et al., 1998). Reductive amination produces 3-amino-allosucrose (3-amino-3-deoxy-α-D-allopyranosyl-β-D-fructofuranoside) which can be acylated by fatty acids to yield surfactants; substitution by methacrylic acid derivatives produces polymer building blocks for the synthesis of polymers bearing sucrose-derived side chains (polyvinyl saccharides); coupling with amino acids gives the corresponding conjugates with selective coupling to the 3-position of sucrose (Pietsch et al., 1994; Stoppok and Buchholz, 1999; Buchholz et al., 2001; Anders, 2002).

Pyranose oxidases provide access to various keto-sugars and keto-sugar acids which may serve as chiral intermediates and synthons in chemo-enzymatic synthesis. Enzymes acting on 2-, 3-, 4-, or 6-position of different sugars such as D-glucose, -galactose, -xylose or -ribose have been identified (Röper, 1990; Giffhorn et al., 2000). Problems in the technical application here relate to the limited enzyme stability, since hydrogen peroxide is produced as a by-product. A range of modifications via oxidoreductases requiring coenzymes and their regeneration, where membrane systems provide advantadges, will be mentioned subsequently (see Section 5.4).

Further important enzymatic routes to different products are dealt with in other sections of this book, including amino acid synthesis (Chapter 3, Sections 3.2.1.1 and 3.2.2.6, Sections 5.4 and 6.4.3), semisynthetic antibiotics (Sections 2.7 and 9.2), and a wide range of synthetic routes (Chapter 3).

5.4
Membrane Systems and Processes

Enzymes can be retained in, or recycled to, a reactor by means of semipermeable membranes, corresponding to immobilization (Figs. 5.9 and 5.10). Ultrafiltration membranes are commercially available for such purposes with 5 or 10 kDa cut-offs, and these are impermeable to proteins of higher molar masses.

Membrane systems offer advantages for cofactor-dependent reactions, when the coenzyme must be regenerated. In general, two enzymes (E_1 and E_2) are required which catalyze a coupled redox reaction:

$$AH_2 + NAD^+ \xrightarrow{E_1} A + NADH + H^+$$

$$B + NADH + H^+ \xrightarrow{E_2} BH_2 + NAD^+$$

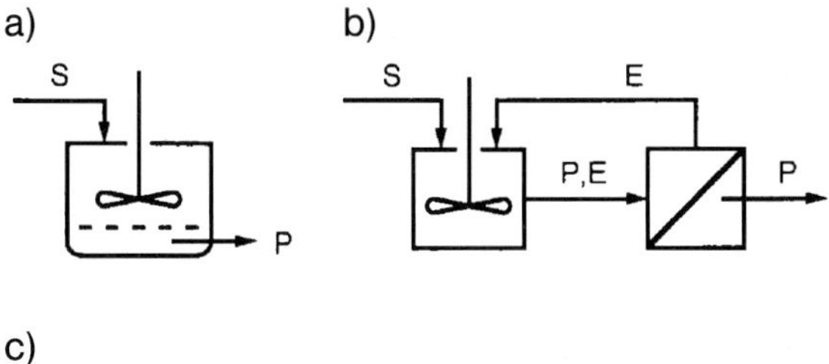

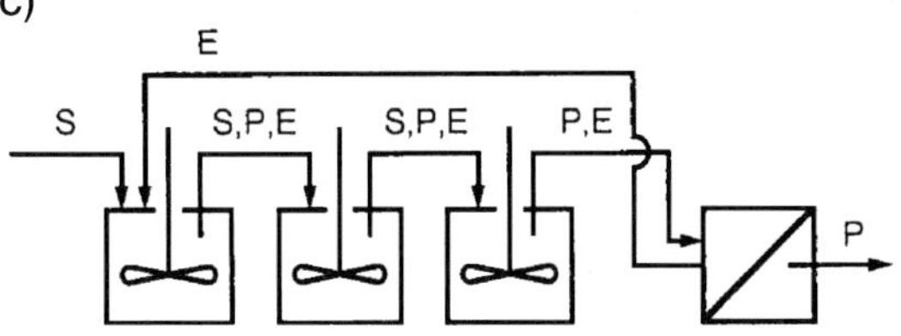

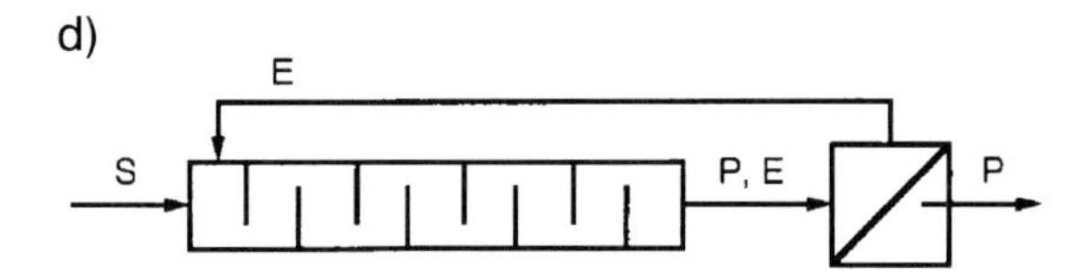

Fig. 5.9 Principles of membrane systems. (a) Membrane reactor with integrated semipermeable membrane; (b) continuous stirred tank reactor; (c) cascade; (d) tubular reactor, each system with membrane module/unit and recycling of enzyme.

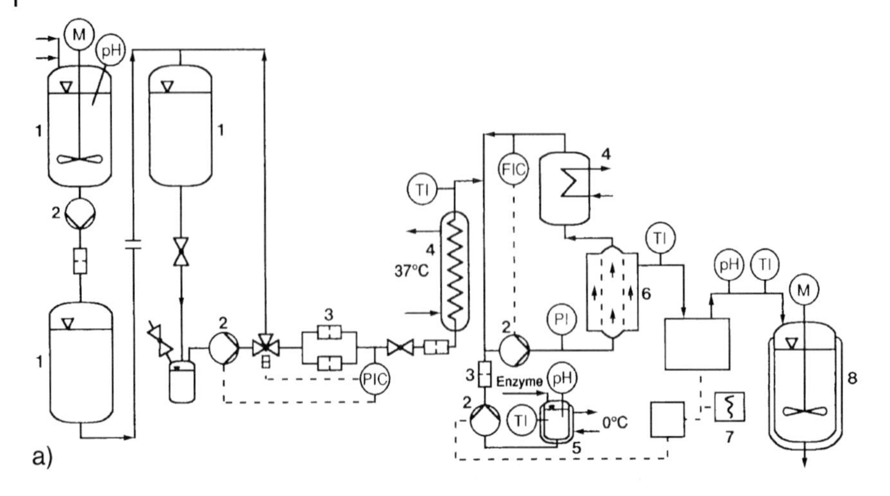

Fig. 5.10 (a) Scheme of production system with membrane unit, with mixing and storage tanks (1), pumps (2), filters (3), heat exchangers (4), enzyme tank (5), ultrafiltration module (6), analytical monitors (7), product vessel (8), control units for pH, temperature (TI), pressure (PIC), volumetric flow (FIC), (regulated control ... —). (b) Unit for amino acid production (reproduced with kind permission of Degussa, Hanau, Germany).

Both enzymes and the cofactor must be retained in, or recycled to, the reactor for cost reasons. The efficiency of cofactor regeneration and recycling should be, for NADH (with a price in the range of 1 € g^{-1} and a product of moderate price, <25 € kg^{-1}), better than 99.9 %. The total turnover number of such systems must be >1000 for high-price products, such as L-*tert*-leucine, and >10 000 for low-price products, such as L-leucine, in order to provide for economic conditions (Kula and Wandrey, 1987; Kragl et al., 1996). The covalent immobilization of such systems with two enzymes and a cofactor, which must access both active centers, is difficult and tedious, and thus economically not feasible.

The method of choice for coenzyme-dependent, two- or multi-enzyme systems is the application of membrane reactors which retain or recycle all catalytically active components. The low molecular-weight cofactors therefore must be coupled to polymers of sufficiently high molecular weight. An example is presented in Figure 5.11. Polyethyleneglycol with a mean molar mass of 20 kDa serves as the soluble macromolecular carrier (Riva et al., 1986). All types of reactor systems with membrane module may be applied (see Fig. 5.9), but in order to provide for a favorable enzyme efficiency, a cascade of continuous stirred tanks or a tubular reactor with a membrane module which recycles the enzymes and the cofactor are appropriate systems (Fig. 5.9c and d). Combinations of stirred and tubular reactors with a membrane module should also be considered. A scheme of a production unit with peripheral vessels, valves, pumps and control units is shown in Figure 5.10.

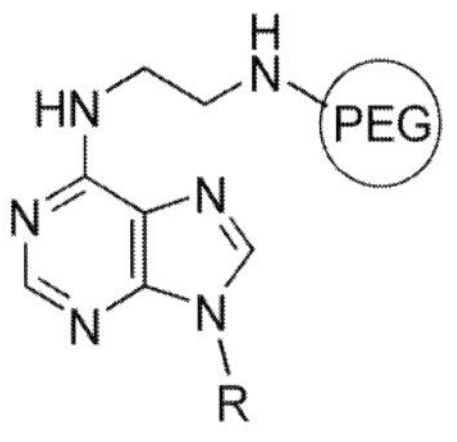

Fig. 5.11 Modified high molecular-weight derivative of a cofactor: PEG-N-6-C2-aminoethyl-NAD^+ (PEG: Polyethylene glycol) (Riva et al., 1986).

Several membrane systems are available, with different membrane structures and module configurations (Fig. 5.12). Details of these were discussed by Kula and Wandrey (1987) and Tutunjian (1985). A considerable number of developments, notably with oxidoreductases, or dehydrogenases (DH), respectively, have been published, and some have been applied on a technical scale (Liese et al., 2000) (see Chapter 3, Section 3.2.1).

For cofactor-dependent reactions a cosubstrate for cofactor regeneration is required. This should be soluble, non-toxic and not inhibitory, and be readily available at low price. The product should be easy to separate and/or represent a valuable by-product. For the regeneration of NAD^+, formic acid or its salts are suited as it is cheap, the enzyme required (formiate DH, FDH from *Candida boidinii*) is readily available, and the product (CO_2) can be easily separated (see Chapter 3, Section 3.2.1, Fig. 3.6, and Table 3.2).

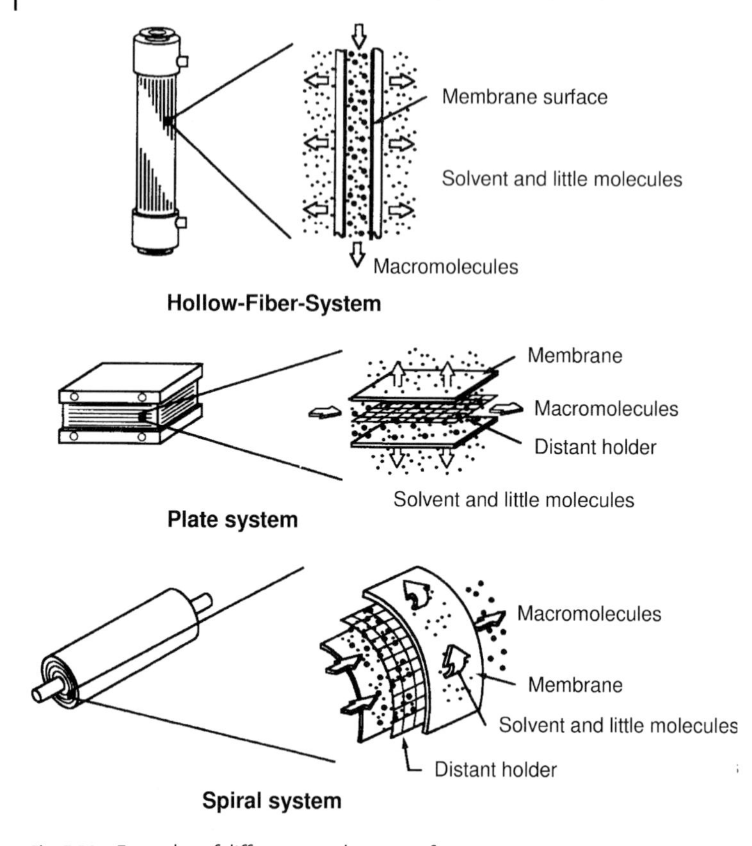

Fig. 5.12 Examples of different membrane configurations (Tutunjian, 1985).

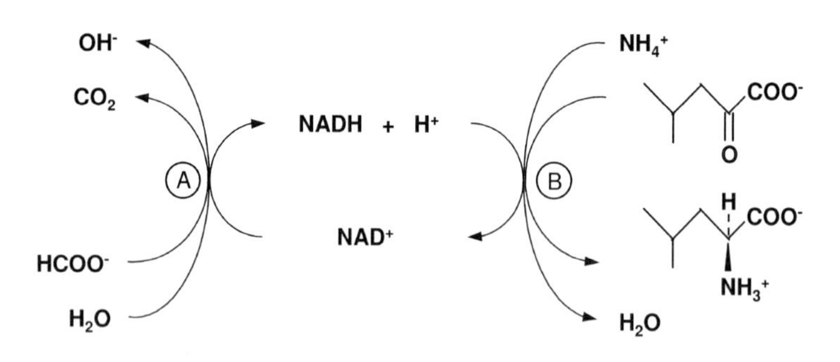

Fig. 5.13 Reaction scheme for continuous production of L-leucine from a keto acid (α-keto isocaproate) with leucine DH (B), formiate DH (A) and polymer-bound cofactor.

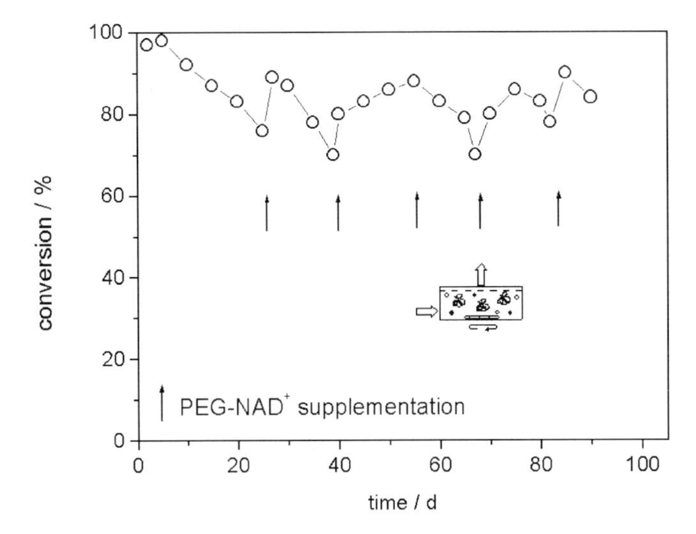

Fig. 5.14 Continuous production of L-leucine in a single enzyme membrane reactor; conditions (feed concentrations): 100 mM α-ketoisocaproate, 400 mM ammonia, and 400 mM formic acid; LeuDH 2.5 U mL^{-1}, FDH 2.5 U mL^{-1}; 0.2 mM PEG-NAD^+; Residence time 1.25 h. The arrows indicate supplementation of PEG-NAD^+ when the cofactor concentration decreased to 0.1 mM (Kragl et al., 1996).

Considerable data have been published relating to the production of L-leucine from a keto-acid (α-keto isocaproate) with leucine DH (LeuDH) in a membrane reactor (Kula and Wandrey, 1987; Kragl et al., 1996) (Figs. 5.13 to 5.15 and Table 5.9): the derivative of the coenzyme (PEG-NADH) remains in the reactor with high degree of retention ($R > 0.999$ %), where the definition of R is:

$$R = (C_r - C_f)/C_r$$

where C_r and C_f are the concentrations in the retentate and filtrate, respectively, index $_0$ for $t = 0$); the loss by wash-out follows from:

$$C_r / C_{r,0} = \exp - [(1 - R)/\tau)$$

where τ is the mean residence time.

The amounts of enzyme consumed and product obtained as a function of process time are shown in Figure 5.15. Results including concentrations, conversion, space-time yield, turnover number and consumption of enzymes per product unit are listed in Table 5.9. On examination, the continuous operation with high efficiency over an extended time scale of 90 days was seen to be remarkable.

For further examples, see Chapter 3, Section 3.2.1.1. Hydroxy acids can be produced following the same principle. (*R*)-2-Hydroxy-4-phenyl-butyric acid is produced industrially from 2-oxo-4-phenyl-butyric acid using (*R*)-lactate-NAD oxido-

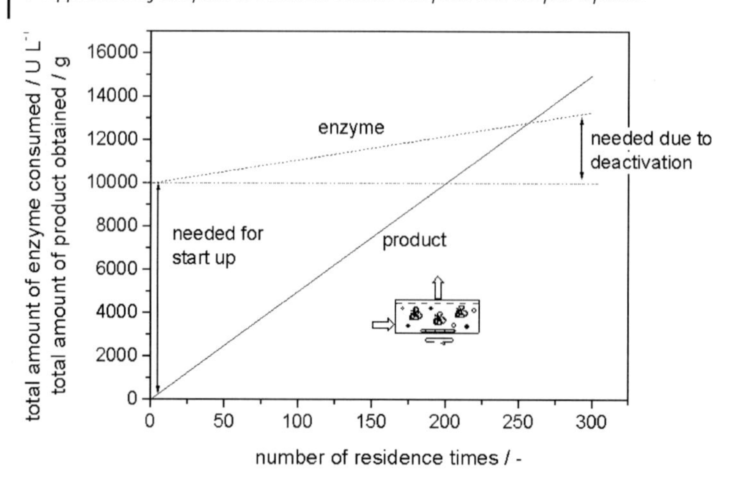

Fig. 5.15 Total amounts of enzyme consumed and product obtained as a function of process time; example used for calculation: substrate 500 mM, enzyme deactivation 3 % per day, molecular weight of product 100 g mol^{-1}. Residence time 1 h (Kragl et al., 1996).

Table 5.9 Conditions and results for the production of L-leucine (Kragl et al., 1996).

Feed concentrations	
• Ketoisocaproate	100 mM
• Ammonia	400 mM
• Formic acid	400 mM
Reactor concentrations	
• PEG-NAD$^+$	0.2 mM
• LeuDH	2.5 U mL^{-1}
• FDH	2.5 U mL^{-1}
Reactor volume	10 mL
pH	8.0
Temperature	25 °C
Residence time	1.25 h
Mean conversion	85 %
Space-time yield	214 g L^{-1} d^{-1}
Total turnover number	80 000
Consumption LeuDH	261 U $kg^{-1}_{product}$
Consumption FDH	298 U $kg^{-1}_{product}$

reductase. Formate DH and formic acid are used for cofactor regeneration. The reaction is carried out in a continuous stirred tank reactor with an ultrafiltration membrane (cut-off 10 kDa; c.f. Fig. 5.9b) with 4.6 hours residence time. At a conversion of 91 % with 99.9 % ee, the enzyme consumption is 150 U kg^{-1} lactate DH and 150 U kg^{-1} FDH (Liese et al., 2000, pp. 113–115). The stereochemistry, empirical rules and further examples have been discussed earlier in Chapter 3, Section 3.2.

The advantages and disadvantages of membrane processes are summarized in Table 5.10, and further examples will be referred to in the following section. Different amino acids can be obtained following the principle mentioned earlier, or by stereospecific hydrolysis of acetyl derivatives of the D,L-mixture with acylases. Separation of the L-amino acid and the acetylated D-derivative proceeds subsequently by ion-exchange columns. An elegant variant is the cofactor regeneration by choice of an appropriate substrate: by oxidation of an α-hydroxy acid followed by reductive amination of the resulting keto-acid to give the amino acid.

The regio- and stereospecific transformation of carbohydrates by redox enzymes can also be carried out in membrane systems, an example being the stereospecific formation of D-tagatose and D-xylulose (Haltrich et al., 2000).

However, for cases of high-cost enzymes, or limited availability and systems of high complexity (more than two enzymes with cofactors), the use of recombinant microorganisms or cells, optimized by metabolic engineering, may well be more convenient as compared to membrane systems with isolated enzymes. Examples of this are the stereospecific mannitol formation from fructose by cells of *Leuconostoc mesenteroides*, with cofactor regeneration by glucose oxidation, and a new system of bacteria designed for the production of vitamin C (Anderson, 1985; Grindley, 1988; Kaup et al., 2003).

Table 5.10 Advantages and disadvantages of membrane processes.

Advantages	***Disadvantages***
No loss and cost for immobilization	Higher investment and more complex periphery
Production with constant conversion due to continuous enzyme dosage	Continuous pumping required
Easy exchange of substrate/enzyme systems	Tubular reactor characteristics require cascade reactor or column with internals
Homogeneous catalysis	Soluble enzymes are in general less stable as compared to immobilized
No mass transfer limitation	
Efficient use of multi-enzyme systems	

5.5 Exercises

5.5.1 Exercise 1: Production of Invert Sugar by Invertase

Invert sugar is to be produced in a technical process from sucrose by hydrolysis with means of invertase (E):

$$\text{Sucrose} + H_2O \xrightarrow{E} \text{Glucose} + \text{Fructose}$$

(A) Reaction temperature: 20 °C (no consideration of enzyme inactivation)

Conversion of 600 kg sucrose per m^3 (1.75×10^3 mol m^{-3}) by 95 % to give a final concentration of 30 kg m^{-3} (0.0877×10^3 mol m^{-3}) in 10 h (3.6×10^4 s).

Calculation of the conversion by means of Eq. (5.2):

$$[S]_0 - [S] + K_m \cdot \ln([S]_0/[S]) = V_{max} \cdot t$$
$$K_m = 0.01 \text{ mol/l} \triangleq 10 \text{ mol/m}^3$$

Result with the data given:

$$(1.75 - 0.0877)\times10^3 + 10 \ln 20 = V_{max}\times3.6\times10^4$$

(obviously the term '10 ln 20 = 30' can be neglected; that means the reaction is zero-order in the range considered here)

$$1.66\times10^3 = V_{max}\times3.6\times10^4$$
$$V_{max} = 4.6\times10^{-2} \text{ (kat m}^{-3}) = 2.76\times10^6 \text{ U m}^{-3}$$

2.76×10^6 U (1 kat = 6×10^7 U) corresponds to the enzyme amount required per m^3 of reaction volume.

Inactivation can be neglected at 20 °C.

(B) Reaction temperature: 50 °C (consideration of inactivation)

Reaction conditions (Westphal, G., et al. *Acta Biotechnol.* 1988, *8*, 357–365):

Activation energy of the reaction: $E_A = 39.1$ kJ mol^{-1}

Activation energy of the inactivation reaction: $E_I = 404.4$ kJ mol^{-1}

Rate constant of inactivation: $\ln k_i = -48610 \cdot (1/T) + 143.33$
(Temperature, K; $T/K = \vartheta/°C + 273.15$)

Result for the rate constant of inactivation: k_i (50 °C) = 8.29×10^{-4} (min^{-1}).

Residual activity ($[E]/[E]_0$) after 10 h: $\ln([E]_0/[E]) = k_i \cdot t$; $[E]_0/[E] = 1.64$, which corresponds to a residual activity of 60.8 %.

Calculation of the mean enzyme activity considering inactivation (integral over reaction time t, divided by t):

$$V_{max} = V_{max,0} \frac{(1 - e^{-k_i \cdot t})}{k_i \cdot t}$$

From this follows the activity required at the start of the reaction:

$$V_{max,0} = V_{max} \frac{k_i \cdot t}{(1 - e^{-k_i \cdot t})}$$

For the example discussed (50 °C) the effective mean activity is

$$V_{max,m} = 0.788\ V_{max,0}$$

The rate constant k_2 at 50 °C is higher as compared to k_1 at 20 °C, corresponding to:

$$k_2/k_1 = (\exp - E_A/RT_2)/(\exp - E_A/RT_1)$$

$$\ln (k_2/k_1) = \frac{E_A}{R}\left(\frac{1}{T_1} - \frac{1}{T_2}\right)$$

$$\ln (k_2/k_1) = \frac{39.1}{8.315 \times 10^{-3}}\left(\frac{1}{293.15} - \frac{1}{323.15}\right) = 1.489$$

$$k_2/k_1 = 4.43$$

It follows that only 1/4 of the enzyme amount or activity, respectively, is required at 50 °C, as compared to that at 20 °C, in order to obtain the same reaction rate. However, due to inactivation the effective amount required is:

$$V_{max,0} = 1/0.788\ V_{max,m} = 1.27\ V_{max,m}$$

The total amount required therefore is:

$$V_{max,0} = \frac{1}{4.43} \cdot 1.27 \cdot 2.76 \cdot 10^6\ \mathrm{U\ m^{-3}}$$

$$V_{max,m} = 0.79 \cdot 10^6\ \mathrm{U\ m^{-3}}$$

That corresponds to 29 % of the amount required at 20 °C.

For the productivity, it follows that:

$P(t) = 600 \times 0.95$ kg (invert sugar/m^3)/13.17×10^{-3} (kat/m^3) $\approx$ 43.3 tons/kat, or 0.72 g U^{-1} (for t = 10 h).

For an enzyme activity of 300 U mg^{-1} this corresponds to ca. 216000 kg invert sugar per kg enzyme. For a technical enzyme with an activity of 10 U mg^{-1}, this would correspond to ca. 7200 kg product per kg enzyme.

Questions

1. Would a further increase of temperature to 55 °C be advantageous? An increased temperature would be favorable in order to reduce the risk of contamination by infections.

2. Calculate the amount of enzyme required (kat m^{-3}) in order to convert sucrose (60 kg m^{-3}) to dextran by dextran-sucrase with 90 % yield in 7 days; must substrate inhibition (K_i = 1.5 mol L^{-1}) be taken into account?

   $K_m = 2.4\times10^{-2}$ mol L^{-1}; $k_i = 7.1\times10^{-4}$ h^{-1} (Inactivation, 20 °C)

Activation energies: reaction: 40 kJ mol^{-1}; inactivation: 200 kJ mol^{-1} (values assumed). Reaction temperatures: a) 20 °C; b) 40 °C.

5.5.2 Exercise 2: Calculation of Enzyme Inactivation under Different Conditions

Data for inactivation kinetics were obtained for two thermostable amyloglucosidases (GA), including stabilization by sorbitol as a model polyol (Feng et al., 2002): *Thermoanaerobacterium thermosaccharolyticum* GA was stable up to 65 °C, k_i (min^{-1}) being 0.0043 (7.17×10^{-5} s^{-1}) at 70 °C. For *A.niger*, GA k_i values at 60, 65, and 70 °C were 0.018, 0.062, and 0.088 (3.0×10^{-4} s^{-1}, 1.03×10^{-3} s^{-1}, and 1.47×10^{-3} s^{-1}), respectively.

Questions

1. Compare the stability of the two enzymes at 70 °C, in terms of inactivation time for a 20 % residual activity; give results in minutes, or hours, as appropriate.

   Result: Inactivation time for *Thermoanaerobacterium* GA: 374 min, or 6.2 h;
   for *A. niger* GA: 18.3 min.

   Sorbitol stabilized *A. niger* GA considerably at 65 °C, with k_i-values of 0.020, and 0.0080 (3.33×10^{-4} s^{-1} and 1.33×10^{-4} s^{-1}) at 1 M and 2 M sorbitol, respectively, while 4 M sorbitol gave complete stability over 40 min of incubation.

2. Calculate the factor of stabilization of *A. niger* GA by 2 M sorbitol from the above half-life times at 65 °C.

   Result: no sorbitol: half-life of 11.2 min
   2 M sorbitol: half-life of 86.6 min
   Stabilization factor: 7.7

3. For a peptidase ('glutenase'), inactivation data were given with the product description, where times for 90 % inactivation in the presence of wheat gluten were ~220 min at 60 °C and 15 min at 70 °C, whereas inactivation was rapid at 80 °C, as required in the baking process (Novo Nordisk, 1995).
   What does "inactivation was rapid at 80 °C" mean in quantitative terms (minutes)?

Calculate k_i for 60 and 70 °C, the respective activation energy, k_i at 80 °C, and the time for 90 % inactivation at that temperature.

Result: k_i at 60 °C: 0.0105 min^{-1}; at 70 °C: 0.153 min^{-1}.
Activation energy $E_A = 248$ kJ mol^{-1}; time for 90 % inactivation at 80 °C is 1.23 min.

5.5.3
Exercise 3: Hydrolysis of Penicillin with Penicillin Amidase

Penicillin G (300 mmol) is hydrolyzed by penicillin amidase (PA) at 37 °C and pH 7.8 ($K_m = 10$ µM) in 120 min to yield 95 % of 6-APA and phenylacetic acid. Enzyme inactivation may be neglected if the pH is kept constant. Phenylacetic acid is a competitive inhibitor ($K_i = 20$ µM).

Questions

1. How many units (U) of PA are required per kg of 6-(APA)? (prices: 6-APA ~100 € kg^{-1} (~2 € per 100 U).
2. Is the process meaningful for biotechnical purposes? How many times must the enzyme be reused in order to bring the enzyme cost below 5 % of the production cost?

The amount of enzyme required for a given conversion of substrate in a given time can be calculated from Eqs. (2.34) and (2.36) for different cases. The integration of the Michalis–Menten equation for competitive product inhibition gives (per L reaction volume)

$$V_{max} = \frac{([S]_0 - [S])\left(1 - \frac{K_m}{K_i}\right) + \left(K_m + \frac{K_m [S]_0}{K_i}\right) \ln([S]_0/[S])}{t}$$

$$= \frac{(0.3 - 0.015)(1 - 0.5) + (10^{-5} + 0.15)\ln(20)}{120} = \frac{0.6\ 10^6}{120}\left[\frac{\mu M}{L\ min}\right]$$

$$= 5\ 10^3\ [U/L]$$

Amount of enzyme required for 1 kg of 6-APA:

$5 \times 10^3\ (0.285\ 0.216)^{-1} = 5000/0.062 = 80\,600$ U

Enzyme cost per kg 6-APA: ~2000 €.

Consequences: For a cost range of 5 € kg^{-1}, the enzyme must be reused more than 400 times in order to make the process economically competitive. The enzyme must therefore be applied as an immobilized bioctalyst for industrial purposes.

Literature

General information

Atkinson, B., Mavituna, F., *Biochemical Engineering and Biotechnology Handbook*, 2nd edn., pp. 529–546. MacMillan Publisher Ltd., New York, **1991**

Ballesteros, A., Plou, F.J., Iborra, J.L., Halling, P.J. (Eds), *Stability and Stabilization of Biocatalysts*. Elsevier, Amsterdam, **1998**

BRENDA – the Enzyme Database, via www.empproject.com/links/

Godfrey, T., West, S.I. (Eds), *Industrial Enzymology; Introduction to industrial enzymology*, pp. 1–8. Macmillan Press, London, **1996**

Laane, C., Tramper, J., Lilly, M.D. (Eds), *Biocatalysis in Organic Media; Studies in Organic Chemistry*, Vol. 29. Elsevier, Amsterdam, **1987**

Liese, A., Seelbach, K., Wandrey, C., *Industrial Biotransformations*. Wiley-VCH, Weinheim, **2000**

Straathof, A.J.J., Adlercreutz, P., *Applied Biocatalysis*. Harwood Academic Publishers, Amsterdam, **2000**

Uhlig, H., *Industrial enzymes and their Applications*. John Wiley & Sons, Inc., New York, **1998**

Whitaker, J.R., Voragen, A.G.J., Wong, D.W.S. (Eds), *Handbook of Food Enzymology*. Marcel Dekker, New York, **2003**

References

Adler-Nissen, J., *Enzymic hydrolysis of food proteins*, Appendix p. 355, Elsevier Publishers, Barking, **1986**

Aleshin, A.E., Hoffman, C., Firsov, L.M., Honzatko, R.B., Refined crystal structures of glucoamylase from *Aspergillus awamori* var. X100, *J. Mol. Biol.* **1994**, *238*, 575–591

Aleshin, A.E., Stoffer, B., Firsov, L.M., Svensson, B., Honzatko, R.B., Crystallographic complexes of glucoamylase with maltooligosaccharide analogs: relationship of stereochemical distortions at the nonreducing end to the catalytic mechanism, *Biochemistry*, **1996**, *35*, 8319–8328

Anders, J., Synthese neuartiger Polyvinylsaccharide auf der Basis von 3-Aminosaccharose. PhD thesis, Technical University, Braunschweig, **2002**

Anderson, S., Berman, C., Lazarus, M.R., et al., Production of 2-keto-L-gulonate, an intermediate in L-ascorbate synthesis, by a genetically modified *Erwinia herbicola*, *Science* **1985**, *230*, 144–149

Athes, V., Combes, D., Effect of high hydrostatic pressure on enzyme stability, in: Ballesteros, A., Plou, F.J., Iborra, J.L., Halling, P.J. (Eds), *Stability and Stabilization of Biocatalysts*, pp. 205–210. Elsevier, Amsterdam, **1998**

Bajpai, P., Application of enzymes in the pulp and paper industry, *Biotechnol. Prog.*, **1999**, *15*, 147–157

Bentley, I.S., Williams, E.C., Starch conversion, in: Godfrey, T., West, S.I. (Eds), *Industrial Enzymology*, pp. 339–357. Macmillan Press, London, **1996**

Bergmeyer, H.U. (Ed.), *Methods of enzymic analysis*, 3rd edn. Verlag Chemie, Weinheim, **1983**

Beier, L., Svendsen, A., Andersen, C., Frandsen, T.P., Borchert, T.V., Cherry, J.R., Conversion of the maltogenic α-amylase Novamyl into a CGTase, *Protein Eng.*, **2000**, *13*, 509–513

Benen, J.A.E., Voragen, A.G.J., Visser, J., Pectic Enzymes; and Benen, J.A.E., Alebeek, G.W.M., Voragen, A.G.J., Visser, J., Pectic Esterases; Benen, J.A.E., Visser, J., Polygalacturonases, in: Whitaker, J.R., Voragen, A.G.J., Wong, D.W.S. (Eds), Handbook of Food Enzymology, pp. 845–848, 849–856, 857–866. Marcel Dekker, New York, **2003**

Böker, M., Jördening, H.-J., Buchholz, K., Kinetics of Leucrose formation from sucrose by dextransucrase, *Biotech. Bioeng.*, **1994**, *43*, 856–864

Bommarius, A.S., Drauz, K., Hummel, W., Kula, M. R., Wandrey, C., Some new developments in reductive amination with cofactor regeneration, *Biocatalysis*, **1994**, *10*, 37–47

Bryjak, J., Noworyta, A., Storage stabilization and purification of enzyme by water-soluble synthetic polymers, *Enzyme Microb. Technol.*, **1994**, *16*, 616–621

Buchholz, K., Glümer, A., Skeries, B., Schmalbruch, B., Warn, S., Yaacoub, E., Saccharidpolymere und -copolymere auf der Basis von Zukkerbausteinen. In: Moderne Polymere-Kohlenhydrate und Pflanzenöle als innovative Rohstoffe, 27-45, Fachagentur Nachwachsende Rohstoffe (Ed.). Gülzow, Germany, **2001**

Buchholz, K., Seibel, J., Isomaltooligosaccharides, in: Eggleston, G., Côté, G.L. (Eds), *Oligosaccharides in Food and Agriculture.* ACS Symposium Series 849, pp. 63–75. American Chemical Society, Washington, **2003**

Cheetham, P.S.J., Production of isomaltulose using immobilized microbial cells; in: Mosbach, K. (Ed.), *Methods Enzym.*, Vol. 136, S432–S454, **1987**

Cordt, S. de, Vanhoof, K., Hu, J., Maesmans, G., Hendrickx, M., Tobback, P., Thermostability of soluble and immobilized α-amylase from *Bacillus licheniformis, Biotech. Bioeng.*, **1992**, *40*, 396–402

Coutinho, P.M., Reilly, P.J., Structure–function relationships in the catalytic and starch binding domains of glucoamylase, *Protein Eng.*, **1994**, *7*, 393–400

Curtis, R.A., Ulrich, J., Montaser, A., Prusnitz, J.M., Blanch, H.W., Protein-Protein Interactions in Concentrated Electrolyte Solutions, *Biotechnol. Bioeng.* **2002**, *79*, 367–380

Declerck, N. et al., Probing structural determinants specifying high thermostability in *Bacilluslicheniformis* alpha-amylase, *J. Mol. Biol.*, **2000**, *301*, 1041–1057

Demuth, B., Jördening, H.-J., Buchholz, K., Modelling of oligosaccharide synthesis by dextransucrase, *Biotech. Bioeng.*, **1999**, *62*, 583–592

Denner, W.H.B., Gillanders, T.G.E., The legislative aspects of the use of industrial enzymes in the manufacture of food and food ingredients, in: Godfrey, T., West, S.I. (Eds), *Industrial Enzymology*, pp. 397–411. Macmillan Press, London, **1996**

Eriksen, N., Detergents; in: Godfrey, T., West, S.I. (Eds), *Industrial Enzymology*, pp. 187–207. Macmillan Press, London, **1996**

Feng, P.-H., Berensmeier, S., Buchholz, K., Reilly, P.J., Production, purification, and characterization of *Thermoanaerobacterium thermosaccharolyticum* glucoamylase, *Starch/Stärke*, **2002**, *54*, 328–337

Foster, K.A., Frackman, S., Jolly, J.F., Production of enzymes as fine chemicals, in: Reed, G., Nagodawithana, T.W. (Eds), *Biotechnology*, Vol. 9, pp. 73–120. VCH, Weinheim, **1995**

Fullbrock, P.D., Practical limits and prospects (Kinetics), in: Godfrey, T., West, S.I. (Eds), *Industrial Enzymology*, pp. 503–540. Macmillan Press, London, **1996**

Gaertner, A., Barnett, C., Dale, D., Enhanced stability in cleaning applications through improved formulations. Lecture, 91st Annual AOCS (American Oil Chemists' Society) meeting, April **2000**, San Diego, California, USA

Giardina, T., Günning, A.P., Juge, N., Both binding sites of the starch-binding domain of *Aspergillus niger* glucoamylase are essential for inducing a conformational change in amylose, *J. Mol. Biol.*, **2001**, *313*, 1149–1159

Gibson, T.D., Hulbert, J.N., Woodward, J.R., Preservation of shelf life of enzyme based analytical systems using a combination of sugars, sugar alcohols and cationic polymers or zinc ions, *Anal. Chim. Acta*, **1993**, *279*, 185–192

Giffhorn, F., Köpper, A., Huwig, A., Freimund, S., Rare sugars and sugar-based synthons by chemo-enzymatic synthesis. *Enzym. Microbiol. Technol.*, **2000**, *27*, 734–742

Godfrey, T., Textiles, in: Godfrey, T., West, S.I. (Eds), *Industrial Enzymology*, pp. 359–371. Macmillan Press, London, **1996**

Gottschalk, T.E., et al., Structure, function and protein engineering of starch-degrading enzymes, *Biochem. Soc. Trans.*, **1998**, *26*, 198–204

Grassin, C., Fauquembergue, P., Fruit Juices, in: Godfrey, T., West, S.I. (Eds), *Industrial Enzymology*, pp. 225–264. Macmillan Press, London, **1996**

Grindley, J.F., Payton, M.A., van den Pol, H., Hardy, K.G., Conversion of glucose to 2-keto-L-gulonate, an intermediate in L-ascorbate synthesis, by a recombinant strain of *Erwinia citreus, Appl. Environ. Microbiol.*, **1988**, *54*, 1770–1775

Haltrich, D., Leitner, B., Nidetzky, B., Kulbe, K.-D., Pyranose oxidase for the production of carbohydrate- based food ingredients, in: Bielecki, S., Tramper, J., Polak, J. (Eds), *Food Biotechnology*, pp. 137–149. Elsevier Science, Amsterdam, **2000**

Harris, E.M.S., Aleshin, A.E., Firsov, L.M., Honzatko, R.B., Refined structure for the complex of 1-deoxynojirimicin with glucoamylase from *Aspergillus awamori* var. X100 to 2.4-Å resolution, *Biochemistry*, **1993**, *32*, 1618–1626

Hendrix, M., Wong, C.-H., Enzymatic Synthesis of Carbohydrates, in: *Bioorganic Chemis-*

try: Carbohydrates, Hecht, S.M. (Ed.), Oxford University Press, **1999**, 198–243

Ikeda, M., Clark, D.S., Molecular cloning of extremely thermostable esterase gene from hyperthermophilic Archeon *Pyrococcus furiosus* in *E. coli*, *Biotechnol. Bioeng.*, **1998**, *57*, 624–629

Illanes, A., Altamirano, C., Aillapán, A., Packed bed reactor performance with immobilized lactase under thermal inactivation, *Enzym. Microbiol. Technol.*, **1998**, *23*, 3–9

Kaup, B., Bringer-Meyer, S., Sahm, H., Metabolic engineering of *Escherichia coli*: construction of an efficient biocatalyst for d-mannitol formation in a whole-cell biotransformation, *Appl. Microbiol. Biotechnol.*, **2003**, 1432–0614 (Online)

Kragl, U., Kruse, W., Hummel, W., Wandrey, C., Enzyme engineering aspects of biocatalysis: cofactor regeneration as examples, *Biotechnol. Bioeng.*, **1996**, *52*, 309–319

Kula, M.-R., Wandrey, C., in: Mosbach, K. (Ed.), *Methods in Enzymology*, Bd. 136, S9 S21, **1987**

Ladenstein, R., Antranikian, G., Proteins from hyperthermophiles: stability and enzymatic catalysis close to the boiling point of water, *Adv. Biochem. Eng.*, **1998**, *61*, 37–85

Lieker, H.P., Thielecke, K., Buchholz, K., Reilly, P.J., High-performance anion -exchange chromatography of saturated and unsaturated oligogalacturonic acids. *Carbohydrate Res.*, **1993**, *238*, 307–311

Maarel, van der, M.J., van der Veen, B., Uitdehaag, J.C., Leemhuis, H. Dijkhuizen, L., Properties and applications of starch-converting enzymes of the α-amylase family, *J. Biotechnol.*, **2002**, *94*, 137–155

Marchal, L.M., et al., Monte Carlo simulation of the α-amylolysis of amylopectin potato starch, *Bioprocess Biosystems Eng.*, **2001**, *24*, 163–170

Mirza, O., Skov, L.K., Remond-Simeon, M., Crystal structures of amylosucrase from *Neisseria polysaccharea* in complex with D-glucose and the active site mutant Glu328Gln in complex with the natural substrate sucrose, *Biochemistry*, **2001**, *40*, 9032–9039

Monsan, P., Combes, D., Effect of water activity on enzyme action and stability, *Ann. N. Y. Acad. Sci.*, **1984**, *434*, 48–60

Morgan, J.A., Clark, D.S., Salt-Activation of Nonhydrolase Enzymes for Use in Organic Solvents, *Biotechnol. Bioeng.* **2004**, *85*, 456–459

Novo Nordisk: Glutenase, B 836a-D 2000, June **1995**

Okkels, J.S., Method for preparing polypeptide variants. PCT WO 97/07205 Novo Nordisk A/S, **1995**

Olsen, H.S., Use of enzymes in food processing, in: Reed, G., Nagodawithana, T.W. (Eds), *Biotechnology*, Vol. 9, pp. 663–736, VCH, Weinheim, **1995**

Olsen, K., et al., Reaction mechanisms of Trp120→Phe and wild-type glucoamylases from *Aspergillus niger*. Interactions with maltooligodextrins and acarbose, *Biochemistry*, **1993**, *32*, 9686–9693

Pierre, F., Perrin, P., Champ, M., Bornet, F., Meflah, K., Menanteau, J., Short-chain fructo-oligosaccharides reduce the occurrence of colon tumors and develop gut-associated lymphoid tissue in mice, *Cancer Res.*, **1997**, *57*, 225–228

O'Rourke, T., Brewing, in: Godfrey, T., West, S.I. (Eds), *Industrial Enzymology*, pp. 103–131. Macmillan Press, London, **1996**

Pedersen, S., personal communication, **2002**

pedro web: *http://afmb.cnrs-mrs.fr/—pedro* (see Reilly, **1999**)

Pietsch, M., Walter, M., Buchholz, K., Regioselective synthesis of new sucrose derivatives via 3-ketosucrose, *Carbohydrate Res.* **1994**, *254*, 183–194

Poulson, Private communication, Enzyme Engineering Conference, Potsdam, Germany, **2001**

Prenosil, J.E., Dunn, I.J., Heinzle, E., in: Rehm, H.-J., Reed, G. (Eds), *Biotechnology*, Bd. 7a, pp. 489–545. Verlag Chemie, Weinheim, **1987**

Quax, W.J., Mrabet, N.T., Luiten, R.G.M., Schuuhuizen, P.W., Staussens, P., Lasters, I., Enhancing the thermostability of Glucose Isomerase by protein engineering. *Bio/Technology*, **1991**, *9*, 723–742

Rastall, R.A., Hotchkiss Jr., A.T., Potential for the development of prebiotic oligosaccharides from biomass, in: Eggleston, G., Côté, G.L. (Eds), *Oligosaccharides in Food and Agriculture*. ACS Symposium Series 849, pp. 44–53. American Chemical Society, Washington, **2003**

Reddy, B.S., Prevention of colon cancer by pre- and probiotics: evidence from laboratory studies, *Br. J. Nutr.*, **1998**, *80*, 219–230

Reilly, P.J., Protein engineering of glucoamylase to improve industrial performance – a review. *Starch/Stärke* **1999**, *51*, 269–274

Reilly, P., Glucoamylase, in: Whitaker, J.R., Voragen, A.G.J., Wong, D.W.S. (Eds), *Handbook of Food Enzymology*, pp. 727–738. Marcel Dekker, New York, **2003**

Riva, S., Carrea, C., Veronese, F.M., Bückmann, A.F., Effect of Coupling Site and Nature of the Polymer on the Coenzymatic Properties of Water-Soluble Macromolecular NAD Derivatives with Selected Dehydrogenase Enzymes, *Enzyme Microb. Technol.* **1986**, *8*, 556–567

Robyt, J.F., *Essentials of Carbohydrate Chemistry*. Springer, New York, **1998**

Röper, H., Selective oxidation of D-glucose: Chiral intermediates for industrial utilization. *Starch/Stärke*, **1990**, *42*, 342–349

Sauer, J., et al., Glucoamylase: structure/function relationships, and protein engineering, *Biochim. Biophys. Acta*, **2000**, *1543*, 275–293

Schmidt, E., et al., Optimization of a process for the production of (*R*)-2-hydroxy-4-phenyl-butyric acid, *J. Biotechnol.*, **1992**, *24*, 3215–3217

Schols, H.A., Voragen, A.G.J., Pectic polysaccharides, in: Whitaker, J.R., Voragen, A.G.J., Wong, D.W.S. (Eds), *Handbook of Food Enzymology*, pp. 829–843. Marcel Dekker, New York, **2003**

Schomburg, D., Stephan, D. (Eds), *Enzyme Handbook*. Springer, Berlin, **1996**

Schomburg, D., Schomburg, I. (Eds), *Springer Handbook of Enzymes*. Springer, Berlin, **2002** (http://www.springer.de/enzymes)

Sicard, P.J., Leleu, J.-B., Tiraby, G., Toward a new generation of glucose isomerase through genetic engineering, Starch **1990**, *42*, 23–27

Sigurskjold, B.W., Christensen, T., Payre, N., Cottaz, S., Driguez, H., Thermodynamics of binding of heterobidentate ligands consisting of spacer-connected acarbose and beta-cyclodextrin to the catalytic and starch-binding domains of glucoamylase from *Aspergillus niger* shows that the catalytic and starch-binding sites are in close proximity in space, *Biochemistry*, **1998**, *37*, 10446–10452

Sierks, M.R., Ford, C., Reilly, P.J., Svensson, B., Site-directed mutagenesis at the active site Trp120 of *Aspergillus awamori* glucoamylase, *Protein Eng.*, **1989**, *2*, 621–625

Sierks, M.R., Svensson, B., energetic and mechanistic studies of glucoamylase using molecular recognition of maltose OH groups coupled with site-directed mutagenesis, *Biochemistry*, **2000**, *39*, 8585–8592

Stoppok, E., Buchholz, K., The production of 3-keto derivatives of disaccharides, in: Bucke, C. (Ed.), *Methods in Biotechnology – Carbohydrate Biotechnology Protocols*, pp. 277–289. Humana Press, Totowa, New Jersey, USA, **1999**

Suvd, D., Fujimoto, Z., Takase, K., Matsumura, M., Mizuno, H., Crystal structure of *Bacillus stearothermophilus* alpha-amylase: possible factors of the thermostability, *J. Biochem. (Tokyo)*, **2001**, *129*, 461–468

Svendsen, A., *Biochim. Biophys. Acta*, **2000**, *1543*, 223–238

Svensson, B., Frandzen, T.P., Matoni, J., Mutational analysis of catalytic mechanism and specificity in amylolytic enzymes; in: Petersen, S.B., et al. (Eds), *Carbohydrate Bioengineering*, pp. 125–145. Elsevier, Amsterdam, **1995**

Svensson, B., Bak-Jensen, K.S., Mori, H., The engineering of specificity and stability in selected starch-degrading enzymes; in: Gilbert, H.J. (Ed.), *Recent Advances in Carbohydrate Bioengineering*, pp. 272–281. The Royal Society of Chemistry, Cambridge, **1999**

Syldatk, C., Altenbuchner, J., Mattes, R., Siemann-Herzberg, M., Microbial hydantoinases. *Appl. Microbiol. Biotechnol.*, **1999**, *51*, 293–309

Timme, V., Buczys, R., Buchholz, K., Kinetic investigations on the hydrogenation of 3-ketosucrose, *Starch/Stärke*, **1998**, *50*, 29–32

Tolan, J.S., Pulp and paper, in: Godfrey, T., West, S.I. (Eds), *Industrial Enzymology*, pp. 327–338. Macmillan Press, London, **1996**

Tramper, J., Poulson, P.B., Enzymes as processing aids and final products; in: Straathof, A.J.J., Adlercreutz, P. (Eds), *Applied Biocatalysis*, pp. 55–91. Harwood Academic Publishers, Amsterdam, **2000**

Tutunjian, R.S., in: Moo-Young, M. (Ed.), *Comprehensive Biotechnology*, Bd. 2, S411–S437. Pergamon Press, Oxford, **1985**

Uhlig, H., *Enzyme arbeiten für uns*, pp. 287–306. Hanser, München, **1991**

Uhlig, H., *Industrial enzymes and their applications*, John Wiley & Sons, Inc., New York, **1998**

Vind, J., Optimization of fungal lipase for diverse applications, *Engineering Enzymes*, preprints p. 31, Paris, **2002**

Weemaes, C., et al., High pressure, thermal, and combined pressure-temperature stabilities of α-amylases from *Bacillus* species, *Biotech. Bioeng.*, **1996**, *50*, 49–56

Ye, W.N., Combes, D., Monsan, P., Influence of additives on the thermostability of glucose oxidase, *Enzyme Microbiol. Technol.*, **1988**, *10*, 498–502.

6
Immobilization of Enzymes (Including Applications)

Why immobilize biocatalysts?	Method for reuse and stabilization of biocatalysts
Immobilization of intracellular enzymes: in cells or as isolated enzymes?	For high enzyme density in a cell or when cofactor regeneration is required: immobilization in cells (see Chapter 7); otherwise immobilization of isolated enzymes
Important enzyme properties for immobilization	Protein surface properties (functional groups, ionic charge, and hydrophobic groups; see Sections 2.2 and 6.1.1)
Carriers for immobilization, properties	Porous systems, high internal surface area for adsorption or covalent binding; functionalization of carrier surface
Methods for immobilization on different carriers	• Adsorption: equilibrium (T, pH, I), prevention of desorption by cross-linking • Covalent binding: activation of the carrier, reaction with functional enzyme groups
Relevant aspects	Yield of active immobilized enzyme, kinetics and efficiency, stability
Examples	Relevant technical applications

Biocatalysts and Enzyme Technology. K. Buchholz, V. Kasche, U.T. Bornscheuer

ISBN: 3-527-30497-5

6.1
Principles

Insoluble enzymes offer all the advantages of classical heterogeneous catalysis: convenient separation for reuse after the reaction by filtration, centrifugation, etc.; application in continuous processes, in fixed-bed, or fluidized-bed and stirred-tank reactors provided with a filter system for retention.

Continuous processes combine the generally more simple technical equipment with the potential for easier process control, automation and convenient coordination with up- and downstream processing, including product recovery and purification.

Today, more than 15 processes of major importance are available on stream, in addition to a considerable number of further, specialized applications. Selected examples of industrial application are listed in Table 6.1, and these range from a production scale of several million down to a few hundred tons per year, albeit with high value added, notably chiral, products (further examples are provided in Chapter 1, Section 1.3, Table 1.2, p. 13). The most important of these processes are to be found in the food and pharmaceutical industries, although the production of basic chemicals such as acrylamide is also included. For details of sales and costs, the reader is advised to consult Godfrey and West (1996), West (1996), Cheetham (2000), and Liese et al. (2000).

The most obvious reason for immobilization is the need to reuse enzymes, if they are expensive, in order to make their use in industrial processes economic. The benefits and shortcomings of continuous processing are summarized in Table 6.2. Limitation by mass transfer in heterogeneous biocatalysts, and thus reduced efficiency of the enzyme, is another important topic that requires consideration and optimization (see Chapter 8, Section 8.4).

The contact of an enzyme with the carrier surface may induce changes due to forces that cause interaction of surface active groups, both of the matrix and the protein. Conformational changes and inactivation have been observed in many cases. Recently, an analysis including adsorption isotherms has been published (Norde and Zoungrana, 1998).

Numerous investigations have been reported on the stabilization of enzymes by their immobilization on insoluble carriers. In general, the effects are not understood on a molecular basis, but two are clear: (1) to reduce inactivation by protease hydrolysis, as immobilization decreases the access of peptidases to the targets of proteolysis; and (2) to provide stabilization of the enzyme's tertiary structure by a rigid matrix, which restricts unfolding to a certain degree. Otherwise, this subject is governed by empirical analysis and procedures. In recent years, continued efforts have resulted in commercial biocatalysts of high operational stability, such as glucose isomerases with productivities in the range of 20 000 kg (product) kg^{-1} (biocatalyst), and penicillin acylases with productivities of 200 to 2000 kg (product) kg^{-1} (biocatalyst), and operating lifetimes in excess of 1000 h. The range of enzymes available is dependent on technical conditions such as temperature, conversion of substrate, and residual activity when the biocatalyst is replaced.

Table 6.1 Examples of immobilized enzymes used in major commercial processes (estimates) (Poulsen, 1984; West, 1996; Cheetham, 2000; Liese et al., 2000; Schmid et al., 2002).

Enzyme (amount of immobilized biocatalyst used)	***Product (amount)***	***Companies (enzyme suppliers; applicants)***
Glucose isomerase (xylose isomerase) (1500 t a^{-1})	Glucose-fructose syrup ca. 10 million t a^{-1}	Novozymes, Degussa, DSM, Danisco/Genencor, Nagase; Archer Daniels Midland, A. E. Staley, Cargill, CPC International and others
Sucrose mutase	Isomaltulose[a)] ca. 100 000	Cerestar, Mitsui Seito Co., Südzucker
β-Galactosidase (Lactase)	Glucose-galactose syrup, >6000 t a^{-1}	Central del Latte, Snow Brand Milk Prod., Sumimoto Chemical Industries, Valeo/Alko
Penicillin acylase (or amidase)[b)] (several t a^{-1})	6-APA (6-amino penicillanic acid) >10 000 t a^{-1}	DSM, Asahi Chemical Ind. Co., Beecham, Biochemie, Bristol Myers, Toyo Jozo, Pfizer
1. D-amino acid oxidase; 2. Glutaryl amidase[b)]	7-ACA (7-amino cephalosporanic acid)	Asahi Chemical Ind. Co., Aventis, Biochemie, DSM, Toyo Jozo
Thermolysin	Aspartame[c)] ca. 10 000 t a^{-1}	DSM, Holland Sweetener Company
Nitrilase	Acrylamide[d)] >95 000 t a^{-1} Nicotinamide >5000 t a^{-1}	Nitto Chemicals, Lonza
Aminoacylase (ca. 10 t a^{-1})	L-Amino acids, e.g., alanine, leucine, phenylalanine, Dopa, *tert*- leucine; several 1000 t/a	Amino, Degussa, DSM, Fluka, Tanabe Seiyaku, and others
Hydantoinases	D-amino acids, e.g., *p*-hydroxy phenylglycine[e)], phenylalanine, tryptophan, valine ca. 1000 t a^{-1}	Ajinomoto, Bayer, Degussa, DSM, Kanegafuchi
Lipases[e)]	Pharmaceutical and agrochemical intermediates, e.g., (*S*)-alcohols, (*R*)-1-phenylethylamine multi-kg; >1000 t a^{-1}	Novozymes, DSM, Genzyme Co., BASF, Bristol Myers Squibb, DSM, Fluka, Schering Plough, Sumitomo Chemical Co., Tanabe Seiyaku Co., and others

a) Sucrose isomer: α-D-glucosyl-1,6-D-fructoside.
b) Mainly for semisynthetic penicillin and cephalosporin antibiotics (ampicillin, amoxycillin) with sales $>3 \times 10^9$ € a^{-1} (Cheetham, 2000, pp. 120–122); see Section 9.2.
c) Sales ca. 850×10^6 € a^{-1} (Cheetham, 2000, pp. 113–117).
d) OECD, 2001
e) See Sections 3.2.2.1 and 3.2.2.6.

Table 6.2 Reasons and limitations for enzyme immobilization.

Reasons	*Limitations*
Reuse of enzyme, reducing cost	Cost of carriers and immobilization
Continuous processing • Facilitated process control • Low residence time (high volumetric activity) • Optimization of product yield	Mass transfer limitations Problems with cofactors and regeneration Problems with multienzyme systems
Easy product separation and recovery Stabilization by immobilization	Changes in properties (selectivity) Activity loss during immobilization

Enzymes may be immobilized by transformation into an insoluble form and retained – as a heterogeneous biocatalyst – inside a reactor by mechanical means, or by inclusion in a definite space (Fig. 6.1). The second principle – inclusion by a semipermeable membrane – was discussed in Chapter 5, Section 5.4. Further inclusion in microcapsules should also be mentioned, whereby small particles are utilized, but the enzyme remains soluble. To date, these applications have not been successful, and subsequently the principles applied most frequently – namely binding to carriers – will be dealt with here in more detail.

One-step reactions are the domain of immobilized enzymes, if not other factors, such as high molecular-weight substrates, represent serious problems. The classical principles of binding enzymes onto carriers include physical adsorption, ionic binding to ion exchangers, and covalent coupling to an insoluble matrix (Fig. 6.1). The last two procedures dominate in technical applications. Important examples applied on the large scale are glucose isomerization, acrylamide production, penicillin hydrolysis, and the production of enantiomerically pure amino acids (see Table 6.1).

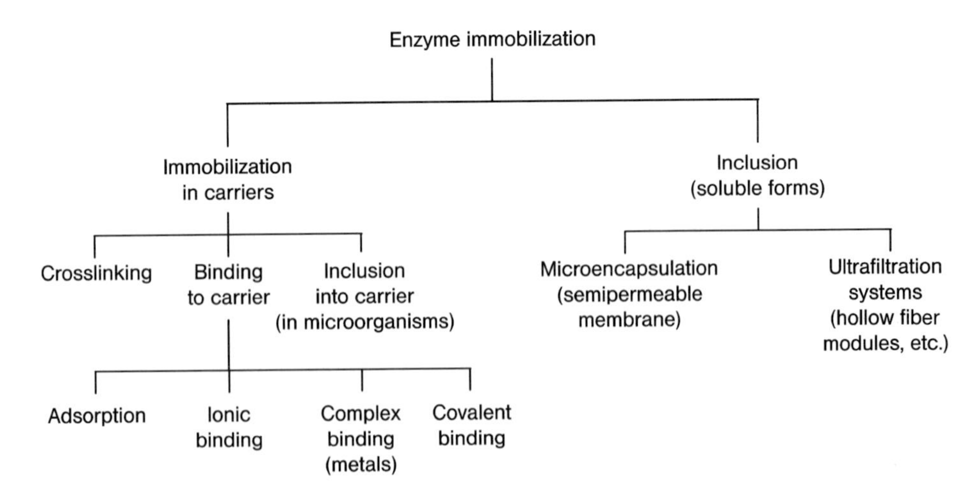

Fig. 6.1 Principles of enzyme immobilization.

Intracellular enzymes are either soluble or membrane bound. Immobilization of the cells clearly includes the enzymes listed (see Chapter 7), and this provides advantages over the immobilization of isolated enzymes for complex transformations which require more than one active enzyme and/or cofactor regeneration.

6.1.1 Parameters of Immobilization

As for other physical and chemical processes, the rate and yield of immobilization depend on the parameters involved, notably the type of carrier, method of immobilization, concentration, pH, temperature, and reaction time (Monsan, 1977/78; Buchholz et al., 1979).

Binding to insoluble porous carriers is the standard method for both laboratory and industrial applications. The properties of the external protein surface and the functional groups accessible play a major role in binding to carriers. Adsorption is dependent upon the hydrophilic and hydrophobic characteristics of the surface regions. The dominating ionic groups and their interaction depend on the amino acids, with pH-dependent charge and their density, which in turn determines the overall surface charge, which is influenced by the pH of the solution (Table 6.3). Most (up to 95 %) of the titratable residues are located on the surface of soluble proteins, and about 40 % are hydrophobic residues (Petersen, 1998; Creighton, 1993). Recently, the protein surface has been modified by recombinant methods in order to introduce amino acids that favor the interaction with surfaces, or to shift the isoelectric point (Laurents et al. 2003). This may, however, influence the structure of the active site, i.e. the activity, and stability (see Chapter 2, Section 2.11).

For covalent binding, several functional groups may be utilized, if they are accessible at the protein surface (Table 6.3). However, only a few are used in practice, notably the amino groups of lysine and arginine, and the carboxyl groups of aspartic and glutamic acid.

It is important to consider these strong interactions between the enzyme and the carrier surface with regard to ionic, hydrophilic and/or hydrophobic, and hydrogen bonding, and also with respect to enzyme stability. Several strong interactions (and their cooperative effects) may result in unwanted irreversible adsorption at the carrier surface and loss of enzyme activity. This may also be due in part to induced conformational changes of the tertiary structure of the protein. These effects have been observed notably for multiple interactions at rigid carrier surfaces.

In some cases the introduction of surface groups – known as "spacers" – in an appropriate density may be advisable, as these protect the enzyme against an aggressive surface which is rendered inaccessible and thus avoids inactivation. Similarly, the adsorption of a certain amount of inactive, cheap protein (e.g., bovine serum albumin) prior to enzyme immobilization has been found to be useful as a protective step.

The specific functionality of the amino and carboxyl groups, and notably binding due to their reactivity and density at many protein surfaces (provided that they are not involved in the catalytic step), are apparent from the data listed in Table 6.3.

Table 6.3 Functional groups in proteins relevant for immobilization.

Functional groups	*pK$_a$*-[a]	*Mean content in proteins %*	*Trypsin tot.(access.)*[b] *number*	*Chymo-trypsin tot. (access.)*[b]	*Gluco amylase*
Amino-, Guanidino-:					
Lysine-NH_2	10.5		13 (>11)	13 (>11)	8
Arginine-NH_2		7	3.8	2	3
Carboxyl-:					
Aspartate-COO^-	3.9	4.8	24 (7)	21 (7–8)	38
Glutamate-COO^-	4.1	4.8			
Other:					
Tyrosine-OH		3.4	10 (3)	4	12
Tryptophan-NH		1.2	4	7–8	12
Histidine-	ca. 7	3.4/2.2			4
Mercapto-					
Cystine-SH					
Oligosaccharides[c]					

[a] Values depend to some degree on the microenvironment.
[b] tot: total content; (access.): accessible at the protein surface.
[c] For enzymes from eucaryotes only.

6.2 Carriers

Here, only a few methods taken from the overwhelming body of literature will be presented, according to their practical relevance. Methods have been extensively reported and summarized earlier (see General literature; notably Mosbach, 1987; Bickerstaff, 1997; Buchholz et al., 1979). Very few methods have found broad application both in the laboratory and on a technical scale.

The first option for the immobilization of enzymes may be realized by crosslinking, and this includes the recently developed techniques of crosslinking enzyme crystals (or the protein at the crystal surface) (CLEC®) and enzyme aggregates (Tischer and Kasche, 1999; Partridge et al., 1996; Cao et al., 2000). In addition, crosslinking of the microorganism which might provide the carrier, thereby leaving the singular enzyme required in an active form, has been successfully applied to glucose isomerase.

Immobilization by matrix inclusion into polymeric networks is effective only with whole microorganisms, and is one of the most convenient methods available (see Chapter 7). Here, the enzyme has a too small diameter and generally leaks out of the particles (an exception is dextran sucrase when bound to high molecular-weight dextran; Reischwitz et al., 1995).

The standard method for enzyme immobilization is binding to insoluble porous carriers, and this will form the focus of this section. Adsorption on the (internal) carrier surface, binding to ion exchangers and covalent binding are all options used in the laboratory as well on the industrial scale. Some examples will be discussed in more detail in order to illustrate the potential of these methods.

The immobilization of an enzyme requires certain specific functions, technical applications and on the other hand further qualities. A summary of the main properties of matrices which must be considered and serve specific purposes is listed in Table 6.4. Recently, a range of commonly used carriers was characterized with respect to particle size distribution, porosity surface area, and skeletal density (Barros et al., 2000).

Interaction of the carrier surface with the protein must be appropriate in so far as no unwanted side effects or reactions may occur (e.g., inactivation, often observed on glass surfaces). The accessible surface inside the pores must be sufficiently large in order to accommodate the amount of enzyme required in a monomolecular layer providing for good activity of the biocatalyst. On the other hand, the pore diameter must be sufficiently large in order to allow the protein to diffuse into the internal vol-

Table 6.4 Properties of matrices relevant for enzyme immobilization.

Properties	***Desired***
Chemical	
• Hydrophilicity/hydrophobicity	
• Swelling properties	Low
• Chemical stability	High
• Microbial stability	Good
Morphological	
• Particle diameter, particle distribution	0.2–1 mm/narrow distribution
• Pore size[a]	30–60 nm
• (Inner) surface for adsorption/binding	large
Mechanical	
• Resistance to pressure/compressibility	Good/low
• Elasticity	Sufficient (no abrasion by stirrer)
General	Food grade (for application in food manufacture) Low cost

[a] Pore diameter and corresponding inner surface (example: porous glass):

	Lower limit:	Upper limit:
Pore diameter	20–40 nm	80–160 nm
Inner surface:	250–125 $m^2\ g^{-1}$	45–25 $m^2\ g^{-1}$

Trypsin with a molar mass of 23 800 Da requires a surface of ca. 15 nm^2

ume (the diameter of small proteins being in the range of 3 to 5 nm, more than the threefold size of the pore being required for unrestricted diffusion). As increasing pore diameter correlates with decreasing internal surface (Table 6.4), a compromise is required with respect to the capacity and transport by diffusion (see Chapter 8, Section 8.4). The capacity of a carrier for the immobilization of an enzyme by adsorption may be tested by measuring the adsorption isotherm under optimal conditions (pH, T, buffer, ionic strength). In many cases, a Langmuir-type isotherm, suggesting a monomolecular protein layer on the internal matrix surface, has been observed.

Furthermore, the particles should preferentially exhibit a regular shape (in general spherical, Fig. 6.2) and a narrow particle size distribution. This provides for optimal flow in a fixed-bed reactor, or suspension in a fluidized-bed reactor, respectively. Fluid flow and pressure drop in fixed-bed reactors depend strongly on these properties, as the pressure drop clearly increases as the void volume decreases, this being due either to the presence of small particles and/or to a broad particle size distribution (see Chapter 9, Section 9.3.5).

The different types of carriers used may be classified according either to the basic material, origin or source, or to their structure. Furthermore, the functional groups accessible on the inner (pore) surface also play a major role. In this respect, frequently used carriers are (Table 6.5):

- inorganic;
- organic from natural sources; and
- organic synthetic materials.

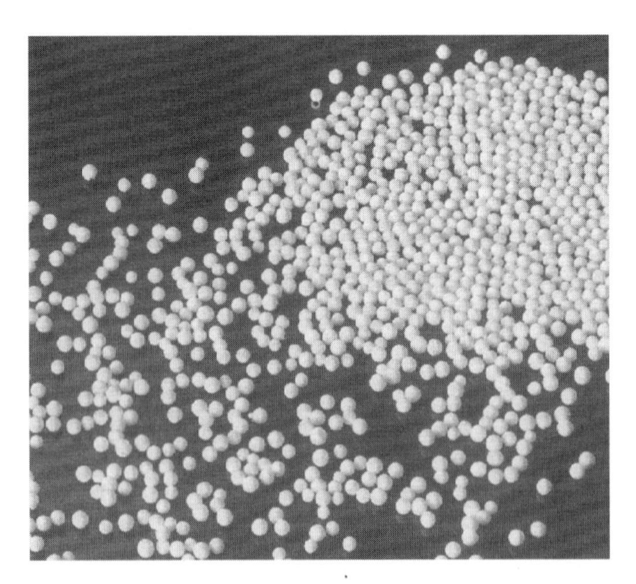

Fig. 6.2 Particles of regular shape and narrow size distribution (Weidenbach et al., 1984).

Table 6.5 Common carriers for enzyme immobilization.

Inorganic carriers	***Application***[a]	***Functional groups***[b]	***Manufacturer***[c]
Porous glass SiO_2	L	-OH	Corning (USA) Waters (USA) Schuller (D)
Porous silica SiO_2 (Deloxan®)	I	-OH	Grace (USA) Solvay (D) Degussa (D)
Alumosilica		-OH	
Organic carriers, natural origin		(Polyol-functions)	
Polysaccharides:			
Agarose (derivatives)	L	-OH	Amersham-Pharmacia (S)
Crosslinked dextrans (Sepharose®)	L	-OH	Amersham-Pharmacia (S)
Cellulose	L,I	-OH	
Proteins:			
Gelatin			
Biomass (e.g., bacteria)[d]	I		
Organic synthetic carriers			
(Meth-) Acrylate-Derivates (Copolymers) (Eupergit®)	L,I	—CH(O)CH$_2$ (epoxide)	Bio Rad (USA) Koch-Light (USA) Degussa (D)
Acrylamide derivatives (Copolymers)			
Vinylacetate derivatives (Copolymers)	L		
Maleic acid anhydride-derivatives	I	—C(=O)–O–C(=O)— (anhydride)	
Polyamides (Nylon)		-NH_2[e]	
Polystyrene derivatives		—C$_6$H$_4$—NH$_2$	
Polypropylene			Akzo (NL)
Sepabeads			Mitsubishi (J)

[a] Application: L: laboratory mainly; I: industrial scale.
[b] Original functional group.
[c] Selected commercial suppliers.
[d] Chemically crosslinked microbial mass with one (remaining) active enzyme.
[e] After reduction of the -NO_2^- group.

6.2.1
Inorganic Carriers

Inorganic carriers exhibit high pressure stability, but may undergo abrasion in stirred vessels; organic materials from natural sources in most cases offer favorable compatibility with proteins; organic synthetic carriers in general exhibit high chemical stability. SiO_2-based carriers can be simply functionalized by introducing amino groups bound to spacers, for example by treating with aminopropyl triethoxysilane (APTS) (Weetall, 1976) (see below, Covalent binding). A complete functionalization leading to a high density of functional groups is essential to produce a stable carrier in aqueous solutions. The covalent binding of enzymes is carried out using glutaraldehyde activation of the carrier. These materials, such as porous glass, silica and alumina-based materials, are commercially available and have been used extensively at both laboratory scale and in technical applications. Extensive research and developmental studies have led to the provision of carriers with good properties for all basic types of materials. The applications of these have included immobilized β-galactosidase (lactase) for lactose hydrolysis.

Celite is a convenient carrier used for the adsorption and stabilization of enzymes in organic media (Balco et al., 1996; Andersson et al., 1999). Bentonite also has excellent adsorption capacity (up to 1.5 g protein g^{-1} bentonite) for enzymes such as penicillin amidase, without substantial loss of activity. It is used for enzyme isolation by adsorption and desorption on a large scale. Crosslinking with glutaraldehyde prevents desorption, while the carrier may be entrapped in alginate to provide biocatalyst particles of an appropriate size (Dauner-Schütze et al., 1988).

6.2.2
Polysaccharides

A range of polysaccharides and derivatives have been used for enzyme immobilization (Fig. 6.3). Polysaccharides exhibit a typical wide network structure (Fig. 6.4) but, due to their hydrophilic properties, they exert weak interactions with proteins with rather no inactivation.

Cellulose derivatives have been used both on laboratory and technical scales. Typical functional groups in commercial carriers are DEAE- (diethylaminoethyl-) and CM- (carboxymethyl-) groups (Fig. 6.3). Thus, a composite material of polystyrene, titanium dioxide and DEAE-cellulose has been applied on an industrial scale for the immobilization of glucose isomerase (see below, Fig. 6.9) (Antrim and Auterinen, 1986). The enzyme is bound by ion exchange with high stability. After inactivation of the biocatalyst (having produced in the range of 10–15 t of product syrup per kg biocatalyst), the carrier can be regenerated and will bind new enzyme molecules.

Dextran (Fig. 6.3) has been used widely for enzyme immobilization, notably in the gel chromatography of proteins. Dextran is water-soluble and is crosslinked for application as an insoluble carrier material with either a narrow or wide network structure, depending on the range of application. A wide network structure exhibits a rather low mechanical and pressure stability, but this is improved by having a narrow structure. A wide range of commercial preparations of dextran (e.g., Sephadex®;

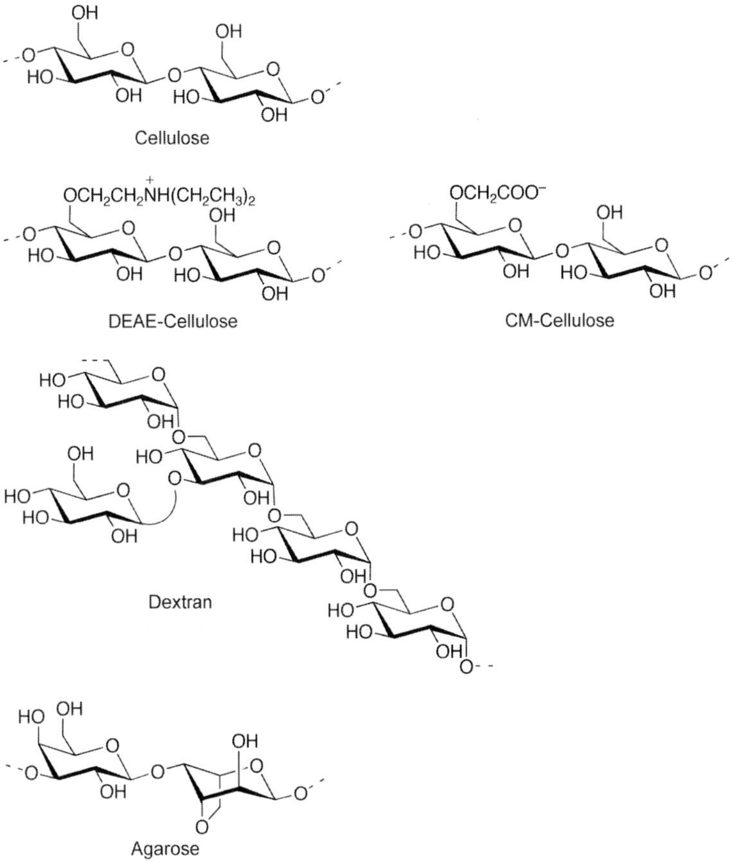

Cellulose

DEAE-Cellulose

CM-Cellulose

Dextran

Agarose

Fig. 6.3 Structures of polysaccharides used as carriers for enzymes.

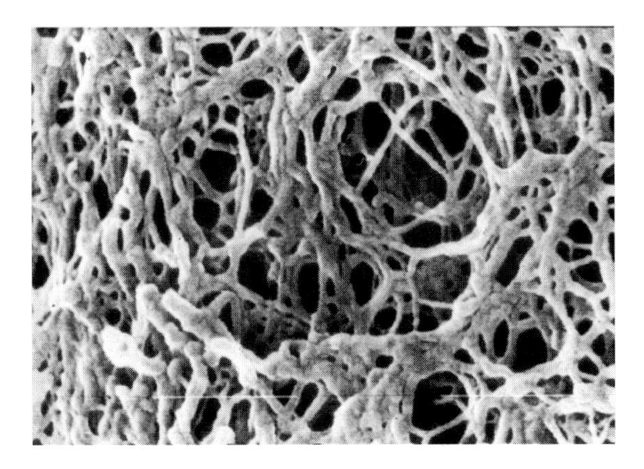

Fig. 6.4 Pore structure of agarose. Scanning electron microscopy (SEM) picture of a porous network (2 % agarose, white line 500 nm, total figure about 2500 nm broad) (Medin, 1995; courtesy of J.C. Jansson).

Table 6.6 Carriers, functional groups, protein functional groups, and reactions.

Original functional group of matrix	*Activation reagent*	*Active intermediate*	*Reactive group of matrix*	*Reactive group of enzyme*	*Coupling reaction, binding*
SiO_2 ⊢〰NH_2 Aminoalkyl	$OHC(CH_2)_3CHO$	⊢$N{=}CH(CH_2)_3CHO$[a]	–CHO Aldehyde	$-NH_2$	Schiff base[a]
⊢$CONH_2$ Acrylamide	$OHC(CH_2)_3CHO$	⊢$CON{=}CH(CH_2)_3CHO$[a]	–CHO Aldehyde	$-NH_2$	Schiff base[a]
⊢CH_2–CH–CH_2 (epoxide, bridged by O) Glycidylmethacrylate	X		–O–CH–CH_2 (epoxide, bridged by O) = NH, O, S	$-NH_2$ –OH –SH	Alkylation
⊢COOH Acrylic acid Methacrylic acid	R–N=C=N–R′, H^+		–COO–C(–NH–R)(=$^+$NH–R′)	$-NH_2$	Peptide bond
SiO_2 ⊢〰COOH Acid	R–N=C=N–R′, H^+		O-Acyliso-urea	$-NH_2$	Peptide bond
⊢CO–O–CO⊣ (cyclic) Methacrylic acid anhydride			–C(=O)–O–C(=O)– Acid anhydride	$-NH_2$	Peptide bond

├OH ├OH Polysaccharide	CNBr	$-O\backslash C{=}NH / -O$ [a]	$-O\backslash C{=}NH / -O$ Imidocarbonate		Isourea Imidocarbonate Carbamate
Polystyrene [b]			├C_6H_4–$N_2^+Cl^-$ Diazonium salt	$-NH_2$ $-SH$ –C_6H_4–OH	Diazo bond
├$CONH_2$ Acrylamide	H_2NNH_2; HNO_2		$-CH_2CON_3$ Acyl azide	$-NH_2$ $-SH$ –C_6H_4–OH	Peptide bond
├C_6H_4–NH_2	Cl_2CO		–C_6H_4–NCO Isocyanate	$-NH_2$	Urea derivative
├R–NH_2	Cl_2CS		–R–NCS	$-NH_2$	Thiourea derivative

[a] simplified
[b] Copolymer with amino groups

Amersham-Pharmacia (S)) are available. Activation for protein binding is conventionally performed using cyanogen bromide (see Table 6.6) (Scouten, 1987; Kohn and Wilchek, 1982), but the limited mechanical stability, coupled with the toxicity of reagents used for carrier activation have restricted the use of dextran mainly to the laboratory. Other polysaccharides such as agarose, starch, pectin, and chitosan (Fig. 6.3) have been used, but have achieved only limited success. Their mechanical stability can be increased by crosslinking the polymer chains.

6.2.3 Synthetic Polymers

Ion-exchange materials have proven to be an economic and technically appropriate solution, as a wide range of carriers is currently available offering good capacity for enzyme immobilization, as well as properties that are relevant to industrial-scale processing. These polymers have also found widespread application in commercial processes, mainly due to the cost effectiveness of the carriers and simplicity of the method (see Table 6.7). They have also been shown to be very useful in the immobilization of lipases for reactions in nonaqueous systems (Balcao et al., 1996).

Polystyrene is a common chemical basis for ion exchange and adsorbent materials, and is crosslinked with divinylbenzene with different contents and thus polymer net-

Table 6.7 Commercial ion-exchange resins.

Chemical basis[a]	***Functional groups***
Polystyrene-derivatives, crosslinked	Weak anion exchangers: $-NR_2H^+$
Polyacrylester-derivatives, crosslinked	Weak cation exchangers: $-COO^-$
Commercial suppliers of ion-exchange resins[a]	Types[c]
Amberlite (Polyacrylester-derivatives, Rohm & Haas)	e.g., IRA 68; IRA 94 S
Lewatit (Bayer)	e.g., E 2001/85
Duolite (Polystyrene-, Polyacrylester-derivatives, Diamond Shamrock)	e.g., A561, A568, S761
Dowex (Dow Chemical)	
Diaion (Mitsubishi)	e.g., HP 20; HPA 25; HPA 75; WA 30
Recommended parameters for immobilization[a, b]	
Carrier particle size	0.1–0.8 mm
Pore size	10–200 nm
Internal surface	10–100 $m^2\ g^{-1}$
Enzyme load	0.1–10 g (protein) L^{-1} wet resin
Adsorption time; temperature	0.5–20 h; T: 10–60 °C

a) Ullmann (1989).
b) Mitsubishi Kasei Corp. (1988).
c) Price range: 5–25 € L^{-1}.

work densities. Particles exhibiting both gel- or macroporous structures are available. Carriers with a macroporous structure are particularly suited as they offer pores that are sufficiently wide for protein diffusion inside the particle, and the high mechanical strength that is required in technical reactors. Functional groups may be introduced either via copolymerization, for example with maleic anhydride carrying carboxyl groups, or by derivatization of the polymer via nitration and reduction to yield amino groups, or by sulfonation. Acrylic-type polymers and derivatives with functional groups are also available within a wide range of commercial products (see Table 6.7).

Special carriers for enzyme immobilization have also been developed, based on materials such as polymethacrylate and polyvinylacetate copolymers. A broad range of copolymers with different characteristics, including hydrophilicity and hydrophobicity balance, functional groups with carboxyl-amino-, epoxy (oxiran-), also including spacers, have been designed, and some are currently available commercially (e.g., Eupergit®; see www.roehm.com; www.degussa.com; www.pharma-polymers.com) (Fig. 6.5). Immobilization to carriers with an epoxide function have also been

Fig. 6.5 Synthetic polymer carrier; macroporous particle structure. Scanning electron microscopy (SEM) pictures of Eupergit C 250 L. The spherical shape and porous structure is easily recognizable. (Reproduced courtesy of S. Menzler, Degussa/Röhm GmbH; www.roehm.com.)

investigated, including their appropriate reaction conditions (Burg et al., 1988; Katchalski-Katzir and Krämer, 2000; Chikere et al., 2001; Martin et al., 2003).

Synthetic resins, including the so-called XAD adsorbents with hydrophobic characteristics, are suitable for the immobilization of hydrophobic enzymes, such as lipases. A wide range of immobilization systems and their application in synthetic reactions has been reported. Results including carriers and binding methods used, reactions with optimal conditions have been summarized by Balcao et al. (1996).

Optimal conditions for enzyme adsorption and, potentially, crosslinking must be established empirically by trial and error, screening of different carriers and parameters for immobilization (for examples of appropriate immobilization conditions, see Exercises 6.5).

6.3
Binding Methods

6.3.1
Adsorption

The adsorption of enzymes onto silica or clay particulate materials is both simple and cost-effective. Pedersen and Christensen (2000) developed a straightforward, economic process of lipase adsorption, agglomeration, drying and sieving to produce an active and stable biocatalyst of definite mechanical strength and particle diameter (Pedersen and Christensen, 2000). Clay is used in a similar principle, as it can adsorb enzymes with a high activity and yield (Dauner-Schütze et al., 1988).

Adsorption onto ion-exchange materials has proven to be an efficient, technically appropriate and economic method as a broad range of carriers is currently available at a reasonable price (5–25 € L^{-1}). A selection of commercial ion exchangers is listed in Table 6.7. The selection of an appropriate carrier for an enzyme must be performed experimentally, as also must the identification of optimal conditions for immobilization, such as the ratio of enzyme and carrier amounts (or concentration, respectively), pH, buffer, and temperature. In order to stabilize an adsorbed enzyme, cross-linking with glutaraldehyde is a favored option when the adsorption is reversible (see below and Exercises).

The adsorption of lipases onto synthetic resins has been used on many occasions, and successfully. Adsorption was reported to be tight, with desorption not occurring to any significant extent when using established procedures. The effect of crosslinking on adsorption has also been investigated. Mechanisms of partial unfolding in the context of multiple binding at the carrier surface have been discussed, and a wide range of investigations including binding yield, operational conditions, stability, and water activity have been reported and summarized (Bornscheuer, 2003; Balco et al., 1996). An example of the adsorption of a lipase onto a silica carrier was provided by Pedersen and Christensen (2000).

6.3.2
Covalent Binding

An analysis of 125 protein families revealed that most (up to 95 %) of polar, titratable residues are located at the surface of soluble proteins (Petersen et al., 1998). It should be kept in mind that exceptions such as lipases have hydrophobic surfaces. The carrier functional groups, as well as their density, play a major role in binding both of the overall protein and in the activity yield. A general conclusion drawn from the literature is that, for every biocatalyst and system, the optimal conditions must be found by empirical investigation.

An example of the functionalization of a silica carrier and enzyme binding is shown in Figure 6.6. Porous silica or glass is treated with aminopropyl triethoxysilane in order to introduce amino groups. Subsequently, the carrier is activated with glutaraldehyde (GA), followed by enzyme binding. A detailed study of the parameters involved in coupling to an amino group activated carrier via GA was published by Monsan (1977/78). Until now, this method has been the most popular due to its simplicity, low cost and effectiveness. It has been applied manyfold both at laboratory and technical scales. However, it must be kept in mind that GA in solution consists of a mixture of oligomers which may also react, and that different reaction mechanisms may occur to an extent which is not quite clear. The condensation reaction yields double bonds which can be recognized by a light color; hydrogenation is advisable providing for bonds which are chemically more stable. The reaction conditions essentially allow for good yields (>90 % of active immobilized enzyme can be achieved with trypsin, for example) notably pH, *T*, *t*, amount of carrier, and concentrations of GA and

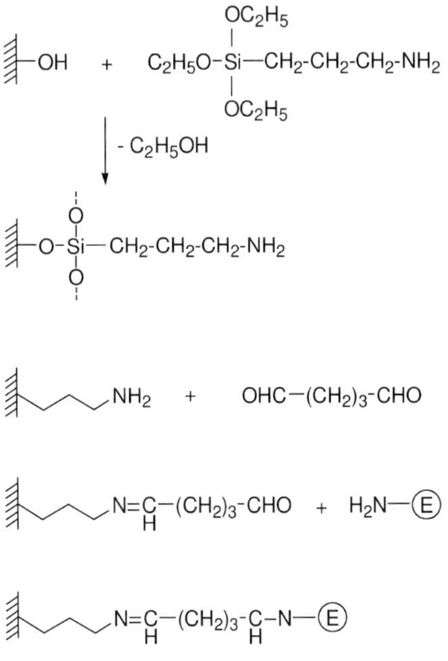

$$\text{–OH} + C_2H_5O\text{–Si}(OC_2H_5)_2\text{–}CH_2\text{-}CH_2\text{-}CH_2\text{-}NH_2 \xrightarrow{-\,C_2H_5OH} \text{–O–Si}(\text{O–})_2\text{–}CH_2\text{-}CH_2\text{-}CH_2\text{-}NH_2$$

$$\text{~~}NH_2 + OHC\text{–}(CH_2)_3\text{-}CHO$$

$$\text{~~}N{=}CH\text{–}(CH_2)_3\text{-}CHO + H_2N\text{–(E)}$$

$$\text{~~}N{=}CH\text{–}(CH_2)_3\text{-}CH\text{–}N\text{–(E)}$$

Fig. 6.6 Derivatization of a silica carrier; activation with glutaraldehyde, and binding of an enzyme by a free amino group (e.g., lysine).

Table 6.8 Immobilization of trypsin via glutaraldehyde, range of conditions (from Monsan, 1977/78).

Carrier:	
Silica (SiO_2) amino-functionalized (Sherosil); amount	100 g
Internal surface range	6–180 $m^2\ g^{-1}$
Pore size	75–2800 nm
Enzyme:	Dissolved under slightly acidic conditions in order to prevent autolysis
Trypsin	
Conditions:	
Activation reaction	pH 8.6 (phosphate buffer, 50 mM), 15 h, 25 °C
Concentration range of glutaraldehyde	0.5–5 %
Coupling reaction	2 g L^{-1} protein in phosphate buffer pH 8.6, 50 mM, 20 mL, 2 h, 4 °C

enzyme. Optimal conditions for trypsin immobilization on a silica carrier were published by Monsan (1977/78). The ranges of favorable results is reproduced in Table 6.8. A two-step procedure with activation of the carrier in the first step and subsequent enzyme immobilization is recommended. An example is given in Section 6.5. The reaction is rather quick, with coupling proceeding in the range of minutes (at 25 °C) or a few hours (at 4 °C) (Buchholz et al., 1979, pp. 169–181).

Details of further investigations of the binding to glass carriers with different activation methods, functional groups and spacers (including GA and succinate) and immobilization yields were published by Mannens et al. (1987). Covalent multipoint attachment, including recombinant approaches, contributes to the stabilization of enzymes in carriers (Fernandez-Lafuente et al., 1995; Rosell et al., 1995; Allard et al., 2002). Binding of an enzyme (L-aminoacylase) to different carriers has also been described by Toogood et al. (2002). The data include derivatization of XAD resins with amino groups and activation by GA, as well as optimal conditions for high activity yield after immobilization. A range of parameters for the activation of carriers (silica, agarose) with tresyl chloride and the immobilization of enzymes, including coupling reagent, pH, temperature, were investigated by Nilsson and Mosbach (1987).

Another reaction which has been used successfully at both laboratory and technical levels is protein binding by epoxide groups. These react with nucleophilic functions such as amino-, thiol- and hydroxyl groups (which are less reactive) under formation of stable C–N-, C–S- or C–O- bonds, respectively (Fig. 6.7). Yields of actively bound enzymes are high when the procedure is optimized. The reaction rates are low, such that effective coupling may require many hours or even days (Krämer, 1979). A commercial carrier of this type is shown in Figure 6.5. Various enzymes, including lipoxygenases, hydrolases, penicillin amidase, hydantoinase, β-galactosidase and lipases have successfully been immobilized and, in part, applied at the technical scale (Chikere et al., 2001; Ferrer et al., 2002).

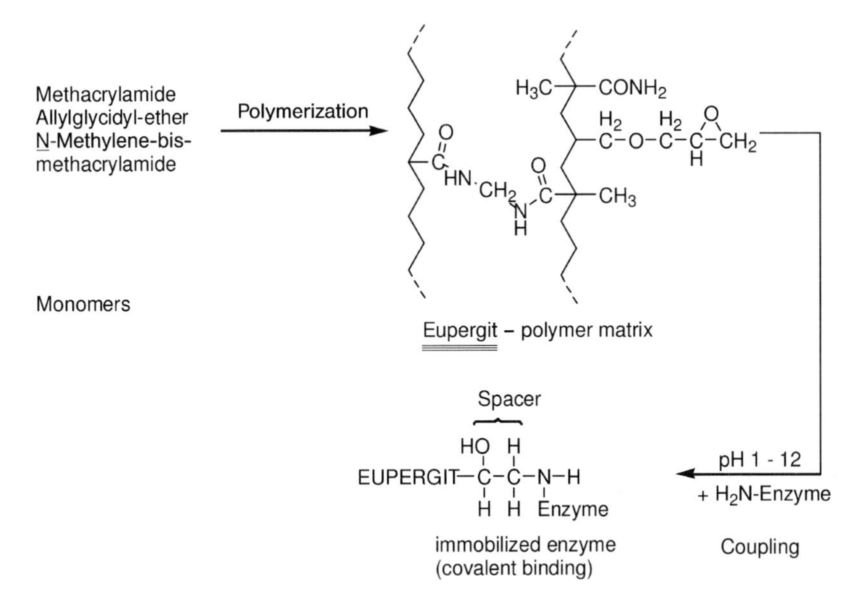

Fig. 6.7 Covalent binding of penicillin amidase to a carrier matrix (Eupergit®) by epoxide groups (Katchalski and Krämer, 2000).

A popular laboratory method has been the activation of polysaccharide carriers with cyanogen bromide and subsequent coupling to yield different types of rather instable bonds; this has resulted in substantial leakage of enzyme from the matrix. Moreover, the reagents and byproducts are highly toxic, and scale-up is difficult. Consequently, this approach was not applied industrially to any major degree. The activation of carboxyl groups by carbodiimide derivatives and binding to amino groups to yield amide bonds has also been used in many laboratories; application in technical systems however is limited due to the expensive and toxic reagents. More extensive overviews of the many now traditional methods have been published by Scouten (1987) and Buchholz and Kasche (1997, pp. 141–166).

Several different methods have been developed for coupling in enzyme sensors, including affinity interaction, an example being biotinylated protein binding to adsorbed avidin (Polzius et al., 1996).

Crosslinked enzyme crystals (West, 1996; Partridge et al., 1996; Schoevart et al., 2004) (CLECs®) consist of pure enzymes that are crosslinked, for example, with GA. The diffusion of substrates and products occurs through microscopic channels of 2–5 nm diameter. Due to the high activity and small particle diameter (d_p 7.5 μm), diffusion limitation at high substrate conversion must be taken into account. Thus, the activity for penicillin G amidase was reported to be 16 000 U mL^{-1} (particle d_p 7.5 μm), resulting in an effectiveness of 0.27 for CLECs, as compared to 170 U mL^{-1} (particle d_p 80 μm) and an effectiveness of almost 1 for Eupergit 250L, at 10 mM substrate concentration (Tischer and Kasche, 1999). CLECs are insoluble, and exhibit high operational stability in both aqueous and organic systems. Systematic investigations have shown that residual water bound to the enzyme molecules – which depends on the drying procedure as well on operational conditions in organic solvents – plays a crucial role (Partridge et al., 1996; West, 1996; Grim, 2001).

A similar method has been developed by Cao et al. (2000). With appropriate precipitants, such as ammonium sulfate, *tert*-butyl alcohol or PEG, enzymes can be precipitated while retaining most (up to 100 %) activity, followed by crosslinking with GA, to form insoluble crosslinked enzyme aggregates (CLEAs). Thus, penicillin amidase could be immobilized with about 80 % hydrolytic activity. Moreover, the specific activity in the synthesis of ampicillin was comparable to that of the free enzyme. In addition, the preparation maintained its synthesis activity in a wide range of organic solvents, whereas the native enzyme exhibited a very low stability.

Crosslinking may also be carried out on the basis of the whole microorganism containing the enzyme in sufficiently high concentrations, so that the cells provide the matrix, or carrier, respectively. The microorganism is then no longer viable (it is dead), and no side reactions can occur to produce toxic or unwanted byproducts. This procedure has been applied to glucose isomerase immobilization (see Section 6.4), and can be carried out in such a way that the enzyme is stabilized and sufficient mechanical stability is provided.

For the immobilization procedure, it is essential to have a balance of enzyme activity introduced and recovered in the immobilized biocatalyst. The role of a spacer at the matrix surface may be important in order to provide flexibility for the enzyme (Manecke et al., 1979; see also Buchholz and Kasche, 1997, pp. 141–166).

Furthermore, the efficiency of immobilized enzymes plays a crucial role, as discussed in Chapter 8, Section 8.4.2. Special procedures which provide improved efficiency by coupling enzymes in the shell of a particle near the external surface have been developed and applied, an example is penicillin hydrolysis (Carleysmith et al., 1980; Borchert and Buchholz, 1984). A sophisticated technique to analyze the enzyme distribution inside a carrier has been developed using confocal laser microscopy (Spieß and Kasche, 2001), while the binding and stabilization of enzymes by multipoint attachment was investigated by Guisán et al. (1997).

6.4
Examples: Application of Immobilized Enzymes

By far the most important processes are glucose isomerization for application as sweeteners in food and penicillin hydrolysis in the pharmaceutical sector. Both of these processes will be treated in more detail in the context of process technology in Chapter 9; hence, only some general information will be presented in this section.

6.4.1
Hydrolysis and Modification of Carbohydrates

High-fructose corn syrup is currently produced at a scale of about 10 million tons (t) per year, using about 1500 tons of immobilized enzyme worth 50 Mio €. The product is a syrup containing about 53 % glucose and 42 % fructose and 5 % other products, or a syrup containing about 40 % glucose and 55 % fructose (it is upgraded in sweetness to that of sucrose by addition of fructose from chromatographic separation). This has the advantage that at an equal sweetener level it is some 10–20 % cheaper than sucrose and, as a technical advantage, exhibits a lesser tendency to crystallize in a wide range of food products. It is largely used in drinks such as cola, lemonades, nectars, fruit juices, confectionary, and sauces, mainly in the USA. Sweetness and taste are almost equal to that of sucrose.

The isomerization of glucose to fructose is an equilibrium-controlled reaction (Fig. 6.8b), with fructose equilibrium concentrations of 48 % at 45 °C and 55 % at 85 °C. Processing at 85 °C however is not possible due to the limited enzyme operational stability and the instability of fructose at temperatures beyond 65 °C due to the Maillard reaction. The mechanism of isomerization by glucose isomerase (GI) consists of ring opening, isomerization through a hydride shift from C2 of the open-form glucose to C1 of the product fructose mediated by two Mg^{2+} ions, followed by ring closure to give the α-ketol (Fig. 6.8a) (Lavie et al., 1994; Reilly and Antrim, 2003).

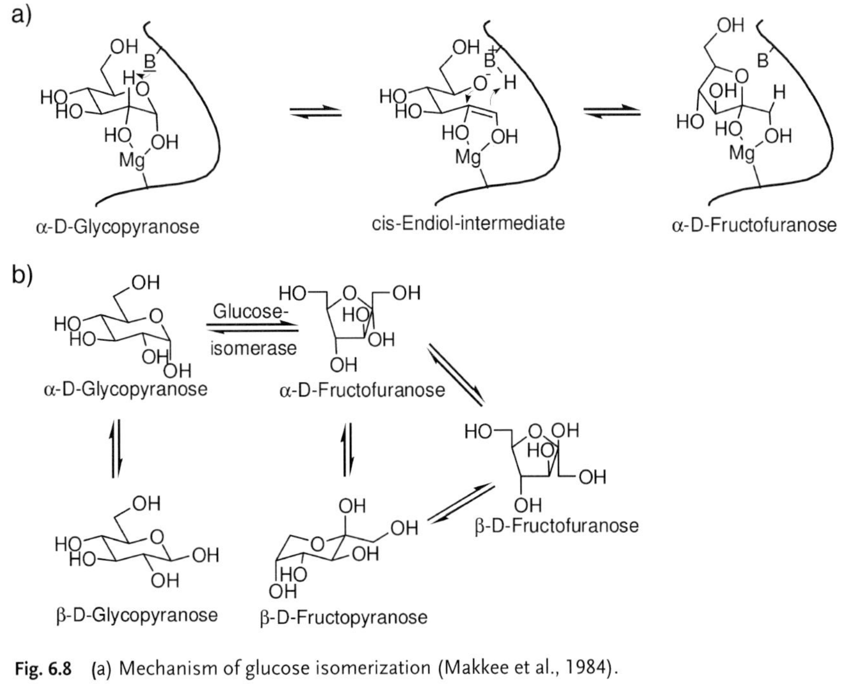

Fig. 6.8 (a) Mechanism of glucose isomerization (Makkee et al., 1984). (b) Chemical equilibria in solution.

Many GI (which basically is a D-xylose ketol-isomerase, EC 5.3.1.5) have been characterized and their tertiary structures are now known. The amino acid sequence alignment has been presented by Misset (2003). Most GIs are homotetramers composed of two tightly bound dimers. Monomers have $(\beta,\alpha)_8$ barrels, with eight β-strands being surrounded by eight helices (Lavie et al., 1994; see also overviews: Misset, 2003; Reilly and Antrim, 2003; SWISS-PROT). GI has been successfully genetically engineered with respect to thermal stability (see Section 2.11.3, Table 2.15) and tight binding of Mg^{2+}. This provides a significant advantage insofar as two steps up- and downscale – to provide Mg^{2+} ions to, and remove them from – the product syrup by ion exchange are no longer necessary. Furthermore, GI has been modified with reference to substrate binding and catalytic activity on glucose and xylose, exhibiting a higher V_{max} and a lower K_m on glucose, which provides advantages for glucose isomerization. Rational protein design has succeeded in increased flexibility of the active site to accommodate the larger substrate (Genencor International, 1990).

The productivity of GI is in the range of 12–20 t (dry substance, d.s.) kg^{-1} biocatalyst, with a half-life of 80 to 150 days (Misset, 2003; Swaisgood, 2003). Reactor dimensions typically are 1.5 m diameter and 5 m height for fixed-bed bioreactors, and the operation is carried out at 58–60 °C. Several different procedures have been developed, and different commercial biocatalysts are utilized at the technical scale. One established method of immobilization is crosslinking of the cells of *Streptomyces murinus* with GA while maintaining almost all of the GI activity (Fig. 6.9a). In another procedure, the enzyme is isolated and purified by chromatography and crystallization, and finally adsorbed to an ion-exchange matrix (Fig. 6.9b) (Pedersen and Christensen, 2000; Antrim and Auterinen, 1986). Commercial biocatalysts are listed in Table 6.9. Process data are provided in Chapter 9, Section 9.4.2.

A considerable number of further processes using immobilized biocatalysts to convert sugars and polysaccharides into new products, and to synthesize oligosaccharides and derivatives, mostly as sweeteners and functional foods, have been applied industrially, and several more are currently under development. These comprise hydrolytic as well as synthetic reactions by hydrolases and glycosyltransferases.

Enzymatic processes fully exhibit their potential in the carbohydrate field in terms of regio- and stereospecificity, and this makes them superior to chemical and classical catalytic reactions. The potential to build up different linkages via α- and β-bonds, and to different regio positions (e.g., 1 to 6 in hexose sugars) makes the synthesis of oligosaccharides dramatically complex. In general, enzymes easily form a single bond, or one main product in high yield (mostly >80 %) with low byproduct formation of isomers. The synthesis of gluco- and fructo-oligosaccharides based on sucrose by glycosyltransferases furthermore utilizes the advantage of the high glycosidic bond energy which is almost equal to that of nucleotide activated sugars (–23 kJ Mol^{-1}). This drives the synthesis to high yields when water is used as the solvent, which is a precondition for food-grade and economic processing. (For regulatory issues in the EU and the USA, see Praaning (2003) and Whitaker (2003, pp. 67–76), respectively.)

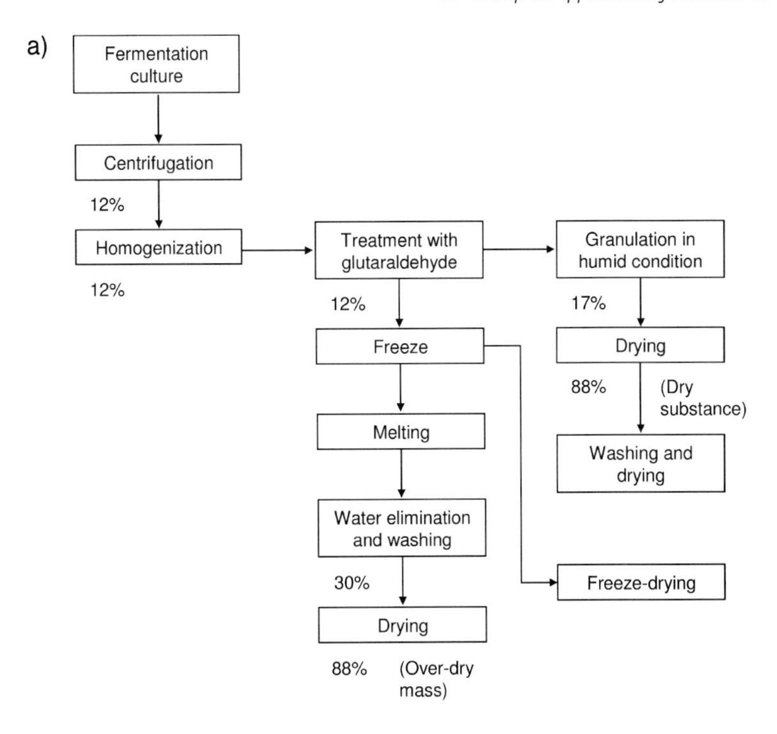

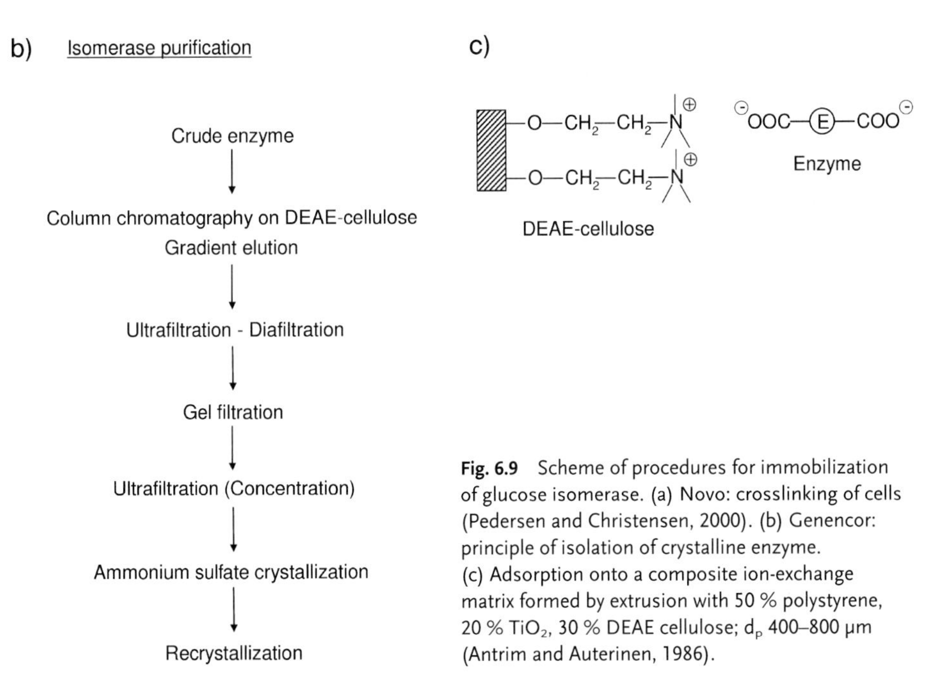

Fig. 6.9 Scheme of procedures for immobilization of glucose isomerase. (a) Novo: crosslinking of cells (Pedersen and Christensen, 2000). (b) Genencor: principle of isolation of crystalline enzyme. (c) Adsorption onto a composite ion-exchange matrix formed by extrusion with 50 % polystyrene, 20 % TiO_2, 30 % DEAE cellulose; d_p 400–800 μm (Antrim and Auterinen, 1986).

Table 6.9 Examples of commercial immobilized glucose isomerases (Pedersen and Christensen, 2000).

Manufacturer	*Trade name*	*Enzyme source*	*Immobilization method*	*Typical values of initial space velocities at 60 °C (bed vol. h^{-1})*
CPC (Enzyme Bio-Systems)	G-zyme G 994	*S. olivochromogenes*	Purified GI adsorbed on an anion-exchange resin	6
Genencor International	Spezyme (600 IGIU g^{-1})	*S. rubiginosus*	Purified GI adsorbed on an anion exchange resin consisting of DEAE-cellulose agglomerated with polystyrene and TiO_2	3.9
Nagase	Sweetase	*S. phaechromogenes*	Binding of heat-treated cells to an anion-exchange resin, granulated	1.4
Novozymes A/S	Sweetzyme T (300 IGIU g^{-1})	*S. murinus*	Crosslinking of cell material with glutaraldehyde, extruded	2.1
UOP	Ketomax 100	*S. olivochromogenes*	PEI-treated ceramic alumina with glutaraldehyde crosslinked, purified GI	5.2

Figure 6.10 provides some selected examples of established processes, and further examples and data for glycosyltransferases are listed in Table 6.10. The hydrolysis of sucrose is accomplished by invertase (immobilized, for example, on Eupergit®) (Uhlig, 1991, p. 204). This process competes with others based on hydrolysis by ion-exchange resins, which are well established. The enzymatic process offers the advantage of less byproduct (e.g., hydroxymethylfurfurol) formation. Lactose in milk and whey is hydrolyzed by β-galactosidase (lactase) into glucose and galactose, exhibiting a greater sweetness and improved fermentation behavior when used as a substrate. The principal reason for hydrolyzing lactose in milk and whey products, however, is to overcome the problem of lactose intolerance, which is widespread among certain populations. Whey syrup with 60 % solids is treated at a scale of several 1000 t a^{-1}. A typical biocatalyst is based on Eupergit®, with a half-life of 20 months, and a productivity of 2000 kg per kg biocatalyst. Other companies purchasing biocatalysts, or applying this process, include Corning, Valio, Sumimoto, and Snam Progetti (Uhlig, 1991, 1998; Swaisgood, 2003).

α-Galactosidase has been used in Japan for the hydrolysis of raffinose into sucrose and galactose in concentrated sugar syrups and molasses from sugar beet processing. This process provides for improved sugar crystallization (where raffinose acts as an inhibitor) and a higher sucrose yield. The enzyme is crosslinked inside the microorganism (mycelium of *Mortierella vinacea*), thus gaining considerably improved operational stability (Fig. 6.11) (Linden, 1982). The Hokkaido Sugar Co. has been treating ca. 300 t of molasses with 1–5 % raffinose per day (Hokkaido Sugar Co., 1990).

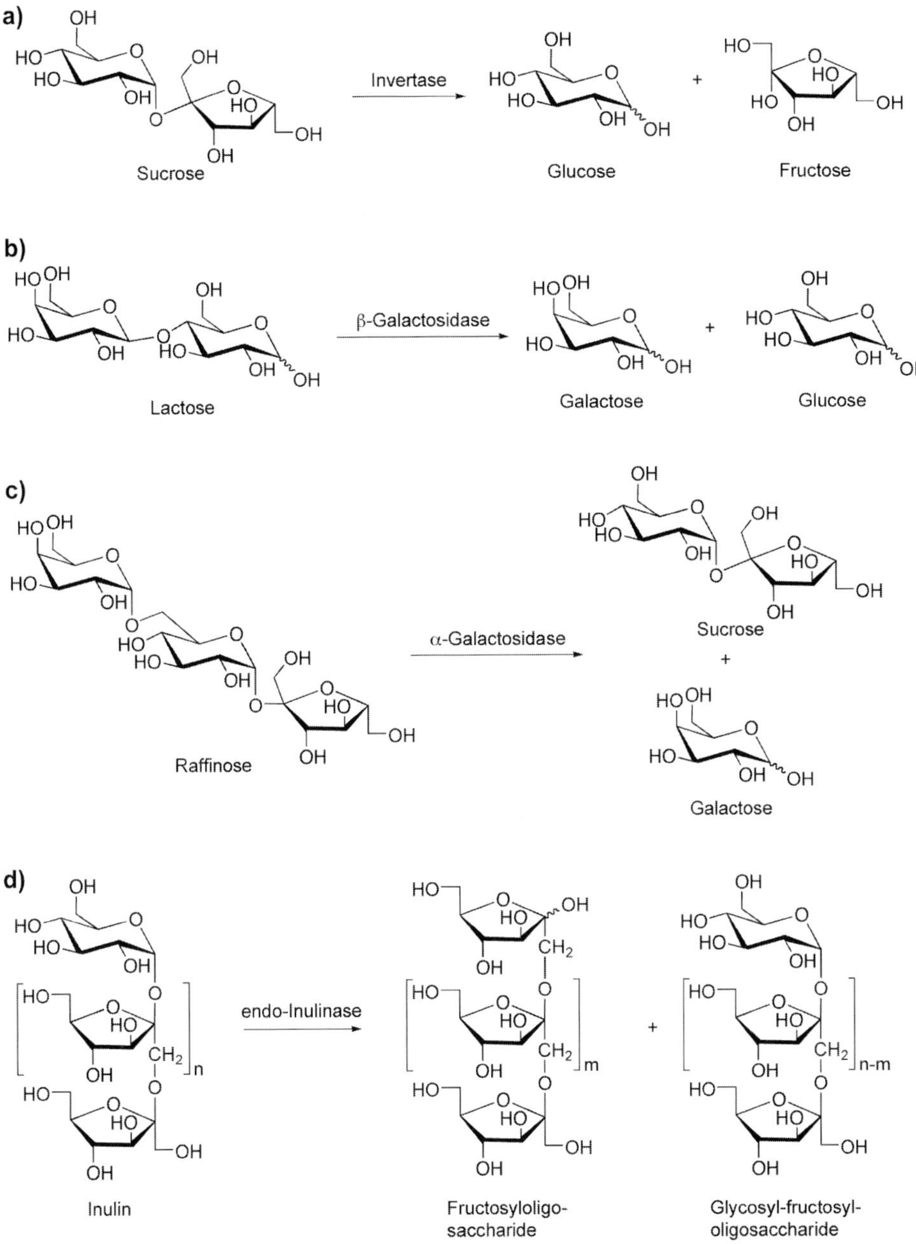

Fig. 6.10 Hydrolytic reactions. (a) Sucrose with invertase; (b) lactose with β-galactosidase; (c) raffinose with α-galactosidase; (d) inulin with inulinase.

A range of oligosaccharides is produced by glycosyltransferases, including both glucosyl- and fructosyltransferases, and using sucrose or starch, or dextrins, respectively, as convenient, high-quality and low-price commercial substrates. The alternative of using glycosyl hydrolases at high concentration in order to favor the condensation reaction over hydrolysis has not succeeded in application, mainly due to limited yields and regioselectivity.

The production of isomaltulose (α-D-glucosyl-1,6-D-fructose) by isomerization of sucrose (α-D-glucosyl-1,2-β-D-fructose) (Fig. 6.12a) is established on the large scale, with about 60000 tons produced annually (Südzucker AG, Cerestar, Germany; Mitsui Seito Co., Japan). The enzyme sucrose mutase is a glycosyltransferase which effects isomerization by an intramolecular rearrangement. For enzyme immobilization, the cells (*Protaminobacter rubrum*) with enzymatic activity are crosslinked and entrapped in alginate beads. Most remarkable is the much increased stability of the biocatalyst at high sugar concentrations, this being further improved under operational conditions in sucrose solution (1.6 M; half-life of over 8000 h) (Cheetham, 1987). The yield of isomaltulose critically depends on the kinetic control of the reaction, with increasing formation of trehalulose as a byproduct at extended reaction (residence) time. Most of the product is further processed by catalytic hydrogenation to a mixture of α-D-glucosyl-sorbitol and -mannitol (Isomalt®) which is used as an alternative sweetener in many food products (Schiweck et al., 1991).

Another disaccharide which may serve as an alternative sweetener is Leucrose (α-D-glucosyl-1,5-β-D-fructose), which can be produced in an analogous rearrangement of sucrose by dextransucrase under appropriate conditions (Schwengers, 1991) (Fig. 6.12a).

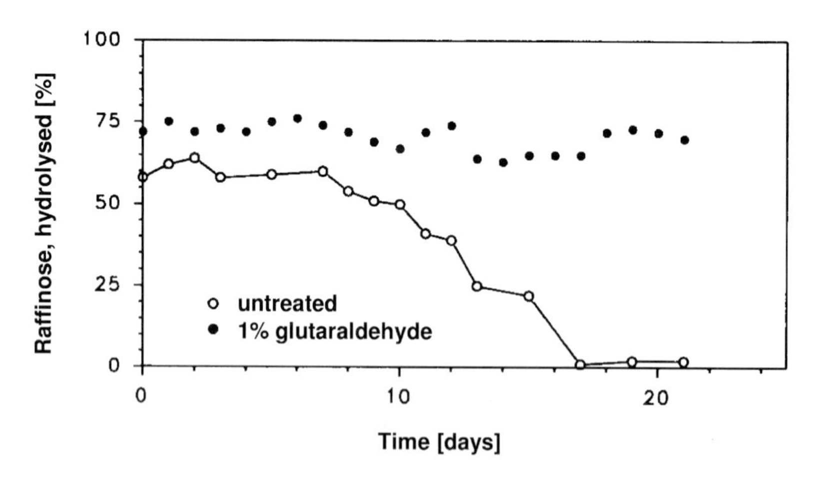

Fig. 6.11 Stabilization of α-galactosidase by crosslinking of the cells with glutaraldehyde (GA), and degree of raffinose hydrolysis, (○) by untreated mycelium, (●) by mycelium crosslinked with 1 % GA (1.5 h, pH 7.0, 20 °C); reaction in molasses with ca. 30 % dm, 2 h batches, 50 °C, pH 5.2). (From Saimura, according to Linden, 1982).

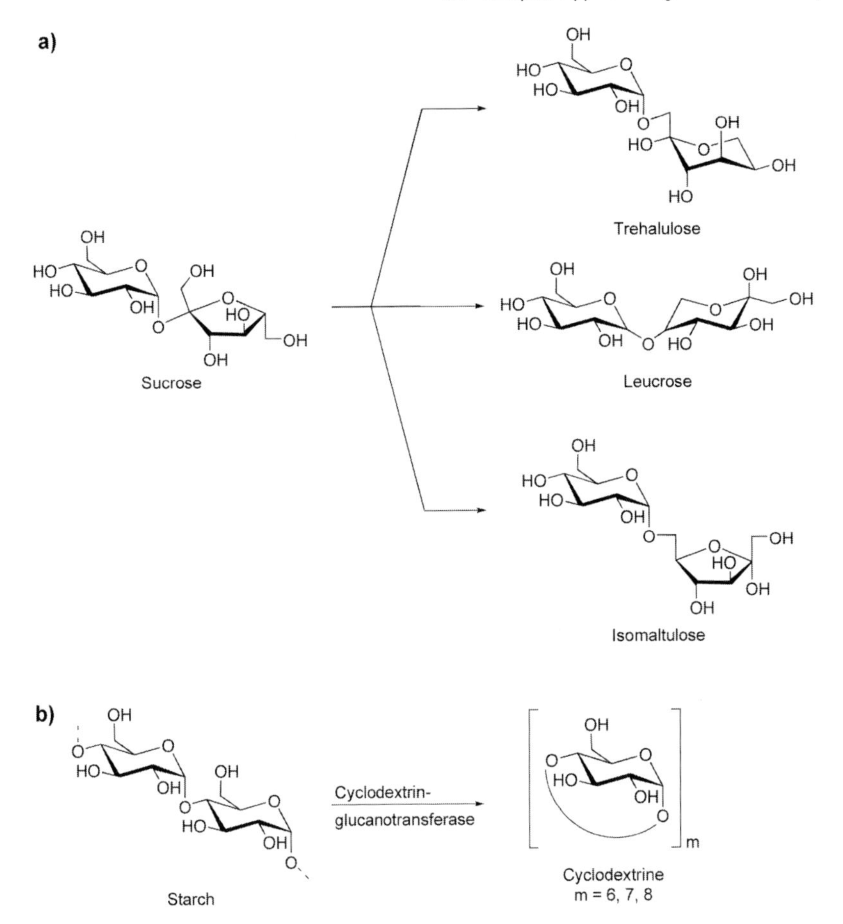

Fig. 6.12 Transglycosidations by glycosyltransferases. (a) Sucrose isomerization; (b) cyclodextrin synthesis.

For recent overviews on enzymes and oligosaccharides in food, see Whitaker et al. (2003) and Eggleston and Côté (2003). Isomalto-oligosaccharides are produced by companies in Japan and Europe for use in food, feed and dermocosmetics (Paul et al., 1992; Remaud-Siméon et al., 1994; Buchholz and Monsan, 2003). Raw materials are starch or sucrose with an acceptor, such as glucose or maltose, to which a glucosyl unit from sucrose is transferred.

Fructo-oligosaccharides (FOS) have gained major attention during recent years as functional food ingredients and medical foods (Rastall et al., 2003; Tungland, 2003). They are manufactured either by partial hydrolysis of inulin, or by glycosyltransferases using sucrose as the substrate (Yun and Kim, 1999; Yun and Song, 1999; Tungland, 2003). An alternative method to produce FOS is by fructosyltransferases acting on sucrose, with the transfer of a fructosyl unit from sucrose to another sucrose linking the fructosyl units by β-(2→1) bonds to the fructosyl moiety of sucrose.

Several fructosyltransferases – mainly from lactic acid bacteria – have recently been cloned, expressed and characterized (van Hijum et al., 2001, 2002). These may be of both the inulosucrase type forming β-(2→1) bonds, and of the levansucrase type, forming β-(2→6) bonds. The product mixture obtained with inulosucrases is composed mainly of glucose, sucrose and tri- to pentasaccharides, with one glucose and two to four fructosyl units linked by β-(2→1) bonds (kestose, etc.) exhibiting prebiotic properties. Physiological functionalities attributed to FOS are prevention of dental caries, dietary fiber, stimulation of bifidobacteria, prevention of diarrhea and constipation, reduction of serum cholesterol, further provision of texture, mouthfeel, taste improvement, and fat replacement (Yun and Song, 1999; Tungland, 2003; Rastall and Hotchkiss, 2003). The production of total oligosaccharides (except isomaltulose; see Table 6.10) is estimated to be in the range several tens of thousands of tonnes annually. A selection of commercial products and their producing companies is listed in Table 6.10.

The price of oligosaccharides in the food sector ranges from 2 to 5 € kg^{-1}; thus, these products (except for cyclodextrins) represent bulk low-price commodities requiring cheap raw materials (sucrose, starch, inulin), together with economic processing with a low cost for the biocatalyst, which must be immobilized and exhibit high productivity.

Table 6.10 Applications of immobilized glycosyltransferases (Fuji and Komoto, 1991).

Enzyme (reference)	*Product*	*Market (estimated, t a^{-1})*	*Company*
Sucrose mutase (1)	Isomaltulose	100 000	Südzucker (D)
Cyclodextrin transferase	Malto-oligosyl-saccharides	5000	Hayashibara (J)
α-Glucosidase and Glucosyltransferase (4)	Isomalto-oligo-saccharides	4800	Hayashibara (J) and others
Dextransucrase	Isomalto-oligo-saccharides		BioEurope, Solabio (F)
Dextransucrase[a]	Dextran	200	Pharmacia (S)
β-Galactosidase	Galacto-oligo-saccharides	7000	Yacult (J)
Fructosyltransferase (2)	Neosugar	4000	Beghin-Meji Industries (F,J)
Cyclodextrin transferase (3)	β-Cyclodextrin and other cyclodextrins	3000	Wacker-Chemie (D)

a) Single example where the product is not made by an immobilized enzyme; Pharmacia is now a Pfizer company.

Reference sources:
(1) Schiweck (1991); (2) Ouarne and Guibert (1995); (3) Wimmer (1999); (4) Buchholz and Monsan (2003).

Cyclodextrins (Fig. 6.12b) are products of major importance in the fields of food, pharmaceuticals (drug protection, slow release) and commodities; for example, they are used in textile drying, for odor removal and as a perfume carrier. In food, their function may be aroma complexation and slow release, stabilization of flavors, and the reduction of bitterness. They are also used to remove cholesterol from milk and egg products. Cyclodextrins are torus-shaped molecules of cyclic α-(1→4)-linked gluco-oligosaccharides. They form inclusion complexes with a wide variety of hydrophobic guest molecules, leading to the applications mentioned. The enzymes used for their manufacture from starch or dextrins are cyclodextrin glycosyltransferases; several have been crystallized, and their structures elucidated and characterized in detail, including the reaction mechanism (Dijkhuizen and van der Veen, 2003).

6.4.2 Hydrolysis and Synthesis of Penicillins and Cephalosporins

The second example of high economic importance is the hydrolysis and synthesis of penicillin and cephalosporin antibiotics. An overview of this is provided in Chapter 9 (Section 9.2). The first large scale enzyme processes developed using immobilized biocatalysts were the hydrolysis of penicillin G and V (Fig. 6.13; see also Section 1.5, Fig. 1.7), during the 1970s. Thermodynamic data for these processes are provided in Section 2.8 (see Table 2.7 and Fig. 2.20), and these are used to determine the process window for the dominating hydrolysis process – that is, the hydrolysis of penicillin G (see Exercise, 2.12). Most of the immobilized biocatalysts used in these processes are produced by the companies that use them for hydrolysis, and few data have been published on these proprietary supports (Liese et al., 2000, pp. 281–295). Some companies that have produced (or still produce) immobilized enzymes for commercial purpose did, however, publish more extensive data on their immobilized biocatalysts (Katchalski-Katzir and Krämer, 2000; *http//:biochem.roche.com/indbio/ind/pga/f_p0.htm*).

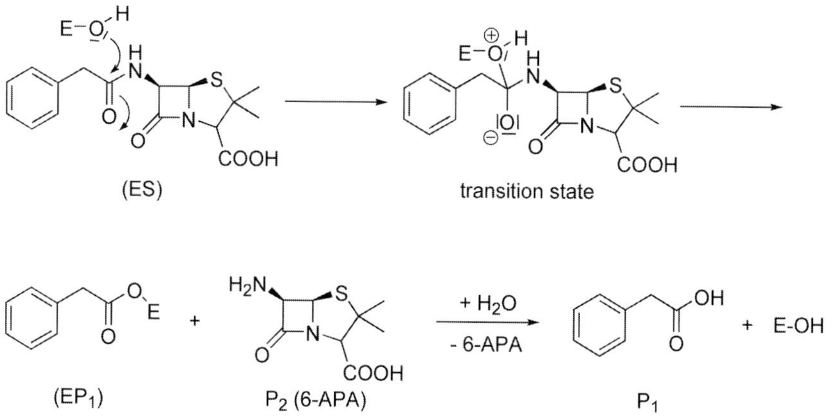

Fig. 6.13 Penicillin hydrolysis by penicillin amidase.

The penicillin amidases used are from *Alcaligenes faecalis, Bacillus megaterium,* and *Escherichia coli.* Due to the better pH stability under alkaline conditions, the enzyme from *A. faecalis* is the best for this process, since at pH 9 almost quantitative hydrolysis is possible. The enzyme has been covalently immobilized to different supports, and some data for this enzyme process are listed in Table 6.11.

The data published for the process (pH and *T*) are within the process window derived in Chapter 2 (Exercises 2.12). pH control is essential since at pH <7.5 the equilibrium conversion is too low, and at pH >8 both the substrate and product degrade too rapidly. Efficient mixing in technical reactors is important in order to prevent enzyme inactivation by the base (ammonia) added. The conversion is 98–99 %, and the yield is in the range of 86–93 % (enzyme productivity decreases with increasing yield). The enzyme productivity is in the range of 1000–2000 kg product kg^{-1} biocatalyst (or in the range of 5 g product U^{-1}). The enzyme cost per product is about 2.5 € kg^{-1} (Poulson, 1984; Cheetham, 2000; Liese et al., 2000, pp. 281–295).

Another enzyme process – the hydrolysis of cephalosporin C to 7-amino cephalosporanic acid – that was developed and introduced on an industrial scale during the 1990s is detailed in Chapter 9, Section 9.2.

Semisynthetic cephalosporins and penicillins are produced from the hydrolysis products 6-aminopenicillanic acid (6-APA), 7-aminocephalosporanic acid (7-ACA) and 7-aminodesacetoxycephalosporanic acid (7-ADCA) by the condensation of other

Table 6.11 Reaction conditions for the hydrolysis of penicillin G (Liese et al., 2000, pp. 281–295; D. Krämer, personal communication).

Producer/User of immobilized biocatalyst	***Asahi Chem. Ind. (Jp)*** ***Asahi Chem. Ind. (Jp)***	***DSM (NL)*** ***DSM (NL)***	***Recordati (It)*** ***Several***
Enzyme source	*B. megaterium*	*A. faecalis*	*E. coli*
Immobilization	Covalently to porous polyacrylonitrile fibers	Covalently to a gelatin-based carrier	Covalently to Eupergit®
Yield after immobilization	n.a.	n.a.	60–70 %
Activity	n.a.	≈200 U mL^{-1} wet carrier	100–150 U mL^{-1} wet carrier
Reactor	packed bed	stirred tank or packed bed	stirred tank (10^4 U L^{-1})
Substrate (%, w/w)	0.3 M penicillin G	n.a.	Penicillin-G-K (8 %)
Temperature [°C]	30–36	n.a.	37
Initial pH	8.4	n.a.	7.6–7.8
Operational stability	≈1000 h	n.a.	Up to 1000 cycles, >60 days

n.a.: Not available.

side chains to the amino group (see Chapter 9, Section 9.2.1). The dominating process here is still chemical synthesis. In order to reduce costs due to the large amount of byproducts that are difficult to recycle, enzyme processes have recently been developed and successfully introduced on an industrial scale (Bruggink, 2001).

6.4.3
Further Processes

Amino acid synthesis is a major sector to which continued research and development efforts have been devoted. The main reason is that enantiopure products are required, as is generally the case in the food and pharmaceutical sectors (for more details, see Chapter 3, Section 3.2.1.1). Chemical synthesis is straightforward, flexible and economic to provide a range of precursors of different amino acids, such as hydantoins and caprolactam derivatives, including non-natural varieties (Syldatk et al., 1999) (see Chapter 3, Section 3.2.2.6). For the industrial production of enantiopure amino acids from these precursors, and also from mixtures of L- and D- iso-

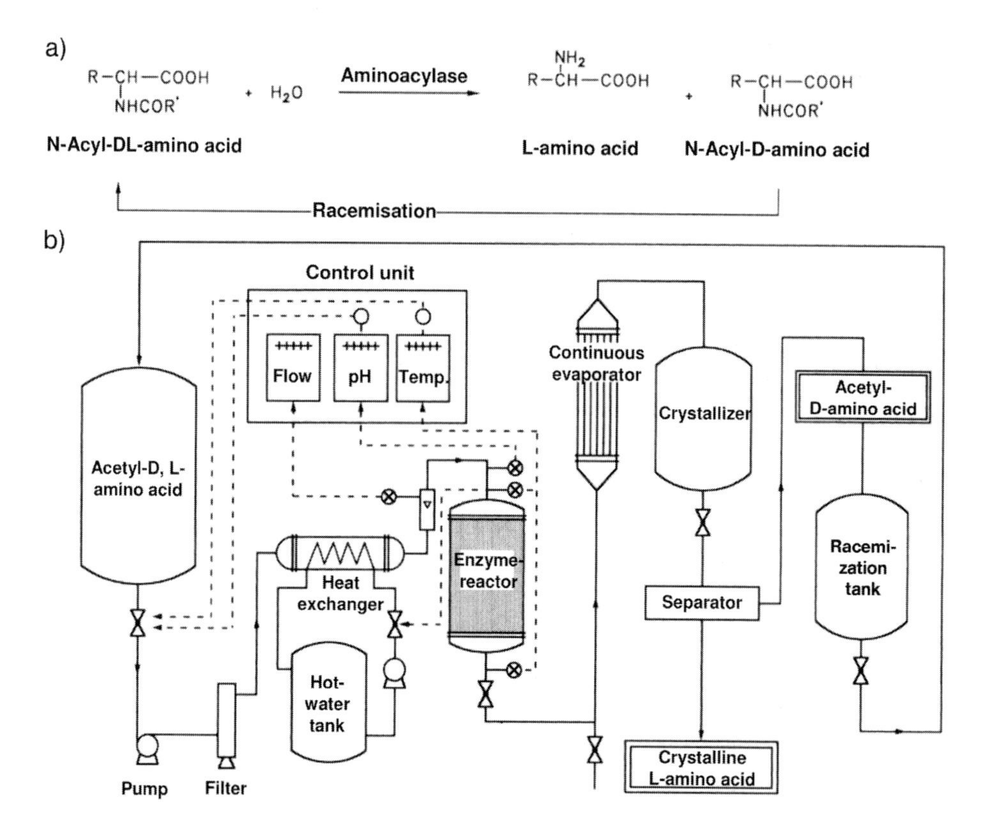

Fig. 6.14 Scheme of the production of L-amino acids by amino acylases (Chibata et al., 1987). (a) Reaction; (b) reactor system and peripheral instrumentation.

mers produced by chemical synthesis, immobilized systems are applied in order to keep enzyme costs low. These processes first were introduced by Chibata and co-workers in Japan in 1969 (Chibata et al., 1987). The classical route is acetylation of the D,L-mixture followed by stereospecific hydrolysis by an amino acylase and subsequent separation by ion exchange and crystallization. The remaining acetylated isomer is recycled after racemization (Fig. 6.14). The amino acylase has been immobilized by adsorption onto an ion-exchange resin, though other methods have also been applied.

Reactions using dehydrogenases are discussed in Chapter 3, Section 3.2.1.1. Further processes use different principles (for an overview, see below). Yet another route to the production of chiral amino acids is via transamination with α-keto or α-hydroxy acids as readily available substrates. This was discussed in Chapter 5 (see Section 5.4).

6.4.3.1 Some Further Principles for Amino Acid Synthesis

- The production of L-aspartic acid by addition of ammonia to fumaric acid (Chibata et al., 1987):

Fumaric acid —(Aspartase, + NH_3)→ L-Aspartic acid

- Decarboxylation of L-aspartic acid by L-aspartate-4-decarboxylase (Chibata et al., 1987):

L-Aspartic acid —(L-Aspartate-4-decarboxylase, - CO_2)→ L-Alanine

- Amino group transfer from L-aspartic acid by a transaminase to a 2-ketoacid as starting material, with subsequent decarboxylation of the byproduct oxaloacetic acid; shift of the equilibrium towards the product wanted:

L-Aspartic acid + 2-Keto acid —(Transaminase)⇌ Oxalacetic acid + L-Amino acid

Oxalacetic acid —(Decarboxylase, - CO_2)→ Pyruvate

- Synthesis of L-tryptophan from L-serine and indole:

L-Serine + Indole → L-Tryptophan synthase → L-Tryptophan

- Stereospecific cleavage of caprolactams (Chibata et al., 1987):

D,L-α-Amino-ε-caprolactam → Hydrolase, + H_2O → L-Lysine + D-α-Amino-ε-caprolactam

α-Amino-ε-caprolactam Racemase

An economically highly relevant peptide synthesis is that of aspartame (Fig. 6.15), a high-intensity sweetener which is used to replace sugar and reduce the caloric content in foods, for example in soft drinks. The synthesis uses protected (*N*-benzylcarboxy-) aspartic acid (Z-Asp) and cheap racemic phenylalanine methyl ester (Phe Me) (see also Section 3.2.2.3). The residual D-Phe Me is racemized and recycled. Thermolysin was found after a broad screening effort to catalyze the synthesis both regio- and enantioselectively, without hydrolyzing the methyl ester. It is applied after immobilization, for example as a CLEC, in stirred-tank reactors using ethyl acetate : water as the solvent. The yield is >95 % at 55 °C, and annual sales are in the range of 850×10^6 € (Cheetham, 2000; Grim, 2001).

A broad range of further processes is performed with immobilized biocatalysts, notably for the production of chiral compounds and intermediates, as well as drugs and agrochemicals using esterases, lipases, and other enzymes (for further details, see Chapter 3).

6.5 Exercises

6.5.1 Immobilization of Enzymes on Ion Exchangers

- Lipase (Novo, European Patent 0382 767, 1988)

Adsorption on ion exchanger: Lewatit E2001/85

Enzyme: *Mucor miehei*-Lipase 2.6 g (112 Lipase-U mg^{-1}) in 25 mL solution, addition to carrier: 8.5 g, pH 6.1, 21 h.

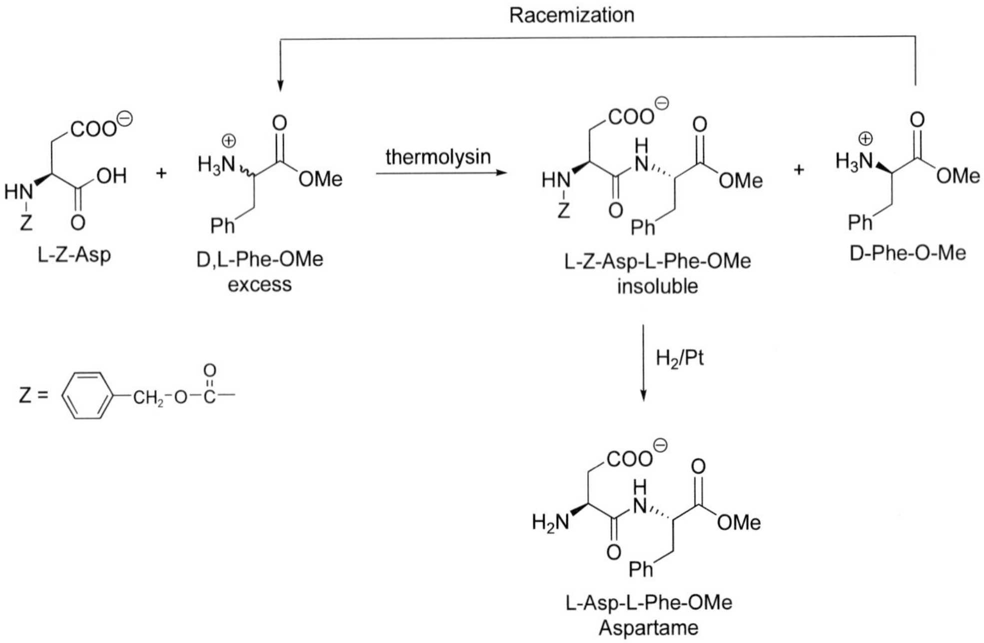

Fig. 6.15 Synthesis of aspartame using thermolysin.

Immobilized activity: 171 BIU g^{-1} (batch inter-esterification unit, 60 °C).

Application in molten substrates without water or solvents.

- Transglycosidase and Pullulanase (Cerestar, European Patent 875 585, 1997; further examples: European Patent 336 376, Mitsubishi Kasei Corp., 1988)

Co-immobilization of two enzymes on Duolite A 568; adsorption and crosslinking.

Enzymes:

1. Transglycosidase (Amano) 11 mg mL^{-1}, 103 TGU mL^{-1} (solution).

Addition of 1 mL solution to 10 mL Duolite, 4 h, 20 °C.

2. Pullulanase (400 PU mL^{-1} solution), addition of 15 mL solution; 4 h.

Crosslinking: addition of 5 mL glutaraldehyde solution (1 %; 0.2 % final concentration), 12 h, 20 °C, washing.

A procedure for lipase immobilization on a silica carrier was reported by Pedersen and Christensen (2000).

- Amyloglucosidase, porous silica as carrier

Reagents and equipment

- Amyloglucosidase (AG) (may be kept for some time as dry preparation in the refrigerator; test for activity before use);
- Porous silica, or porous glas, derivatized with amino groups (see Table 6.5 and Section 6.3.2 for different properties);
- Solution of glutaraldehyde (GA), 5 %;
- Sörensen phosphate buffer, pH 7, 0.01 M (add a droplet of toluene as antimicrobial agent);
- Sörensen phosphate buffer, pH 5, 0.01 M (add a droplet of toluene as antimicrobial agent);
- Thermostated glass vessel, about 50 mL, with frit, or sieve, and stirrer;
- Thermostat, 50 °C;
- UV Photometer, cuvette (10 mm);
- Peristaltic pump;
- Glucose and/or ethanol analysis by enzymatic test, or GC, HPLC, or glucose analyser.

Immobilization

Suspend 3 g of carrier in 20 mL buffer (pH 7), degass; add 5 mL of GA solution (5 %), if needed add another 10 mL of buffer; the reaction should proceed for 30 min while stirring and pumping the solution through a bypass (from the lower outlet of the glass vessel and reintroducing it from the top); take off the solution, wash two times with water.

Add solution of AG (corresponding to 0.2 g, or 1000 U) in 5 mL of buffer (pH 7); measurement of extinction at 280 nm every 15 min (dilution by 1:20), let reaction proceed for 45 min; take off the solution (test for residual enzyme activity), wash 3 times with water, suspend in 50 mL of buffer (pH 5).

Measurement of activity in suspended solution of immobilized enzyme (with a sample of suspended carrier) and with a solution of dextrin as substrate (low molecular weight; high molecular weight substrates hydrolysis is limited by diffusion); glucose analysis by enzymatic test, or HPLC, or glucose analyzer. The concentrations of glucose and/or ethanol are graphically recorded as a function of time, the catalytic activity (g (glucose/ethanol) per g (yeast) per h) determined from the slope; compare the reaction with the same amount (activity) of free enzyme; calculate balance of active immobilized enzyme.

- Amyloglucosidase, Eupergit C® as carrier

Reagents and equipment

- Amyloglucosidase (AG);
- Eupergit C® from Degussa (Röhm), Darmstadt (Germany);
- Potassium phosphate buffer, pH 5, 0.01 M;
- Citrate buffer 0.1 M (pH 4,5);
- Glass vessels, 10 or 20 mL volume.

Immobilization

Prepare a solution of 50 mg (or 300 U) of AG in 50 mL of buffer (pH 6) in 4 glass vessels; add different amounts of carrier (see the following table) and let react while shaking the closed vessels for 24 h at 25 °C. After sedimentation of the carrier the supernatant is removed with a pipette. The carrier with immobilized enzyme is washed 3 to 4 times with citrate buffer 0.1 M (pH 4,5).

The supernatants of the different preparations are tested for residual enzyme activity. The immobilized enzyme preparations are suspended in 25 mL of citrate buffer (pH 4,5); samples are tested for activity in shaked glass vessels (see above).

Amounts for immobilization

Number	***Amyloglucosidase (g)***	***Eupergit (g)***	***Ratio of enzyme to carrier***
1	0.05	0.25	0.20
2	0.05	0.50	0.10
3	0.05	0.75	0.067
4	0.05	1.00	0.05

Literature

General Literature

Bickerstaff, G.F. (Ed.), *Methods in Biotechnology, Vol. 1: Immobilization of Enzymes and Cells*, Humana Press Inc., Totowa, NJ, USA, 1997.

Buchholz, K. (Ed.), Characterization of immobilized biocatalysts; in: *DECHEMA Monograph* **84**, VCH, Weinheim, 1979.

Godfrey, T., West, S. (Eds), *Industrial Enzymology*, Macmillan Press, London, 1996.

Liese, A., Seelbach, K., Wandrey, C., *Industrial Biotransformations*, Wiley-VCH, Weinheim, 2000.

Mosbach, K. (Ed.), *Meth. Enzymol.*, Vol. 135, Academic Press, Orlando, 1987.

Pedersen, S., Christensen, M.W., Immobilized biocatalysts, in: Straathof, A.J.J., Adlercreutz, P. (Eds), *Applied Biocatalysis*, pp. 213–228, Harwood Academic Publishers, Amsterdam, 2000.

Tischer, W., Kasche, V., Immobilization enzyme crystals, in: *Trends Biotechnol.*, **1999**, *17*, 326–335.

SWISS-PROT, *http://www.kr.expasy.org/enzyme/*

References

Allard, L., Cheynet, V., Oriol, G., et al., Versatile Method for Production and Controlled Polymer-Immobilization of Biologically Active Recombinant Proteins, *Biotechnol. Bioeng.* **2002**, *80*, 341–348.

Andersson, M., Samra, B., Holmberg, H., Adlercreutz, P., Use of Celite-immobilised chloroperoxidase in predominantly organic media, *Biocatal. Biotransform.*, **1999**, *17*, 293–303.

Antrim, R.L., Auterinen, A.-L., A new regenerable immobilized glucose isomerase, *Stärke*, **1986**, *38*, 132–137.

Balcao, V.M., Paiva, A.L., Malcata, F.X., Bioreactors with immobilized lipases: State of the art, *Enzyme Microb. Technol.*, **1996**, *18*, 392–416.

Barros, R.J., Wehtje, E., Garcia, F., Adlercreutz, P., Physical characterization of porous materials, *Biocat. Biotrans.*, **2000**, *16*, 67–85.

Böker, M., Jördening, H.-J., Buchholz, K., Kinetics of Leucrose formation from sucrose by dextransucrase, *Biotechnol. Bioeng.*, **1994**, *43*, 856–864.

Borchert, A., Buchholz, K., Improved biocatalyst effectiveness by controlled immobilization of enzymes, *Biotechnol. Bioeng.*, **1984**, *26*, 727–736.

Bornscheuer, U.T., Immobilizing enzymes: how to create more suitable biocatalysts, *Angew. Chem. Int. Ed. Engl.*, **2003**, *42*, 3336–3337.

Bruggink, A. (Ed.), *Synthesis of β-lactam antibiotics*, Kluwer Academic Publishers, Dordrecht, 2001.

Buchholz, K., Duggal, SK., Borchert, A., Characterization of immobilized biocatalysts; in: DECHEMA Monograph **84**, pp. 1–48, 169–181, VCH, Weinheim, 1979.

Buchholz, K., Kasche, V., *Biokatalysatoren und Enzymtechnologie*, VCH, Weinheim, 1997.

Buchholz, K., Monsan, P., Dextransucrase, in: Whitaker, J.R., Voragen, A.G.J., Wong, D.W.S. (Eds), *Handbook of Food Enzymology*, pp. 589–603, M. Dekker, New York, 2003.

Burg, K., Mauz, O., Noetzel, S., Sauber, K., Neue synthetische Träger zur Fixierung von Enzymen, *Angew. Makrom. Chem.*, **1988**, *157*, 105–121.

Cao, L., van Rantwijk, F., Sheldon, R.A., Cross-linked enzyme aggregates: a simple and effective method for the immobilization of penicillin acylase, *Org. Lett.*, **2000**, *2*, 1361–1364.

Carleysmith, S.W., Dunnil, P., Lilly, M.D., Kinetic behaviour of immobilizal penicillin acylase, *Biotech. Bioeng.*, **1980**, *22*, 735–756.

Cheetham, P.S.J., Production of isomaltose using immobilized microbial cells, in: Mosbach, K. (Ed.), *Methods Enzymol.*, Vol. 136, pp. 432–454, 1987.

Cheetham, P., Case studies in the application of biocatalysts; in: Straathof, A.J.J., Adlercreutz, P. (Eds), *Applied Biocatalysis*, pp. 93–152, Harwood Academic Publishers, Amsterdam, 2000.

Chibata, I., Tosa, T., Sato, T., Application of immobilized biocatalysts in pharmaceutical and chemical industries, in: Rehm, H.J., Reed, G. (Eds), *Biotechnology*, Vol. 7a, pp. 653–684, Verlag Chemie, Weinheim, 1987.

Chikere, A.C., Galunsky, G., Schünemann, V., Kasche, V., Stability of immobilised soybean lipoxygenases: influence of coupling conditions of the ionization state of the active site Fe, *Enzyme Microb. Technol.*, **2001**, *28*, 168–175.

Creighton, T.E., *Proteins: Structures and Molecular Properties*, W.H. Freeman, New York, 1993.

Dauner-Schütze, C., Brauer, E., Borchert, A., Buchholz, K., Development of a high capacity adsorbent for enzyme isolation and immobilization; in: Moo-Young, M. (Ed.), *Bioreactor immobilized enzymes and cells*, pp. 63–70, Elsevier Applied Science, London, 1988.

Demuth, B., Jördening, H.-J., Buchholz, K., Modelling of oligosaccharide synthesis by dextransucrase, *Biotechnol. Bioeng.*, **1999**, *62*, 583–592.

Dijkhuizen, L., van der Veen, B., Cyclodextrin glycosyltransferase, in: Whitaker, J.R., Voragen, A.G.J., Wong, D.W.S. (Eds), *Handbook of Food Enzymology*, pp. 615–627, M. Dekker, New York, 2003.

Eggleston, G., Côté, G.L., Oligosaccharides in Food and Agriculture, ACS Symposium Series 849, American Chemical Society, Washington, 2003.

Fernandez-Lafuente, R., Rosell, C.M., Rodriguez, V., Guisan, J.M., Strategies for enzyme stabilisation by intramolecular crosslinking with bifunctional reagents, *Enzyme Microb. Technol.*, **1995**, *17*, 517–523.

Ferrer, M., Plou, F.J., Fuentes, G., Cruces, M.A., Andersen, L., Kirk, O., Christensen, M., Ballesteros, A., Effect of the immobilisation method of lipase from *Thermomyces lanuginosus* on sucrose acylation, *Biocatal. Biotransform.*, **2002**, *20*, 63–71.

Fuji, S., Komoto, M., Novel carbohydrate sweeteners in Japan, *Zuckerindustrie*, **1991**, *116*, 197–200.

Genencor International, Glucose isomerases having altered substrate specificity. European Patent 1 264 883, 1990.

Godfrey, T., West, S., Introduction to industrial enzymology, in: Godfrey, T., West, S. (Eds), *Industrial Enzymology*, pp. 1–8, Macmillan Press, London, 1996.

Grim, M.D., Cross linked enzyme crystals: Biocatalysts for the organic chemist, in: Kirst, H.A., Yeh, W.-K., Zmijewski, M.J. (Eds), *Enzyme Technologies for Pharmaceutical and Biotechnological Applications*, pp. 209–226, Marcel Dekker, New York, 2001.

Guisán, J.M., Rodriguez, V., Rosell, C.M., Stabilization of immobilized enzymes by chemical modification with polyfunctional macromolecules, in: Bickerstaff, G.F. (Ed.), *Methods in Biotechnology, Vol. 1: Immobilization of Enzymes and Cells*, pp. 289–298, Humana Press Inc., Totowa, NJ, USA, 1997.

van Hijum, S.A.F.T., van Geel-Schutten, G.H., Rahaoui, H., van der Maarel, M.J.E.C., Dijkhuizen, L., Characterization of a novel fructosyltransferase from *Lactobacillus reuteri* that synthesizes high-molecular-weight inulin and inulin oligosaccharides, *Appl. Environ. Microbiol.*, **2002**, *68*, 4390–4398.

van Hijum, S.A.F.T., Bonting, K., van der Maarel, M.J.E.C., Dijkhuizen, L., Purification of a novel fructosyltransferase from *Lactobacillus reuteri* 212 and characterization of the levan produced, *FEMS Microbiol. Lett.*, **2001**, *205*, 323–328.

Hokkaido Sugar Co. Ltd., Tokyo, Procedure of Enzymatic Hydrolysis of Raffinose (Jap.), 1990. See also Obana, J., Hashimoto, S., Enzyme applications in the sucrose industry, *Sugar Technol. Rev.*, **1976/77**, *4*, 209–258.

Katchalski-Katzir, E., Krämer, D.M., Eupergit C, a carrier for immobilization of enzymes, *J. Mol. Catal. B Enzymol.*, 2000, *10*, 157–176.

Kohn, J., Wilchek, M., A new approach (cyanotransfer) for cyanogen bromide activation of sepharose at neutral pH, which yields activated resins free of interfering nitrogen derivatives, *Biochem. Biophys. Res. Commun.*, **1982**, *107*, 878–884.

Krämer, D.M., Characterization of immobilized biocatalysts; in: Buchholz, K. (Ed.), DECHEMA Monograph **84**, p. 168, VCH, Weinheim, 1979.

Laurents, D.V., Huyghues-Despointes, B.M.P., Bruix, M., Thurlkill, R.L., Schell, D., Newsom, S., Grimsley, G.R., Shaw, K.L., Trevino, S., Rico, M., Briggs, J.M., Antosiewicz, J.M., Scholtz, J.M., Pace, C.N., Charge-Charge interactions are key determinants of the *pK* values of ionizable groups in ribonuclease SA (pI = 3.5) and a basic variant (pI = 10.2), *J. Mol. Biol.*, **2003**, *325*, 1077–1092.

Lavie, A., Allen, K.N., Petsko, G.A., Ringe, D., X-ray crystallographic structures of D-xylose isomerase-substrate complexes position the substrate and provide evidence for metal movement during catalysis, *Biochemistry*, **1994**, *33*, 5469–5480.

Linden, J.C., Immobilized α-D-galactosidase in the sugar beet industry, *Enzyme Microb. Technol.*, **1982**, *4*, 130–136.

Makkee, M., Kieboom, A.P.G., van Bekkum, H., Glucose Isomerase Catalyzed D-Glucose-D-Fructose Interconversion, *Recl. Trav. Chim. Pays-Bas*, **1984**, *103*, 361–364.

Manecke, G., Ehrenthal, E., Schlüsen, J., in: Buchholz, K. (Ed.), Characterization of Immobilized Biocatalysts, *Dechema-Monographien*, Bd. **84**, S49–72, 1979.

Mannens, G., Slegers, G., Lambrecht, R., Immobilization of acetate kinase and phosphotransacetylase on derivatized glass beads, *Enzyme Microb. Technol.*, **1987**, *9*, 285.

Martin, M.T., Plou, F.J. Alcalde, M., Ballesteros, A., Mol, J., Immobilisation on Eupergit C of cyclodextrin glucosyltransferase (CGTase) and properties of the immobilised biocatalyst, *Catal. B: Enzymatic*, **2003**, *21*, 299–308.

Medin, A.S., Studies on structure and properties of agarose, Acta Universitatis Uppsaliensis, Uppsala, 1995.

Misset, O., Xylose (glucose) isomerase, in: Whitaker, J.R., Voragen, A.G.J., Wong, D.W.S. (Eds), *Handbook of Food Enzymology*, pp. 1057–1077, Marcel Dekker, New York, 2003.

Mitsubishi Kasei Corp., European Patent 336 376, 1988.

Monsan, P.J., Optimization of glutaraldehyde activation of a support for enzyme immobilization, *J. Mol. Catalysis*, **1977/78**, *3*, 371–384.

Nilsson, K., Mosbach, K., Tresyl Chloride-Activated Supports for Enzyme Immobilization, in: Supports for Enzyme Immobilization, *Meth. Enzymol.*, **1987**, *135*, 65–78.

Norde, W., Zoungrana, T., Activity and structural stability of adsorbed enzymes, in: Ballesteros, A., Plou, F.J., Iborra, J.L., Halling, P.J. (Eds), *Stability and Stabilization of Biocatalysts*, pp. 495–504, Elsevier, Amsterdam, 1998.

Ouarne, F., Guibert, A., Fructo-oligosaccharides: enzymic synthesis from sucrose, *Zuckerindustrie*, **1995**, *120*, 793–798.

Partridge, J., Hutcheon, G.A., Moore, B.D., Halling, P.J., Exploiting hydration hysteresis for high activity of cross-linked subtilisin crystals in acetonitrile, *J. Am. Chem. Soc.*, **1996**, *118*, 51.

Paul, F., Lopez Munguia, A., Remaud, M., Pelenc, V., Monsan, P., U.S. Patent 5,141,858, 1992.

Pedersen, S., Christensen, M.W., Immobilized biocatalysts, in: Straathof, A.J.J., Adlercreutz, P. (Eds), *Applied Biocatalysis*, pp. 213–228, Harwood Academic Publishers, Amsterdam, 2000.

Petersen, S.B., Jonson, P.H., Fojan, P., Protein engineering the surface of enzymes, *J. Biotechnol.*, **1998**, *66*, 11–26.

Polzius, R., Schneider, T., Bier, F.F., Bilitewsky, U., Koschinski, W., Optimization of Biosensing using grating couplers: immobilization on tantalum oxide waveguides, *Biosensors Bioel.*, **1996**, *11*, 503–514.

Poulson, P.B., Current applications of immobilized enzymes for manufacturing purposes, *Biotechnol. Gen. Eng. Rev.*, **1984**, *1*, 121.

Praaning, D.P., Regulatory issues of enzymes used in foods from the perspective of the E.U. market, in: Whitaker, J.R., Voragen, A.G.J., Wong, D.W.S. (Eds), *Handbook of Food Enzymology*, pp. 59–65, M. Dekker, New York, 2003.

Rastall, R.A., Hotchkiss, Jr., A.T., Potential for the development of prebiotic oligosaccharides form biomass, ACS Symposium Series, **2003**, *849*, 44–53.

Reilly, P., Antrim, R.L., Enzymes in grain wet milling, in: *Ullmann's Encyclopedia of Industrial Chemistry*, Wiley-VCH, Weinheim, 2003.

Reischwitz, A., Reh, K.D., Buchholz, K., Unconventional immobilization of dextransucrase with alginate, *Enzyme Microb. Technol.*, **1995**, *17*, 457–461.

Remaud-Siméon, M., Lopez Munguia, A., Pelenc, V., Paul, F., Monsan, P., Production and use of glycosyltransferases from *Leuconostoc mesenteroides* NRRL B-1299 for the synthesis of oligosaccharides, *Appl. Biochem. Biotechnol.*, **1994**, *44*, 101–117.

Rosell, C., Fernandes-Lafuente, R., Guisan, J.M., Modification of enzyme properties by

the use of inhibitors during their stabilisation by multipoint covalent attachment, *Biocatal. Biotransform.*, **1995**, *12*, 67–76.
Sauber, K., Lessons from industry, in: van den Tweel, W.J.J., Harder, A., Buitelaar, R.M. (Eds), *Stability and Stabilization of Enzymes*, Proceedings of an International Symposium Elsevier Science Publishers B.V., 1993.
Schiweck, H., Munir, M., Rapp, K.M., Schneider, B., Vogel, M., New developments in the use of sucrose as an industrial bulk chemical, in: Lichtenthalter, F.W. (Ed.), *Carbohydrates as Organic Raw Materials*, pp. 57–94, VCH, Weinheim, 1991.
Schmid, A., Hollmann, F., Park, J.B., Bühler, B., The use of enzymes in the chemical industry in Europe, *Curr. Opin. Biotechnol.*, **2002**, *13*, 359–366.
Schoevaart, R., Wolbers, M.W., Golubovic, M., Preparation, Optimization, and Structures of Cross-Linked Enzyme Aggregates (CLEAs), *Biotechnol. Bioeng.* **2004**, *87*, 754–761.
Schwengers, D., Leucrose, a ketodisaccharide of industrial design; in: Lichtenthaler, F.W. (Ed.), *Carbohydrates as Organic Raw Materials*, pp. 183–195, Verlag Chemie, Weinheim, 1991.
Scouten, W.H., A survey of enzyme coupling techniques, in: Mosbach, K. (Ed.), *Methods in Enzymology*, Vol. 135, pp. 30–65, Academic Press, Orlando, 1987.
Spiess, A., Kasche, V., Direct measurement of pH profiles in immobilised enzymes during kinetically controlled synthesis using CLSM, *Biotechnol. Prog.*, **2001**, *17*, 294–303.
Swaisgood, H.E., Use of Immobilized Enzymes in the Food Industry, *Handbook of Food Enzymology*, Whitaker, J.R. et al. (Eds.), Marcel Dekker, Inc. New York, 2003, 359–366.
Syldatk, C, Altenbuchner, J., Mattes, R., Siemann-Herzberg, M., Microbial Hydantoinases, *Appl. Microbiol. Biotechnol.* **1999**, *51*, 293–309.
Tischer, W., Kasche, V., Immobilization enzyme crystals, *Trends Biotechnol.*, **1999**, *17*, 326–335.
Toogood, H.S., Taylor, I.N., Brown, R.C., Taylor, S.J.C., McCague, R., Littlechild, J.A., Immobilisation of the thermostable L-aminoacylase from *Thermococcus litoralis* to generate a reusable industrial biocatalyst, *Biocat. Biotransform.*, **2002**, *20*, 241–249.
Tungland, B.C., Fructooligosaccharides and other fructans, in: Eggleston, G., Côté, G.L. (Eds), *Oligosaccharides in Food and Agriculture*, ACS Symposium Series 849, American Chemical Society, Washington, 2003.
Uhlig, H., *Enzyme arbeiten für uns*, Hanser, München, 1991; Industrial enzymes and their applications, John Wiley & Sons Inc., New York, 1998.
Ullmann's Encyclopedia of Industrial Chemistry, Vol. A14, pp. 393–459, VCH, Weinheim, 1989.
Weetall, H.H., Covalent coupling methods for inorganic support materials, in: Mosbach, K. (Ed.), *Methods in Enzymology*, Vol. 44, pp. 134–148, Academic Press, Orlando, 1976.
Weidenbach, G., Bonse, D., Richter, G., Glucose isomerase immobilized on SiO_2-carrier with high productivity; *Stärke*, **1984**, *36*, 412–416 (s. auch Beaded Catalysts and Carriers, Prospekt, Fa. Kali-Chemie, Hannover, Germany).
West, S., Chemical biotransformations, in: Godfrey, T., West, S. (Eds), *Industrial Enzymology*, pp. 155–175, Macmillan Press, London, 1996.
Whitaker, J.R., Voragen, A.G.J., Wong, D.W.S. (Eds), *Handbook of Food Enzymology*, M. Dekker, New York, 2003.
Wimmer, T., personal communication, Wacker.-Chemie, München, 1999.
Yun, J.W., Song, S.K., Enzymatic production of fructooligosaccharides from sucrose, in: Bucke, C. (Ed.), *Methods in Biotechnology – Carbohydrate Biotechnology Protocols*, pp, 141–151, Humana Press, Totowa, New Jersey, USA, 1999.
Yun, J.W., Kim, D.H., Enzymatic production of inulooligosaccharides from inulin, in: Bucke, C. (Ed.), *Methods in Biotechnology – Carbohydrate Biotechnology Protocols*, pp, 153–163, Humana Press, Totowa, New Jersey, USA, 1999.

7
Immobilization of Microorganisms and Cells

Why immobilize microorganisms?	• No enzyme isolation required • Application of complex reactions • Application of multienzyme systems • Intracellular cofactor regeneration • Continuous processing
Methods of immobilization	• Organism as carrier only (see Chapter 6) • Inclusion in ionotropic gels • Adhesion
Technical applications	• Acetic acid (vinegar) production • Alcohol production • Beer maturation • Anaerobic waste water treatment • Exhaust gas purification
Examples and exercises	

7.1
Introduction

The focus of this chapter is on the immobilization of living, or viable microorganisms and cells. The background of the application of such systems is the observation that the enzyme activities of immobilized cells remain active for longer periods compared to that in isolated cells, this being due to the (re)synthesis of enzymes and cofactors as well as their regeneration. This is also essentially true for resting cells. Remarkably, as long ago as 1823 the cells of *Acetobacter*, when adsorbed onto wood chips, were used in the technical production of acetic acid from ethanol (Knapp, 1847, pp. 480–486). The basic principles of this process were later laid down by Hattori and Fusaka in 1959, and by K. and R. Mosbach in 1966. Living bacteria in mixed

Biocatalysts and Enzyme Technology. K. Buchholz, V. Kasche, U.T. Bornscheuer

ISBN: 3-527-30497-5

cultures have been applied successfully to environmental technologies since the 1970s and 1980s (Jördening and Buchholz, 1999; Lettinga et al., 1999). Currently, cells with multienzyme systems and/or intracellular cofactor regeneration – notably recombinant cells with designed reaction pathways (Section 7.6) – have gained much interest. These were first used in the production of intermediates and high added-value compounds which otherwise might be difficult to produce. When compared to classical fermentation processes, immobilized microorganisms offer the advantage of a short residence time, and high volumetric productivity in continuous processes in different types of reactors (e.g., tubular and fluidized-bed reactors) is possible. The immobilized systems may function with resting cells that do not grow (appropriate growth limitation by controlling C-, N- or P- sources), but nevertheless can regenerate cofactors; on the other hand, the growth of microorganisms may be possible under appropriate conditions. The immobilization of growing cells is applicable when the formation or regeneration of enzyme activities is dependent upon growth, and if a new biomass is formed and/or delivered to the medium.

In addition to microorganisms, plant or animal cells can also be immobilized. All of these biocatalysts offer the advantages of reuse and continuous application. Further advantages of immobilized microorganisms and cells include:

- no requirements for enzyme isolation and purification;
- reaction sequences or transformations requiring multienzyme systems;
- formation/production of secondary metabolites (eventually coupled to growth);
- cofactor regeneration in the native system;
- application of anchorage-dependent cells that grow only when attached to surfaces (e.g., mammalian cells);
- application of immobilized mixed cultures with synthrophic growth or metabolism; and
- protection against shear forces, and in part also against toxic substances (e.g., oxygen, metal ions, aromatics).

Thus, complex reaction sequences (e.g., alcohol formation from sugar) can only be performed by viable microorganisms (bacteria, yeasts) that are capable of synthesizing enzymes and regenerating cofactors. The immobilization of microorganisms is also applicable if the enzyme(s) are difficult to isolate or show low stability and activity outside the cell, as is the case for nitrile hydratase in acrylamide production. (The topic of enzyme immobilization is discussed in Chapter 6).

Problems relating to the application of viable, immobilized microorganisms and cells result from a variety of different conditions, including:

- insufficient stability;
- mass transfer limitation due to growing cells in the catalyst matrix;
- side reactions, degradation of products; and
- byproducts from the lysis of cells or traces of toxic metabolites (notably in pharmaceutical and food applications).

One general shortcoming of immobilized microorganisms compared to singular immobilized enzymes is that of low volumetric activity. Indeed, the above-mentioned problems have considerably limited the use of immobilized microorganisms and cells until recently (Freeman and Lilly, 1998). Thus, the continuous production of alcohol by immobilized yeast was relatively unsuccessful, despite major efforts being made in the development and scale-up of the system. The immobilization of *Gluconobacter oxidans* for oxidative biotransformations was also hindered by the high oxygen demand, which led to mass transfer limitations and a loss in activity after only one biotransformation cycle (Schedel, 2000).

Thus, the main areas for the application of immobilized microorganisms and cells include complex transformations that require the regeneration of enzymes and cofactors and/or cell growth. These include wastewater treatment, notably with anaerobic cultures, and exhaust gas purification (Jördening and Buchholz, 1999; VDI, 1996; Klein and Winter, 2000).

A review on the research and development of immobilized microorganisms has been written by Brodelius and Vandamme (1987), whilst basic research on their use has been compiled by Wijffels et al. (1996) and Wijffels (2001). This has incorporated topics such as adhesion, details of immobilization procedures, viable cells, physiology, influence of the microenvironment, mass transfer and dynamic modeling, oxygen concentration profiles and mechanical stability. Junter et al. (2002) later presented data on the physiology and protein expression that support the existence of a specific metabolic behavior in the immobilized state. These authors focused on the proteomic approach as a complementary tool to gene-level investigations.

For details of immobilized animal cells, the reader is referred to reviews by Tokayashi and Yokoyama (1997), Doyle and Griffiths (1998) and Waugh (1999), and for details on immobilized plant cells to Dörnenburg and Knorr (1995) and Ishihara et al. (2003).

The most important principles for the immobilization of microorganisms and cells are their inclusion in polymer matrices, adsorption or adhesion onto carriers, and flocculation and/or aggregation. The most important method for inclusion into ionotropic gels is discussed in detail in Section 7.3, while adhesion to carriers – based on the interaction of surfaces of microorganisms and solid surfaces – is detailed in Section 7.5. Further methods are summarized in Section 7.2.

In industrial applications, three types of immobilized viable cell systems are used currently: (1) beer maturation with yeast cells; (2) anchorage-dependent mammalian cells immobilized on microcarriers (which is a rather important technology for the production of vaccines); and (3) environmental technology using mixed cultures. Despite much effort being made into research investigations (up to pilot plant level for the continuous production of alcohol), no other industrial applications have yet been reported which utilize immobilized viable cells. An outlook presents new perspectives for such applications (Section 7.6).

7.1.1
Fundamental Aspects

The immobilization of microorganisms leads to modifications in their microenvironment, with consequential concerns relating to mass transfer phenomena that affect substrate and product gradients and metabolism. At present, few details are available of these phenomena and analytic problems, mainly due to their complexity.

Mass transfer phenomena are described more fully in Chapter 8, Section 8.4. Currently, little is known about the influence of external mass transfer, which may be insignificant in many cases (Radovic, 1985) (c.f. Chapter 9, Section 9.3.4). Rather, more has been published with respect to pore diffusion and the efficiency of immobilized microorganisms (for surveys, see Radovic, 1985; Westrin and Axelsson, 1991). Systematic studies with varying particle diameters and densities of cells within the matrix have been reported by Klein and Vorlop (1985).

As a general rule, and if no cell growth occurs, when microorganisms are immobilized in ionotropic gels the diffusion of low molecular-weight substrates is not significantly slower than diffusion in water (Table 7.1). The effective diffusion coefficients are mainly in the range of 50–100 % of that in free solution. Similar conditions may be valid for biofilms growing on surfaces. However, under conditions of significant growth and increasing biofilm thickness (>100 μm), mass transfer can be reduced severely, the consequence being a much lower effective diffusion of substrates into the biofilm. Omar (1993) has reported on oxygen diffusion into gels in the presence of microorganisms, while mass transfer and coupled intracellular phenomena have been analyzed in pellets of *Aspergillus niger* and biofilms (Bössmann et al., 2003).

Table 7.1 Comparison of diffusion coefficients (D) of substrates in water and in particles of Ca^{2+}-alginate at 30 °C (Tanaka et al., 1984; Berensmeier et al., 2004), and in different matrices at 37 °C (ag: agarose; low a: low-density alginate; high a: high-density alginate) (Lundberg and Kuchel, 1997).

Substrate	*MW [g mol^{-1}]*	*D [10^{-6} cm^2 s^{-1}], or [10^{-10} m^2 s^{-1}] in water*	*D [10^{-6} cm^2 s^{-1}], or [10^{-10} m^2 s^{-1}] in Ca^{2+} alginate (2 %)*
L-Tryptophan	204	6.7	6.7
α-Lactalbumin	15 600	1.0	1.0
Albumin	69 000	0.70	no diffusion
(Tanaka et al., 1984)			
Glucose	180	7.0	5.6 ± 0.2
Sucrose	342	5.23	4.7 ± 0.2
(Berensmeier et al., 2004)			
Glucose	180	9.6 ± 0.2	
– ag			6.8 ± 0.1
– low a			7.6 ± 0.1
– high a			6.7 ± 0.3
Glycine	75	13.8 ± 0.4	
– ag			10.4 ± 0.3
– low a			12.5 ± 0.3
– high a			9.7 ± 0.1
(Lundberg and Kuchel, 1997)			

Diffusion–reaction–growth correlations were investigated both experimentally and by modeling using *Escherichia coli* as a model organism (Lefebvre and Vincent, 1995). Heterogeneity in the biomass distribution inside gel-immobilized cell systems was examined in detail, and the phenomena responsible included substrate diffusion and accumulation of inhibiting products (see also Fig. 7.7). Heterogeneity is greatly enhanced when the substrate concentration is decreased, and when the membrane thickness is increased (Lefebvre and Vincent, 1997).

Immobilized-cell physiology, which relates to the fundamental aspects of correlation of immobilization and the response of cell growth and productivity, was reviewed by Junter et al. (2002). Information has been collected on a variety of organisms such as *Pseudomonas* sp., *Bacillus* sp., *Saccharomyces cerevisiae* and *Candida* sp., together with details of the matrices used for immobilization (e.g., alginate, agarose, carrageenan beads). The data reported have included the growth and productivity of secondary metabolites such as ethanol, enzymes, and antibiotics. A variety of growth rates was reported for immobilized cells – whether decreased, unchanged, or enhanced – compared to free cell systems. In most cases, growth was limited due to nutrient and/or oxygen mass transfer, with the immobilized biomass concentrating in the peripheral areas of the particles. Growth-promoting effects were attributed to protection against inhibiting or toxic substances or metabolites (e.g., ethanol, see below). Many examples have shown unchanged or lower specific productivities compared to those of suspended cultures (Junter et al., 2002). In contrast, an enhanced rate of ethanol formation – by 40–50 % – has been found for immobilized yeast as compared to suspended cells, and a growth rate reduction of 45 % and lower intracellular pH values were also observed (Doran and Bailey, 1986; Galazzo et al., 1987).

More detailed analysis revealed a high resistance of several bacteria (e.g., *Pseudomonas aeruginosa*) against antibiotics, and parameters other than diffusion limitation were clearly involved in these phenomena. It could be shown for *E. coli* that cell permeability was changed due to an alteration in the porins as channels for nonspecific diffusion. Subsequent two-dimensional electrophoresis (2-DE) of the total cellular proteins from suspended and immobilized *E. coli* cells revealed notable qualitative and quantitative differences. Those proteins for which amounts varied according to the growth mode represented about 20 % of the total cellular proteins detected. In addition, indications were identified which related some of these differences to stress response (Junter et al., 2002).

7.2 Immobilization by Aggregation/Flocculation

Several methods for the immobilization of microorganisms and cells are presented in Figure 7.1. For most applications, aggregation or flocculation (Fig. 7.2), adsorption and/or adhesion as well as entrapment in polymeric networks – and notably inclusion into ionotropic gels (mainly alginate) – have gained broad acceptance, and these methods will be treated in more detail in the following sections. Those methods in which the whole microorganism is immobilized, but only a singular enzyme

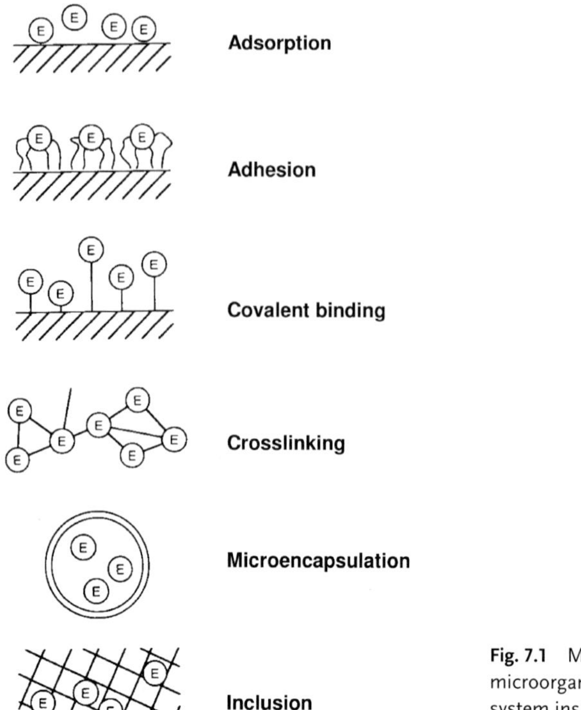

Fig. 7.1 Methods of immobilization of microorganisms (E signifies the enzyme system inside the immobilized cells).

activity is used and the organism itself serves as the matrix (in general after crosslinking with glutaraldehyde), are described in Chapter 6.

The aggregation or flocculation of microorganisms and cells – that is, the formation of agglomerates – can be regarded as a natural and simple method of immobilization (Scott, 1987). The basic phenomena and physiology of these processes have been discussed in detail elsewhere (Marshall, 1984; Wingender and Flemming, 1999).

Flocs and biofilms usually contain mixed populations of bacteria, where filamentous morphologies may play a major role, but they also represent a habitat for eucaryotic microorganisms. Maximal concentrations of bacteria range from 10^{10} to 10^{12} cells per mL, and this results from both direct and indirect interactions of the organisms. Cell-surface polymers and extracellular polymeric substances are of major importance for the development and structural integrity of flocs and biofilms (Wingender and Flemming, 1999).

Flocculating or aggregating anaerobic bacteria (including other material such as $CaCO_3$) have been applied successfully in biological wastewater treatment (Sam-Soon et al., 1987; Lettinga et al., 1999). Under appropriate conditions, pellets with settling (sedimentation) properties are formed which contain mixed bacterial populations of high density, but which may also include inorganic material. These systems are well known as either "upflow anaerobic sludge bed" (UASB) and "expanded granular sludge bed" (EGSB). The aggregates or pellets usually have a diameter of 1–4 mm and, under favorable conditions (e.g., inclusion of $CaCO_3$), exhibit high sedimentation velocities (up to 0.17 m min^{-1}). In mixed cultures, filamentous (e.g., *Methanotrix soehngenii*) and rod-type morphological forms have been observed; basically, the aggregate should contain all species required for degradation of the substrate, including hydrolytic, acetogenic, and methanogenic bacteria.

At the laboratory scale, reaction rates of chemical oxygen demand (COD) degradation of up to 50 kg m^{-1} d^{-1} could be obtained, where the biomass exhibited a specific activity of about 2 kg COD kg^{-1} organic dry matter (odm essentially biomass) and day (Hulshoff et al., 1983). In technical scale reactors with volumes of 2000–5000 m^3, the reaction rates are lower by factors between 2 and 4, and criteria for flocculation or granulation have been established for such systems (Table 7.2). The low cost, enhanced efficiency and operational stability for certain types of wastewater make this method of immobilization particularly attractive.

Flocculating yeast species have been investigated for the continuous production of ethanol, with improved productivity in tubular reactors and stirred-tank reactors fitted with a subsequent settler system (Scott, 1987; Netto et al., 1985). However, the use of this method on a technical scale often shows problems of operational stability of the biocatalyst, as has been observed for example with *Zymomonas* sp. in ethanol fermentation.

The conditions for *stable* flocculation have been investigated for several examples (see below), but are known neither generally nor completely. Multivalent ions (e.g., Ca^{2+}) and polyelectrolytes (e.g., chitosan) may favor the aggregation of microorganisms. The substrate, pH, ionic strength and temperature each play a major role; relevant characteristics are listed in Table 7.3.

Table 7.2 Characteristics, parameters and criteria for flocculation or granulation of anaerobic mixed cultures (Hulshoff et al., 1983; Sam-Soon et al., 1987).

Parameters	***Criteria***
Type and availability of substrates	Beneficial substrates: Low fatty acids (besides acetic acid) N-source (not limited) High partial pressure of hydrogen Unfavorable substrates: Lipids
Temperature	30–37 °C (practical application range)
pH	6.5–7.8
Ionic strength, ratio of mono- and divalent ions	Ratio of Ca^{2+} to K^+-ions: >5 : 1
Control of start-up	Selection pressure for aggregating species

Table 7.3 Characteristics of life in microbial aggregates (Wingender and Flemming, 1999).

Characteristics	*Associated phenomena*
Formation of stable microconsortia	High cell density
Cell-to-cell communication	Facilitated gene transfer
Long retention time	
Restriction of mass transfer	Formation of gradients in pH, oxygen, nutrients
	Aerobic, microaerophilic and anaerobic habitats in close proximity
	Improved tolerance to toxic substances

Aggregation is also an important phenomenon in fungi, an example being *Aspergillus niger*. The morphological development is thought to start with the aggregation of conidia, immediately after inoculation. A model for conidial aggregation has been proposed which consists of two separate aggregation steps. The first takes place immediately after inoculation (time scale 5–8 h), and the second is triggered by germination and hyphal growth (time scale about 10 h). A model of these steps is shown in Figure 7.2 (Grimm et al., 2004).

Product formation of mycelial organisms is intimately linked with their morphology. Pellet morphology and glucoamylase formation were investigated under different agitation intensities of *A. niger* AB 1.13. For pellet formation, it is necessary to

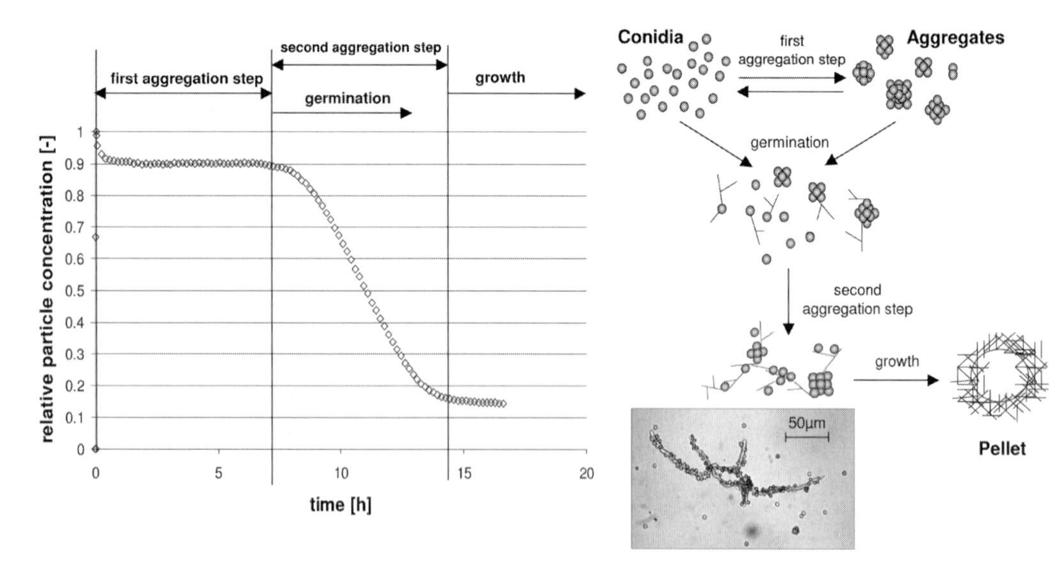

Fig. 7.2 Kinetic model of conidial aggregation of *Aspergillus niger* AB 1.13 based on experimental results.

apply low energy dissipation rates, which also favor product formation (Kelly et al., 2004).

In further studies, *A. niger* has been used as an expression system for glucose oxidase (GOD),and a pellet model has been developed, assuming both suspended hyphae and a fixed number of pellets. Although pellet density was identified as a key parameter, total pellet number was also seen to be important as it influenced the pellet surface and hence the turnover of substrate and product formation (Rinas et al., 2004).

7.3 Immobilization by Entrapment

7.3.1 Entrapment in Polymeric Networks

This method of immobilization has recently dominated research and development studies, whilst adhesion has been applied more to environmental technologies. The many and diverse methods for producing polymeric porous networks allows the design of different solutions with complete retention of the cells to be immobilized, yet with almost unlimited transport of substrates and products (except for polymers). It is important to select nontoxic components under appropriate conditions (pH, temperature, solvents) in order to maintain the catalytic potential of the cells. Thus, methods using organic solvents have not been widely used.

Ionotropic gel formation with polymeric polyelectrolytes is most important when compared to other inclusion methods, as it combines the use of low-cost materials (most are based on natural polysaccharides) and gentle conditions, together with a broad range of matrices and designs (see Section 7.3.2).

A simple and gentle method is that of *gelation*, which involves a phase transition in a polymer–solvent mixture due to a temperature shift. For this purpose, the cells are suspended at elevated temperature in the polymer (e.g., agar or agarose) solution; subsequent cooling causes the formation of a three-dimensional network – the gel – which includes the cells. The nature of the polysaccharides used allows for a smooth immobilization, and this is particularly suited for sensitive plant cells (Hulst and Tramper, 1989)

Polymers based on polyvinylalcohol (PVA) have been used successfully as matrices for the immobilization of different organisms (Jahnz et al., 2001). Gel formation can be performed at room temperature under gentle conditions, with high rates of survival of active organisms. The polymer is heated to produce a solution, which is then cooled to room temperature and mixed carefully with the suspension of microorganisms. Droplets are deposited onto a surface, followed by partial drying (15–20 min); a stabilizer solution is then overlaid to complete the hardening. The resultant particles have a flat, lens-like shape, with a diameter of 3–4 mm and a thickness of 200–400 μm – and therefore short diffusion path for substrates. The process has been automated to yield reproducible catalysts (LentiKats®). Typical microorganisms immobilized in this way include *Saccharomyces cerevisiae* and *Leuconostoc oenos* for alcohol production, slow-growing *Nitrosomonas* and *Nitrobacter* sp. for the oxidation of ammonia to nitrate in water, and *E. coli* for the production of tryptophan. The system can also be operated under nonsterile conditions, as the matrix is not degraded by fungi or other organisms (Jahnz et al., 2001).

Chemically resistant and mechanically stable networks can be formed via polycondensation and polyaddition reactions. Thus, bifunctional oligomeric prepolymers were reacted with appropriate multifunctional components. Polyurethanes based on prepolymers with two terminal isocyanate groups have proven to be flexible systems, notably for the conversion of hydrophobic substrates (Fukui and Tanaka, 1982; Kawamoto and Tanaka, 2003). Prepolymers for photoinducible radical polymerization with varying properties (hydrophilic, hydrophobic) have also been developed (Fukui et al., 1987). However, due to toxic effects and the rather elaborate chemistry involved these systems were not used to any great extent. The immobilization of yeast for alcohol production was developed up to the pilot scale, however. The system could be run at high productivity, even under nonsterile conditions, for over 350 days (Nojima and Yamada, 1987).

Basic considerations of processing, including the effect of processing parameters, have been summarized by Freeman and Lilly (1998). This review is restricted to synthetic applications, where the supply of oxygen and nutrients, high specific productivity, stabilization with operation times of at least 100 h and feasibility for large-scale operation were considered. Biotransformations by means of oxidoreductases, biosynthesis of organic acids, antibiotics, enzymes or other proteins are reported.

Processing parameters relevant for high productivity and operational stability include:

- immobilization method,
- mode of operation (e.g., repeated batch versus continuous),
- aeration and mixing,
- bioreactor configuration,
- medium composition (including feeding of substrates, precursors, or additional nutrients),
- temperature, pH,
- *in situ* product and/or excess biomass removal.

The authors stress the relevance of understanding cell physiology and correlations with productivity and operational stability, which has limited application for viable immobilized microorganisms in synthesis (Freeman and Lilly, 1998).

A new strategy to widen the scope of cells – for example, towards new substrate utilization – has been developed by Ueda et al. (2003), namely the genetic modification of yeast cells which display enzymes on their surface.

7.3.2
Entrapment in Ionotropic Gels

7.3.2.1 Principle

Water-soluble polyelectrolytes can form solid polymeric networks (gels) by crosslinking with either polycations or polyanions. Polysaccharides with carboxy- or sulfonyl groups (alginate, pectin, carrageenan) or amino groups (chitosan obtained from chitin by deacetylation) are specifically suited for this. These materials are available at low or reasonable cost, and are also nontoxic. Furthermore, this method offers diverse optimization strategies with different procedures and conditions; several counterions can be applied for crosslinking to form a variety of secondary structures. The most frequently used ions are Ca^{2+} and K^{+} as cations, and polyphosphates as anions (Table 7.4). In principle, Al^{3+} and Fe^{3+} can also be used, though restrictions with regard to their use in the food and pharmaceutical industries limit their application.

The structural elements of the most popular polysaccharides are shown in Figure 7.3. The secondary and tertiary structures which lead to the formation of networks (gels) are shown schematically in Figure 7.4. Inside these networks, the microorganisms, cells and parts of cells (e.g., microsomes) can be entrapped, both gently and efficiently. Enzymes usually cannot be immobilized using this method as their dimensions are smaller by several orders of magnitude, and consequently they diffuse easily out of the gel (Reischwitz et al., 1995).

The structural elements are of major importance with regard to the mechanical properties of the gel and particles formed.

Alginic acid is a polyuronic acid extracted from seaweeds, and is composed of varying proportions of 1→4-linked β-D-mannuronic (M) and α-L-guluronic acids (G) (see Fig. 7.3). The residues occur in varying proportions depending on the source, and are arranged in block patterns comprised of homopolymeric regions (MM blocks and GG blocks) interspersed with alternating regions of heteropolymeric regions (MG blocks). The gel strength is related to the G content, which varies from 20 to 75 % depending on the seaweed source. The GG blocks have preferential binding sites for divalent cations, such as Ca^{2+}. The resultant gel is biochemically inert and

Table 7.4 Ionic polysaccharides and counterions for crosslinking.

Fixed matrix ion	*Carrier matrix*	*Counter-ion*
$R\text{-}COO^-$	Alginate	Ca^{2+} (Al^{3+}, Zn^{2+}, Fe^{2+}, Fe^{3+})
	Pectin	Ca^{2+} (Al^{3+}, Zn^{2+}, Mg^{2+})
	Carboxymethyl cellulose	Ca^{2+} (Al^{3+})
$R\text{-}SO_3^{2-}$	Carrageenan	K^+, Ca^{2+}
$R\text{-}NH_3^+$	Chitosan	Polyphosphates

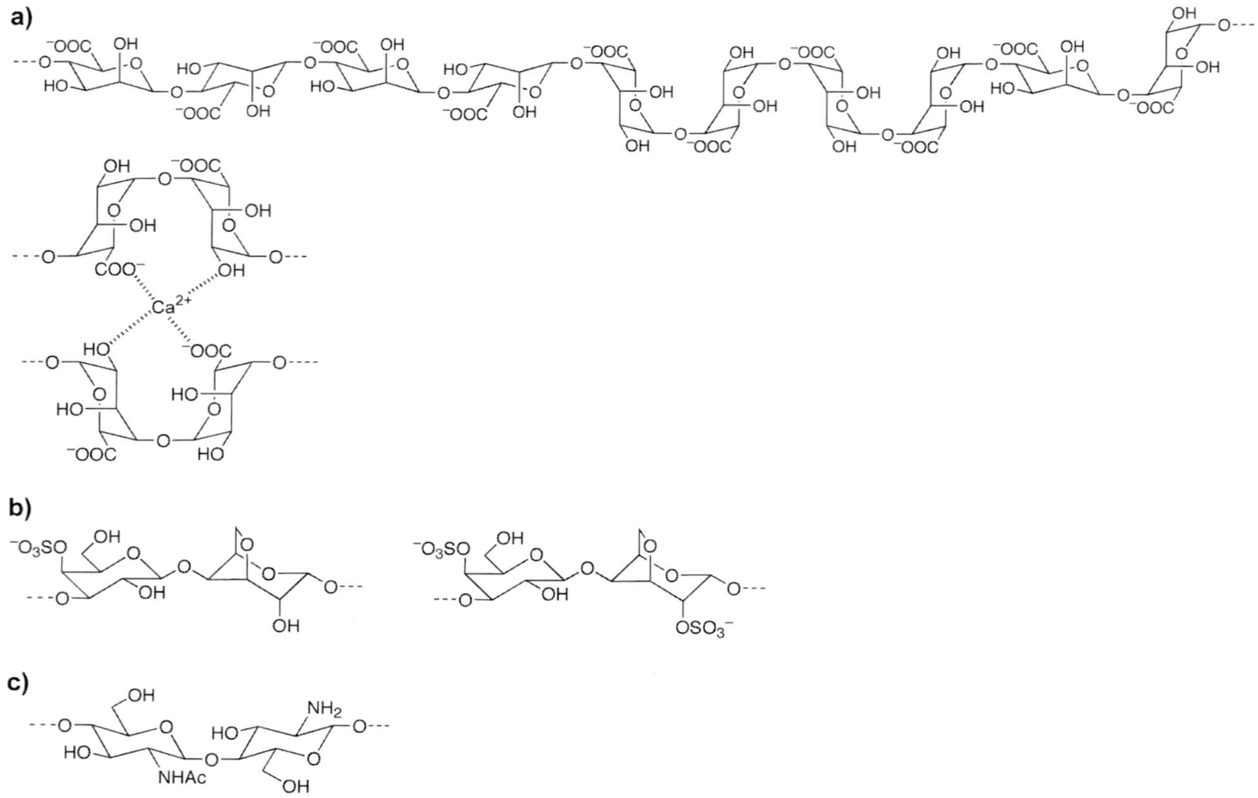

Fig. 7.3 Structures of ionic polysaccharides. (a) Alginates with D-mannuronic-, L-guluronic-, and mixed structural elements; (b) κ- (R = OH) and ι-(R = OSO_3^-) carrageenan; (c) Chitosan.

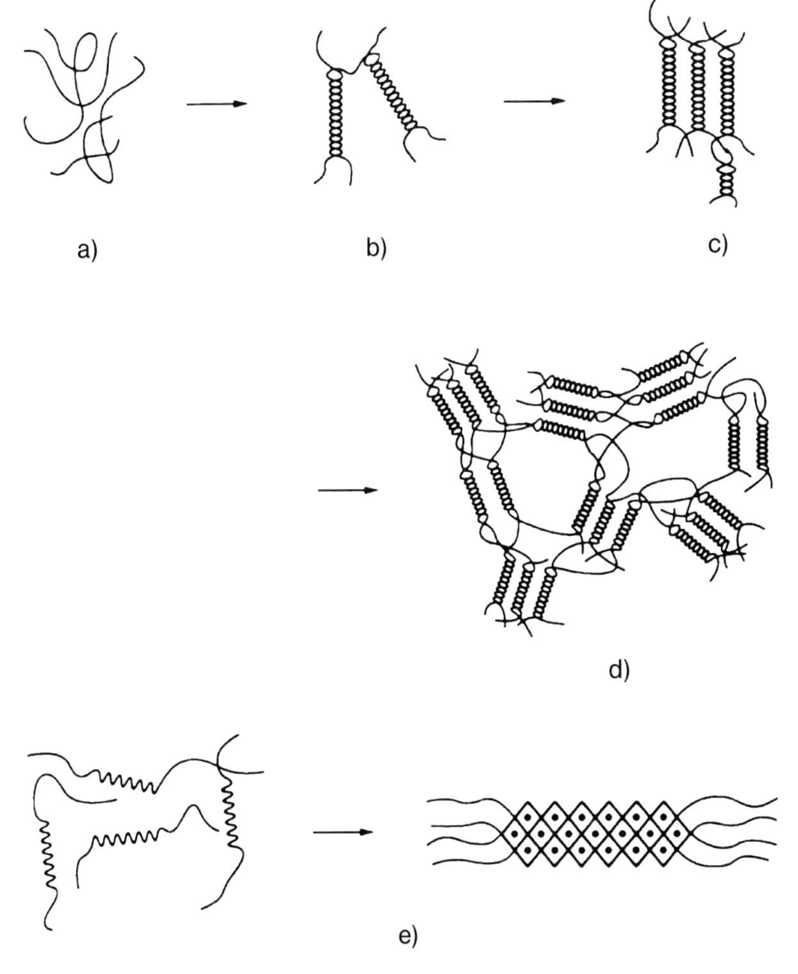

Fig. 7.4 Mechanisms of gel formation. (a) Statistical polymer network in solution; (b) double/twin helices; (c) bundle/packet of double helices and (d) their tertiary/supramolecular structure; (e) network formation in alginates due to crosslinking by multivalent cations (•).

mechanically stable, with interstitial spaces that are suitable for cell immobilization (Fraser and Bickerstaff, 1997).

Carrageenans are able to build different networks with other types of binding, which are formed also with K^+ ions. These polyanion systems are unstable at elevated pH, whereas the polycation systems tend to dissolve at low pH (Brodelius and Vandamme, 1987).

A considerable number of publications have demonstrated the merits and shortcomings as well as the flexibility of gel entrapment methods, notably with alginate and carrageenan (see surveys in Mosbach, 1987; Nojima and Yamada, 1987; Nagashima et al., 1987; Freeman and Lilly, 1998; Dörnenburg and Knorr, 1995; Bickerstaff, 1997; and Junter et al., 2002).

7.3.2.2 Examples

The entrapment of cells in alginate is one of the simplest methods of immobilization. Alginates are commercially available as water-soluble sodium alginates, and have been used for more than 65 years in the food and pharmaceutical industries as thickening, emulsifying, film-forming, and gelling agents. Entrapment in insoluble calcium alginate gel is recognized as a rapid, nontoxic, inexpensive, and versatile method for the immobilization of enzymes and cells (Fraser and Bickerstaff, 1997). This has also been the most popular method, with numerous applications for immobilizing both microorganisms and plant cells. Major developments of this procedure were reported by Klein and Vorlop (1985), Kierstan and Bucke (1977) and Jahnz et al. (2001).

To prepare immobilized cells, a solution of sodium alginate, typically 2–4 % (w/v), and an appropriate buffer solution is mixed with the cells to be immobilized and passed dropwise through a nozzle (syringe) into a vessel containing a solution of calcium chloride (200 mM) under gentle stirring (Fig. 7.5). In this way particles of about 2 mm diameter (range: 1–4 mm) are obtained. The particle size can be reduced in part by drying. The density of the biomass can thus be increased significantly by up to 30 % (relative to wet weight) (see also Exercise 7.5.1).

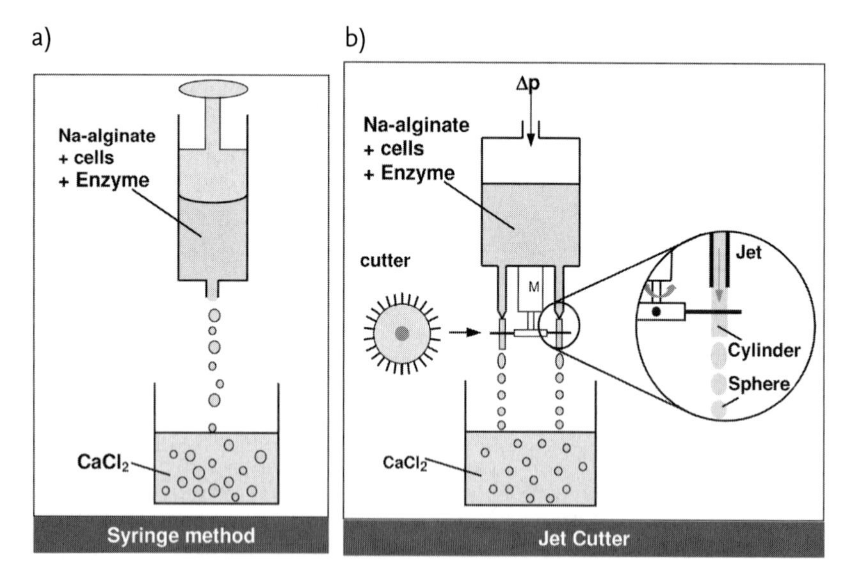

Fig. 7.5 Immobilization in alginate. (a) Syringe method; (b) Jet cutter method with alginate-cell suspension, pressure application Δp, rotating cutter, jet-nozzle, cylindrical solution leaving the nozzle, cutting to yield spherical droplets falling down into the $CaCl_2$ solution for crosslinking (Pruesse et al., 2000).

Specific methods have been developed in order to control the particle diameter, and notably to prepare small particles with optimal mass transfer characteristics. In a system developed by Prüße et al. (1998), the fluid is pressed through a nozzle as a liquid jet. Beneath the nozzle, the jet is cut into cylindrical segments by a rotating cutting tool made of small wires fixed in a holder. The segments form spherical beads while falling into a 200 mM $CaCl_2$ solution (Fig. 7.5), the particle diameter being dependent on the speed of rotation. In this way, particles of 0.5 mm and narrow size distribution can be obtained (Fig. 7.6). Scale-up to produce biocatalysts in the kilogram scale has been achieved (Prüße et al., 1998). Furthermore, particles could be obtained which are appropriate for use in fluidized-bed reactors by the additional

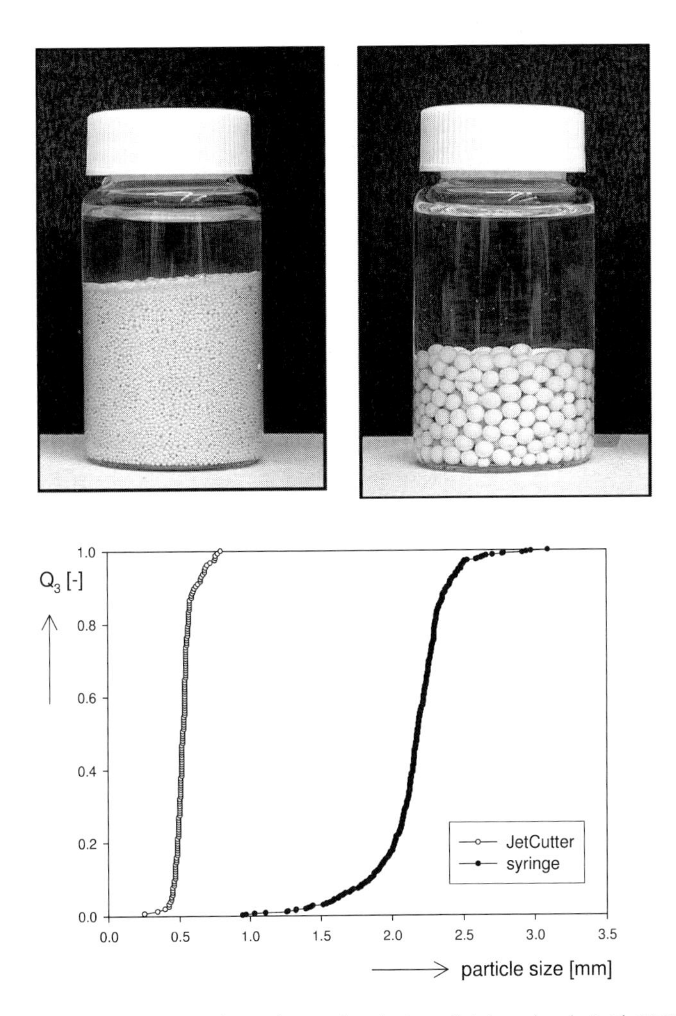

Fig. 7.6 Particles and particle size distribution of alginate beads (with 30 % (w/w) sand for improved settling properties) prepared by syringe or Jet cutter (Berensmeier et al., 2004).

entrapment of titanium dioxide or sand particles, thus providing high sedimentation rates due to an enhanced density (Berensmeier et al., 2004).

The destabilization of calcium alginate is promoted by chelating agents (phosphate, citrate, EDTA) that remove calcium ions and by other mono- (K^+-, Na^+-) or divalent cations that can exchange with Ca^{2+}. Operational stability during the conversion of substrate solutions requires a ratio of monovalent and Ca^{2+} ions of <5 or 20 (maximal ratio) in general. This is due to the exchange and stationary equilibrium of all ions present in solution. Leakage can occur from the gel beads, and this is influenced by initial alginate concentration, mechanical treatment of the beads, and cell productivity, if dividing cells have been immobilized. Electron microscopy of a 2 % calcium alginate bead showed pores that ranged from 5 to 200 nm in diameter. After gelation with Ca^{2+} ions, alginate particles may be further crosslinked with chitosan, polyacrylic acid, polyvinylalcohol, or polyethylenimine to produce more stable, lower-porosity complexes with improved leakage characteristics (Fraser and Bickerstaff, 1997).

Many investigations and major developmental efforts have been devoted to ethanol production by immobilized yeast, or bacteria (e.g., *Zymomonas*) up to the pilot scale with several m^3 reactor volumes (Nagashima et al., 1987; Junter et al., 2002). For fluidized-bed application, which offers advantages for technical substrates with fine suspended particles, alginate beads including fine TiO_2 particles co-immobilized with yeast cells proved advantageous. The particles exhibited improved mechanical stability and settling velocity. The activity was high due to growth of the yeast which was originally immobilized and distributed homogeneously inside the matrix, such that the active cells concentrated near the particle surface. Thus, the inhomogeneous distribution of yeast cells resulted in favorable conditions for mass transfer, with only a short diffusion path for substrate and products (Fig. 7.7).

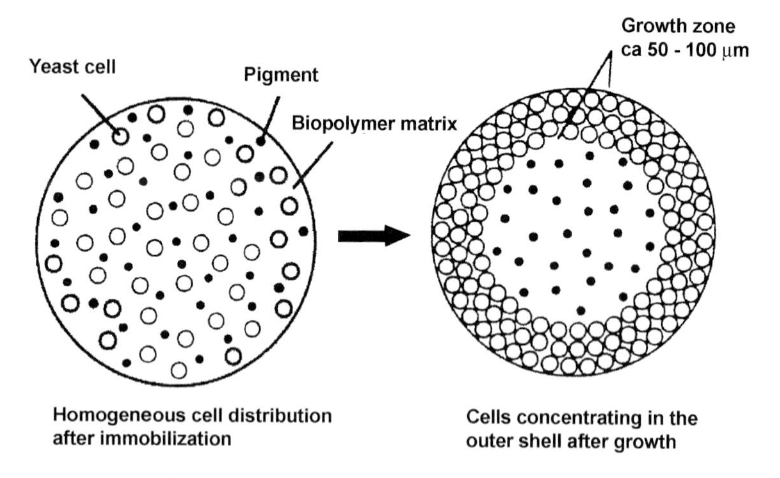

Fig. 7.7 Biocatalyst with homogeneous (after immobilization, left) and inhomogeneous distribution of yeast cells (after growth, right) (pigment: fine TiO_2 particles). (From Buchholz, 1989; with kind permission of Lurgi GmbH, Frankfurt).

Owing to the industrial importance of this system, several studies have focused on the metabolic responses (Doran and Bailey, 1986; Galazzo et al., 1987; Norton and D'Amore, 1994; Junter et al., 2002). A stimulation of active metabolism on immobilization, with increased specific rates of substrate (essentially glucose) uptake and product (essentially ethanol) secretion were observed for different types of immobilized systems. Enhanced production efficiencies as compared to suspended cells were frequently reported. As highlighted by Norton and D'Amore (1994), however, careful interpretation should be made of published data on productivities as results are often reported on a volumetric scale, which is of interest from a biochemical engineering aspect, but does not characterize the intrinsic behavior of the immobilized cells. Despite numerous efforts however, no industrial process for ethanol production with immobilized yeast has yet been established. An example of this, namely the final stage in beer production, or maturation, is provided in Section 7.4.

The first industrial utilization of immobilized microorganisms was based on the method with entrapment in κ-carrageenan developed by Chibata et al. (1987) for the production of amino acids (single enzyme step, nonviable cells). In principle, gelation occurs by dissolving the potassium salt of κ-carrageenan (2–5 % solution) at 70–80 °C, suspending the cells at 40 °C in the polysaccharide solution and subsequently cooling. For particle stability however, it proved advantageous to drop the suspension into a solution of potassium chloride (e.g., 0.1 M). In a two-step procedure a small amount of cells is suspended in the polysaccharide solution, after which the cells are grown in the gel by incubation in a nutrient medium. In this way, high yeast cell densities of up to 5×10^{12} cells L^{-1} could be obtained. Details of the procedures are provided by Chibata et al. (1987).

Particles with carrageenan can be further stabilized and hardened by several methods, such as crosslinking by hexamethylenediamine, polyethyleneimine, or glutaraldehyde. A survey summarizing the developments of gel immobilized microorganisms is provided in Table 7.5.

Table 7.5. Examples of immobilized microorganisms in different supports (Chibata et al., 1987; Freeman and Lilly, 1998; Junter et al., 2002).

Product	*Organism*	*Carrier*	*Reactor*[a]
Ethanol	*Saccharomyces cerevisiae*	Alginate	
Ethanol	*Zymomonas mobilis*	Alginate	
2,3-Butanediol	*Enterobacter aerogenes*	Alginate	
L-Isoleucine	*Serratia marcescens*	Alginate	
11β-Hydroxy-progesterone	*Acetobacter phoenicis*	Alginate	
Prednisolone	*A. globiformis*	Polyacrylamide	RB
L-DOPA	*Erwinia herbicola*	Carrageenan	PB
Thienamycin	*S. cattleya*	Celite	BC
Glucoamylase	*Aspergillus niger*	Alginate	SF/RB
(Conversion of toluene)	*Pseudomonas fluorescens*	Anion-exchange resin	
	Bacillus subtilis	Carrageenan	

[a] RB: repeated batch; PB: packed bed; SF: shake flask.

Higher specific production rates and/or yields with immobilized as compared to suspended cells have been observed, for example in the production of enzymes and secondary metabolites such as antibiotics. However, other examples exist whereby cultures in the immobilized state displayed unchanged or even lower specific productivities as compared to suspended cultures, mainly due to mass transfer limitations (Junter et al., 2002).

Industrial applications with microorganisms immobilized in polymer matrices have been rare, however, and consequently no technical application for ethanol production has been reported. Redox reactions which require multienzyme systems with cofactors and their regeneration appear to be carried out mainly with suspended whole-cell systems and batch-type reactions (see Liese et al., 2000; Cheetham, 2000; Kirst et al., 2001).

Plant cells can also be immobilized in alginate or carrageenan (Rosevear, 1988; Hulst and Tramper, 1989; Dörneburg and Knoll, 1995). Due to the complex and specific requirements of the cells, and also the high cost of cell culture, these methods are limited to the formation of products of high added value, such as pharmaceutical specialties, components of aromas and fragrances, and secondary metabolites in general. Thus, the red pigment shikonin was produced industrially using a plant cell culture, while major efforts were also undertaken for the production of rosmarinic acid, berberine, and ginseng (Rosevear, 1988). De-novo synthesis, the biotransformation of precursors (e.g., the transformation of digitoxin to digoxin), and the production of indole alkaloids are in principle possible (Hulst and Tramper, 1989).

A further example relates to the immobilization of cells of *Taxus* sp. plant cells (Tang and Mavituna, 2001). Roisin et al. (1996) presented factorial designs which were developed to analyze the metabolic response of *Gibberella fjikuroi* in the production of secondary metabolites such as gibberellins and carotenes.

A review by Dörneburg and Knoll (1995) outlined progress into the biotransformation of exogenous substrates by cultured plant cells, and highlighted the potential of regio- and stereoselective hydroxylation, oxidoreduction, hydrogenation, glycosylation, and hydrolysis for a variety of organic compounds. The development of techniques was also described elsewhere (Ishihara, 2003).

Immobilized plant cells offer, in principle, the same merits as those of immobilized microorganisms, namely repeated or continuous application of the biocatalyst, simple product recovery, high cell density and thus high volumetric productivity, protection of the cells against shear forces, and decoupling of the growth and production phases.

The immobilization of *animal* and *human cells* has also attracted attention over many years, notably for the production of monoclonal antibodies with hybridoma cells, therapeutic proteins by recombinant cells, and vaccines. Immobilization can compensate for low growth rates and productivity if high cell densities are realized. Indeed, cell densities between 10^7 and 10^8 cells mL^{-1} have been obtained in fixed and fluidized-bed reactors, compared to about 10^6 cells mL^{-1} in suspended systems. Microcarrier systems are now well-established both at laboratory and technical scales (see Section 7.4).

7.4 Adsorption

Acetic acid (vinegar) has been produced by adsorbing *Acetobacter* sp. onto wood chips since 1823, and even today 60 m^3 reactors filled with curled beechwood shavings are used as so-called "trickling generators" with ethanol and oxygen as substrates. A pump circulates an ethanol/water/acetic acid mixture from a storage reservoir to a distributor at the top of the tank. A blower is used to force air through the packing. A good conversion efficiency is 88–90 % of the theoretical value, and a correctly operated generator has a packing life of 20 years (Atkinson and Mavituna, 1991, pp. 1171–1173).

Although the use of immobilized yeast cells has been widely studied for beer production, the only industrial production process known to date with viable cells is the final stage in beer manufacture, where aroma formation occurs during maturation. Immobilized yeast cells (*Saccharomyces cerevisiae*) were immobilized by adsorption onto porous silica particles ("Siran", Schott, D) with a high internal surface (up to 0.4 m^2 g^{-1}, or 21–90 m^2 L^{-1}, d_p 2–3 mm). Two reactors, each of 2.5 m^3, were installed for a production volume of 400 000 hL (1 hL = 100 L) of beer. During maturation and aroma formation, the essential step of α-acetolactate transformation into acetoine and diacetyl has been optimized with respect to lower investment (due to a small reactor volume), with a much reduced residence time (2 h compared to 10–20 days in the conventional process), continuous processing with easy process automation, less purification chemicals (surfactants), and less wastewater (Breitenbücher, 1996).

Further developments involved the production of low-alcohol beer (Champagne, 1996), and alcohol-free beer production with immobilized *Saccharomyces cerevisiae* also promoted much interest. Details of the flavor formation and cell physiology were investigated by Iersel et al. (1999). Yeast cells were attached to the surface of a carrier consisting of polystyrene coated with DEAE-cellulose. During operation in a packed-bed reactor, changes in the cell physiology were observed, notably with respect to alcohol acetyl transferase and the formation of ethyl acetate and isoamyl acetate, which depended significantly on the temperature being in the range of 2 to 12 °C. In fact, it was found that an optimal and constant flavor profile could be obtained by introducing regular aerobic periods to stimulate yeast growth.

The growth of mammalian cells on microcarriers is well established, and has been used successfully on a production scale for the manufacture of monoclonal antibodies with hybridoma cells, therapeutic proteins by recombinant cells, and vaccines. Thus, microcarrier culture – that is, the growth of anchorage-dependent cells on small particles (usually spheres with a diameter range of 100–300 μm) suspended in stirred culture medium – has made major progress and impacted heavily on protein production. The availability of human diploid cell lines allowed a rapid expansion in the manufacture of human vaccines (Doyle and Griffith, 1998). Following considerable development work, a range of suitable microcarriers became available based on gelatin, collagen, crosslinked dextran, cellulose, glass, and polystyrene. The key criteria were to prepare the surface both chemically and electrostatically appropriate for cell attachment, spreading, and growth. Thus, 1 g of Cytodex microcarrier (Pharmacia) has a surface area of 6000 cm^2 and, when used at a concentration of 2 g L^{-1} produces an area of 12 000 cm^2 L^{-1}. Adherent CHO (Chinese hamster ovary, including recombinant) cells could be immobilized efficiently on (e.g., gelatin) modified, silica-based carriers. Modified cellulose was used efficiently for the immobilization of hybridoma cells and hepatocytes (Pörtner et al., 1995; Tokayashi and Yokoyama, 1997; Doyle and Griffith, 1998).

Requirements for cell immobilizing carriers have been summarized as follows:

- Material where cells can easily attach to proliferate on the surface, innoxious to the cells.
- Density and size adequate for fluidization by mild mixing.
- Large surface area per unit volume.
- Mechanical strength and durability.
- In porous support: appropriate dimensions for cell accommodation and nutrient and oxygen diffusion, the latter being one of the critical parameters.
- Scale-up potential and commercially available at a reasonable price (Tokayashi and Yokoyama, 1997).

Scale-up has been performed using reactors of 1–5 m^3 in size, albeit with low cell densities, usually ca. 2×10^6 cells mL^{-1} (Tokayashi and Yokoyama, 1997). Stirred reactors, fixed-bed and fluidized-bed reactors have also been investigated (Lüllau et al., 1994; Pörtner et al., 1995; Waugh, 1999).

7.5 Adhesion

The adsorption of microorganisms at, and their growth on, surfaces is a well-known and common phenomenon in nature. Typical matrices include stones, sand, clay minerals, plant materials, teeth, as well as metal and polymer/plastic surfaces. Adsorption is based on a physical – and, in principle – reversible, interaction. Often, however, a subsequent step follows when microorganisms produce and secrete secondary metabolites that cause a strong, partially irreversible adhesion to the primary matrix, in many cases by formation of a secondary matrix of polymers formed as metabolites. In the context of this chapter this process is termed "adhesion", and it is most often encountered with biofilm-producing bacteria.

For technical applications, immobilization based on adhesion is important because it forms a very stable and biocompatible attachment of the microorganisms to the carrier. It represents an elegant and powerful method of immobilization, as simple, convenient and cheap carriers (e.g., sand, pumice, charcoal, lava, wood chips, straw) can be used in general, without any special additives. As reactors, tubular configurations with packed beds, as well as fluidized beds with carriers exhibiting good settling properties, are convenient.

A range of processes has gained importance in the field of environmental technology; indeed, this sector is considered to be the most relevant inasmuch as it has considerable economic importance for the application of living immobilized microorganisms. For example, the annual overall cost for biological wastewater treatment in Germany alone is estimated as over 10×10^9 €. Diverse mixed microorganism cultures are used, after being adapted to the conditions of the different wastewaters (which clearly cannot be closed systems operating under sterile conditions). Both aerobic cultures for wastewater and exhaust air purification, and anaerobic cultures for the treatment of high organic load wastewater of the food and chemical industries are used, and in this respect several hundred functional plants are now on stream world wide.

7.5.1
Basic Considerations

The interaction of microorganisms or cells and a carrier depends essentially on the surface properties of each, respectively. This is notably true for the first step, the primary adsorption – that is, before extracellular metabolites, which may modify the surface, are formed and secreted.

Whereas many matrices (e.g., glass, silica, clay minerals, cellulose) have been well-characterized, less is known about the surface properties of microorganisms, which in turn depend on the growth conditions and/or the surrounding medium. X-ray–photoelectron-spectroscopy has proven to be the most promising method to provide information on the elements and types of bond in the external molecular layers down to a 10-nm depth from the surface (Brodelius and Vandamme, 1987; Amory and Rouxhet, 1988; Mozes et al., 1988). This method allowed for an analysis of the elemental composition of the cell surfaces of different bacteria and yeasts. Furthermore, the electrophoretic mobility provides information on the surface charge, and hydrophobic chromatography on the hydrophobicity of the surface (van Loosdrecht et al., 1987; Amory and Rouxhet, 1988). Correlation of these physico-chemical data led to the following conclusions:

- Phosphate groups play a dominant role with respect to the surface charge; the N/P-concentration of the surface correlates with the surface charge.
- The bacteria investigated exhibited higher N- and P-concentrations as compared to yeasts. It must be considered in this respect that the elemental composition for a certain strain depends on the fermentation conditions (e.g. the substrates) and the fermentation time (cell age).
- Proteins of the cell surface correlate with the hydrophobicity. For yeasts, the hydrophobicity is proportional to the N/P-ratio at the cell surface. In contrast, the hydrophobicity of bacteria exhibits an inverse correlation with the N/P-ratio and the oxygen concentration; here, hydrophobicity is proportional to the hydrocarbon percentage of the cell surface.
- The main interactions of the surfaces of microorganisms and carriers are van der Waals forces, dipole–dipole, and electrostatic interactions. In many cases, repulsive forces must also be taken into account as both microorganisms (e.g., *Saccharomyces, Acetobacter, Bacillus* sp.) and carrier materials (such as glass- or silica-

surfaces) carry negative surface charges. In such cases the surface charge of the carrier can be modified or changed (reversed) by appropriate ionic components (Fe^{3+}, chitosan, etc.) in order to facilitate the primary adsorption.

Flocculation, where microorganisms aggregate to form larger particles, may be considered as a special form of immobilization by adsorption and/or adhesion. In such cases bacteria and yeasts form granules which may include other material such as calcium carbonate, as is the case in anaerobic wastewater treatment. These granular particles can be readily retained, for example in fluidized-bed reactors. For flocculating yeasts further specific interactions than those mentioned before could be identified (Kihn et al., 1988).

Adhesion and flocculation in general require selection processes and also conditions for operational stability that are not (yet) known. Processes occuring during the adsorption of microorganisms may be divided phenomenologically into the following steps (Fig. 7.8) (Rouxhet, 1990; Wanda et al., 1990):

1. (Primary) adsorption.
2. Adhesion (secondary steps, building of bridges, for example by cell organelles, pili, or secondary metabolites such as extracellular proteins and/or polysaccharides.
3. Growth of the adhering cells (formation of a biofilm and/or a secondary matrix of microorganisms, secondary metabolites and further components such as calcium carbonate) up to a limiting condition where growth and detachment by shearing reach a stationary equilibrium.

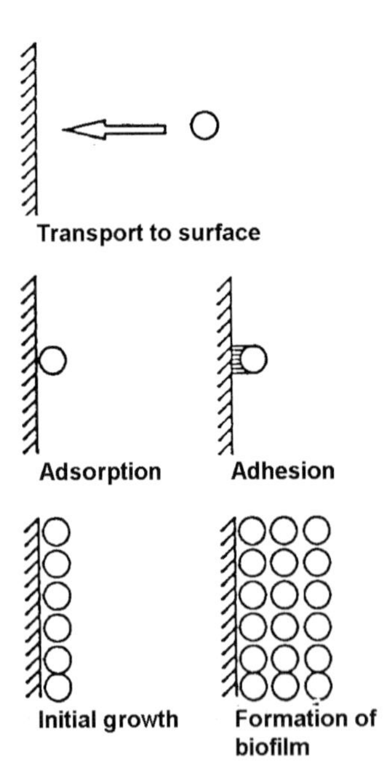

Fig. 7.8 Scheme of the different steps of adsorption, adhesion, and growth of microorganisms on a surface (Rouxhet, 1990).

Figure 7.9 illustrates the single steps of adsorption and adhesion taken from an example of growth of an anaerobic mixed culture on glass surfaces. It is clear that the primary adsorption proceeds quickly (within several hours), whereas the formation of a biofilm takes a much longer time (several days or weeks) (Fig. 7.9 a and b). Formation of the biofilm depends critically on the available carbon source; this provides the stimulus for the formation of extracellular products (such as polysaccharides; Fig. 7.9 c) which form the secondary matrix on the carrier surface. For anaerobic organisms, complex media with different carbon sources – notably medium-chain fatty acids – have proven to be superior with respect to the formation of a polysaccharide matrix and thus a biofilm with high stability and activity, as compared to media with a single carbon source (Wanda et al., 1990).

The (secondary) adhesion can provide for mechanically stable biofilms which, in fluidized-bed reactors, resist abrasion. It can be initiated by selection pressure on the bacterial mixed culture. For this, the carrier is kept in the reactor (e.g., with a fixed or fluidized bed) which is provided with wastewater carrying a mixed culture of broad species distribution. The medium residence time should be below the reciprocal growth rate of the microorganisms, so that all organisms that are unable to adhere to the surface are eluted. Both aerobic (with a hydraulic residence time τ of ca. 45 min.) and anaerobic (τ = 1–4 h) biomass can be immobilized by adhesion via this method, thereby implementing a selection pressure (Heijnen et al., 1991). The growth to high density of active biomass on the carrier may require between a few days and several weeks, but it can be accelerated by the application of an inoculum with carrier-fixed (preselected) microorganisms (which are transferred by a low rate of detachment of the inoculum and subsequent attachment to the new carrier) (Jördening et al., 1991). The analysis of profiles in bioaggregates and biofilms has been developed considerably in recent years (Bössmann et al., 2003).

Several conditions (which are known only partially for different applications) must be maintained in order to obtain an immobilized bacterial culture with high activity and stability:

- Stable pH: in large reactors this requires sufficient mixing and buffer capacity.
- Control of temperature: a convenient and cheap source of energy (< 50 °C) and heat exchangers are required in order to maintain the optimal temperature.
- Minimal substrate requirements: carbon source, minerals (N-, P-, and sources of other growth requirements); constant concentrations are favorable, large variations may cause problems; starvation and sudden changes of substrates can lead to a breakdown of cultures which are not adapted; notably, anaerobic cultures may require long time (weeks) in order to form new active biomass.
- No substances which are toxic to the microorganisms.

Many different materials have been applied as carriers: for example, particles (granules, spherical, porous or compact carriers) can be used in fluidized beds preferentially, but only partially in fixed-bed reactors due to problems of occlusion; internals can be used in fixed-bed reactors.

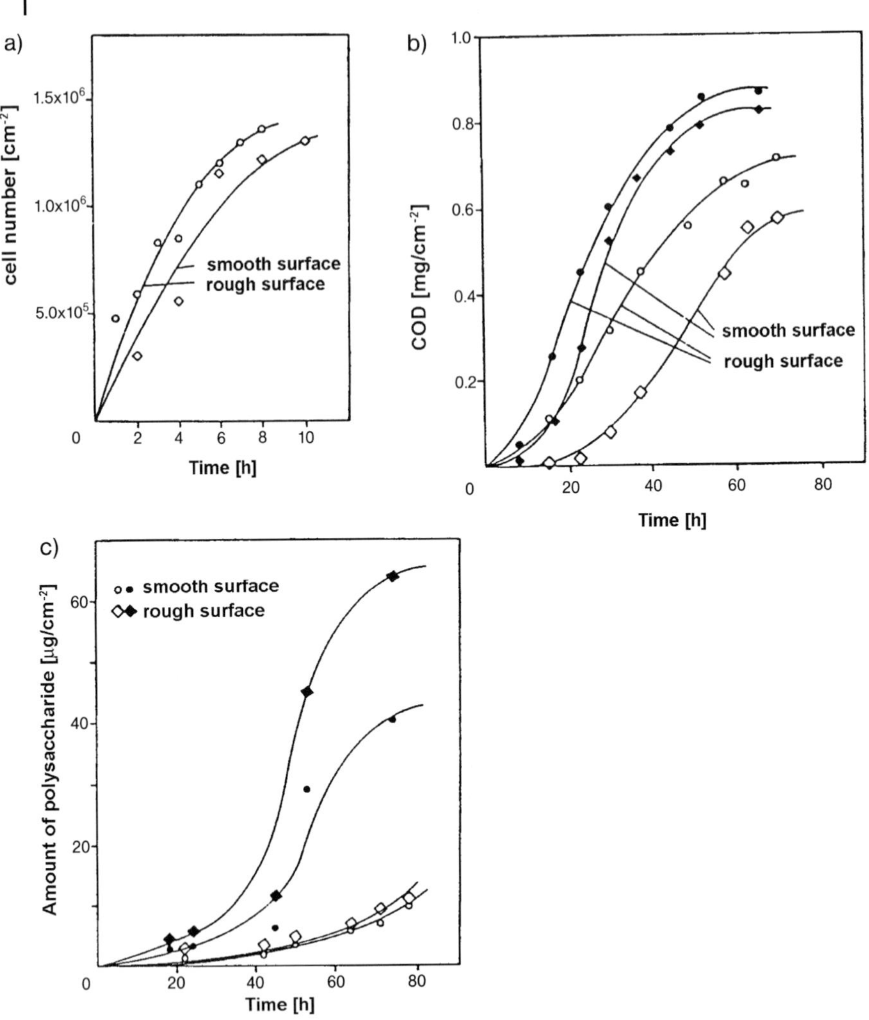

Fig. 7.9 Processes during growth of an anaerobic mixed culture on either smooth and even or, in contrast, raw glass surfaces (Wanda et al., 1990). (a) Rate of primary adsorption (measurement of cell number per cm^2) with acetic acid as carbon source. (b) Formation of a biofilm on a glass surface (measurement via COD (chemical oxygen demand) per cm^2) with different carbon sources (○, ◊ reactor A: acetic acid; ●, ◆ reactor B: acetic and butyric acid). (c) Polysaccharide content of the system shown in (b).

The type of carrier has a major influence on the density of biomass, and thus on the reactor space–time-yield which can be obtained. When compared to nonporous carriers (e.g., sand), porous materials (e.g., porous silica, pumice) allow for a much higher density of active biomass per volume, which in turn provides for a much higher degradation potential (Keim et al., 1989, 1990; Jördening et al., 1991; Aivasidis and Wandrey, 1992). A further advantage of porous carriers is their lower density and thus lower energy requirements for maintaining fluidization, when compared to nonporous carriers. Thus, the fluidization of pumice requires only about 5 W m^{-3}, which is one-tenth of that required for sand (50 W m^{-3}, both at 50 % bed expansion and similar particle size; see Table 7.6b) (Jördening et al., 1991).

The formation of a biofilm in the initial phase depends on the inoculum and favorable conditions (substrate, pH, temperature). The growth rates for anaerobic mixed cultures, including methane bacteria, are in the range of 0.04 to 0.15 d^{-1}. In pilot reactors an increase in performance of 10–20 % per day during the exponential growth phase could be obtained (Jördening et al., 1991), while the thickness of the biofilm on a smooth surface may be in the range of 0.02 to 0.2 mm. Extended investigations on the dynamics in the biofilm have been undertaken by Trulear and Characklis (1982).

Porous carriers with sufficient large pores are favorable for an optimal density of microorganisms. Messing (1982) found that a pore diameter five times larger than the cell dimension represents an optimum, when the cells should be able to grow (Fig. 7.10). In the case under investigation (ceramic carrier), a cell density of 10^8 to 10^9 anaerobic cells per gram carrier was obtained (carrier: Cordierit; Corning Glass, New York; particle diameter 2–6 mm, average pore diameter 3 µm).

A selection of carriers and their properties is presented in Table 7.6. Materials used for applications include inorganic (glass or porous glass, respectively, silicates, ceramics, lava, etc.), synthetic organic (polyurethanes, polyethylene, polypropylene, charcoal), and natural organic carriers (peat, chips of bark). The most important criteria for carrier materials are their mechanical, chemical and biological stability, the specific weight (or density), the surface per reactor volume (m^2 per m^3) available for attachment of microorganisms, and the void volume available for fluid flow. In addition, it should be noted that several additives can favor immobilization, notably calcium carbonate precipitates or ions, which either ease or enhance adsorption.

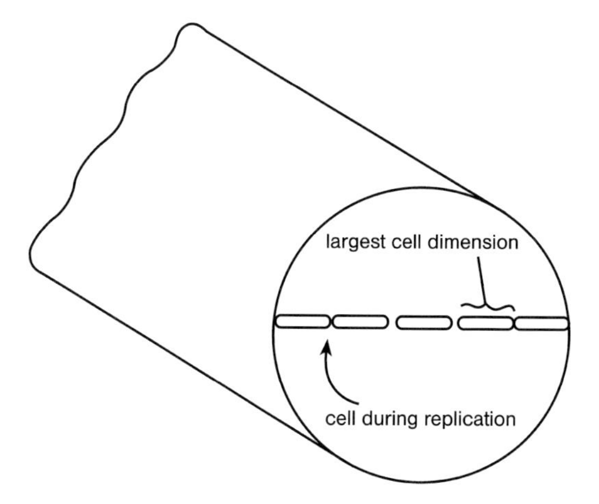

Fig. 7.10 Schematic view of cell division inside a pore of optimal dimensions (Messing, 1982).

Table 7.6 a Typical supports for fixed-bed reactors (Jördening and Buchholz, 1999).

Support	*Diameter range [mm]*	*Surface [m^2 m^{-3}]*	*Bed porosity*	*Eq. pore D.*[a]	*Reference*
Raschig rings	10.16	45–49	0.76–0.78		Carrondo et al. (1983)
Pall rings	90.90	102	0.95	20	Young and Dahab (1983)
	Corrugated modular blocks	98	0.95	46	
Pall rings	25	215			Schultes (1998)
Hiflow 90®	90	65	0.965	>30 mm	Weiland et al. (1988)
Plasdek C. 10®	modular blocks	148	0.96	>30 mm	Weiland and Wulfert (1986)
Flocor R®	corrugated rings	320	0.97	>30 mm	
Ceramic Raschig rings	25	190	0.74		Andrews (1988)

[a] Equivalent pore diameter.

Tab. 7.6 b Support material for anaerobic fluidized-bed systems (Jördening and Buchholz, 1999).

Support	*Diameter [mm]*	*Density [kg m^{-3}]*	*Surface [m^2 m^{-3}]*	*Porosity*	*Upflow Velocity [m h^{-1}]*	*Biomass [kg m^{-3}]*	*Reference*
Sand	0.5	2540	7100[a]	0.41	30	4–20	Anderson et al. (1990)
Sepiolite	0.5	1980	20300[b]			32	Balaguer et al. (1992)
GAC	0.6					34	Chen et al. (1995)
Biomass granules					2–6.5		Franklin et al. (1992)
Sand	0.1–0.3	2600			16	40	Heijnen (1985)
Biolite	0.3–0.5	2000			5–10	30–90	Ehlinger (1994), Holst et al. (1997)
Pumice	0.25–0.5	1950	$2.2 \cdot 10^6$	0.85	10		Jördening (1996), Jördening and Küster (1997)

[a] Calculated from the given data with the assumption of total sphericity.
[b] Calculated with data given in Sanchez et al. (1994).

7.5.2 Applications

Microorganisms immobilized by adhesion are applied mainly in areas of environmental technology. Many fluidized-bed reactors with viable anaerobic mixed cultures either with carrier or as granular bacterial aggregates are utilized in the food industry for the purification of high-load wastewater. Likewise, trickle bed systems to immobilize bacteria performing nitrification for the elimination of nitrogen-containing substances, as well as biofilters, are used for exhaust gas purification, mainly for the elimination of foul-smelling organic pollutants, in a large number of different production facilities.

The advantages of systems with immobilized microorganisms are a high density of organisms and thus a high biocatalyst activity, when compared to free suspended microorganisms. Hence, high-capacity reactors may have a much smaller volume and lower costs. Difficult or cost-intensive recycling of slow-growing microorganisms (e.g., anaerobic) is not necessary due to immobilization. The cultures applied are capable of adapting to specific requirements (largely varying different substrates and substrate concentrations), and are often able to convert compounds at very low concentrations.

7.5.2.1 Anaerobic Wastewater Treatment

There are considerable advantages of anaerobic wastewater treatment as compared to aerobic treatment. Typically, the specific productivity of an anaerobic system is much higher, and the engineering is simple. Whereas much energy is needed for aeration in aerobic wastewater treatment plants, anaerobic treatment produces energy in the form of usable biogas. On the other hand, this also means that the growth of anaerobic bacteria is slow, and for this reason in particular the start-up of an anaerobic plant is slow. An overview of the most important reaction pathways of anaerobic degradation of biomass (polysaccharides, proteins, lipids) is presented in Figure 7.11. The so-called "acidifying bacteria" convert carbohydrates, for example, to fatty acids, alcohols, carbon dioxide and hydrogen, and this in general is a rather rapid reaction. These primary fermentation products are further converted by acetogenic bacteria to yield acetate, formiate, CO_2, and H_2 that ultimately are used by methanogenic bacteria to form biogas, mainly CH_4 and CO_2 (McInerny, 1999).

The kinetics of this reaction network are complex. For the growth of new biomass, the essential step during the slow start-up of an anaerobic reactor, the concentration of biomass x can accumulate with a rate μ corresponding to exponential growth

$$\mu = \frac{1}{x} \cdot \frac{dx}{dt} \tag{7.1}$$

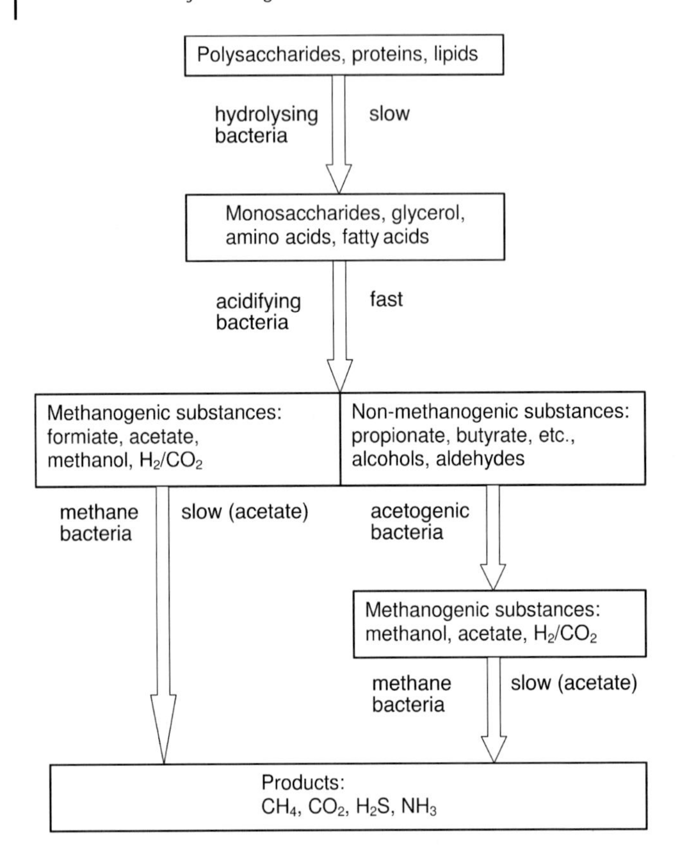

Fig. 7.11 Scheme of anaerobic degradation reactions.

Integration gives the concentration of biomass x at time t.

$$x = x_0 \cdot e^{\mu \cdot t} \tag{7.2}$$

The growth rate μ depends on the substrate concentration [S] in a saturation type curve – that is, μ depends on [S] as follows:

$$\mu = \mu_{max} \cdot \frac{[S]}{K_m + [S]} \tag{7.3}$$

(K_m corresponds to the substrate concentration where the growth rate reaches half of the maximal value.)

From Eqs. (7.1) and (7.3) one obtains for the growth of microorganisms (for start-up with high substrate concentration, i.e., $S \gg K_m$, as is used conventionally):

$$\frac{dx}{dt} = \mu_{max} \cdot \frac{x \cdot [S]}{K_m + [S]} \sim \mu_{max} \cdot x, \quad \text{if} \quad S \gg K_m \tag{7.4}$$

For substrate conversion, the simple equation of Monod, or Michaelis–Menten, respectively can often be used, corresponding to the assumption of one rate-limiting step (in most cases methane formation):

$$-\frac{d[S]}{dt} = \frac{V_{max} \cdot [S]}{K_m + [S]} \quad \text{(approximation)} \tag{7.5}$$

The stationary effectiveness factor η for a system with microorganisms immobilized by adhesion is

$$\eta = \frac{v_{Biofilm}}{v_{Suspension}} \tag{7.6}$$

and can be calculated or estimated according to Chapter 8, Section 8.4.2. In most cases this value is near 1, but at a high film thickness with an active biomass η may be significantly lower. In such cases inactivation and detachment of the microorganisms from the carrier surface may occur due to insufficient nutrient supply. Kinetics and mass transfer in anaerobic granular aggregates of microorganisms (sludge) has been investigated by Gonzales-Gil et al. (2001). It was found that external mass transfer can be neglected, whereas an increase in apparent K_m was observed due to intraparticle mass transfer. A model was used to describe reaction diffusion kinetics.

Integration of Eq. (7.5) gives:

$$[S]_0 - [S] + K_m \cdot \ln\frac{[S]_0}{[S]} = V_{max} \cdot t \tag{7.7}$$

With this equation an estimation of the conversion ($[S]_o - [S]$) or the exit concentration [S], respectively, can be obtained which depend on the concentration at the reactor inlet $[S]_o$, the residence time $\tau = t$ and the values of V_{max} and K_m determined empirically. For this purpose it is essential to control the pH (≥ 7) and the temperature at constant values. It must furthermore be considered that the (apparent) values of V_{max} and K_m may change with stationary conditions (e.g., the concentration or the substrate composition).

As analytical data for the wastewater charge COD, values are conveniently taken. If one acid (mostly acetic acid) is dominating, the COD may serve as an approximation for the substrate concentration [S] (see Eq. 7.5) (1 g acetic acid corresponds to 1.07 g COD). This simple approximation however can only be used if the stationary conditions are well defined and constant.

In cases where this approach cannot be used, the coupled differential equations corresponding to the most important substrates and intermediary products can be established based on simple Michaelis–Menten kinetics (Eq. 7.5) and solved by numerical methods. As an example, the differential equations for the degradation of butyric acid (But) (a rather common case) are given below. In this case, *iso*-butyric acid (i-But) is formed by an equilibrium reaction from butyric acid. Acetic acid (Ace) is formed from butyric acid only (Eqs. 7.8 to 7.10) (Mösche, 1998). These equations can be solved numerically using the Runge–Kutta method and an algorithm for the optimization of the parameters with minimum error.

$$\frac{d[\text{But}]}{dt} = -\left\{\frac{V_{\text{max, But, i-But}}\left(1 - \frac{[\text{i-But}]}{[\text{But}]\,K_{\text{eq}}}\right)}{1 + \frac{K_{\text{m, But, i-But}}}{[\text{But}]}\cdot\left(1 + \frac{[\text{i-But}]}{K_{\text{m, i-But, But}}}\right)}\right\} - \left(\frac{V_{\text{max, But, Ace}}\,[\text{But}]}{K_{\text{m,But, Ace}} + [\text{But}]}\right) \tag{7.8}$$

$$\frac{d[\text{i-But}]}{dt} = \left\{\frac{V_{\text{max, But, i-But}}\left(1 - \frac{[\text{i-But}]}{[\text{But}]\,K_{\text{eq}}}\right)}{1 + \frac{K_{\text{m, But, i-But}}}{[\text{But}]}\cdot\left(1 + \frac{[\text{i-But}]}{K_{\text{m, i-But, But}}}\right)}\right\} \tag{7.9}$$

$$\frac{d[\text{Ace}]}{dt} = \left(\frac{V_{\text{max, But, Ace}}\,[\text{But}]}{K_{\text{m, But, Ace}} + [\text{But}]}\right) - \left(\frac{V_{\text{max, Ace}}\,[\text{Ace}]}{K_{\text{m, Ace}} + [\text{Ace}]}\right) \tag{7.10}$$

K_{eq} is the equilibrium constant of the isomerization of But to i-But; Subscripts: But,i-But and i-But,But refer to the isomerization reaction of butyric and iso-butyric acid, respectively; But,Ace refers to the conversion of butyric to acetic acid; Ace refers to acetate conversion. The first term in Eqs. (7.8) and (7.9) can be obtained from Eq. 2.37 (p. 67) for the rate of a reversible one-substrate reaction, where the ratio of (V_{max}/K_m) of the forward to the reverse reaction equals K_{eq}.

These kinetic models do not describe external deviations such as the instationary occurrence or accumulation of lactic and/or propionic acid – a common phenomenon in wastewater from the food industries. The conversion of these may become a serious bottleneck (if the mixed culture is not adapted), with a drop in pH and inactivation of the methanogens. Typical data for industrial substrates and kinetics are listed in Table 7.7.

Table 7.7 Data for common industrial substrates (wastewater from the sugar industry) and kinetics of immobilized anaerobic systems (laboratory data) (Jördening et al., 1991; Buchholz et al., 1992).

Wastewater composition	***[kg m^{-3}]***	***Kinetic data***	
Organic substances		Active biomass	
COD (total)	5–15	(determined as odm)[a]	15–80 g L^{-1}
Acetate	1–2	max. growth rate μ_{max}	0.03–0.08 day^{-1}
Propionate	0.5–1.5	biofilm thickness	0.02–0.2 mm
Butyrate	1–2	V_{max} for acidification	up to 100 g (COD) L^{-1} day^{-1}
(D-, L-) Lactate	0.2–2.5	V_{max} for methanogenesis (porous glass as carrier)	30–150 g (COD) L^{-1} day^{-1}
		total degradation rate	40–80 g (COD) L^{-1} day^{-1}

[a] odm = organic dry matter.

Growth kinetics for anaerobic bacteria – including slow-growing methanogenic species – are in the range of 0.03 to 0.08 d^{-1}, corresponding to doubling rates (where the active biomass grows to twice the original mass) of 7 to 20 days (Jördening et al., 1991). For the start-up of such systems therefore, about 5–10 weeks are required when sufficient inoculum is available (about 5 % of the final active biomass is required) and optimal strategies are applied. For campaign-dependent systems (operating during certain seasons), which are common in many food industries, it means that a restart at the beginning of a campaign requires 1 to 2 weeks to attain full capacity (assuming a common value of 50 % activity loss during 6–8 months of intercampaign). Further kinetic data for UASB and EGSB systems have been published by Gonzales-Gil et al. (2001).

The anaerobic treatment of wastewaters includes acid formation and methane formation, which differ significantly in terms of nutritional needs, growth kinetics and sensitivity with regard to the environmental conditions, notably pH (Fig. 7.11) (Demirel and Yenigün, 2002). Hence, these two steps are preferentially physically separated, and two-stage systems are considered to be superior, as their performance in terms of stability and space–time-yield will be superior to that of one-stage systems. The first step does not require immobilized bacteria, as those involved exhibit rapid growth, whereas the second step requires immobilized systems for high performance. The load of reactors for the second step with volatile fatty acids (in a system with a separate acidification reactor) may be higher by a factor of 4 to 5, compared to a one-step system and feeding with complex substrates (not acidified completely) (Henze and Harremoes, 1983).

The biocatalyst comprises granular aggregates of microorganisms, or the carrier with adhering microorganisms, extracellular secondary products (such as polysaccharides) and eventually precipitated inorganic compounds (such as $CaCO_3$). Carriers for fixed and fluidized-bed reactors have been mentioned previously (see Tables 7.6 a and b). The carrier (e.g., porous silica, pumice, plastics) should exhibit a large, rough, and – preferentially – a macroporous surface in order to accommodate large numbers of bacteria and provide protection against shear forces. The microorganisms of the second, methanogenic step comprise complex mixed cultures (including propionate-, butyrate-, lactate-transforming and methanogenic bacteria) which may secrete extracellular secondary metabolites such as polysaccharides and proteins to form a secondary matrix that is favorable for adhesion (see Fig. 7.9 and 7.12). The complex reaction sequences (Eqs. 7.8–7.10) can proceed in a single particle as methanogenic and the other species mentioned which form acetate, formiate and hydrogen grow symbiotically, and utilize the products formed immediately, including interspecies hydrogen transfer (McInerny, 1999).

One very broad and successful use of high-performance anaerobic treatment in fixed- and fluidized-bed reactors relates to wastewater from industries based on agricultural and forestry products, with typically high concentrations of organic substrates that are readily degraded by anaerobic bacteria. These offer the advantage of high-load systems, require much less volume and space, and hence less investment as compared to conventional systems. Furthermore, these systems tend to operate more stable under transient conditions, such as fluctuations of substrates and pH.

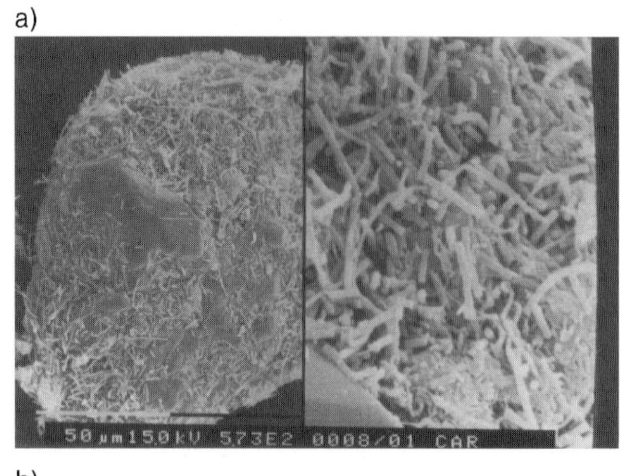

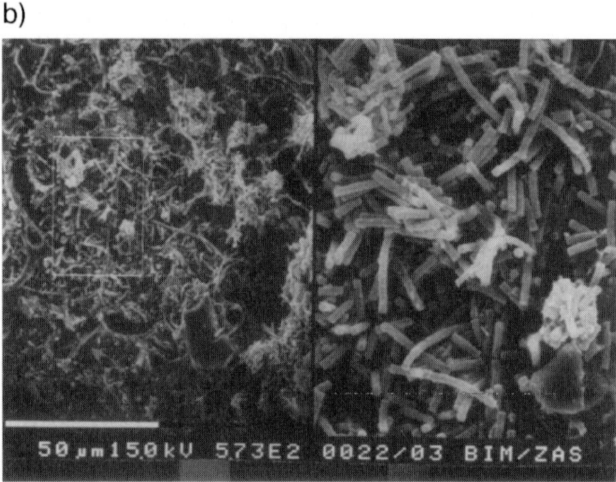

Fig. 7.12 Microorganisms of an anaerobic mixed culture adhering to (a) sand and (b) pumice particles at two different magnifications (scale bar = 50 µm).

In fact, a large number of different types of plants with immobilized anaerobic bacteria are currently on stream, including over 50 fixed-bed and about 20 fluidized-bed reactors with carriers. Dominant, however, are about 200 of the expanded granular sludge bed (EGSB) and about 700 upflow anaerobic sludge blanket (UASB) systems which operate without a carrier, but rather utilize granular bacteria aggregates (Lettinga et al., 1999; Macarie, 2001; Frankin, 2001).

Substrates typically result from raw material washing procedures, blanching, extraction, fermentation, or enzymatic processing. Original substrates are in most cases carbohydrates such as sugar, starch, cellulose, and hemicellulose, proteins and fats, all of which readily undergo bacterial degradation to fatty acids, mainly acetic, propionic, and butyric and lactic acids. The majority of installations are in the potato, starch and sugar industries, in fruit, vegetable and meat processing, in cheese, yeast, alcohol, citric acid, pectin manufacturing, and in the paper and pulp industries. The concentrations of substrates are typically in the range of 5 to 50 kg (COD) m^{-3}, and these are diluted by recirculation (loop reactor for control of concentration, pH, and to provide suspension of particles; see Chapter 9, Section 9.1) down to less than 2 kg (COD) m^{-3}.

The most widely used system worldwide is the UASB-system (Lettinga et al., 1999; McCarty, 2001). The incoming wastewater is distributed equally over the cross-section by a system of tubes. At low superficial upflow velocities (1–2 m h^{-1}) the wastewater flows vertically through a bed of granular particles of aggregated bacteria (sludge bed). In the upper part of the reactor, a three-phase separation unit is integrated, which avoids disintegration of the sludge pellets. A disadvantage of this system is a lack of information with regard to the formation of granular particles which are essential for success of the system. A further development of the system resulted in EGSB-plants with a significantly higher loading potential. The high liquid upflow velocities require specific separators (screens or modified internal lamella separators) (van Lier et al., 2001; Lettinga et al., 1999).

Typical loading rates are in the range of 50 to 90 kg COD m^{-3} d^{-1} for common wastewater from agricultural and food processes with a high conversion (over 80 % of degradable COD) at laboratory and pilot scales (Jördening and Buchholz, 1999). Despite these favorable results on the laboratory scale, the data reported for the performance on a technical scale are in a significantly lower range of 8 to 30 kg m^{-3} d^{-1}, as shown in Table 7.8 for fixed-bed and Table 7.9 for fluidized-bed reactors (see for example, Jördening and Buchholz, 1999; Macarie, 2001; Jördening and Buchholz, 2004). The loading rates for EGSB plants is over 20 kg m^{-3} d^{-1} COD, and for UASB about 10 kg COD m^{-3} d^{-1} (Table 7.9) (Lettinga et al., 1999; Frankin, 2001).

Anaerobically treated wastewater usually cannot be released directly into the environment. The COD-concentration in the effluent is generally not as low as can be

Table 7.8 Data of industrial-scale anaerobic fixed-bed systems.

Wastewater	*Reactor size [m³]*	*Support material*	*Load [kg m⁻³ d⁻¹]*	*Removal [%]*	*Reference*
Meat processing	22[a)]	Porous glass (Siran®)	10–50	up to 80	Breitenbücher (1994)
Dairy	362	Plastics (Flocor® and cloisonyle)	10–12	70–80	Austermann-Haun et al. (1993)
Sugar	1400	Plastic rings (Flocor®)	13		Weiland et al. (1988)
Potato processing	660	Pallrings (100 mm)	3–6		Weiland et al. (1988)
Starch	4300[b)]	Lava	25	>70	Schraewer (1988)
Distillery	13000	Modular plastic blocks	8–12		Weiland et al. (1988)
Chemistry	1900	Plastics	16–20	90	Henry and Varaldo (1988)

a) Two-stage system with acidification (40 m^3) as a first stage.
b) Two-stage system with acidification reactor of 1000 m^3 as the first stage.

Table 7.9 Data for technical-scale anaerobic fluidized bed reactors, UASB and EGSB systems.

Waste[a]	*Company*	*Support*	*Reactor Vol. [m^3]*	*Ratio*[b] *[–]*	*Load [kg m^{-3} d^{-1}]*	*Concen-tration [kg COD m^{-3}]*	*HRT*[c] *[h]*	*Removal [%]*	*Reference*
					Fluidized bed				
Soy bean	Dorr–Oliver	sand	304	2.0	14–21	0.8–10		75–80	Sutton et al. (1982)
Several	Degremont	biolite	210–480		16–21	3.8–5	0.25	50–60	Ehlinger (1994)
Yeast and pharmaceuticals	Gist-Brocades[d]	sand	400	4.4	8–30	1.9–4	0.14–0.41	95–98	Franklin et al. (1992)
Sugar beet	BMA	pumice	500	5.0	20–30	1–5	1.5	90–95	Jördening (1996)
					UASB				
Aspartame	Nutrasweet Co.		2×600		7.8	22			Macarie (2001)
PET[e]	Eastman Chemical		144		12	12		90–95	
Nylon	Rhone Poulenc[f]		990		8	16		80	
					EGSB				
FormAldehyde	Caldic Europort		275		17	40		98	Macarie (2001)
DMT[e]	Kosa		550		13	34			
DMT and PET[e]	Sasa		2×1000		13	6.5			

a) Source of waste water.
b) Height/diameter.
c) HRT: Hydraulic retention time.
d) Currently DSM (NL).
e) PET: polyethylene terephthalate; DMT: dimethylterephthalate.
f) Currently Aventis.

achieved with aerobic treatment, and this often leads to an odorous product. For this reason, high-loaded wastewaters are often treated in a combination of anaerobic and aerobic processes. The second (aerobic) step is then, if necessary, constructed also for nitrogen elimination.

7.5.2.2 **Nitrogen Elimination (Nitrification and Denitrification)**

Nitrogen elimination from wastewater is a standard requirement, as limiting values by authorities for total nitrogen-containing compounds (notably ammonia) require this step to be taken. It is usually a subsequent process after anaerobic degradation of organic compounds. During the anaerobic degradation of nitrogen-containing compounds, nitrogen is converted to ammonia. Subsequent nitrification and denitrification converts it to nitrogen (N_2) and, to a lesser degree to N_2O, which should be avoided as far as possible. Here, only a short summary of the reaction and process steps is given, as general and basic aspects were dealt with in the preceding section (for details, see Dorias et al., 1999; Kayser, 2004).

Two groups of bacteria catalyze the oxidation of ammonia to nitrate; the ammonium-oxidizing group (e.g., *Nitrosomonas* sp.) form nitrite, which subsequently is oxidized to nitrate by nitrite-oxidizing bacteria (e.g., *Nitrobacter* sp.):

$$NH_4^+ + 1.5\ O_2 \xrightarrow{\textit{Nitrosomonas}} NO_2^- + 2\ H^+ + H_2O \quad +275\ \text{kJ mol}^{-1} \tag{7.11}$$

$$NO_2^- + 0.5\ O_2 \xrightarrow{\textit{Nitrobacter}} NO_3^- \quad +75\ \text{kJ mol}^{-1} \tag{7.12}$$

Sum of the two steps:

$$NH_4^+ + 2\ O_2 \longrightarrow NO_3^- + 2\ H^+ + H_2O \quad +350\ \text{kJ mol}^{-1} \tag{7.13}$$

Nitrate subsequently is reduced under anoxic conditions by denitrifying bacteria via nitrite to yield H_2O and N_2. The denitrification with acetic acid as a cosubstrate proceeds as follows:

$$8\ NO_3^- + 5\ CH_3COOH \longrightarrow 4\ N_2 + 10\ CO_2 + 6\ H_2O + 8\ OH^- \tag{7.14}$$

Kinetic data for nitrification and denitrification have been published by Verstraete and van Vaerenberg (1986), and by Dorias et al. (1999).

Technical systems with immobilized bacteria in mixed cultures are well established, and two main systems are used: (1) trickling filters, and (2) rotating biological contactors. Conditions for high process efficiency and general guidelines can be found in Dorias et al. (1999) and EPA (1993). In trickling filters, the main portion is the inserted packing material, which may be either rocks, lava, or plastic material. It must allow for both free flow of the wastewater and free air flow (for aerobically operated filters), and the removal of biological surplus sludge with the water flow. Therefore, rocks or slags with a particle size of 40–80 mm and 80–150 mm, respectively, are used. A large variety of plastic media has been specifically designed with a specific surface in the range of 200 $m^2\ m^{-3}$ and a void space of approximately 95 %. Three different process methods are applied: simultaneous denitrification in the trickling filter with nitrate return; pre-denitrification in an anoxic zone; or post-denitrification, which may operate as trickling filters. Rotating biological contactors with submerged drums, specially equipped with rotating discs made of plastic material placed halfway into the water, are used mainly in smaller wastewater treatment plants.

7.5.2.3 **Exhaust Gas Purification**

Immobilized mixed cultures are frequently used in exhaust gas treatments, as a wide variety of organic substances (Table 7.10) and even some inorganic odor compounds can be eliminated from the waste gas (VDI 1991, 1996, 2002). Microbiological and engineering fundamentals have been summarized in Klein and Winter (2000). This principle has found successful application in the food industries, including animal-, meat-, fish-, yeast-, and aroma production, as well as in the chemical industries, with diverse examples such as the degradation of hydrocarbons (aliphatic and aromatic), phenols, sulfides, amines, amides, and even halogenated compounds, in foundries and for spray cabins in automobile production (Table 7.10). The elimination of acrylonitrile and styrene mixtures has been described recently (Lu et al., 2002).

Reactor types with immobilized mixed cultures of microorganisms include the bioscrubber or biofilter, and trickling filter. Examples of their application have been summarized by Fischer (2000) and Plaggemeier and Lämmerzahl (2000), and in VDI (1989, 2002). The advantages of biological exhaust gas treatment are first, operation at low temperature, and second, the efficiency of the elimination of toxic and odorous substances at low concentrations in the waste gas stream, at low cost.

The conditions for efficient treatment are good mass transfer – that is, the sufficiently rapid transfer of the components from the gas into the fluid phase – and their biological degradability. If these conditions are satisfied, then mixed cultures can adapt to the substrate if an inoculum with a sufficiently broad bacterial consortium is present. Important parameters are the concentration of the components, pH and buffer capacity of the fluid phase, temperature, available surface, and the residence time distribution of the phases. Substances with toxic properties towards microorganisms can inhibit the system, however.

Table 7.10 Examples for groups of readily degradable substances in exhaust gas (VDI, 1991, 1996).

Group of substances	*Examples*
Aliphatic hydrocarbons	hexane, ethylene
Aromatic hydrocarbons	toluene, xylene
Alcohols	methanol, butanol
Ethers	tetrahydrofuran
Aldehydes	formaldehyde, acetic aldehyde
Carboxylic acids	butyric acid
Amines	trimethylamine

The phenomena governing the efficiency of a biofilter or trickling filter are mass transfer, and the kinetics of degradation. The transfer of the substances to be degraded from the gas phase to the fluid phase (biofilm) depends on the solubility and the equilibria for both phases, and the kinetics of mass transfer. The concentrations in equilibrium (c^*) of gas (g)- and fluid (l)- phases are correlated by the Henry coefficient, He:

$$c^*_{\mathrm{g}} = \mathrm{He} \cdot c_{\mathrm{l}} \tag{7.15}$$

The rate of mass transfer can be obtained from the mass balance for the gas phase:

$$Q_{\mathrm{g}}\,(c_{\mathrm{go}} - c_{\mathrm{ge}})\,V_{\mathrm{R}}^{-1} = k_{\mathrm{L}}\,a\,\Delta c \tag{7.16}$$

where Q_g is the volumetric gas flow rate ($m^3\,h^{-1}$), the indices o and e signify ingoing and outcoming flow, and V_R is the reactor volume; k_L is the mean mass transfer coefficient and a the interface area; Δc is the logarithmic mean of the concentration difference of both phases.

The mass transfer in the fluid phase is rate-limiting in general. The mass transfer coefficient k_L can be calculated approximately using the correlation of particle size d_p and effective diffusion coefficient D'_s, or the corresponding correlation of Sherwood (Sh), Reynolds (Re), and Schmidt (Sc) numbers (for further correlations of dimensionless numbers, see Chapter 9, Section 9.3):

$$\mathrm{Sh} = k_{\mathrm{L}}\,d_{\mathrm{p}}\,(D'_{\mathrm{s}})^{-1} = 0.054\,\mathrm{Re}_{\mathrm{l}}^{0.6}\,\mathrm{Re}_{\mathrm{g}}^{0.85}\,\mathrm{Sc}^{0.33} \tag{7.17}$$

(with indices l and g for the liquid and gas phases, respectively). The surface a is difficult to determine, but can be estimated from the carrier surface. For compounds which are easily degraded the mass transfer is considered to be rate-limiting. In contrast, for compounds which are degraded slowly the rate of the biological reaction will be rate-limiting; often, the reaction rate then follows Michaelis–Menten kinetics (after an initial phase):

$$v = k\,x_{\mathrm{b}}\,[\mathrm{S}]_{\mathrm{b}}\,(K_{\mathrm{m}} + [\mathrm{S}]_{\mathrm{b}})^{-1} \tag{7.18}$$

(with concentration of microorganisms x_b and substrate $[S]_b$ in the biofilm).

The degradation routes are more or less complex, depending on the compounds and the microorganisms, which are mostly bacterial and fungal species. Part of the substrate (up to 50 %) is utilized for the growth of biomass, and this must be removed from the system. The oxidative degradation in general yields CO_2 and water; for example, in the case of propionic aldehyde:

$$C_2H_5CHO + 4\,O_2 \rightarrow 3\,CO_2 + 3\,H_2O$$

Inorganic odorous substances can also be transformed in part, and thus be removed from the waste gas. Ammonia is oxidized after absorption to nitrite and nitrate, while hydrogen sulfide can be oxidized by some bacterial species to sulfuric acid.

The characteristics of the waste gas stream are relative humidity, temperature, waste gas flow rate, chemical pollutant identity and concentration, and odor concentration. Essential parameters are the relative humidity (>95 %) and the temperature of the gas (<60 °C), surface loading rate (m^3 m^{-2} h^{-1}), the mass loading rate (g m^{-3} h^{-1}), the elimination capacity (g m^{-3} h^{-1}) and the removal efficiency (%). For odor elimination, the olfactrometric measurement is essential (VDI 1991, 2002; Waweru et al., 2000).

Reactors used frequently are the biofilter (this is the most common, with many hundreds of installations) and the trickling filter (trickle-bed air biofilter), which achieves higher specific rates of treatment. Biofilters are fixed-bed reactors where the waste gas flows through a bed of carrier material (mostly natural fiber material) with bacterial mixed cultures adhering to the surface.

Trickling filters are another standard method for the purification of waste gas, and are capable of degrading more contaminants per unit volume, as compared to biofilters, and of treating up to 100000 m^3 h^{-1}. In a trickle-bed reactor, an aqueous phase is continuously circulated through a bed of inert material. The packing can consist of inert bulk material (Raschig rings, saddles, etc.) or of structured packings. The material can be plastics, such as polyethylene, poplypropylene, or polyurethane foam.

The surface should be appropriate for adhesion of the biofilm. The waste air flows either concurrently or countercurrently with a recirculated liquid through the packing (Fig. 7.13). A uniform distribution of the fluid over the cross-section is important for sufficient mass transfer. The pressure drop must be low (10 to 50 Pa m^{-1} at a flow

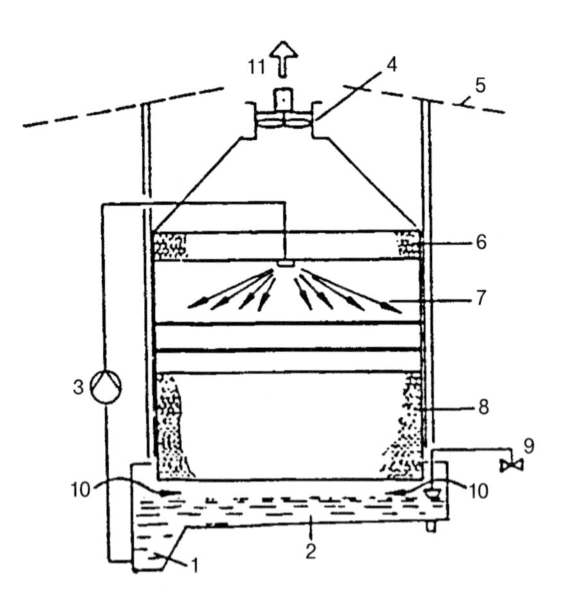

Fig. 7.13 Scheme of a biological trickling filter, with water reservoir (1), collecting trough (2), pump (3), ventilator (4), root (5), packings for droplet removal (6), water distribution system (7), bed with packings (8), outflow control (9), waste gas inflow (10), purified gas (11).

Table 7.11 Ranges of pollutant concentrations and operating parameters for biofilters and trickling bed reactors (VDI, 1996; Waweru et al., 2000).

	Biofilter	***Trickling bed reactor***
Pollutant concentration [g m^{-3}]	<1	<0.5
Henry coefficient [–]	<10	<1
Surface loading rate [m^3 m^{-2} h^{-1}]	50–200	100–1000
Mass loading rate [g m^{-3} h^{-1}]	10–160	<500
Empty bed contact time [s]	15–60	1–10[a] ; 30–60[b]
Volumetric loading rate [m^3 m^{-3} h^{-1}]	100–200	
Elimination capacity [g m^{-3} h^{-1}]	10–160	
Removal efficiency [%]	95–99	

[a] For easy degradable compounds.
[b] For slowly degradable compounds.

rate of 200 m h^{-1}), and it must be ensured that excess sludge can be removed. The volume of such reactors may be several hundred of cubic meters, and the height of the packed bed in the range of 2 to 10 m. Ranges of pollutant concentrations and operating parameters are summarized in Table 7.11 (VDI 1996, 2002; Waweru et al., 2000).

7.6
Outlook: Designed Cells

Biotransformation by designed recombinant whole-cell systems represents a field of high actual interest, and a range of examples has been developed successfully for several multi-step reactions. These are currently mostly used as suspended cells. The potential of using such systems continuously as an immobilized biocatalyst remains an open question. One bottle-neck of this seems to be catalyst stability, notably with respect to cofactor regeneration and resynthesis. Metabolic engineering is a key tool in this context in order to identify limitations (e.g., in redox equivalents) and for improving yields by optimized metabolic fluxes (Edwards et al., 1999; Segre et al., 2002).

A whole-cell biotransformation system for the conversion of D-fructose to D-mannitol was developed in *E. coli* by constructing a recombinant oxidation/reduction cycle (Kaup et al., 2003). First, the *mdh* gene, encoding mannitol dehydrogenase of *Leuconostoc pseudomesenteroides* ATCC 12291 (MDH), was expressed. To provide a source of reduction equivalents needed for D-fructose reduction, the *fdh* gene from *Mycobacterium vaccae* N10 (FDH), encoding formate dehydrogenase, was functionally co-expressed. FDH generates the NADH used for D-fructose reduction by dehydrogenation of formate to carbon dioxide. The introduction of a further gene, encoding the glucose facilitator protein of *Zymomonas mobilis*, allowed the cells efficiently to take up D-fructose (Fig. 7.14). Biotransformations conducted under pH control by formic acid addition yielded D-mannitol at a concentration of 362 mM within 8 h.

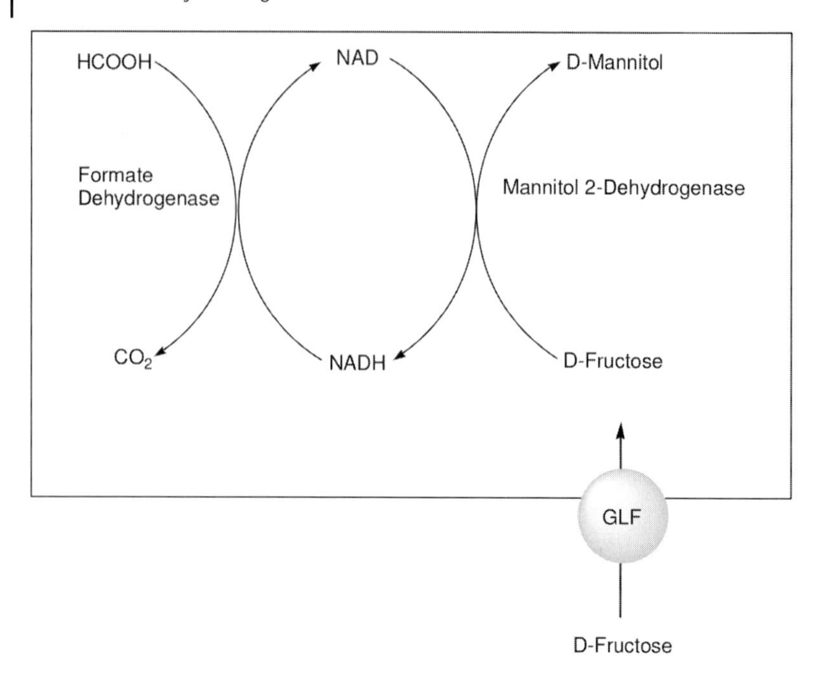

Fig. 7.14 Biocatalyst for D-mannitol formation from fructose in a whole-cell biotransformation (GLF, glucose facilitator) (from Kaup et al., 2003).

The yield of mannitol (relative to fructose) was 84 mol%. An extension of this system might include the integration of glucose isomerase so that glucose can be used as a substrate (Kaup et al., 2003).

$$\text{D-fructose} + \text{NADH} \xrightarrow{\text{MDH}} \text{D-mannitol} + \text{NAD}^+$$

$$\text{HCOOH} + \text{NAD}^+ \xrightarrow[\text{FDH}]{} \text{CO}_2 + \text{NADH}$$

Several whole-cell processes are currently used on the industrial scale (Schmid et al., 2002). Thus, (*R*)-2-(4′-hydroxyphenoxy) propionic acid is manufactured at BASF by the oxidation of (*R*)-2-phenoxypropionic acid (Fig. 7.15) (for basic aspects and details of reactions catalyzed by oxygenases, see Section 3.2.1.2).

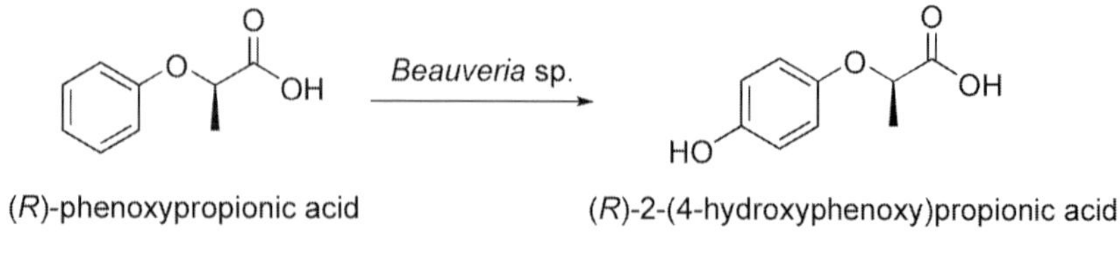

Beauveria sp.

(*R*)-phenoxypropionic acid → (*R*)-2-(4-hydroxyphenoxy)propionic acid

Fig. 7.15 Whole-cell biotransformation: Oxidation of (*R*)-2-phenoxypropionic acid to (*R*)-2-(4'-hydroxyphenoxy) propionic acid (BASF) (from Schmid et al., 2002).

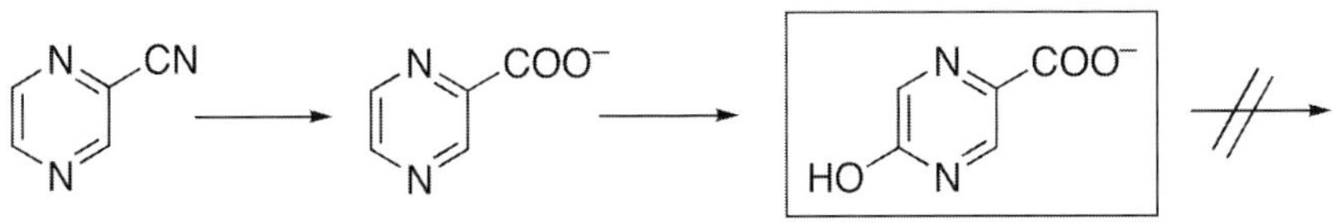

Fig. 7.16 Whole-cell biotransformation for the production of 5-hydroxypyrazine carboxylic acid (Lonza) (from Schmid et al., 2001).

Another process combines the sequence of two enzymes, nitrile hydratase and amidase, catalyzing the formation of (*S*)-2-phenylpropionic acid from racemic 2-phenylpropionitrile (Lonza, 2004; Schmid et al., 2002) (for details of reactions catalyzed by nitrilases and nitrile hydratases, see Section 3.2.2.5). A further step is integrated in a process where the nitrile group of 2-cyanopyrazine is hydrolyzed to pyrazine carboxylic acid, followed by the regioselective hydroxylation to 5-hydroxypyrazine carboxylic acid by the same bacterial cells (Fig. 7.16) (Schmid et al., 2001). An-

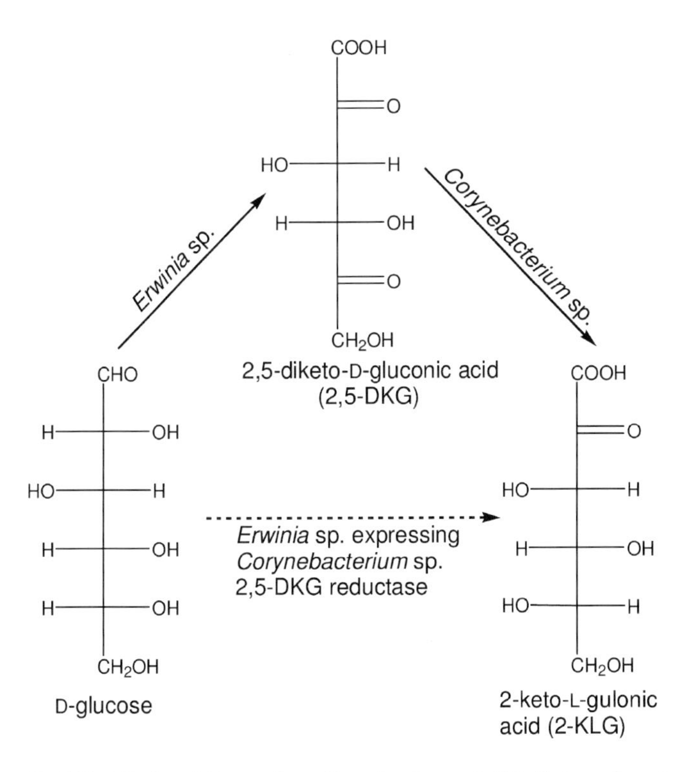

Fig. 7.17 Whole-cell biotransformation of D-glucose to 2,5-diketo-D-gluconic acid (DKG) and 2-keto-L-gulonic acid (KLG) by a recombinant strain of *Erwinia* sp. (in solution, glucose and sugar acids are present in equilibrium as cyclic pyranose and lactones, respectively) (from Grindley et al., 1988).

other process utilizes the xylene degradation pathway for the manufacture of 5-hydroxypyrazine carboxylic acid from 2,5-dimethylpyrazine (Schmid et al., 2001).

Much effort has been devoted to new routes for L-ascorbate synthesis, and particularly by the construction of designed recombinant microorganisms. Thus, an efficient system was developed with a recombinant strain of *Erwinia* sp. which oxidizes D-glucose to 2,5-diketo-D-gluconic acid. Moreover, this strain encodes a recombinant reductase from *Corynebacterium* sp. producing 2-keto-L-gulonic acid, which easily can be rearranged by cyclization to vitamin C (Fig. 7.17). When a culture of the *Erwinia* strain was fed with glucose to a total of 40 g L^{-1}, 49.4 % of the glucose was converted to 2-keto-L-gulonate during a 72-h bioconversion (Grindley et al., 1988). Anderson et al. (1985) succeeded in constructing a similar concept. BASF has realized a different concept on the industrial scale using a mixed culture of two microorganisms. Sorbitol as the substrate is oxidized to L-sorbose followed by oxidation to 2-keto-L-gulonic acid, which then is rearranged to vitamin C; however, details of this pathway have not been published.

Yet another product of commercial interest is indigo, one of the world's largest-selling textile dyes. A biotransformation route has been designed by cloning genes for a multicomponent hydroxylase from *Acinetobacter sp.* in a solvent-resistant *E. coli* mutant encoding a NADH regeneration system. This strain was able to produce indigo from indole in an organic solvent-water two-phase culture system (Doukyu et al., 2003).

Steroid pathway intermediates, formed during sterol catabolism, are widely recognized as pharmaceutically important precursors for drug synthesis. Different approaches have been developed to overcome difficulties due to the limited capabilities of microorganisms in such pathways. The group of Dijkhuizen focused on the bioengineering of molecularly defined mutant strains of *Rhodococcus* species blocked at the level of steroid δ^1-dehydrogenation and steroid 9-α-hydroxylation. A molecular toolbox has been developed including characterization of the steroid catabolic pathway, cloning and expression vectors for *Rhodococcus sp.* for use in genomic library construction, functional complementation and gene expression. The application of this toolbox has allowed rational construction of *Rhodococcus* strains with optimised properties for sterol/steroid bioconversions (van der Geize et al., 2002).

A challenge for the application of several of these approaches is the development of stable systems, which can be used repeatedly or continuously. This implies cofactor regeneration and enzyme resynthesis in the protein turnover cycles of the cells in order to maintain productivity over extended time scales and thus provide economic synthesis of the products.

7.7 Exercises

7.7.1 Entrapment of Yeast in Calcium Alginate for Ethanol Production

Equipment

- Syringe (20 or 100 mL); sieve or filter funnel; thermostated shaker; thermostated column bioreactor with mesh (ca. 10–100 mL).
- Glucose and/or ethanol analysis by enzymatic test, or GC, HPLC, or glucose analyzer.
- Sodium alginate solution: sodium alginate solutions can be sterilized by autoclaving, or by membrane filtration. Prepare well before required. A solution of alginate, typically 2–4 % (w/v), is prepared by dissolving solid sodium alginate in distilled water or a buffer solution appropriate to the enzyme or cells to be immobilized. The dissolving process is slow, and normally requires stirring for up to 5 h. The solution may be stored at 4 °C, but should be prepared freshly each week.
- Calcium chloride ($CaCl_2 \cdot 6H_2O$) solution: prepare liter volumes in concentrations of 0.15 M. The solution may be stored at room temperature.
- Dried baker's yeast *Saccharomyces cerevisiae:* suspend 1.0 g in 5 mL distilled water and mix in a small pestle and mortar until a smooth paste is formed.
- Yeast cell-alginate mixture: mix 1 mL of the yeast paste with 50 mL of alginate solution and stir the mixture thoroughly to ensure complete mixing. The maximum workable concentration of sodium alginate solution is ca. 4–5 % (w/v), depending on the quality of sodium alginate (this is due to high viscosity).
- Buffer: 0.1 M sodium phosphate buffer, pH 7.0.

Immobilization

The set-up for the formation of beads is shown in Figure 7.5. All the immobilization steps by ionotropic gel formation with sodium alginate may be carried out at 4 °C. Transfer the alginate mixture with yeast cells to a dropping device, such as a syringe (100 mL; a 0.55-mm internal diameter (i.d.) needle is appropriate; the number and size of the alginate beads are influenced by the size of the opening through which the alginate is allowed to pass). Place a beaker containing 1 L of $CaCl_2$ solution below the dropping device and magnetically stir the solution to produce a light vortex. Adjust the flow control to allow dropwise flow, from a height of about 10 cm, into the $CaCl_2$ solution. Allow the flow to continue until the desired number of beads have been formed.

Allow a further 20–30 min stirring before collecting the beads using either a Büchner funnel or a large filter funnel. Care must be taken when using suction with Büchner or other funnels to ensure that the beads are maintained in liquid at all times. Wash the beads on the funnel with 1 L of buffer appropriate to the cells (for yeast cells, use distilled water). Store the beads in a minimal amount of distilled water until required (Fraser and Bickerstaff, 1997).

Supplements (titandioxide or fine sand particles) can be added to the alginate solution prior to the gelation process in order to adjust the bead density, sedimentation rate and mechanical strength (Berensmeier et al., 2004).

Dissolution of the beads can be achieved by incubating them at room temperature with 0.1 M sodium phosphate buffer, pH 7.0, for a few hours, depending on the original alginate concentration, until complete.

Test: Conversion of Glucose and Formation of Ethanol

Discontinuous Experiment

Fractions of beads (5–10 g) with different diameters are placed into Erlenmeyer flasks; 50 mL of glucose solution (5 g L^{-1}) are added, and the flasks put into a shaker at 37 °C; concentrations of glucose and/or ethanol are determined at 10- and 20-min intervals during 2 h (conversion expected, ca. 80 %). Concentrations of glucose and/or ethanol are recorded graphically as a function of time, the catalytic activity (g (glucose/ethanol) per g (yeast) per h) is determined from the slope; the activity is compared for biocatalyst beads of different diameter; from these data the efficiency may be estimated (see Chapter 8, Sections 8.3 and 8.4).

The theoretical yield is 2 mol ethanol from 1 mol glucose (or 0.51 g ethanol from 1 g glucose). About 10 % of the glucose is consumed for maintenance metabolism.

Continuous Experiment

Place 100 g biocatalyst (with 10–30 % immobilized yeast) into a column with a sieve to obtain a packed bed. A glucose solution (5 g L^{-1}) is pumped through the bed of immobilized yeast at rates of 0.4 up to 4 L h^{-1} (residence time in the range of 30 to 3 min, calculated for 200 mL bed volume). Between three and five residence times are required until stationary conditions are approached. Take samples for the analysis of glucose and/or ethanol every 10 or 20 min. Calculate the conversion, space–time yield, and catalyst productivity and compare for different flow rate and conversion.

Note: A technical process should operate at near-maximal concentrations, which is about 10–12 % for ethanol, due to toxic effects limiting the productivity of yeast. The glucose concentration therefore would be approximately 200 g L^{-1}. At high conversion – as would be required for economic reasons – product inhibition and inactivation of the biocatalyst must be taken into consideration.

7.7.2
Characterization of an Anaerobic Fluidized-Bed Reactor

The technical properties of a fluidized-bed reactor with two different carriers are to be investigated; the performance of an anaerobic fluidized-bed reactor shall be characterized.

Equipment

Two glass reactors (tubes with a sieve at the bottom, volume 7 L) with recirculation and two pumps (for introducing the substrate and for recirculation), flow meter, conductometer, pH measurement and regulation, equipment for COD- or TOC-measurement, gas chromatograph (GC) or HPLC, or ion chromatograph.

Reagents

Sand (d_p 0.1–0.2), porous glass, or pumice (d_p 0.2–0.4), carrier with adhering anaerobic mixed culture, substrate (acetate, propionate, butyrate, or a mixture of acids).

Pressure Drop in the Fluidized Bed

For a reactor without a carrier, with sand, and with porous glass or pumice, 20 different superficial upflow velocities shall be established, the pressure drop (Δp) and the bed expansion should be measured as a function of the flow rate (Q). The point of beginning expansion is determined from the correlation of Q and Δp.

Performance of an Anaerobic Fluidized-bed Reactor

The kinetic parameters V_{max} and $K_{m\,(apparent)}$ should be determined in a fluidized-bed reactor with adhering anaerobic mixed culture. The carrier material is expanded by ca. 50 %, with an appropriate flow rate. A substrate solution with a concentration of 1 g (COD) is introduced into the reactor. Samples are taken every 30 min, and COD or TOC (Total Organic Carbon) are determined, respectively. The experiment is finished when no more gas is formed.

Anaerobic COD Degradation

The maximal reaction rate and the apparent Michaelis–Menten-constant $K_{m\,(apparent)}$ as well as the transformation of different acids and intermediates shall be investigated by means of the degradation and intermediate formation of organic acids. Measurements are performed by taking samples from a continuous fluidized-bed reactor (carrier: porous glass or pumice with adhering anaerobic mixed culture including acetogenic and methanogenic bacteria; T: 37 °C; pH: 6.8–7.3; substrate: butyric acid at the start of the experiment; samples every 15 min). Concentrations of butyric and acetic acids and, if possible, intermediates such as isobutyric acid, are determined by GC or HPLC. The kinetic parameters are determined according to Michaelis–Menten kinetics, where the reversible isomerization is considered (Eqs. 7.5 to 7.10).

Literature

General Literature

Bickerstaff, G.F. (Ed.), *Methods in Biotechnology, Vol. 1: Immobilization of Enzymes and Cells*, Humana Press Inc., Totowa, NJ, USA, 1997.

Doyle, A., Bryan Griffiths, J. (Eds), *Cell and Tissue Culture: Laboratory Procedures in Biotechnology*, John Wiley & Sons, New York, 1998.

Hauser, H., Wagner, R. (Eds), *Mammalian Cell Biotechnology in Protein Production*, Walter de Gruyter, Berlin, New York, 1997.

Jördening, H.-J., Winter, J., *Environmental Biotechnology*, Wiley-VCH, 2004.

Klein, J., Winter, J. (Vol. Eds), *Environmental Processes III. Biotechnology*, Vol. 11c, Wiley-VCH, Weinheim, 2000.

Marshall, K.C. (Ed.), *Microbial Adhesion and Aggregation*, Springer, Berlin, 1984.

Mosbach, K. (Ed.), *Methods Enzymol.*, Vol. 135, Academic Press, Orlando, USA, 1987.

Wijffels, R.H., Buitelaar, R.M., Bucke, C., Tramper, J., *Immobilized Cells: Basics and Applications*, Elsevier, Amsterdam, 1996.

Wijffels, R.H., *Immobilized Cells*, Springer, Berlin, 2001.

Winter, J. (Vol. Ed.), *Environmental Processes I. Biotechnology*, Vol. 11a, Wiley-VCH, Weinheim, 1999.

References

Aivasidis, A., Wandrey, C., Reaktionstechnische Optimierung und Maßstabsvergrößerung von anaeroben Festbett-Umlauf- und Wirbelschichtreaktoren, *Chem.-Ing.-Tech.*, 1992, *64*, 374–375.

Amory, D.E., Rouxhet, P.G., Surface properties of *Saccharomyces cerevisiae* and *Saccharomyces carlsbergensis*: chemical composition, electrostatic charge and hydrophobicity, *Biochim. Biophys. Acta*, 1988, *938*, 61–70.

Anderson, S., Berman Marks, C., Lazarus, R., et al., Production of 2-keto-L-gulonate, an intermediate in L-ascorbate synthesis, by a genetically modified *Erwinia herbicola*. *Science*, 1985, *230*, 144–149.

Atkinson, B., Mavituna, F., *Biochemical Engineering and Biotechnology Handbook*, 2nd edn., pp. 529–546, 1171–1173, MacMillan Publisher Ltd., New York, 1991.

Austermann-Haun, U., Kunst, S., Saake, M., Seyfried, C. F., Behandlung von Abwässern, in: *Anaerobtechnik* (Böhnke, B., Bischofsberger, W., Seyfried, C. F., Eds), pp. 467–696, Springer-Verlag, Berlin, 1993.

Berensmeier, S., Ergezinger, M., Bohnet, M., Buchholz, K., Design of immobilised dextransucrase for fluidised bed application, *J. Biotechnol.*, 2004, in press.

Bössmann, M., Staudt, C., Neu, T.R., Horn, H., Hempel, D.C., Investigation and modeling of growth, structure and oxygen penetration in particle supported biofilms. *Chem. Eng. Technol.*, 2003, *26*, 219–222.

Breitenbücher, K., Hochleistung durch mehr Biomasse, *UTA Umwelttechnik Aktuell*, 1994, 5, 372-374.

Breitenbücher, K., Schott information 79, 1996.

Brodelius, P., Vandamme, E.J., Immobilized cell systems, in: Rehm, H.J., Reed, G., Kennedy, J.F. (Eds), *Biotechnology*, Vol. 7a, pp. 405–464, VCH Verlagsgesellschaft, Weinheim, 1987.

Buchholz, K., Immobilisierte Enzyme – Kinetik, Wirkungsgrad und Anwendung, *Chem. Ing. Tech.*, 1989, *61*, 611–620.

Buchholz, K., Diekmann, H., Jördening, H.-J., Pellegrini, A., Zellner, G., Anaerobe Reinigung von Abwässern in Fließbettreaktoren; *Chem. Ing. Tech.*, 1992, *64*, 556–558.

Champagne, C.P., Immobilized cell technology in food processing, in: Wijffels, R.H., Buitelaar, R.M., Bucke, C., Tramper, J. (Eds), Immobilized Cells: Basics and Applications, pp. 633–640, Elsevier, Amsterdam, 1996.

Cheetham, P.S.J., Case studies in the application of biocatalysts for the production of (bio)chemicals, in: Straathof, A.J.J., Adlercreutz, P. (Eds), Applied Biocatalysis, pp. 93–152, Harwood Academic Publishers, Amsterdam, 2000.

Chibata, I., Tosa, T., Transformations of organic compounds by immobilized microbial cells, *Adv. Appl. Microbiol.*, 1977, *22*, 1–27.

Chibata, I., Tosa, T., Sato, T., Takata, I., Immobilization of cells in carrageenan, in: Mosbach, K. (Ed.), *Methods in Enzymology*, Vol. 135, pp. 189–198, Academic Press, Orlando, USA, 1987.

Demirel, B., Yenigün, O., Two-phase anaerobic digestion processes: a review, *J. Chem Technol. Biotechnol.* 2002, *77*, 743–755.

Doran, P.M., Bailey, J.E., Effects of Immobilization on Growth, Fermentation Properties, and Macromolecular Composition of

Saccharomyces cerevisiae Attached to Gelatin, *Biotechnol. Bioeng.*, 1986, *28*, 73–87.
Dorias, B., Hauber, G., Baumann, P., Design of nitrification/denitrification in fixed growth reactors, in: Winter, J. (Vol. Ed.), *Biotechnology*, Vol. 11a, pp. 335–348, Wiley-VCH, 1999.
Dörnenburg, H., Knorr, D., Strategies for the improvement of secondary metabolite production in plant cell cultures, *Enzyme Microb. Technol.*, 1995, *17*, 674–684.
Doukyu, N., Toyoda, K., Aono, R., Indigo production by *Escherichia coli* carrying the phenol hydroxylase gene from *Acinetobacter sp.* strain ST-550 in a water-organic solvent two-phase system, *Appl. Microbiol. Biotechnol.*, 2003, *60*, 720–725.
Doyle, A., Bryan Griffiths, J. (Eds), Cell and tissue culture: Laboratory procedures in biotechnology, Chap. 5.8, pp. 262–267: Microcarriers – Basic Techniques. John Wiley & Sons, New York, 1998.
Edwards, J.S., Ibarra, R.U., Palsson, B.O., Metabolic flux balance analysis, in: Papoutsakis, E.T. (Ed.), *Metabolic Engineering*, pp. 13–57, PUBLISHER, 1999.
Ehlinger, F., Anaerobic biological fluidized beds: operating experiences in France, pp. 315–323, *Proc. 7th Int. Symp. Anaerobic Digestion*, Cape Town, South Africa, 1994.
EPA (U.S. Environmental Protection Agency), Manual Nitrogen Control, EPA/625/R-93/010, 1993.
Fischer, K., Biofilters, in: Klein, J., Winter, J. (Vol. Eds), *Environmental Processes III. Biotechnology*, Vol. 11c, pp. 321–332, Wiley-VCH, Weinheim, 2000.
Flitsch, S.L., Watt, G.M., Enzymes in carbohydrate chemistry: formation of glycosidic linkages, *Biotechnology*, 2000, *8b*, 243–274.
Franklin, R.J., Full-scale experiences with anaerobic treatment of industrial wastewater, *Water Sci. Technol.*, 2001, *44*, 1–6.
Franklin, R.J., Koevoets, W.A.A., van Gus, W.M.A., van der Pas, A., Application of the biobed upflow fluidized bed process for anaerobic waste water treatment, *Water Sci. Technol.* 1992, *25*, 373–382.
Fraser, J. E., Bickerstaff, G. F., Entrapment in calcium alginate, in: Bickerstaff, G.F. (Ed.), *Methods in Biotechnology, Vol. 1: Immobilization of Enzymes and Cells*, Humana Press Inc., Totowa, NJ, USA, 1997.
Freeman, A., Lilly, M.D., Effect of processing parameters on the feasibility and operational stability of immobilized viable microbial cells, *Enzyme Microb. Technol.*, 1998, *23*, 335–345.
Fukui, S., Tanaka, A., Immobilized microbial cells, *Annu. Rev. Microbiol.*, 1982, *36*, 145–172.
Fukui, S., Sonomoto, K., Tanaka, A., Entrapment of biocatalysts with photo-crosslinkable resin prepolymers and urethane resin prepolymers, in: Mosbach, K. (Ed.), Methods in Enzymology, Vol. 135, pp. 230–252, Academic Press, Orlando, USA, 1987.
Galazzo, J.L., Shanks, J., Bailey, J.E., Comparison of Suspended and Immobilized Yeast Metabolism using ^{31}P Nuclear Magnetic Resonance Spectroscopy, *Biotechnol. Techn.*, **1987**, *1*, 1–6.
Geize, van der, R., Hessels, G.I., van Gerwen, R., et al., Molecular and functional characteriation of *kshA* and *kshB*, encoding two components of 3-ketosteroid 9-alpha-hydroxylase, a class IA monooxygenase, in *Rhodococcus erythropolis* strain SQ1, *Mol. Microbiol.*, **2002**, *45*, 1007–1018.
Geize, van der R., Hessels, G.I., Dijkhuizen, L., Molecular and functional characterization of the *kst*D2 gene of *Rhodococcus erythropolis* SQ1 encoding a second 3-ketosteroid δ^1-dehydrogenase isoenzyme, *Microbiology*, **2002**, *148*, 3285–3292.
Gonzales-Gil, G., Seghezzo, L., Lettinga, R., Kleerbezem, R., Kinetics and mass transfer in anaerobic sludge, *Biotechnol. Bioeng.*, 2001, *73*, 125–134.
Grimm, L.H., Kelly, S., Hengstler, J., Göbel, A., Krull, R., Hempel, D.C., Kinetic studies on the aggregation of *Aspergillus niger* conidia, *Biotechnol. Bioeng.*, 2004, *87*, 213–218.
Grindley, J.F., Payton, M.A., van den Pol, H., Hardy, K.G., Conversion of glucose to 2-keto-L-gulonate, an intermediate in L-ascorbate synthesis, by a recombinant strain of *Erwinia citreus*, *Appl. Environ. Microbiol.*, 1988, *54*, 1770–1775.
Heijnen, J.J., Mulder, A., Weltevrede, R., Hols, J., van Leeuwen, H.L.J.M., Large-scale anaerobic-aerobic treatment of complex industrial waste water using biofilm reactors, *Water Sci. Technol.*, 1991, *23*, 1427–1436.
Henry, M., Varaldo, C., Anaerobic digestion treatment of chemical industry wastewaters at the Cuise-Lamotte (Oise) plant of Société Française Hoechst 479, in: *Anaerobic Digestion 1988* (Hall, E.R., Hobson, P.N., Eds), pp. 479-486. Pergamon Press, Oxford, 1988.
Henze, M., Harremoes, P., Anaerobic treatment of wastewater in fixed film reactors – a literature review, *Water Sci. Technol.*, 1983, *15*, 1–101.

Hulshoff, L.W., de Zeeuw. W.J., Velzeboer, C.T.M., Lettinga, G., Granulation in UASB-Reactors, *Water Sci. Technol.*, **1983**, *15*, 291–304.

Hulst, A.C., Tramper, J., Immobilized plant cells: A literature survey, *Enzyme Microb. Technol.*, 1989, *11*, 546–558.

Iersel, van M.F.M., van Dieren, B., Rombouts, F.M., Abee, T., Flavor formation and cell physiology during the production of alcohol-free beer with immobilized *Saccharomyces cerevisiae, Enzyme Microb. Technol.*, 1999, *24*, 407–411.

Ishihara, K., Hamada, H., Hirata, T., Nakajima, N., Biotransformation using plant cultured cells, *J. Mol. Catalysis B: Enzymatic*, 2003, *23*, 145–170.

Jahnz, U., Wittlich, P., Pruesse, U., Vorlop, K.D., New matrices and bioencapsulation processes, in: Hofmann, M., Anne, J. (Eds), *Focus on Biotechnology*, Vol. 4, pp. 293–307, Kluver Academic Publishers, Dordrecht, 2001.

Jördening, H.-J., Scaling-up and operation of anaerobic fluidized bed reactors, *Zuckerindustrie*, 1996, *121*, 847–854.

Jördening, H.-J., Jansen, W., Brey, S., Pellegrini, A., Optimierung des Fließbettsystems zur anaeroben Abwasserreinigung, *Zuckerindustrie*, 1991, *116*, 1047–1052.

Jördening, H.-J., Küster, W., Betriebserfahrungen mit einem anaeroben Fließbettreaktor zur Behandlung von Zuckerfabriksabwasser, *Zuckerindustrie*, 1997, *122*, 934–936.

Jördening, H.-J., Buchholz, K., Fixed film stationary-bed and fluidized-bed reactors, in: Winter, J. (Vol. Ed.), *Biotechnology*, Vol. 11a, pp. 493–515, Wiley-VCH, 1999.

Jördening H.-J., Buchholz, K., High-rate anaerobic waste water treatment, in: Jördening, H.-J., Winter, J. (Eds), *Environmental Biotechnology*, pp. 135–162, Wiley-VCH, 2004.

Junter, G.-A., Coquet, L., Vilain, S., Jouenne, T., Immobilized-cell physiology: current data and the potentialities of proteomics, *Enzyme Microb. Technol.*, 2002, *31*, 201–212.

Kaup, B., Bringer-Meyer, S., Sahm, H., Metabolic engineering of *Escherichia coli:* construction of an efficient biocatalyst for D-mannitol formation in a whole-cell biotransformation, *Appl. Microbiol. Biotechnol.*, 2004, *64*, 333–339.

Kawamoto, T., Tanaka, A., Entrapment of biocatalysts by prepolymer methods, in: Whitaker, J.R., Voragen, A.G.J., Wong, D.W.S. (Eds), *Handbook of Food Enzymology*, pp. 331–341, M. Dekker, New York, 2003.

Kayser, R., Activated sludge processes, in: Jördening, H.-J., Winter, J. (Eds), *Environmental Biotechnology*, pp. 79–120, Wiley-VCH, 2004.

Keim, P., Luerweg, M., Aivasidis, A., Wandrey, C., Entwicklung der Wirbelschichttechnik mit dreidimensional kolonisierbaren Trägermaterialien aus makroporösem Glas am Beispiel der anaeroben Abwasserreinigung, *Korresp. Abwasser*, 1989, *36*, 675–687.

Keim, P., Luerweg, M., Striegel, B., Aivasidis, A., Wandrey, C., Einsatzbeispiele und Scale-up von Wirbelschichtreaktoren mit Kugeln aus porösem Sinterglas in der anaeroben Abwasserreinigung, *Chem.-Ing.-Tech.*, 1990, *62*, 336–337.

Kelly, S., Grimm, L.H., Hengstler, J., Schultheis, E., Krull, R., Hempel, D.C., Agitation effects on submerged growth and product formation of *Aspergillus niger, Bioprocess Biosystems Eng.*, 2004, *26*, 315–323.

Kierstan, M., Bucke, C., The Immobilization of Microbial Cells, Subcellular Organelles and Enzymes in Calcium Alginate Gels, *Biotechnol. Bioeng.*, **1977**, *19*, 387–397.

Kihn, J.C., Masy, C.L., Mestdagh, M.M., Yeast flocculation: competition between nonspecific repulsion and specific bonding in cell adhesion, *Can. J. Microbiol.*, 1988, *34*, 773–778.

Kirst, H.A., Yeh, W.-K., Zmijewski, Jr., M.J. (Eds), *Enzyme Technologies for Pharmaceutical and Biotechnological Applications*, Marcel Dekker, New York, 2001.

Klein, J., Vorlog, K.-D., Immobilization Techniques-Cells, Comprehensive Biotechnology, Ed. Moo-Young, M., Pergamon Press, Oxford, **1985**, 2, 203–224.

Knapp, F., Lehrbuch der chemischen Technologie, Vol. *2*, 1847, F. Vieweg und Sohn, Braunschweig, 1847.

Lefebvre, J., Vincent, J.-C., Diffusion-reaction-growth coupling in gel-immobilized cell systems: Model and experiment, *Enzyme Microb. Technol.*, 1995, *17*, 276–284.

Lefebvre, J., Vincent, J.-C., Control of the biomass heterogeneity in immobilized cell systems. Influence of initial cell and substrate concentrations, structure thickness, and type of bioreactors. *Enzyme Microb. Technol.*, 1997, *20*, 536–543.

Lettinga, G., Hulshof Pol, L.W., van Lier, J.B., Zeemann, G., Possibilities and potential of anaerobic waste water treatment using anaerobic sludge bed (ASB) reactors.

In: Winter, J. (Vol. Ed.), *Biotechnology*, Vol. 11a, pp. 517– 526, Wiley-VCH, 1999.

Liese, A., Seelbach, K., Wandrey, C., *Industrial Biotransformations*, Wiley-VCH, Weinheim, 2000.

Lonza (2004): http://www.lonzabiotec.com/biotec/en/areas/0.html

Lu, C., Lin, M.-R., Wey, I., Removal of acrylonitrile and styrene mixtures from waste gases by a trickle bed air biofilter, *Bioprocess Biosyst. Eng.*, **2000**, *25*, 61–67.

Lüllau, E., Biselli, M., Wandrey, C., Growth and metabolism of CHO-cells in porous glass carriers, in: Spier, R.E., Griffiths, J.B., Berthold, W. (Eds), *Animal Cell Technology: Products of Today, Prospects of Tomorrow*, pp. 152–255, Butterworth-Heinemann, 1994.

Lundberg, P., Kuchel, P.W., Diffusion of solutes in agarose and alginate gels: ^{1}H and ^{23}Na PFGSE and ^{23}Na TQF NMR studies, *Magnet. Reson. Med.*, 1997, *37*, 44–52.

Macarie, H., Overview of the application of anaerobic treatment to chemical and petrochemical wastewaters, *Water Sci. Technol.*, 2001, *44*, 201–214.

McCarthy, P.L., The development of anaerobic treatment and its future, *Water Sci. Technol.*, 2001, *44*, 149–156.

McInerny, M.J., Anaerobic metabolism and its regulation, in: Winter, J. (Vol. Ed.), *Biotechnology*, Vol. 11a, pp. 455–478, Wiley-VCH, 1999.

Messing, R.A., High-rate, continuous waste processor for the production of high BTU gas using immobilized microbes; in: Chibata, I., Fukui, S., Wingard, L.B., Jr. (Eds), *Enzyme Engineering*, Vol. 6, pp. 173–180, Plenum Publishing Corp., 1982.

Mösche, M., Anaerobe Reinigung von Zuckerfabriksabwasser in Fließbettreaktoren: vom Pilotmaßstab zum 500-m^3-Reaktor, Fortschrittsberichte VDI, Nr. 201, pp. 49–53, Düsseldorf, 1998.

Mozes, N., Léonhard, A.J., Rouxhet, P.G., On the relation between the elemental surface composition of yeast and bacteria and their charge and hydrophobicity, *Biochim. Biophys. Acta*, 1988, *945*, 324–334.

Nagashima, M., Azuma, M., Nogushi, S., Large scale Preparation of Alginate immobilized Yeast cells and its application to Industrial Ethanol Production, in: Mosbach, K. (Ed.), *Meth. Enzymol.*, Vol. 136, pp. 394–405, Academic Press, Orlando, USA, 1987.

Netto, C.B., Destruhaut, A., Goma, G., Ethanol Production by Flocculating Yeast: Performance and Stability Dependence on a Critical Fermentation Rate, *Biotechnol. Lett.*, **1985**, *7*, 359–360.

Nojima, S., Yamada, T., Large scale production of Photo-Cross-Linkable Resin-Immobilized Yeast, in: Mosbach, K. (Ed.), *Meth. Enzymol.*, Vol. 136, pp. 380–394, Academic Press, Orlando, USA, 1987.

Omar, S.H., Oxygen diffusion through gels employed for immobilization. Part 2. In the presence of microorganisms, *Appl. Microbiol. Biotechnol.*, 1993, *40*, 173–181.

Plaggemeier, T., Lämmerzahl, O., Treatment of Waste Gas Pollutants in Trickling filters, in: Klein, J., Winter, J. (Vol. Eds), *Environmental Processes III. Biotechnology*, Vol. 11c, pp. 333–344, Wiley-VCH, Weinheim, 2000.

Pörtner, R., Shimada, K., Matsumura, M., Märkl, H., High density culture of animal cells using macroporous cellulose carriers, in: Beuvery, E.C., et al. (Eds), *Animal Cell Technology: Developments Towards the 21st Century*, pp. 835–839, Kluwer Academic Publishers, The Netherlands, 1995.

Prüße, U., Dallun, J., Breford, J., Vorlop, K.-D., Production of spherical beads by jet cutting, *Chem. Eng. Technol.*, 2000, *23*, 1105–1110.

Prüße, U., Fox, B., Kirchhoff, M., Bruske, F., Breford, J., Vorlop, K.-D., The jet cutting method as a new immobilization technique, *Biotechnol. Technol.*, 1998, *12*, 105–108.

Radovich, J.M., Mass Transfer Effects in Fermentations Using Immobilized Whole Cells, *Enz. Microb. Technol.* **1985**, *7*, 2–10

Reischwitz, A., Reh, K.D., Buchholz, K., Unconventional immobilization of dextransucrase with alginate, *Enzyme Microb. Technol.*, 1995, *17*, 457–461.

Rinas, U., El-Enshasy, H., Emmler, M., Hille, A., Hempel, D.C., Horn, H., Model-based prediction of substrate conversion and protein synthesis and excretion in recombinant *Aspergillus niger* biopellets, *Biotechnol. Bioeng.*, 2004, in press.

Roisin, C., Bienaimé, C., Nava Saucedo, J.E., Barbotin, J.-N., Influence of the microenvironment on immobilised *Gibberella fujikuroi*, in: Wijffels, R.H., Buitelaar, R.M., Bucke, C., Tramper, J. (Eds), Immobilized Cells: Basics and Applications, pp. 189–195, Elsevier, Amsterdam, 1996.

Rosevear, A., Immobilized plant cells, *Food Biotechnol.*, 1988, *2*.

Rouxhet, P.G., Biocatalysts/interfacial chemistry, *Ann. N. Y. Acad. Sci.*, 1990, *613*, 265–278.

Sam-Soon, P., Loewenthal, R.E., Dold, P.L., Marais, G.R., Hypothesis for pelletisation in the upflow anaerobic sludge bed reactor, *Water SA*, 1987, *13*, 69–80.

Schedel, M., Regioselective oxidation of aminosorbitol with *Gluconobacter oxidans*; Key reaction in the industrial 1-deoxynojirimycin synthesis, *Biotechnology*, 2000, *8b*, 295–311.

Schmid, A., Dordick, J.S., Hauer, B., Kiener, A., Wubbolts, M., Witholt, B., Industrial biocatalysis today and tomorrow, *Nature*, 2001, *409*, 258–267.

Schmid, A., Hollmann, F., Park, J.B., Bühler, B., The use of enzymes in the chemical industry in Europe, *Curr. Opin. Biotechnol.*, 2002, *13*, 359–366.

Schraewer, R., Das Anfahrverhalten von anaeroben Bioreaktoren zur Reinigung hochbelasteter Stärkeabwässer, *Starch/Stärke*, **1988**, *40*, 347–352.

Segre, D., Vitkup, D., Church, G.M. (2002): Analysis of optimality in natural and perturbed metabolic networks, *Proc. Natl. Acad. Sci. USA*, 2002, *99*, 15112–15117.

Sutton, P.M., Li, A., Evans, R.R., Korchin, S., Dorr–Oliver's fixed film and suspended growth anaerobic systems for industrial wastewater treatment and energy recovery, *37th Ind. Conf.*, **1982**, Purdue, West Lafayette, IN, USA.

Tanaka, H., Matsumura, M., Veliky, I.A., Diffusion Characteristics of Substrates in Ca-Alginate Gel Beads, *Biotechnol. Bioeng.*, **1984**, *26*, 53–58.

Tang, C.W., Mavituna, F., Cell immobilisation of taxus media, in: Van Broekhoven, A., et al. (Eds), *Novel Frontiers in the Production of Compounds for Biomedical Use*, pp. 401–407, Kluwer Academic Publishers, Netherland, 2001.

Tokayashi, M., Yokoyama, S., Cell cultivation technology, in: Hauser, H., Wagner, R. (Eds), *Mammalian Cell Biotechnology in Protein Production*, Walter de Gruyter, Berlin, New York, 1997.

Trulear, M.G., Characklis, W.G., Dynamics of biofilm processes, *WPCF*, 1982, *54*, 1288–1301.

Ueda, M., Murai, T., Tanaka, A., Genetic immobilization of enzymes on yeast cell surface, in: Whitaker, J.R., Voragen, A.G.J., Wong, D.W.S. (Eds), *Handbook of Food Enzymology*, pp. 343–357, M. Dekker, New York, 2003.

van Lier, J.B., van der Zee, F.P., Ian, N.C.G., Rebac, S., Kleerebezem, R., Advances in high-rate anaerobic treatment: staging of reactor systems, *Water Sci. Technol.*, 2001, *44*, 15–25.

van Loosdrecht, M.C.M., Lyklema, J., Norde, W., Schraa, G., Zehnder, A.J.B., Electrophoretic mobility and hydrophobicity as a measure to predict the initial steps of bacterial adhesion, *Appl. Environ. Microb.*, 1987, *53*, 1898–1901.

VDI: VDI-Richtlinien 3477, Biological Waste Gas/Waste Air Purification. Biofilters, 1991 (German, English).

VDI: VDI-Richtlinien 3478, Biological Waste Gas Purification. Bioscrubbers and Trickle Bed Reactors, 1996 (German, English).

VDI: VDI-Richtlinien 3477, Biologische Abgasreinigung. Biofilter, 2002 (German).

Verstraete, W., van Vaerenberg, E., Aerobic activated sludge; in: Rehm, H.J., Reed, G. (Eds), *Biotechnology*, Vol. 8, pp. 43–112, VCH, Weinheim, 1986.

Wanda, U., Wollersheim, R., Diekmann, H., Buchholz, K., Adhesion of anaerobic bacteria on solid surfaces, in: de Bont, J.A.M., Visser, J., Mattiassen, B., Tramper, J. (Eds), *Physiology of Immobilized Cells*, pp. 109–114, Elsevier, Amsterdam, 1990.

Waugh, A., Culturing animal cells in fluidized bed reactors, in: Jenkins, N. (Ed.), *Methods in Biotechnology, Vol. 8: Animal Cell Biotechnology*, pp. 179–185, Humana Press Inc., Totowa, NJ, 1999.

Waweru, M., Herrygers, V., Van Langenhove, H., Verstraete, W., Process engineering of biological waste gas purification, in: Klein, J., Winter, J. (Vol. Eds), *Environmental Processes III. Biotechnology*, Vol. 11c, pp. 259–273, Wiley-VCH, Weinheim, 2000.

Weiland, P., Thomsen, H., Wulfert, K., *Entwicklung eines Verfahrens zur anaeroben Vorreinigung von Brennereischlempen unter Einsatz eines Festbettreaktors*, in: Verfahrenstechnik der mechanischen, thermischen, chemischen und biologischen Abwasserreinigung, Part 2: Biologische Verfahren (GVC-VDI, Ed.), pp. 169–186, Düsseldorf: VDI, Düsseldorf, 1988.

Westrin, B.A., Axelsson, A., Diffusion in Gels Containing Immobilized Cells: A Critical Review, *Biotechnol. Bioeng.*, **1998**, *38*, 439–446.

Wingender, J., Flemming, H.-C., Autoaggregation of microorganisms: Flocs and biofilms, in: Winter, J. (Vol. Ed.), *Biotechnology*, Vol. 11a, pp. 65–83, Wiley-VCH, 1999.

8
Characterization of Immobilized Biocatalysts

The characterization is required to answer the following questions:	This is required to answer these questions: Phenomena involved and required data
How much immobilized biocatalyst is required to obtain a desired space-time yield (*STY*)?	The required *STY* and the effectiveness factors (Sections 8.2, 8.3, 8.7)
What other properties influence – in comparison with free biocatalysts – *STY* and productivity with immobilized biocatalysts?	The coupling of reaction and diffusion, particle size and effectiveness factors (Sections 8.3, 8.4)
How can effectiveness factors be calculated? How much can they differ from experimental data?	From relations that describe the coupling of reaction and diffusion (Sections 8.4, 8.7).
Comparison of different continuous reactors (stirred tank or packed bed reactor) – which is better? What effectiveness factor must be used?	Calculation of *STY* for these reactors (Section 8.5)
What properties of immobilized biocatalysts must be known to design an enzyme process? What properties of immobilized biocatalysts are important to improve them for a given enzyme process?	Determination of important mechanical, physical, chemical and catalytic properties of immobilized biocatalysts for an enzyme process (Section 8.6)
Can immobilized biocatalysts be used to carry out enzyme processes with slightly soluble substrates or products in suspensions (emulsions)?	Section 8.8
How can negative effects on enzyme processes due to micro-(concentration or pH-gradients) and nanoenvironmental (pore surface charges) effects in immobilized biocatalysts be minimized ?	Use suitable buffers; select suitable support with small surface charge and diffusion distance for substrates (Section 8.9)

Biocatalysts and Enzyme Technology. K. Buchholz, V. Kasche, U.T. Bornscheuer

ISBN: 3-527-30497-5

8.1 Introduction

Immobilized biocatalysts are used in enzyme technology for analytical, preparative, and industrial purposes. They allow reuse of the enzyme and reduce the biocatalyst cost per unit of produced product (see Chapters 6 and 7). The first studies on the properties of immobilized biocatalysts appeared around 1960. Before this, the importance of immobilized enzymes in natural systems such as cells and tissues or in soil had been recognized and studied (Mclaren and Packer, 1970; Weisz, 1972). Compared with systems using free enzymes, the reaction rate is determined not only by the catalytic properties but also by the mass transfer to, from, and inside the immobilized biocatalysts, and their micro- and nano-environment (pH, ionic strength on a micrometer and nanometer scale). The first experimental and theoretical studies on such heterogeneous systems were performed on physiological systems during the 1920s (Warburg, 1923; Rashevsky, 1940). Phenomenologically similar systems – heterogeneous – chemical catalysts, have been used in chemical engineering since the 1930s, and were initially extensively analyzed quantitatively at that time (Damköhler, 1937; Thiele, 1939; Zeldovich, 1939). The analytical description of both systems, has – once the similarity was recognized – contributed much to the analytical and experimental characterization of immobilized biocatalysts.

These systems provide the basic knowledge for the rational design of enzyme processes with immobilized biocatalysts. For this aim, the quantitative relations for the time t required for a given substrate conversion with a given biocatalyst activity (or *vice versa*), as a function of different system properties must be derived (see Sections 8.2 to 8.5). From this, the space–time yield, the biocatalyst productivity (the amount of substrate converted per unit biocatalyst amount, expressed as activity or weight), and the biocatalyst cost can be obtained. In order to improve the immobilized biocatalysts, properties that are important for their application must be determined, and how they can be influenced by the nano- and microenvironment around the immobilized enzymes, must be studied in detail (see Sections 8.6 to 8.9).

8.2 Factors Influencing the Space–Time Yield of Immobilized Biocatalysts

The space–time yield (STY) of an enzyme process is (see Chapter 2, Section 2.8):

$$STY = \frac{[S]_0 - [S]_t}{t} \tag{8.1}$$

where t is the time required for the desired change in substrate concentration from $[S]_0$ to $[S]$. For free enzymes, the biocatalyst activity, $V_{max,0}$ required to obtain this STY can be calculated from Eq. (2.33)

$$V_{max,0} = \frac{\int_{[S]_t}^{[S]_0} f([S], [P])\, d[S]}{(1 - e^{-k_i t})/k_i} \tag{8.2}$$

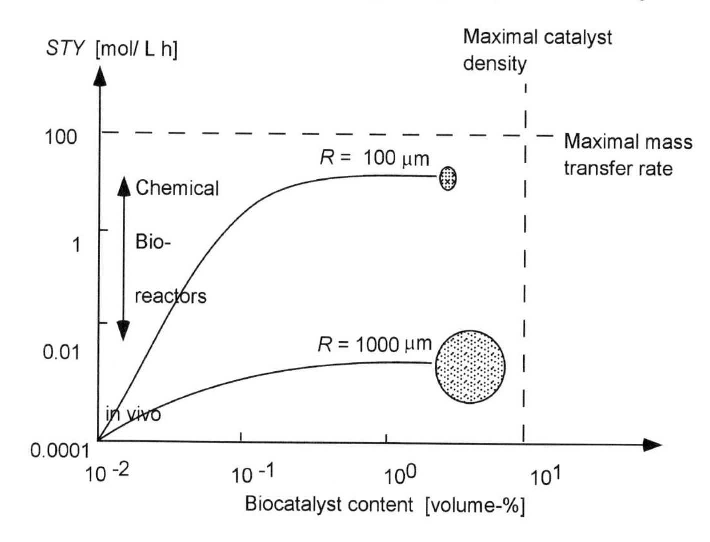

Fig. 8.1 Limiting system properties for the maximal space–time yield (*STY*) in reactors with immobilized biocatalysts as a function of the biocatalyst content (as % of reactor volume) and particle radius *R*.

where k_i is the rate constant for the first-order inactivation of the biocatalyst. In reactors with immobilized biocatalysts with the same catalyst activity, it cannot be assumed that all catalysts are used simultaneously. Some, localized in the inner part of the particle are not accessible for substrate, when it has been converted in the outer part of the particle. To consider this, an effectiveness factor η (degree of catalyst utilization ≤1) must be introduced in Eq. (8.2). This gives the following relationship:

$$V_{\max,0} = \frac{\int_{[S]_t}^{[S]_0} f([S],[P])\, d[S]}{\eta\,(1 - e^{-k_i t})/k_i} \tag{8.3}$$

from which $V_{\max,0}$ can be calculated, once η is known. The definition and calculation of effectiveness factors for different reactors is outlined in Sections 8.3 to 8.5.

The *STY* as a function of $V_{\max,0}$ is limited by the following system properties (Fig. 8.1):

- The maximum amount of substrate that can be transported to the particle with immobilized biocatalysts per unit time.
- The maximal biocatalyst density that can be obtained in the particles with immobilized enzymes or cells.

They can be influenced by engineering (particle size, mass transfer rate) or biochemical/biological (catalyst content) means. Besides this, the *STY* can be influenced by:

- Properties of the particle in which the biocatalyst is immobilized:
  - chemical, mechanical and thermal stability;
  - concentration and type of functional groups (charges, hydrophobic residues) on the inner (in pores) and outer surface per unit particle volume;
  - porosity, pore size;
  - particle density.
- Properties of the immobilized biocatalyst (enzyme or cell):
  - biocatalytic properties of the immobilized biocatalyst (see Chapter 2, Sections 2.6 and 2.7.1);
  - turnover number k_{cat}, Michaelis–Menten constant K_m, stereoselectivity E, or synthesis/hydrolysis-selectivity in kinetically controlled reactions $(k_T/k_H)_{app}$;
  - stability.
- Concentration gradients of substrates and products inside the particles (especially pH-gradients (Tischer and Kasche, 1999).

8.3 Effectiveness Factors for Immobilized Biocatalysts

The substrate concentration outside particles with immobilized enzyme is shown in Figure 8.2. At steady state, the substrate gradient is time-independent, and the same amount of substrate that is transported to the particle per unit time is converted to product inside the particle.

Outside the particle (cell) a concentration gradient is formed, as substrate must be transported to the particle. This diffusion can only be driven by a concentration gradient that is characterized by the thickness δ of the unstirred diffusion layer (film).

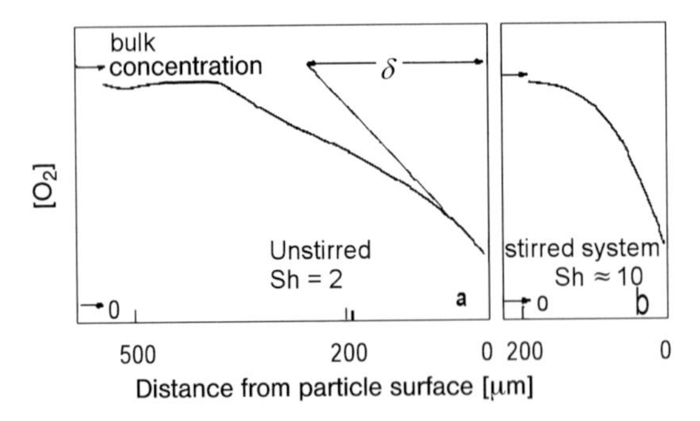

Fig. 8.2 Oxygen concentration gradient outside a spherical particle (radius $R \approx 300$ μm) with immobilized glucose oxidase that catalyzes the oxidation of glucose with oxygen. (a) Without stirring (Sherwood number (Sh) = 2); (b) with stirring (Sh ≈ 10). The oxygen concentration was measured using oxygen microelectrodes with a tip diameter of several micrometers. Note that the thickness of the diffusion layer given by the intersection of the gradient (tangent) at the particle surface and the bulk oxygen content approximately equals the particle radius in the unstirred system. This is in agreement with the theoretical analysis (Kasche and Kuhlmann, 1980).

In this layer the mass transfer occurs only by diffusion. The thickness of the film can be reduced by stirring, which increases the velocity of the particles relative to the solvent (Fig. 8.2). The stirring causes shear forces that may cause abrasion of the particles, and thus the stirring speed has an upper limit in order to avoid this abrasion.

From Figure 8.2 follows that the immobilized biocatalysts are surrounded by a substrate concentration that is lower than the bulk concentration. Two different effectiveness factors – the stationary and the operational – that account for this have been defined (Figs. 8.3 and 8.4).

The *stationary effectiveness factor* η is the ratio of the initial rates (Fig. 8.3), determined under equal conditions and with the same biocatalyst content (V_{max}) and bulk substrate concentration:

$$\eta = v_{imm}/v_f \tag{8.4}$$

that is easy to determine experimentally. The stationary effectiveness factor can be subdivided into an external η_e (for the diffusion layer) and an internal η_i (inside the particle). These are, however, contrary to η, difficult to determine experimentally, and will not be used here.

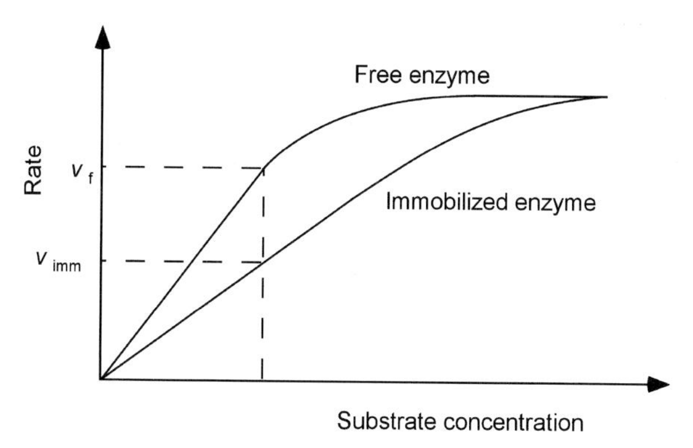

Fig. 8.3 Determination of the stationary effectiveness factor η.

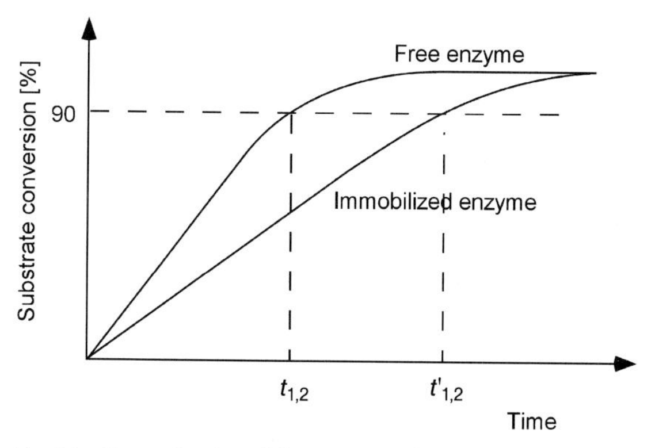

Fig. 8.4 Determination of the operational effectiveness factor $\eta_o = t_{1,2}/t'_{1,2}$ from progress curves.

The stationary effectiveness factor does not provide any information on how rapidly the desired end-point of an enzyme process (see Figs. 1.9 and 2.9) can be reached. During this process the substrate concentration outside the particles and the stationary effectiveness factor are changed. Therefore, the *operational effectiveness factor* η_o was introduced (Kasche, 1983), and is defined as the ratio of the time required to reach the end-point of a process with the same initial substrate concentration, and same amount of free and immobilized biocatalyst per reactor volume unit, under otherwise equal conditions (Fig. 8.4):

$$\eta_o = t_{1/2}/t'_{1/2} \tag{8.5}$$

It can be determined easily from progress curves (substrate concentration as function of time) in Figure 8.4.

8.4 Mass Transfer and Reaction

8.4.1 Maximal Reaction Rate of Immobilized Biocatalysts as Function of Particle Radius

The *STY* or observed rate of the catalyzed reaction v_{obs} as function of the biocatalyst concentration in spherical particles is shown in Figure 8.1. It is limited by the maximum rate of substrate transfer to the particles and the biocatalyst content in the support. When the latter is rate-determining, the rate can be increased by higher biocatalyst concentrations until v_{obs} becomes mass transfer limited.

The optimal biocatalyst concentration, expressed as the activity V'_{max}, can be estimated as follows. At the intersection of the linear part of the curve at low biocatalyst content and the maximal mass transfer rate, given by the horizontal part of the curves in Figure 8.1, the following ratio equals 1

$$\frac{\text{maximal reaction rate in the particle}}{\text{maximal mass transfer rate to the particle}} = \frac{V_p \eta \dfrac{V'_{max}}{f([S],[P])}}{A_p k_L \alpha c_S(\infty)} = 1 \tag{8.6}$$

and can be used to estimate V'_{max}, where V_p and A_p are the particle volume and outer surface area of the immobilized enzyme particle, respectively; η is the stationary effectiveness factor; primed quantities properties of the immobilized biocatalyst; α is the concentration change in the diffusion layer as fraction of the bulk substrate concentration $c_S(\infty)$; and k_L is the mass transfer coefficient. The latter, with the dimension velocity, is a function of the stirring speed in a batch reactor, or of the linear flow rate in a packed-bed reactor. It can be expressed as:

$$k_L = D_S\, \text{Sh}/2R = D_S/\delta \tag{8.7}$$

where D_S is the diffusion coefficient of the substrate, and Sh is the dimensionless Sherwood number (the ratio of particle diameter to the thickness of the unstirred diffusion layer or film). This value is 2 for a spherical particle in an unstirred system (Kasche and Kuhlmann, 1980; see Chapter 9). V'_{max} can be estimated from Eqs. (8.6) and (8.7) with the following assumptions: $c_S(\infty) \gg K'_m$, then $f([S],[P])$ and η in Eq. (8.6) are ≈ 1 (see Chapter 2.7, p. 49). A high initial substrate concentration up to the molar range is desirable in enzyme processes, and this assumption is then justified (see Table 2.4). The expression for V'_{max} derived from Eqs. (8.6) and (8.7) is

$$V'_{max} = (1.5/R^2)\ D_S\ \mathrm{Sh}\ \alpha\ c_S(\infty) \tag{8.8}$$

For these, V'_{max} has been calculated as a function of the particle radius and different bulk substrate concentrations in Figure 8.5, for substrates with MW ≈ 500, D_S is $\approx 6 \times 10^{-6}$ cm^2 s^{-1}, with Sh = 10 and $\alpha = 0.1$. Immobilized biocatalysts that are used in packed-bed reactors should have a radius of more than 100–200 µm to avoid to large back-pressures in these reactors. For stirred batch reactors, the radius can be lower. When the radius is ≤ 10 µm, the particles are difficult to remove from the reaction mixture by filtration or sedimentation. The radius should not be much more than ~100–200 µm, to allow for high rates and *STY* (Fig. 8.5). The question remains as to whether the V'_{max} calculated in this figure can be realized in available supports for biocatalyst immobilization (see Chapter 6). In these particles, up to ~10 g of enzyme can be immobilized per liter of wet support, which provides an enzyme concentration of 100–400 µM for enzymes in the MW-range 25 to 100 kDa. Many enzymes have a turnover number of 100 s^{-1} (see Tables 2.4 and 2.5), and this gives V'_{max} in the range of 0.01 to 0.04 M s^{-1} (or $6–24 \times 10^5$ U L^{-1}) – that is, almost of the same order of magnitude as calculated in Figure 8.5. The value may be smaller due to the assumptions $\alpha = 0.1$ (too large), and $\eta \approx 1$ (too large). In order to analyze this situation, the effectiveness factors and substrate concentration profiles both inside and outside the particles must be calculated (see Section 8.4.2).

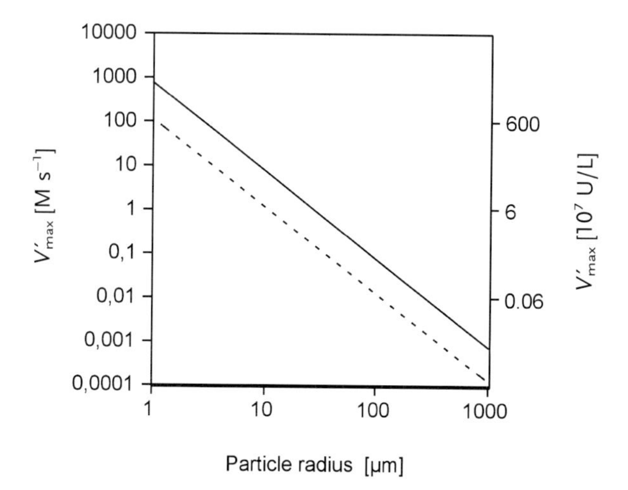

Fig. 8.5 V'_{max} as a function of particle radius calculated from Eq. (8.8) for the initial bulk substrate concentrations 1 M (——) and 0.1 M (------).

8.4.2
Calculation of Effectiveness Factors and Concentration Profiles Inside and Outside the Particles

The concentration profile inside and outside a spherical particle with immobilized biocatalyst can be determined at steady state from the mass conservation relationship in the volume element between $r+dr$ and r (Fig. 8.6). There, the net diffusion equals the conversion of the substrate (simple Michaelis–Menten kinetics without inhibition, properties of the immobilized biocatalyst are primed) or

$$D_S' \cdot A(r+dr)\frac{dc_S(r+dr)}{dr} - D_S' \cdot A(r)\frac{dc_S(r)}{dr} = \frac{V_{max}' \cdot c_S(r)}{K_m' + c_S(r)} A(r) \cdot dr \tag{8.9}$$

where $A(r+dr)$ and $A(r)$ are areas perpendicular to the diffusional flow. From the Taylor series (see mathematical textbooks), it follows that:

$$\frac{dc_S(r+dr)}{dr} = \frac{dc_S(r)}{dr} + \frac{d^2c_S(r)}{dr^2} \cdot dr \tag{8.10}$$

With this, Eq. (8.9) is transformed to

$$D_S'\left(\frac{d^2c_S(r)}{dr^2} + \frac{n}{r} \cdot \frac{dc_S(r)}{dr}\right) = \frac{V_{max}' \cdot c_S(r)}{K_m' + c_S(r)} \tag{8.11}$$

where $n = 0$, 1, or 2 for planar, cylindrical, or spherical geometry, respectively. This equation can be solved with the following boundary conditions:

$$\frac{dc_S(r)}{dr} = 0 \quad \text{at} \quad r = 0$$

$$c_S(R) = c_S(\infty) - \frac{D_S}{k_L} \cdot \frac{dc_S(R)}{dr} \quad \text{for} \quad r = R \text{ (particle surface)}$$

from Eq. (8.7) and Figure 8.6, after which $dc_S(R)/dr$ can then be determined. This provides the substrate transport rate into the particle which, at steady state, equals the rate of substrate conversion in the particle.

It is suitable to transform Eq. (8.11) into a dimensionless form using the variable transformations

$$\gamma = \frac{c_S(\infty)}{K_m'} \qquad z = \frac{r}{R} \qquad c_S(z) = \frac{c_S(r)}{c_S(\infty)}$$

this results in

$$\frac{d^2c_S(z)}{dz^2} + \frac{n}{z} \cdot \frac{dc_S(z)}{dz} = \frac{R^2 \cdot V_{max}'}{D_S' \cdot K_m'} \cdot \frac{c_S(z)}{1 + c_S(z) \cdot \gamma} \tag{8.12}$$

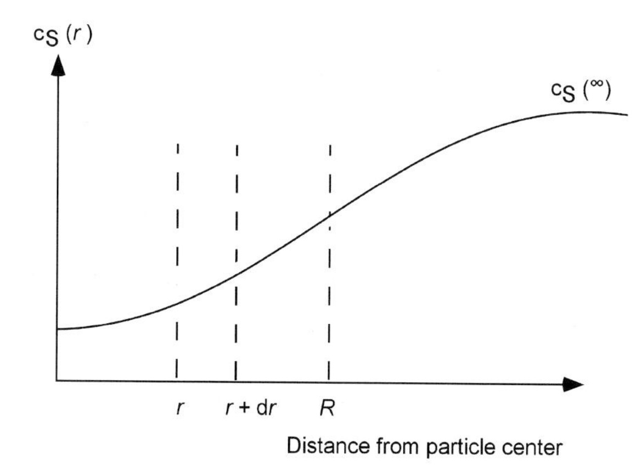

Fig. 8.6 The substrate concentration (c_S) inside and outside a particle (radius R) with immobilized biocatalyst (enzyme, cell) at steady state.

The dimensionless quantity

$$\varphi^2 = \frac{R^2 \cdot V'_{max}}{D'_S \cdot K'_m} \tag{8.13}$$

in Eq. (8.12) is the square of the *Thiele-modulus*, a dimensionless number. It is the ratio of the maximum rate of substrate conversion in the particle, divided by the maximum rate of substrate transport to the particle. Such a ratio characterizing heterogeneous catalysts, was first introduced by Damköhler (1937). It contains properties that influence the activity of immobilized biocatalysts in the support. Of these values, D'_S and R have no influence on the activity of the free biocatalyst. As φ^2 is introduced here it applies for all substrate concentrations – that is, also for first-order ($c_S < K_m$), and zero-order ($c_S \gg K_m$) reactions. (In some textbooks on heterogeneous catalysis, the dimensionless numbers are derived so that they only apply for first- or zero-order reactions.)

The differential equation (Eq. 8.12) can only be solved analytically for first- and zero-order reactions, and must therefore generally be solved by numerical methods. A suitable method here is the collocation method (Villadesen and Michelsen, 1978). When the concentration profile has been calculated, the effectiveness factors can be calculated as follows using the definitions in Section 8.3. The stationary effectiveness factor (Eq. 8.3) is

$$\eta = \frac{3 \cdot D'_S \cdot \dfrac{dc_S(R)}{dr}}{R_m \cdot V(c_S(\infty))} = \frac{3\int_0^1 z^2 \cdot V(c_S(z))\,dz}{V(c_S(1))} \tag{8.14}$$

where R_m is the average particle radius and $V(c_S(\infty)) = V'_{max} \cdot c_S(\infty)/(K'_m + c_S(\infty))$.

Then $t'_{1,2}$ in Eq. (8.5) can be calculated from the integral derived from the expression

$$d(c_S(\infty))/dt = \eta(c_S(\infty)) \cdot V(c_S(\infty)) \cdot n \cdot (4/3)\,\pi R_m^3$$

by variable separation

$$\int_{c_{S1}(\infty)}^{c_{S2}(\infty)} \frac{-3\,\mathrm{d}c_S(\infty)}{\eta(c_S(\infty)) \cdot V(c_S(\infty)) \cdot n \cdot 4R_m^3\pi} = \int_0^{t'_{12}} \mathrm{d}t = t'_{12} \tag{8.15}$$

where n is the particle number per unit volume. The time $t_{1,2}$ in Eq. (8.5) is obtained when Eq. (8.15) is integrated with $\eta = 1$. Then η_o is calculated from Eq. (8.5).

The values of the effectiveness factors calculated for different parameters are given in Figures 8.7 and 8.8. As parameter for the external mass transfer the Sherwood number (Eq. 8.7) is used. In an unstirred system with spherical particles, the Sherwood number is 2, whereas in stirred-tank or packed-bed reactors it can be increased to the range 20–30. At higher values, where more energy is required for stirring and pumping at higher speeds, the effectiveness factors are marginally changed. The values calculated in these figures can be used to estimate the effectiveness factors for immobilized biocatalysts. The comparison with experimental data (see Fig. 8.8) shows a good agreement within experimental error estimated to be in the range of ±20–30 %. This large error results from the many quantities in Eqs. (8.7) and (8.13) that must be determined to estimate the square of the Thiele modulus and the Sher-

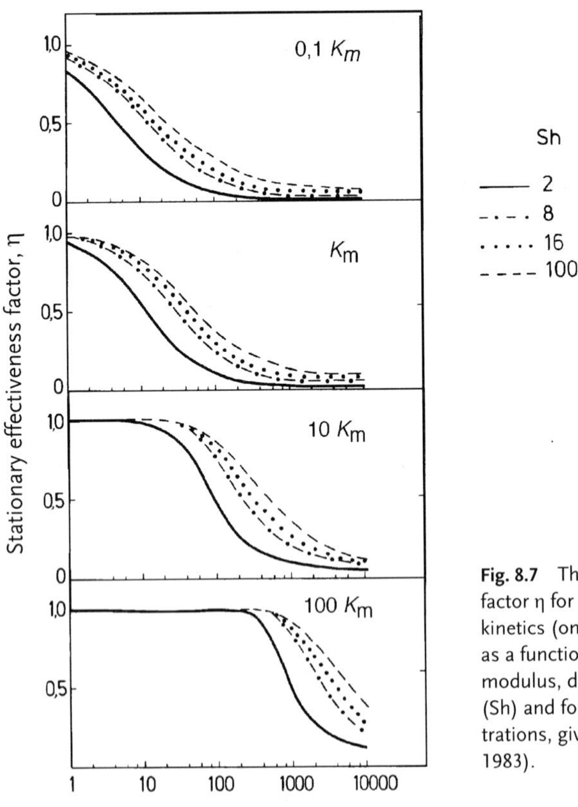

Fig. 8.7 The stationary effectiveness factor η for simple Michaelis–Menten kinetics (one substrate, no inhibition) as a function of the square of the Thiele modulus, different Sherwood numbers (Sh) and for different substrate concentrations, given in units of K_m (Kasche, 1983).

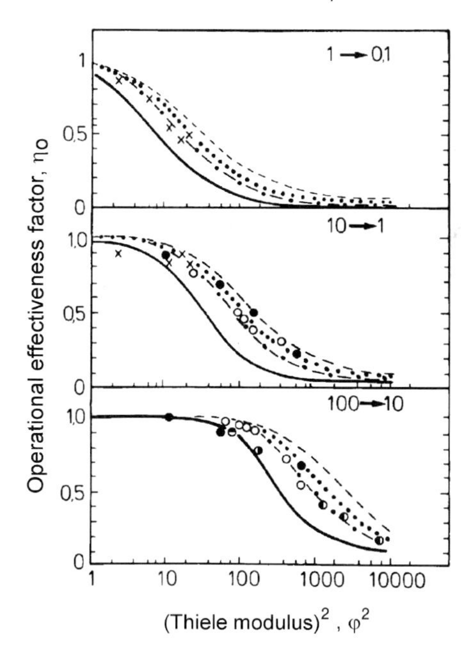

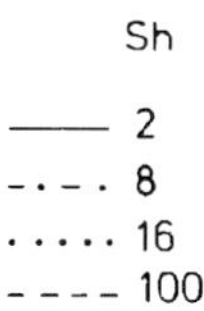

Fig. 8.8 The operational effectiveness factor η_o for simple Michaelis–Menten kinetics (one substrate, no inhibition) as a function of the square of the Thiele modulus, different Sherwood numbers (Sh) and for different initial substrate concentrations, given as γ/K_m, for spherical particles and 90 % substrate conversion. Calculated (curves) and experimental (circles) data for such processes are compared (Kasche, 1983).

wood number. The error in determination of the single quantities is at best ±5 %. For these estimations the K_m-value for the free enzyme is used. From Figure 8.7 follows that the intrinsic molecular property K'_m of the immobilized enzyme can only be determined in systems with $\varphi^2 < 1$, as the rates must also be determined for $[S] \leq K'_m$ (see Section 2.7.1).

The values in Figures 8.7 and 8.8 have been calculated for simple Michaelis–Menten kinetics (one substrate and no inhibition). This does generally not apply for enzyme processes. The effectiveness factors for other cases (one substrate with inhibition; two substrates without and with inhibition) can be calculated as shown here, when the rate equations covering these cases are used. Many enzyme processes, such as hydrolysis reactions, can be considered as reactions with one substrate. Here, however, competitive product inhibition cannot be neglected. For these, K'_m can be estimated to be $K_m(1+[P]/K_i)$, where [P] increases with time (see Table 2.6). For such processes the effectiveness factors can be estimated from Figures 8.7 and 8.8 using the average value $[P] = 0.5\,[S]_0$ – that is, half the initial substrate content.

Thus, these figures can be used to:

- estimate the effectiveness factors for a given immobilized biocatalyst; and
- design immobilized catalysts where the effectiveness factors are ≈1.

8.5
Space–Time Yields and Effectiveness Factors for Different Reactors

In order to design enzyme processes with immobilized biocatalysts, and to evaluate the different reactors used in these, the amount of biocatalyst with a known activity

(V_{max}) required to obtain a given *STY* must be calculated (Eq. (8.3). The relationship between the *STY* and V_{max} will be derived for the most important continuous reactors – the continuous stirred tank (CST) reactor and the packed-bed (PB) reactor – and this will be performed with the following (idealized) assumptions:

- for the CST reactor: complete mixing in the reactor;
- for the PB reactor: plug flow; no radial dispersion.

Both reactors are kept at the same temperature, and contain the same immobilized biocatalyst, the inactivation of which is neglected. The catalyzed reaction is described by simple Michaelis–Menten kinetics (one substrate, no inhibition). This should also answer the question of which effectiveness factor applies to the different reactors.

8.5.1
Continuous Stirred-Tank Reactor

The CST reactor (Fig. 8.9) is characterized as follows: V_T = reactor volume; α_T = immobilized biocatalyst volume as a fraction of the reactor volume; and Q = flow rate. At steady state in the reactor ([S] time-independent) the following relationship applies:

$$[S]_0 \cdot Q - [S] \cdot Q - \frac{\eta([S]) \cdot \alpha_{CST} \cdot V_{CST} \cdot V_{max}[S]}{K_m + [S]} = 0 \tag{8.17}$$

Here, the stationary effectiveness factor (Eq. 8.4) must be used as the substrate concentration in the reactor is constant. With the relative substrate concentrations

$$\gamma_2 = \frac{[S]}{K_m}\ ;\ \gamma_1 = \frac{[S]_0}{K_m} \tag{8.18}$$

Eq. (8.17) can be rewritten as

$$(\gamma_1 - \gamma_2)\,Q = \frac{\eta(\gamma_2) \cdot \alpha_{CST} \cdot V_{CST} \cdot \gamma_2}{(1+\gamma_2)\,K_m} \tag{8.19}$$

The residence time in this reactor $\tau_{12,CST}$ required for the substrate conversion from γ_1 to γ_2 is

$$\tau_{12,CST} = \frac{V_{CST}}{Q} = \frac{\gamma_1 - \gamma_2}{\gamma_2} \cdot \frac{(1+\gamma_2)\,K_m}{\alpha_{CST} \cdot V_{max}\,\eta(\gamma_2)} \tag{8.20}$$

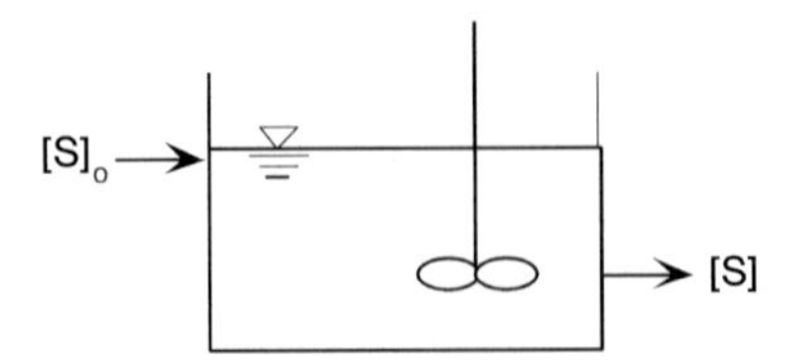

Fig. 8.9 Continuous stirred-tank reactor; operational details.

The *STY* is

$$STY_{CST} = \frac{\text{amount converted substrate (in moles)}}{\text{Unit time} \cdot \text{unit reactor volume}} =$$

$$= \frac{(\gamma_1 - \gamma_2)\, K_m}{\tau_{12,CST}} = \eta\,(\gamma_2) \cdot V_{max} \cdot \frac{\gamma_1 \cdot \alpha_{CST}}{1 + \gamma_2} \tag{8.21}$$

From which the amount of biocatalyst ($\alpha_{CST}\ V_{max}$) per unit reactor volume to obtain this *STY* can be calculated.

8.5.2
Packed-Bed (PB) Reactor or Stirred-Batch Reactor

The reactor (Fig. 8.10) is characterized as follows:

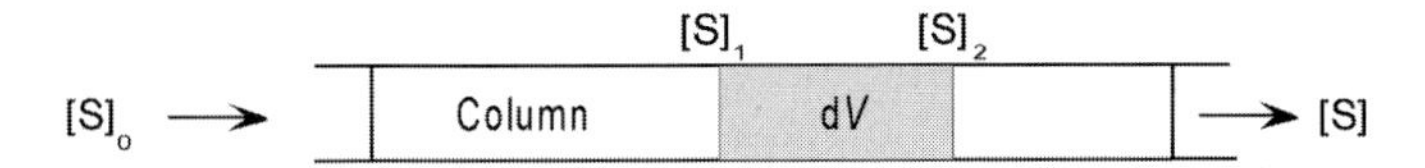

Fig. 8.10 Packed-bed reactor; operational details.

With the change in the subscripts from CST to PB, the same characteristics apply for this reactor. In the volume element d*v* the following relationships apply for the steady state.

$$([S]_1 - [S]_2) \cdot \frac{dV}{dt} - \frac{\eta \cdot \alpha_{PB} \cdot V_{max} \cdot [S] \cdot dV}{K_m + [S]} = 0 \tag{8.22}$$

In the PB reactor the stationary effectiveness factor is not constant, but varies with [S]. With the relative substrate concentrations in Eqs. (8.18) and (8.22) is changed to

$$\frac{d\gamma}{dt} = \frac{\alpha_{PB} \cdot V_{max} \cdot \gamma \cdot \eta\,(\gamma)}{K_m\,(1+\gamma)} \quad \text{or} \quad dt = \frac{K_m\,(1+\gamma)\,d\gamma}{\alpha_{PB} \cdot V_{max} \cdot \eta\,(\gamma) \cdot \gamma}$$

Integration over the packed bed gives the residence time required for the substrate conversion from γ_1 to γ_2. From the definition of the operational effectiveness factor we obtain

$$\tau_{12,PB} = \int_0^{t_{12,PB}} dt = \frac{K_m}{\alpha_{PB} \cdot V_{max}} = \int_{\gamma_2}^{\gamma_1} \frac{(1+\gamma)\,d\gamma}{\eta\,(\gamma) \cdot \gamma} \tag{8.23}$$

$$\tau'_{12,PB} = \frac{1}{\eta_{0,12}\,(\gamma_1)} \cdot \frac{K_m}{\alpha_{PB} \cdot V_{max}} \{\ln\,(\gamma_1/\gamma_2) + (\gamma_1 - \gamma_2)\} \tag{8.24}$$

Equation (8.23) is first integrated with $\eta(\gamma)=1$ (for the free enzyme), and the definition (Eq. 8.5) is then used to derive Eq. (8.24). A more general relationship that also covers product and substrate inhibition when enzyme inactivation can be neglected (see Eq. 2.33) is

$$\tau_{12,PB} = \frac{1}{\eta_{0,12}([S]_0)} \cdot \frac{\int_{[S]_t}^{[S]_0} f([S],[P])\, d[S]}{\alpha_{PB} \cdot V_{max}} \tag{8.25}$$

where f([S], [P]) is given by Eqs. (2.34) to (2.36) (p. 66). It also applies for the discontinuous stirred-batch reactors, for which the operational effectiveness factor must be used. The *STY* for these reactors is analogous to the derivation of Eq. (8.21)

$$STY_{PB} = \eta_{0,12} \cdot V_{max} \cdot \frac{\alpha_{PB}(\gamma_1 - \gamma_2)}{\ln(\gamma_1/\gamma_2) + \gamma_1 - \gamma_2} \tag{8.26}$$

from which the required amount of immobilized biocatalyst ($\alpha_{PB}\ V_{max}$) per unit reactor volume can be calculated.

8.5.3
Comparison of CST and PB Reactors

Equations (8.21) and (8.26) show that the *STY* differ in the effectiveness factors used and the terms expressing the degree of substrate conversion. For the same degree of substrate conversion (≥90 %) the ratio of Eq. (8.26) to Eq. (8.21) is

$$\frac{STY_{PB}}{STY_{CST}} = \frac{\eta_{0,12}(\gamma_1) \cdot \alpha_{PB} \cdot (\gamma_1 - \gamma_2) \cdot (1 + \gamma_2)}{\eta(\gamma_2) \cdot \alpha_{CST} \cdot \gamma_2 \cdot (\ln(\gamma_1/\gamma_2) + \gamma_1 - \gamma_2)} \tag{8.27}$$

a relationship that can be used to compare both reactors. Generally, $\gamma_1 \gg 1$, $\alpha_{PB} > \alpha_{CST}$ and $\eta_{o,12}(\gamma_1) \geq \eta(\gamma)$ (Figs. 8.7 and 8.8). Then the ratio in Eq. (8.27) is >1 – that is, for this case the *STY* is larger for the PB than CST reactor. This also implies that, for the same *STY*, less immobilized biocatalyst and a smaller reactor volume is required in the PB than in the CST reactor. In the latter reactor α_{CST} must be less than 0.05 in order to avoid abrasion due to shear forces and particle–particle collisions. This limits the *STY* that can be obtained in this reactor. In the PB reactor, where α_{PB} is ≥0.5, a much higher *STY* can be obtained at equal reactor volume.

For processes with several substrates, inhibition and enzyme inactivation, Eqs. (8.21) and (8.26) will become more complex. With product inhibition the PB reactor is better than the CST reactor, as the latter is working at a higher product concentration than the former. In some cases, the CST reactor is better than the PB reactor, especially for products that are gases or that must be neutralized, such as H^+. The disadvantageous pH-gradient over the PB reactor can be reduced by dividing it into several consecutive reactors, where the pH is adjusted to the desired value between the reactors. CST and PB reactors with volumes in the range 0.01 to 10 m^3 are used in enzyme processes.

8.6
Determination of Essential Properties of Immobilized Biocatalysts

Immobilized biocatalysts have been developed for a variety of applications in the laboratory, for analytical purposes, and in enzyme technology. In order to allow for their reproducible use, and to select the optimal biocatalyst for a specific application, they must be well characterized – that is, their essential physico-chemical (including mechanical), kinetic properties, and their stability under process conditions must be determined in standardized procedures. These have been proposed in the scientific literature (Buchholz et al., 1979). Properties that are important for large-scale enzyme processes are listed in Table 8.1, but not all of these are important for biocatalyst use on a laboratory or analytical scale. This may explain why immobilized biocatalysts are often insufficiently characterized in the scientific literature, or inadequate information is provided by the immobilized biocatalyst manufacturer.

The characterization of immobilized biocatalysts, as outlined above – and especially comparative studies with the same enzyme immobilized in different supports used for the same enzyme process – can provide important information on the properties of the supports that are important to improve the process. The recent reinvention of nanobiotechnology – that is, molecular interactions on a nanometric scale, as on the pore surface in immobilized biocatalysts or on natural or artificial membrane surfaces – has revived the importance of such studies on immobilized biocatalysts.

When immobilized biocatalysts are used in the food or pharmaceutical industries, health and environmental regulations must be considered. This information can be found on the home pages of the World Health Organization (WHO; www.who.org) or the Food and Drug Administration (FDA; www.fda.gov) (see also Chapter 4, Section 4.1 and Appendix I).

8.6.1
Physico-chemical Properties

The physico-chemical properties of immobilized biocatalysts are listed in Table 8.1. The average particle radius of a swollen particle can be determined using microscopy, or with specialized instruments that determine particle size distribution. A narrow distribution is important for PB reactors as it determines the bed porosity and thus the pressure drop. The measurement of particle size distribution can also be used to measure the abrasion in stirred tank or fluidized bed reactors. To be of technical relevance the abrasion as a function of particle concentration must be performed in reactors where high sheer forces can be realized. This requires baffled vessels and stirrers with a diameter ≥50 mm (see Chapter 9, Section 9.3). As expected, abrasion increases with the concentration of immobilized particles in the reactor, and based on this the upper particle content in these reactors should be less than 5 % of the reactor volume. This limits the space–time yield in these reactors. In enzyme technology, particles with radius R in the range 50 to 500 μm are used, and these have an internal surface area of 10 to 100 $m^2\ g^{-1}$ wet particles. The average pore diameter is generally >20 nm. The pore size distribution can be measured on-

Table 8.1 Basic information or properties that should be given or determined for immobilized biocatalysts, and factors influencing these in order to allow for their reproducible production and to evaluate the support properties that are essential for an enzyme process (Buchholz et al., 1979; EFB, 1983). (Properties in italics must only be determined for immobilized biocatalysts used in industry).

Information/property	***Factors influencing the property***
1. General information	
1.1 Reaction that is catalyzed	
1.2 Enzyme (EC number) and enzyme source (microorganism or tissue); purity, pI (isoelectric point)	
1.3 Support used for the immobilization Composition; functional groups	
2. Immobilization	
2.1 Immobilization method; conditions during immobilization (concentration, pH, I, T); how are unreacted reactive groups on the support inactivated?	pH, I, T, t
2.2 Immobilization yield (with respect to enzyme amount and enzyme activity) and how it is determined	pH, I, T, t [E] at the start of the immobilization, adsorption isotherm for immobilization by adsorption
2.3 For immobilized biocatalysts used in dry solvents: pH and buffer used before they where transferred to the water-free solvent	
3. Physico-chemical characterization of the immobilized biocatalyst	
3.1 Shape, average wet particle radius (R) and its distribution; surface area/g wet weight; average pore radius (R_p) and its distribution	pH, I
3.2 Swelling (wet volume/dry volume)	pH, I
3.3 *Compressibility/pressure drop in fixed beds*	bed height H, R, pressure drop Δp; linear flow rate u
3.4 *Abrasion in stirred tank or fluidized beds reactors*	R, particle density, stirrer diameter and speed
3.5 Fluorescence spectra	pH
3.6 Stationary charges on support (positive or negative) n_c; pore surface: hydrophilic or hydrophobic	pH, I
4. Kinetic characterization	
4.1 Intrinsic properties of the immobilized enzyme (V'_{max}; k'_{cat} and active enzyme content n_E; K'_i; selectivities (stereo- and $(k_T/k_H)_{app}$)	pH, I, T, [S], n_c
4.2 Stationary and operational effectiveness factor at process conditions	pH, I, R, T, V'_{max}, K'_m, D'_S, [S], [P], u for PB, and n, d_i for CST and batch reactor
4.3 Space time yield at process conditions	pH, I, R, T, V'_{max}, K'_m, D'_S, [S], [P]
5. Stability and productivity	
5.1 Storage stability	pH, I, T, t
5.2 *Stability under process conditions (operational stability)*	pH, I, T, k_i, t
5.3 *Productivity* (space time yield integrated over the time the biocatalyst is used) (amount product /per unit amount immobilized enzyme)	pH, I, T, R, V'_{max}, K'_m, D'_S, [S], k_i

ly for dry particles using porosimeters, or from scanning electron micrographs of shock-frozen particles (see Chapter 6).

The particle radius influences the effectiveness factor (see Section 8.4) in all reactors. In PB reactors, the square of the radius determines the pressure drop over the reactor, and this is also influenced by the interstitial volume fraction, ε (or bed porosity). This is about 35–40 % of the bed volume for optimally packed beds of spherical particles. In fluidized bed reactors, the particle radius and density determine the bed expansion as a function of the linear liquid flow rate. In general, the particle radius distribution is heterogeneous, and thus average values must be determined. Note that the simple average (Eq. 8.28),

$$\bar{R}_{\mathrm{p}} = \left(\sum_{\mathrm{i}=1}^{\mathrm{n}} R_{\mathrm{p,i}} \right) / \mathrm{n} \tag{8.28}$$

where n is the number of particles, cannot be used to calculate the overall particle surface that determines the rate of mass transfer to the particles. For this, the following average (Eq. 8.29) must be used

$$\bar{R}_{\mathrm{p,A}} = \left(\sqrt{\sum_{\mathrm{i}=1}^{\mathrm{n}} R_{\mathrm{p,i}}^{2}} \right) / \mathrm{n} \tag{8.28}$$

Some properties must be determined in the reactor used for the enzyme process. In order to minimize the immobilized biocatalyst amount required for this, suitable scaled-down reactors should be used. For packed and fluidized bed reactors, the same bed height but with a much smaller cross-section of the reactor can be used for this purpose. The compressibility is essential in PB reactors, as when the bed height falls by more than 10 % at higher flow rates, it indicates that the particle are deformed at higher pressures, and this causes an increase in the pressure drop. The pressure drop can be determined in a simple column with 1 m bed height, and at least 20 g of particles at different flow rates; a minimum pressure drop of 1 bar must be maintained for 1 h. Compressible particles are not suitable for use in PB reactors. For fluidization see Chapter 9.

The fluorescence spectra of immobilized enzymes can be used to determine the enzyme distribution within the support, with spectral changes due to immobilization being monitored either by spectrofluorimetry or by confocal microscopy (Kasche et al., 1994; Heinemann et al, 2002). When fluorochromes with a pH-dependent fluorescence intensity are co-immobilized with the enzymes, the pH-gradient in the particles during reactions where H^+ are formed or consumed can be determined (Spieß and Kasche, 2001).

One important property of immobilized biocatalysts that frequently is not measured is the stationary charge density, which can be determined by titration. When this has been carried out, net charge densities of more than 10 mM have been observed. This is much larger than net charge densities due to immobilized enzymes, where n_{E} is <0.1 mM. These charges are either formed in the deactivation of functional groups not used for the immobilization of the enzyme, or are due to charged compounds used in the production of the supports (Bozhinova et al., 2004; Kasche et al., unpublished data). As these charges are localized on the surface of the pores

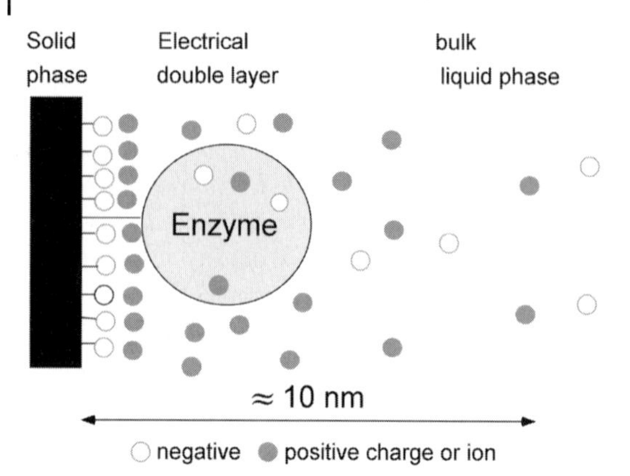

Fig. 8.11 The electrical double layer formed by stationary charges on the surface of pores in particles with immobilized enzymes. In the double layer the pH and ionic strength differs from the corresponding values in the bulk phase. This can change the intrinsic properties of the enzyme that is located in the double layer (Hunter, 1993; González-Caballero and Shilov, 2002).

in the particles, they are distributed in a much smaller volume than the total particle volume. The local charge density at the pore surface can then become more than 1 M. This gives rise to an electric double layer with dimensions in the nm-range with pH- and I-values that differ from the bulk values (Fig. 8.11). This may influence the intrinsic properties of the immobilized enzyme, especially when charged substrate or product molecules are involved in the enzyme process. It can also reduce the immobilization yield when the enzyme has the same charge as the surface. This requires that the isoelectric point (pI) of the enzyme is known, and this can be determined using isoelectric focusing (see Chapter 4, Section 4.5). In this case, immobilization should be carried out at a high ionic strength, in order to reduce repulsion between the support surface and the enzyme, in order to increase the yield (Chikere et al., 2001).

8.6.2
Kinetic Characterization of Immobilized Biocatalysts; Influence of Support Properties on the Nano- and Micrometer Level in Aqueous and other Systems

The intrinsic properties listed in Table 8.1 are required to estimate the amount of biocatalyst needed to obtain a given space–time yield – that is, for the design of an enzyme process. Some of the immobilized enzyme molecules may be inactivated during the immobilization procedure. This can be determined from the mass balance both for concentration and activity (enzyme amount/activity before immobilization – enzyme amount/activity in wash solutions after immobilization) that for pure enzymes can be obtained using spectroscopic measurements. Whether all of the immobilized enzymes also are active can only be determined by measuring the immobilized active site concentration, n_E. As the kinetic properties generally are changed when enzymes are immobilized, this is only possible for enzymes for which the active sites can be titrated. For this aim, substrates that react specifically with the active site irreversibly, and have a colored or fluorescent compound as releasing group, are used. When the substrate is in excess, the concentration of the leaving group equals the number of active sites in the sample of the immobilized enzyme. Alternatively, an inhibitor that irreversibly blocks the active site can be used. By increasing the concentration of this inhibitor until the enzyme becomes inactive, the number of active sites can be titrated. These active site titrations are mainly available for hydrolases such as α-chymotrypsin, trypsin, penicillin amidase, and lipase (Gabel et al., 1974; Svedas et al., 1977; Fujii et al., 2003).

For covalent immobilization, almost quantitative binding yields can be reached. In this case the concentration of immobilized enzyme is limited by the content of reactive groups on the support surface and the total surface area (see Chapter 6, Section 6.2). For noncovalent adsorption, the amount of immobilized enzyme is determined by the adsorption isotherm and concentration of the free enzyme in solution. When peptidases are immobilized, a loss of active enzyme by autoproteolysis must be avoided; this is achieved by dissolving the enzyme in a buffer of a pH at which the enzyme activity is low, and by performing the immobilization at a low enzyme concentration (Kasche, 1983).

The maximal immobilized active enzyme concentration n_E is limited by the cross-sectional area of the enzyme ($\approx 1 - 5 \times 10^{-17}$ m^2 for molecular weight in the range of 10 to 100 kDa) and the total surface area in the particles (10–100 m^2 mL wet support). The concentration is obtained when the pore surface is covered by immobilized enzymes; the maximal values for n_E are then in the range of 1.6 to 16 mM (16–160 mg mL) for a 10-kDa enzyme, and 0.3 to 3 mM (30–300 mg mL) for a 100-kDa enzyme. However, only 10 % of these values are reached in practice (Katchalski-Katzir and Krämer, 2000; Janssen et al., 2002).

8.6.2.1 Determination of V_{max} and k_{cat}

The intrinsic properties of immobilized enzymes can only be determined in systems that are not mass transfer-limited – that is, where the stationary effectiveness factor η equals 1. This applies always when $[S] \gg K'_m$ – that is, for the determination of V'_{max} (see Fig. 8.7). The turnover number k'_{cat} can only be determined for enzymes of which the active sites can be titrated.

A simple method is available to determine η. The particles can be disintegrated by sonication to reduce R and the Thiele modulus to a value where $\eta \approx 1$. By comparing the initial rate of the enzyme-catalyzed reaction with the same initial substrate content for whole and disintegrated particles, η for the whole particles can be determined (Figs. 8.3 and 8.12). Even when $\eta \approx 1$, the turnover number determined may not be an intrinsic property (Fig. 8.12 b). This applies for reactions where acids or bases are either produced or consumed. A pH gradient may then be formed in the particles which perturbs the pH-dependent intrinsic properties (see Chapter 2, Section 2.7). To avoid this, a buffer with high buffering capacity at the pH where measurements are performed should be used. This implies that the pK-value of the buffer should be near this pH-value (Tischer and Kasche, 1999). For such measurements a buffer with an ionic strength of 0.05 M is sufficient (Fig. 8.12).

It is well known that the intrinsic kinetic properties of an enzyme are changed when it has bound a ligand, or is chemically modified in solution. The same is expected for an enzyme which is bound (adsorbed) or chemically modified by covalent

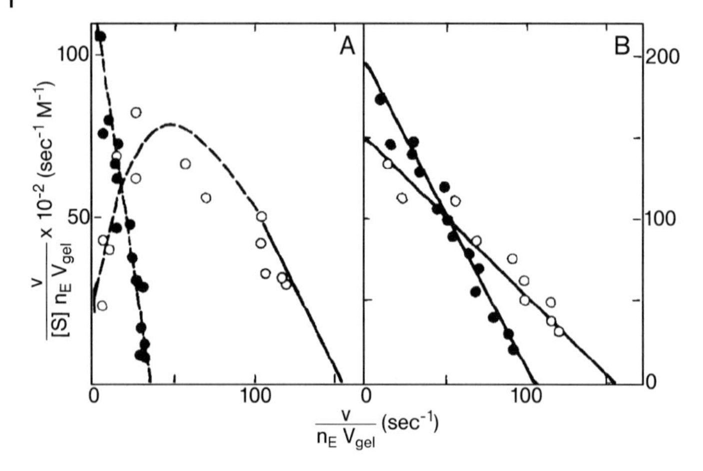

Fig. 8.12 Effects of pH-gradients in stirred-batch reactors for enzyme processes where H^+ is produced. Influence of buffer capacity and particle radius on the determination of k_{cat} and K_m for α-chymotrypsin immobilized in Sepharose 4B particles (n_E = 90 μM), from rate measurements using Eq. (2.13). (A) (R = 60 μm); (B) R = 3 μm (homogenized particles from (A)) at 25 °C and pH 8.0. Total ionic strength 0.25 M of which for buffer 0.05 M (○) or 0.001 M (●). Substrate: *N*-Acetyl-(*S*)-tyrosine-ethyl-ester (Kasche and Bergwall, 1973).

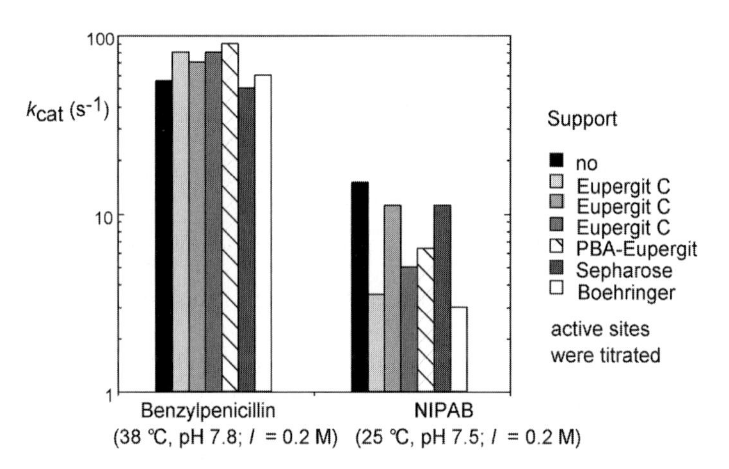

Fig. 8.13 Turnover number for the hydrolysis of penicillin G and 6-nitro-3-phenylacetamido benzoic acid (NIPAB) by penicillin amidase from *E. coli* immobilized in different supports, determined in stirred batch reactors. The ionic strength is only due to the phosphate buffer used (B. Galunsky and V. Kasche, unpublished data).

bonds to the surface of a support. This is shown in Figure 8.13, where the turnover number of penicillin amidase immobilized in different supports is given for two substrates. For one substrate the immobilization led to an increase, and for the other to a decrease in k_{cat}, compared with the value for the free enzyme. The observed changes were up to a factor of 2. The causes for this are still unclear, but these results show that – as expected – the intrinsic properties of enzymes are changed when they are immobilized. This is also observed for enzymes adsorbed to hydrophobic supports (Kasche et al., 1991). In some cases the changes are within experimental error; for example, the turnover number determined for α-chymotrypsin in whole and homogenized particles at high buffer capacity in Figure 8.12 equals k_{cat} for the free enzyme (see Table 2.4).

In many enzyme processes the stationary effectiveness factor η equals 1 at the initial substrate content. In processes where H^+ is either formed or consumed, perturbation of the intrinsic constants due to pH-gradients in the particles cannot be neglected (Spieß and Kasche, 2001). This is shown for the equilibrium-controlled hydrolysis of penicillin and glutaryl-7-amino-cephalosporanic acid in Figure 8.14. Without buffer, the formed acid (p$K \approx$ 4–5) and base (p$K \approx$ 4.5) cause a significant

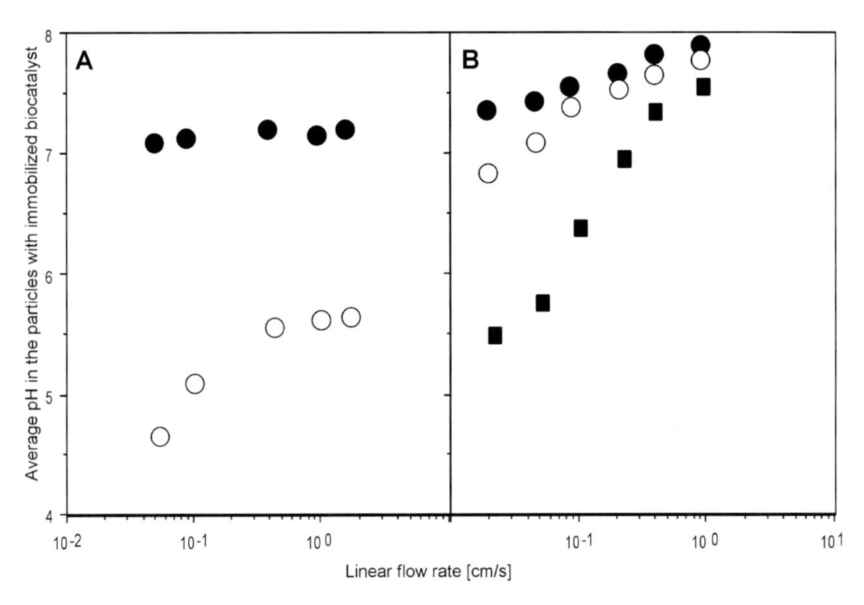

Fig. 8.14 Effects of pH-gradients in immobilized enzyme particles in packed bed reactors for an enzyme process where H^+ is produced. Influence of buffer capacity and linear flow rate on average pH in particles with immobilized biocatalysts. Filled symbols with buffer; open symbols without buffer. The average pH in the particles was determined from the fluorescence intensity of co-immobilized fluorescein, the fluorescence intensity of which is pH-dependent below pH 9. A small packed bed reactor (length 3–5 mm) was used, where the pH decrease over the reactor was less than one pH-unit. The stationary effectiveness factor at the initial substrate content was ≈1; that is, the systems used were not mass transfer controlled. (A) Hydrolysis of 300 mM penicillin by penicillin amidase from *E. coli* immobilized in Eupergit C (n_E = 20 μM; average particle radius 80 μm) with and without phosphate buffer (I = 0.05) at 38 °C. The pH of the substrate solution at the reactor inlet was 7.9. (B) Hydrolysis of 40 mM (circles) glutaryl-7- aminocephalosporanic acid (Glu-7-ACA) with and without phosphate buffer (I = 0.2 M), 135 mM Glu-ACA (squares) with buffering components ammonia, carbonate, and acetate ($I \approx$ 0.6 M) at 25 °C, with glutaryl amidase immobilized in particles similar to Eupergit C produced by former Hoechst AG (n_E = 100 μM; average particle radius 100 μm). The pH of the substrate solution at the reactor inlet was 8.0 (Spieß et al., 1999).

Table 8.2 Intrinsic k_{cat} and K_m-values for free and immobilized penicillin amidase (PA) from *E. coli* and bovine trypsin (TRY) immobilized in different supports for the hydrolysis of different substrates determined in phosphate buffer of different ionic strength (pH 7.5) at 25 °C. The enantioselectivity (Eq. 2.15) for PA is also given (Goldstein et al., 1964; Wiesemann, 1991).

Substrate/enzyme	*Support*	*I* [M]	k_{cat} [s^{-1}]	K_m [mM]	E_{eq}
(*R*)-phenylglycine-amide/PA	None	0.20	35	19	
	Eupergit C	0.20	30	18	
	PBA-Eupergit C	0.20	16	7	
	Sepharose	0.20	22	7	
(*S*)-phenylglycine-amide/PA	None	0.20	6	11	0.30
	Eupergit C	0.20	11	26	0.25
	PBA-Eupergit C	0.20	17	34	0.20
	Sepharose	0.20	14	26	0.15
Benzoyl-Arg-ethylester/TRY	None	0.04	n.d	6.9	
	None	0.50		6.9	
	Ethylene-malate	0.04		0.2	
	Copolymer (negative)	0.50		5.2	

decrease in pH, to about 5 in the particles, but by increasing the linear flow rate to up to 1 cm s^{-1} this can be reduced. A further reduction is achieved by adding a suitable buffer with a pK near the optimum pH for the reaction.

8.6.2.2 K_m and K_i

In order to determine these properties, n_E must not itself be determined, but the effectiveness factor must be equal to 1. As for the turnover number, these intrinsic properties are expected to be changed when an enzyme is immobilized. For the equilibrium constants, a decrease is expected due to steric hindrance caused by the immobilization that reduces the rates of association reactions. Changes in rate constants for the dissociation reaction may counteract this. The K_m-value for the enzyme in the homogenized particles in Figure 8.12 is five-fold larger than the value for the free enzyme (Kasche and Bergwall, 1973). Similar or opposite changes have been observed for other enzymes (Table 8.2).

These results show that the turnover number and the Michaelis–Menten constant can change considerably when an enzyme is immobilized. In most cases – except for the case of adsorption to ion-exchangers shown in Table 8.2 – the causes for this remain unclear. It is probably partly due to conformational changes caused by the covalent binding or adsorption to the surface that, for free enzymes in solution, also leads to changes in k_{cat} and K_m. The nano-environment around the immobilized enzyme also causes changes in the apparent constants (Table 8.2). At low ionic strength the positively charged substrate Benzoyl-Arg-ethylester accumulates at the negatively charged surface of the support, but this does not occur at a higher ionic strength.

8.6.2.3 **Selectivities**

The stereoselectivity E_{kin} and selectivity in the kinetically controlled synthesis of free and immobilized penicillin amidase (see Chapter 2, Section 2.7.1.2) using different supports are illustrated in Figure 8.15. It has been observed that most support for immobilized biocatalysts contain stationary charges (mainly carboxyl or amino groups) whose concentration are orders of magnitude larger than due to the charges on the immobilized enzymes. With charged substrates this can (as shown in Fig. 8.15) cause large changes in the selectivities. The measurements in Figure 8.15 were carried out under conditions where $\eta \approx 1$. This is also verified by comparison of the whole and homogenized supports in Figure 8.15 B. The total rates of synthesis and hydrolysis were not changed when the particles were homogenized. The hydrolysis of (*R*)-phenylglycine-amide by penicillin amidase has been found to be independent of the ionic strength. The reduction in selectivity observed in this figure is due

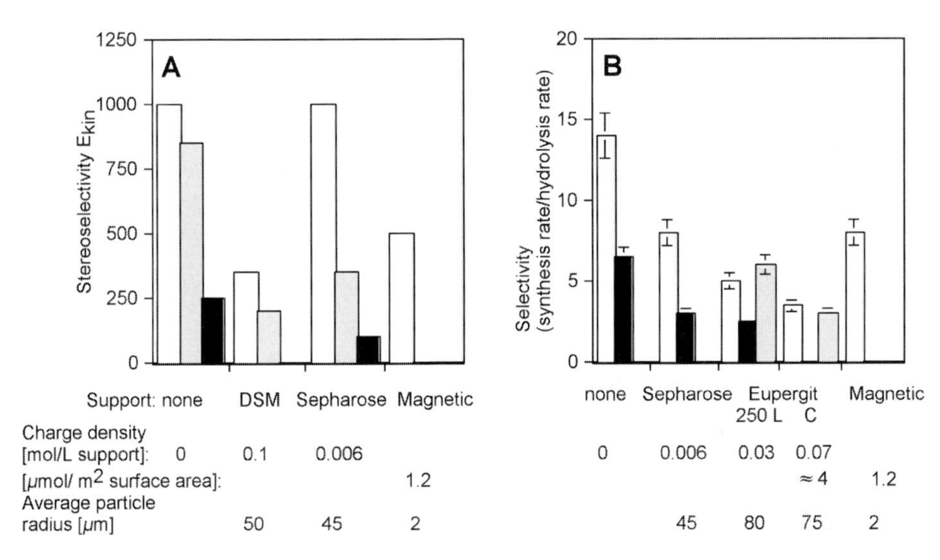

Fig. 8.15 The influence of the support charge on: (A) the stereoselectivity E_{kin} (Eq. 2.16), and (B) selectivity in kinetically controlled synthesis (Eq. 2.1) for reactions with charged substrates catalyzed by free and immobilized penicillin amidase (PA). The DSM support is based on gelatin (Spiess and Kasche, 2001) and the magnetic support on polyvinyl alcohol (Bozhinova, 2004). (A) S_1'-stereoselectivity in the kinetically controlled synthesis of (*R*)-phenylglycil-(*R,S*)-Phe from 20 mM (*R*)-phenylglycine-amide and 100 mM-(*R,S*)-Phe at pH 9.0 and 25 °C, catalyzed by PA from different sources (open bars: from *E. coli*; gray bars: from *A. faecalis*; black bars: from *B. megaterium*). E-values above 1000 cannot be accurately determined, the experimental error in the values below 1000 are ±10 %. (B) Selectivity in the kinetically controlled synthesis of cephalexin from 200 mM (*R*)-phenylglycine-amide and 50 mM 7-aminodesacetoxycephalosporanic acid in phosphate buffer ($I = 0.05$) at pH 7.5 and 5 °C, catalyzed by free and immobilized PA from *E. coli*. The measurements were performed without (open bars) and with black bars addition of 1 M NaCl. For the Eupergit support measurements were also carried without NaCl in homogenized supports grey bars. After homogenization, the average particle size was 2.3 and 1.1 μm for Eupergit 250 L and Eupergit C, respectively.

to the ionic strength in the electric double layer on the support surface where the enzyme is located. This reduces binding of the negatively charged nucleophile (7-ADCA) to its positively charged binding site on the enzyme (Kasche, 1986). The selectivity decreases with the concentration of stationary charges on the support for the immobilized biocatalyst, or the ionic strength for the free enzyme. These results show that the charge density of supports must be reduced in order to improve the selectivity of immobilized biocatalysts for kinetically controlled synthesis of charged substrates.

The changes in stereoselectivity caused by stationary charges on the support in Figure 8.15A are more complex to analyze. The binding of the negatively charged nucleophile (*R,S*)-Phe to the positively charged binding site is reduced by the ionic strength in the double layer. However, this alone cannot cause the reduction in E_{kin}. The p*K* of the N-terminal Ser that is acylated in the enzyme-catalyzed reaction (see Fig. 2.5, p. 38) has been shown to be increased by the negatively charged carboxyl group of (*R*)-Phe (Lummer et al., 1999). This effect can be reduced by the ionic strength in the double layer, and contributes to the decrease in stereoselectivity of the immobilized biocatalysts.

For uncharged substrates the stereoselectivity has been found both to increase and decrease when the enzyme is immobilized in charged and hydrophobic supports where $\eta = 1$ (Kasche et al., 1991; Palomo et al., 2003). Further studies are required to explain the causes for these changes, but the above discussion indicates that nano-environmental effects can cause large changes in selectivities when the enzyme is immobilized.

8.6.2.4 Determinations of Effectiveness Factors

The *stationary effectiveness factor* η can be determined as shown in Figure 8.3, and can also be estimated from the Thiele modulus calculated using the values for the free enzyme, and the initial substrate content using Figure 8.7.

The *operational effectiveness factor* η_o can be determined as shown in Figure 8.4, or estimated from the Thiele modulus calculated as above, using Figure 8.8. It can also be calculated from experimental data for the times required for the same substrate conversion t_{12} and t'_{12} for free (amount per unit reactor volume V_{max}) and immobilized (amount per unit reactor volume V'_{max}) enzyme, respectively using the relationship

$$\eta_{o,12} = (t_{12} \cdot V_{max})/(t'_{12} \cdot V'_{max}) \tag{8.30}$$

assuming that $K_m = K'_m$.

8.6.3 Productivity and Stability under Process Conditions

The biocatalyst concentration V'_{max} and the space–time yield *STY* of immobilized biocatalysts cannot be kept constant under process conditions, due to inactivation of the biocatalyst. Besides the inactivation processes listed in Table 2.9 (p. 81), the following processes contribute to the inactivation of immobilized biocatalysts:

- Abrasion due to shear forces, stirrer and particle-particle collisions in stirred batch reactors.
- Chemical instability of the support or the covalent bond between the support and the enzyme and desorption of adsorbed enzyme (Ulbrich et al., 1991).
- Bimolecular reactions with substrates or products at the high concentrations used in enzyme processes, such as inactivation due to reaction between glucose and amino groups of Lys in glucose isomerase-catalyzed isomerization of glucose (see Section 2.11, p. 98).
- Mass transfer limitations due to adsorption of molecules or microorganisms in or on the support'.
- Desorption of metal ions required for activity and/or stability, such as Zn^{2+} for carboxypeptidase A (Fig. 2.4) or hydantoinase (see Chapter 3, p. 151) or Ca^{2+} for α-amylase, some lipases, penicillin amidase, subtilisin (see Chapter 4, Section 4.4).

The storage, pH- and temperature-stability of immobilized biocatalysts is generally higher than for the free enzymes (Aguado et al., 1995; Monti et al., 2000; Park et al., 2002; Ferreira et al., 2003). The inactivation processes can – as for free enzymes – be characterized by an apparent first-order rate constant, k_i, or a half-life time $t_{1/2}$ – that is, the process time when the biocatalyst activity is 50 % of its initial value. This value is generally determined by the producers or users of immobilized biocatalysts, by determination of the *STY* as a function of time for the specific enzyme process. A high value of $t_{1/2}$ (≥50–100 days) is essential to reduce the biocatalyst cost and to increase biocatalyst productivity. These values have now been reached for important enzyme processes, such as the hydrolysis of penicillin and the isomerization of glucose. The integral of the *STY* over time under process conditions until the immobilized biocatalyst has 20–50 % of the initial *STY*, and is replaced by new biocatalyst, indicates the productivity of the immobilized biocatalyst. This is generally expressed as kg substrate converted per kg immobilized biocatalyst. For different enzyme processes this value varies in the range of 100 to 10 000 kg kg^{-1}. Both the producers and users of immobilized biocatalysts make continuous attempts to increase productivity in order to reduce biocatalyst costs.

8.7
Comparison of Calculated and Experimental Data for Immobilized Biocatalysts

When designing an enzyme process, the immobilized biocatalyst activity V'_{max} required to obtain a desired *STY*, or the time needed for a given substrate conversion with a given V'_{max}, must be either calculated or estimated. This was discussed in Chapter 2, Section 2.8 for free enzymes, but for immobilized enzymes the relationships derived in Section 8.5 can be used. The effectiveness factors calculated in Section 8.4 must also be considered.

Such calculations have been presented for some enzyme processes, such as the hydrolysis of penicillin G, penicillin V, and glutaryl-7-aminocephalosporanic acid (Carleysmith et al., 1980; Haagensen et al., 1983; Tischer et al., 1992; Spiess et al.,

1999). The influence of pH-gradients (Fig. 8.14) and effectiveness factors was only considered in the last of the above references, and the accuracy of these estimations will be illustrated for the hydrolysis of penicillin G (the process window is given in Chapter 2, p. 102, Exercise 14), under conditions used in large-scale processes. This enzyme process, with competitive product inhibition, was previously mainly carried out in batch reactors with some added buffer at a pH ~8.0, and temperatures between 30 and 38 °C. Under these conditions, the end-point of the reaction can be selected to be ≥90 % substrate conversion. For an accurate calculation, all enzyme kinetic constants (V'_{max}, K'_m, and inhibition constants) for the hydrolysis and the back-reaction and their pH- and temperature (T)-dependence of this equilibrium-controlled reaction, must be known. For most enzyme processes these constants (except for V'_{max}) are difficult and time-consuming to determine, and hence these data are rarely published. Such calculations must therefore be performed with simplifying assumptions. For this case they are: constant pH, biocatalyst inactivation, and the back- reaction neglected. The time t required for a substrate conversion from $[S]_0$ to $[S]_t$ for this reaction with competitive product inhibition in a stirred batch reactor is (Eq. (8.3), Eqs. (2.36) and (8.24)).

$$t = \frac{K_m (1 + [S]_0/K_i) \cdot \ln ([S]_0/[S]_t) + (1 - K_m/K_i) ([S]_0 - [S]_t))}{\eta_0 \cdot V_{max,0}} \tag{8.31}$$

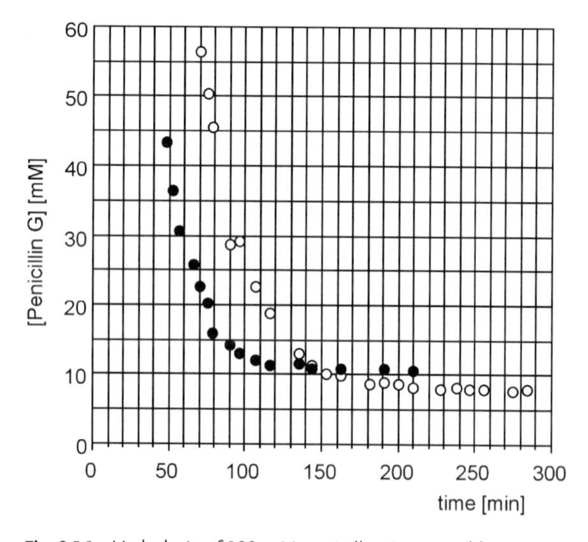

Fig. 8.16 Hydrolysis of 300 mM penicillin G at 38 °C and pH 7.8 (phosphate buffer I = 0.05 M) with free (○, 9000 U L^{-1} reactor) and immobilized (●, 13 000 U L^{-1} reactor) penicillin amidase (PA) from *E. coli* in a stirred batch reactor. The pH in the reaction solution was kept constant by titration with 3 M NH_3. The PA was adsorbed to the bifunctional support (see Table 4.4, p. 188) phenyl-butyl-amine-Eupergit (PBA-Eupergit, average particle radius 80 μm) and crosslinked with glutaraldehyde. The concentration of active immobilized enzyme, determined by active site determination, was 3.4 mg mL^{-1} wet support (or 130 000 U L^{-1} wet support). The penicillin G concentration was determined by HPLC.

With $V_{max,0}$ from the data in Figure 8.16, K_m from Table 2.4 (p. 51), $K_i = 20\ \mu M$ for the competitive product inhibition by phenylacetic acid, $D \approx 5 \times 10^{-6}\ cm^2\ s^{-1}$ the square of the Thiele modulus becomes <1, as for this process the apparent K_m-value $K_m(1+[S]_0/2K_i)$ can be used (see Section 8.4.2, p. 343). From Figure 8.8 it follows that the operational effectiveness factor for 90 % substrate conversion is ≈1. The times calculated with these values are:

- for the free enzyme 41 min (experimental value ≈90 min);
- for the immobilized biocatalyst 29 min (experimental value ≈55 min).

In both cases the calculated value is about a factor of 2 smaller than the experimental value. For the immobilized biocatalyst, this is partly due to the pH-gradient inside the particles (Fig. 8.14A) – that is, the V_{max}-value used in the calculation is too high. Local pH-gradients may also occur with the free enzyme. This can be avoided by using more buffer, but it is not recommended for enzyme processes as the buffer used must ultimately be removed from the product and the waste water. Other causes for an underestimation of the time required for desired substrate conversion are:

- experimental error in the kinetic data used (≥20 % in K_m and K_i (Deranleau, 1969) and ≥10 % for V_{max});
- the assumptions made for the calculation.

With fewer assumptions and more data, the discrepancy between calculated and experimental values can be reduced (Spiess et al., 1999). However, in the practical design of enzyme processes such estimations must be performed (as illustrated above) by using simplifying assumptions, together with personal and published data with considerable experimental errors. Thus, a difference of a factor of 2 in the calculated and experimental data is not unexpected. The additional studies required to determine all necessary data for reducing this difference may take longer than the adjustment of the design based on observed differences between calculated and experimental data.

Today, the hydrolysis of penicillin G is increasingly carried out in several sequential or recirculation PB reactors where the pH is adjusted between the beds or the vessel used in the recirculation reactor (see Chapter 9). In this way, shorter hydrolysis times, in order to reduce loss of the unstable product 6-amino-penicillanic acid, can be obtained (see also Case study 2 in Chapter 9). A higher conversion can be obtained when the hydrolysis is carried out at pH 9 with the more stable penicillin amidase from *A. faecalis* (Chapter 2, p. 81).

8.8 Application of Immobilized Biocatalysts for Enzyme Processes in Aqueous Suspensions

Many substrates and products used or produced in enzyme processes are slightly soluble in water. In Chapter 2, Section 2.9.1 it was shown that processes with such compounds can be carried out with free enzymes and high *STY* in aqueous suspen-

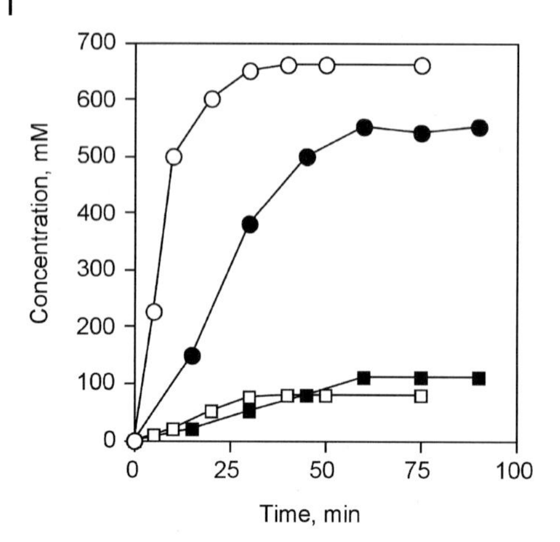

Fig. 8.17 Kinetically controlled synthesis of *N*-acetyl-Tyr-Arg-NH_2 (circles) in aqueous suspensions with precipitated substrate *N*-acetyl-Tyr-ethyl ester (ATEE), with free (open symbols,) and immobilized (closed symbols) α-chymotrypsin at 25 °C. Starting conditions: pH 9.0 (carbonate buffer, $I = 0.2$ M), 750 mM ATEE and 800 mM Arg-NH_2; free enzyme 20 μg mL^{-1}; immobilized enzyme 200 μL^{-1} immobilized in Sepharose (20 μM in the support). The product *N*-acetyl-Tyr (squares) did not precipitate during the reaction. The pH was kept constant at pH 9.0 during the reaction. The free ATEE concentration was less ≤20 mM (Kasche and Galunsky, 1995).

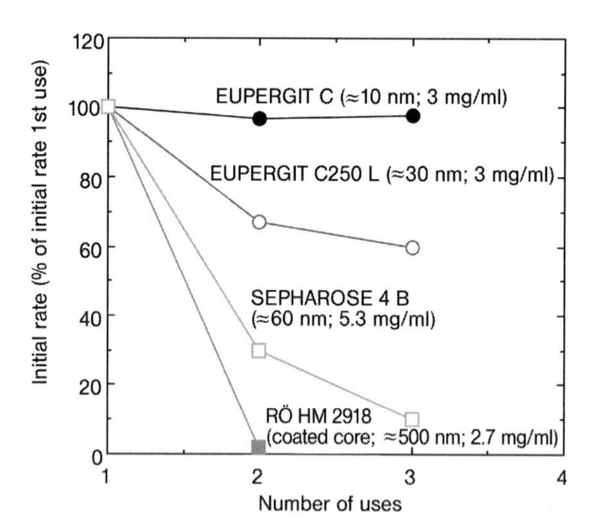

Fig. 8.18 Influence of product precipitation on the initial rate as percentage of rate for the first use for the hydrolysis of 400 mM (*R*)-phenylglycine amide at pH 7.5 and 25 °C catalyzed by penicillin amidase from *E. coli* immobilized in supports with different radii. The support pore radius and the immobilized enzyme concentration is given in the brackets after the support (Kasche and Galunsky, 1995).

sions or emulsions. Figure 8.17 shows that such processes with suspended substrate can also be carried out with immobilized enzymes (Kasche and Galunsky, 1995; Youshko et al., 2002). Indeed, such processes can even be carried out with immobilized enzymes when the product precipitates, although in this situation particles with narrow pores must be used to prevent precipitation of the product in the pores of the immobilized biocatalyst particle, as this would limit reuse of the immobilized biocatalyst (Fig. 8.18, see also Fig. 2.23, p. 73).

Table 8.3 Problems caused by concentration gradients in enzyme processes with immobilized biocatalysts, and how they can be solved

Process	*Problems compared to systems with free biocatalyst*	*Examples*	*How to minimize the problem*
I. Equilibrium-controlled process			Fill out!
1. Hydrolysis involving uncharged substrates/products	Increased product inhibition Reduced rate Increased (equilibrium control) or decreased (kinetic control) formation of byproducts; Electrical double layer with charged supports	Di-, oligo-saccharide hydrolysis	
2. Hydrolysis involving acids and bases	Formation of pH-gradients Decreased rate, biocatalyst stability and yield See 1	Hydrolysis of antibiotics (cephalosporin, penicillin G and V), peptides, proteins, lipids, nucleic acids, etc.	
3. Racemate resolution	See 1 and 2 Decreased steric purity of product	Industrial racemate resolutions	
II. Kinetically controlled process			
Synthesis of condensation products (antibiotics, peptides etc.)	Decreased yield See I: 1 and 2	Synthesis of β-lactam antibiotics (cephalosporin, penicillin), peptides	
Racemate resolution	Decreased steric purity and yield See I: 1–3		

8.9
Improving the Performance of Immobilized Biocatalysts

In Section 8.4 it was shown that concentration gradients of substrates and products are formed in immobilized biocatalyst particles when the effectiveness factors are smaller than 1. In processes where H^+ is formed or produced, a pH-gradient is formed, even in immobilized biocatalysts where the effectiveness factors are ≈1 (Fig. 8.14). The problems caused by such gradients or electric double layers (see Section 8.6.2) in different enzyme processes are summarized in Table 8.3. The reader should try to suggest how the stated problem can be resolved.

The concentration gradients can be reduced by decreasing the Thiele modulus. This can be achieved by using smaller particles or by reducing the diffusion distance between the flowing free solution and the immobilized enzymes. The latter has been achieved in perfusible particles mainly developed for chromatography (Afeyan et al., 1990), though until now these have been used only minimally in enzyme technology.

Recently, however, interest has increased in the study of small, especially non-porous magnetic and non-magnetic particles in the micrometer and nanometer ranges (Jia et al., 2003; Bozhinova et al., 2004). Such non-porous and the perfusible particles can be used for the selective hydrolysis of biopolymers with hydrolases, as the biopolymers can easily diffuse to the enzymes. Examples for their potential application are the proteolytic processing of recombinant pro-proteins to the biologically active protein, such as the processing of recombinant human pro-insulin to insulin.

Magnetic particles can be easily separated from the reaction mixture at the end of the reaction. Larger *STY* can be obtained with smaller porous particles (see Fig. 8.5), but these cannot be used in PB reactors, due to the increased pressure drop. Another limit of small particles for use in PB reactors is shown in Figure 8.19. The total surface area available for convective mass transfer is reduced with decreasing particle radius, due to the increasing fraction of the total outer surface occupied by the particle-particle contact areas, where only diffusive mass transfer is possible. In packed beds there are more than ten such direct particle-particle contact areas (Hinberg et al., 1974; Anderson and Walters, 1986; Renken, 1993).

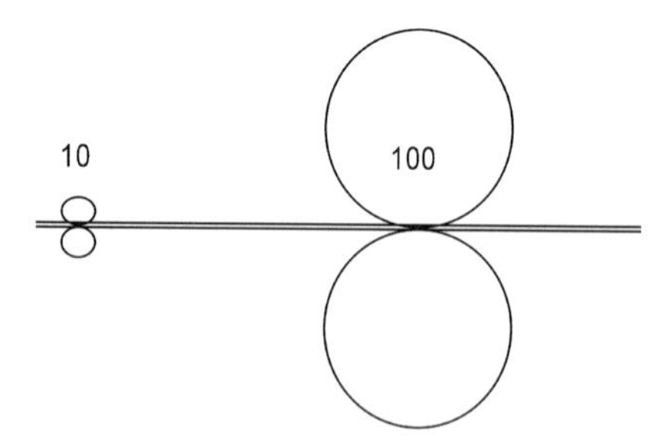

Fig. 8.19 The fraction of a particle–particle contact area with a distance <1 μm (between the horizontal lines) of the total particle surface increases, when the particle diameter decreases. This affects the effective particle surface for convective mass transfer when the particle radius becomes less than 50 μm (Renken, 1993).

8.10 Exercises

1. How can the substrate consumption rate in reactions catalyzed by an immobilized biocatalyst be measured in the free solution?

2. Define the stationary and operational effectiveness factor. How can they be determined?

3. Derive Eq. (8.12) for spherical particles.

4. What particle ranges are suitable for spherical immobilized biocatalysts (see Fig. 8.5 and Section 8.9)? Why?

5. How can you increase the effectiveness factor of a given immobilized biocatalyst?

6. Explain the following results:

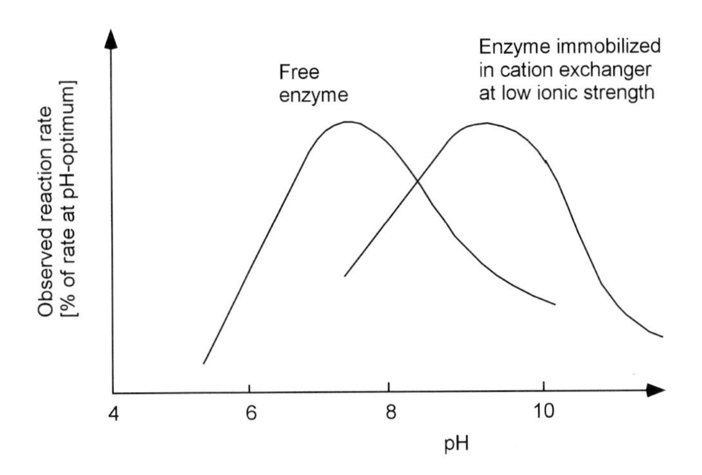

The pH is measured in the solution outside the particles. Why is $\eta > 1$ at some values of pH?

What happens at higher ionic strengths?

7. Try to explain the influence of the effectiveness factor in Eq. (8.26).

8. How are Eqs. (8.20) and (8.24) influenced by competitive product inhibition? How can this be minimized in the case of the product H^+?

9. Why does the space–time yield (*STY*) of immobilized biocatalysts have an upper limit? (Hint: see Fig. 8.1).

10. Explain the following graph for an immobilized biocatalyst:

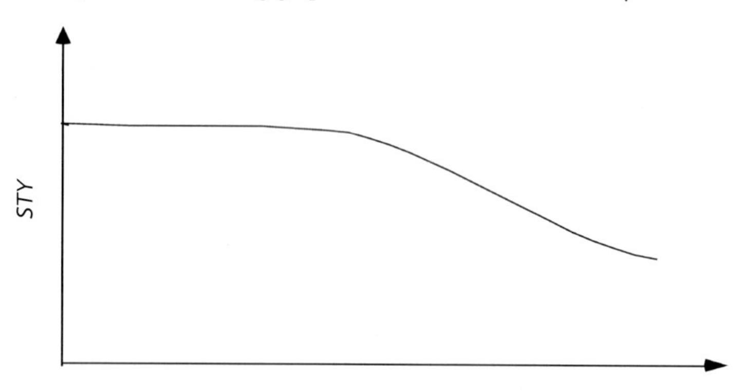

11. How can the *STY* be kept constant in an enzyme process with several reactors? The substrate content and the degree of substrate conversion cannot be changed.

12. Derive Eq. (8.31).

13. Determine the selectivity for the free and immobilized enzyme in Figure 8.15. Why do they differ?

14. In the hydrolysis of maltose by glucoamylase, isomaltose is a by-product formed in the equilibrium-controlled reaction of the product:

    glucose + glucose $\rightleftharpoons$ isomaltose

    In the hydrolysis of lactose by β-galactosidase, oligosaccharides formed in the kinetically controlled reaction between the intermediate product galactosyl-enzyme and the substrate lactose (Fig. 1.9, p. 21) is a by-product. The selectivity of this process is assumed not to be changed upon immmobilization of the enzyme.
    Discuss how the by-product formation is influenced by the immobilization of the enzyme for the above enzyme processes. The same enzyme activity is used in the reactors with free and immobilized biocatalysts.

15. Precursors for the neuropeptide kyotorphin Tyr-Arg can be synthesized in the kinetically controlled process.

    X-Tyr-OEt + Arg-NH_2 (or Arg-OEt) $\rightarrow$ X-Tyr-Arg-NH_2 (or X-Tyr-Arg-OEt) + EtOH

    in aqueous suspension (X = Acetyl, X-Tyr-OEt is suspended (Fig. 2.23, p. 72; Figs. 8.17 and 8.18, p. 360) or in free aqueous solution (X = Maleyl that increases the solubility in water, Chapter 3, Section 3.2.2.3, p. 140) catalyzed by the free or immobilized endo-peptidase α-Chymotrypsin.

    a) Compare and discuss the ad- and disadvantages of these enzyme processes (Hints: How can the products and by-products be isolated from the remaining substrates and the enzyme? Esters are unstable in water at high pH. EtOH destabilizes enzymes).

b) How can kyotorphin be produced from the precursors formed in a)?

c) What properties must an enzyme have to be a suitable biocatalyst for the production of kyotorphin with minimal by-product formation from the unprotected amino acids? (Hints: Figs. 2.3 to 2.5 and Table 2.4). How can the product be isolated from the remaining substrates and the enzyme (Hint: Isoelectric point of kyotorphin, Tyr and Arg (see Chapter 4, Section 4.5)?

Literature

Afeyan, N.B., Gordon, N.F., Mazsaroff, I., Varady, L., Fulton, S.P., Yang, Y.B., Regnier, F.E., Flow-through particles for the HPLC separation of biomolecules: perfusion chromatography, *J. Chromatogr.* **1990**, *519*, 1–29.

Aguade, J., Romero, M.D., Rodríguez, L., Calles, J.A., Thermal deactivation of free and immobilized β-glucosidase from *Penicillium finiculosum*, *Biotechnol. Prog.* **1995**, *11*, 104–106.

Andersson, D.J., Walters, R.R., Equilibrium and rate constants of immobilized concanavalin A determined by high-performance affinity chromatography, *J. Chromatogr.* **1986**, *376*, 69–85.

Bozhinova, D., Galunsky, B., Yueping, G., Franzreb, M., Köster, R., Kasche, V., Evaluation of magnetic polymer micro-beads as carriers for immobilized biocatalysts for selective and stereoselective transformations, *Biotechnol. Lett.* **2004**, *26*, 343–350.

Buchholz, K. (Ed.), *Characterization of immobilized biocatalysts*, DECHEMA Monograph Bd. 84, **1979**.

Carleysmith, S.W., Dunnill, P.D., Lilly, M.D., Kinetic behaviour of immobilized penicillin amidase, *Biotechnol. Bioeng.* **1980**, *22*, 735–756.

Cheetham, P.S.J., Production of isomaltulose using immobilized cells. *Methods Enzymol.* **1987**, *136*, 432–454.

Chikere, A.C., Galunsky, B., Schünemann, V., Kasche, V., Stability of immobilised soybean lipoxygenases: influence of coupling conditions on the ionisation state of the active site Fe", *Enzyme Microb. Technol.* **2001**, *28*, 168–175.

Damköhler, G., Influence of diffusion, fluid flow, and heat transport on the yield in chemical reactors, *Der Chemie Ingenieur* **1937**, *3*, 359–485, translated in *Int. Chem. Eng.* **1988**, *28*, 132–198.

Deranleau, D.A., Theory of the measurement of weak molecular complexes, I. General Considerations, *J. Am. Chem. Soc.* **1969**, *91*, 4044–4049.

Fujii, R., Utsunomiya, Y., Hiratake, J., Sogabe, A., Sakata, K., Highly sensitive active-site titration of lipase in microscale culture media using fluorescent organophosphorus ester. *Biochim. Biophys. Acta* **2003**, *1631*, 197–205.

Ferreira, L., Ramos, M.A., Dordick, J.S., Gil, M.H., Influence of different silica derivatives in the immobilization and stabilization of a *Bacillus licheniformis* protease (Subtilisin Carlsberg), *J. Mol. Catal. B: Enzym.* **2003**, *21*, 189–199.

Gabel, D., Active site titration of immobilized chymotrypsin with a fluorogenic reagent. *FEBS Lett.* **1974**, *49*, 280–281.

Gonzáles-Caballero, B., Shilov, V.N., Electric double layer at a colloid particle, in: *Encyclopedia of Surface and Colloid Science*. Marcel Dekker, New York, **2002**.

Goldstein, L., Levin, Y., Katchalski, E., A water-insoluble polyanionic derivative of trypsin. II. Effect of the polyelectrolyte carrier on the kinetic behavior of the bound trypsin, *Biochemistry* **1964**, *3*, 1913–1919.

Haagensen, P., Karlsen, L.G., Petersen, J., Villadesen, J., The kinetics of penicillin-V deacylation on an immobilized enzyme, *Biotechnol. Bioeng.* **1983**, *25*, 1873–1895.

Heinemann, M., Wagner, T., Doumèche, B., Ansorge-Schumacher, M., Büchs, J., A new approach for the spatially resolved qualitative analysis of the protein distribution in hydrogel beads based on confocal laser scanning microscopy. *Biotechnol. Lett.* **2002**, *24*, 845–850.

Hinberg, I., Korus, R., O'Driscoll, K.F., Gel-entrapped enzymes: kinetic studies of immobilized β-galactosidase, *Biotechnol. Bioeng.* **1974**, *16*, 943–963.

Hunter, R.J., *Foundations of Colloid Science*, Oxford University Press, Oxford, **1993**.

Janssen, M.H., van Langen. L.M., Pereira, S.R.M., van Rantwijk, R., Sheldon, R.A., Evaluation of the performance of immobilized penicillin G acylase using active-site titration, *Biotechnol. Bioeng.* **2002**, *78*, 425–432.

Jia, H., Zhu, G., Wang, P., Catalytic behaviors of enzymes attached to nanoparticles: the effect of particle mobility. *Biotechnol. Bioeng.* **2003**, *84*, 406–414.

Kasche, V., Bergwall, M., in: *Insolubilized Enzymes* (Salmona, M., Saronio, C., Garattini, S., Eds.). Raven Press, New York, **1973**, pp. 77–86.

Kasche, V., Kuhlmann, G., Direct measurements of the thickness of the unstirred diffusion layer outside immobilized biocatalysts. *Enzyme Microb. Technol.* **1980**, *2*, 309–312.

Kasche, V., Correlation of theoretical and experimental data for immobilized biocatalysts. *Enzyme Microb. Technol.* **1983**, *5*, 2–14.

Kasche, V., Galunsky, B., Michaelis, G., Binding of organic solvent molecules influences the P_1'-P_2' stereo- and sequence specificity of α-chymotrypsin in kinetically controlled peptide synthesis, *Biotechnol. Lett.* 1991, *13*, 75–80.

Kasche, V., Mechanism and yields in enzyme catalyzed equilibrium and kinetically controlled synthesis of β-lactam antibiotics peptides and other condensation products, *Enzyme Microb. Technol.* **1986**, *8*, 4–16.

Kasche, V., Gottschlich, N., Lindberg, Å., Niebuhr-Redder, C., Schmieding, J., Perfusible and non-perfusible supports with monoclonal antibodies for biospecific purification of *E. coli* penicillin amidase within its pH-stability range, *J. Chromatogr.* **1994**, *660*, 137–145.

Kasche, V., Galunsky, B., Enzyme-catalyzed biotransformations in aqueous two-phase systems with precipitated substrate and/or product, *Biotechnol. Bioeng.* **1995**, *45*, 261–267.

Katchaski-Katzir, E., Krämer, D., Eupergit® C, a carrier for immobilization of enzymes of industrial potential, *J. Mol. Catal. B: Enzym.* **2000**, *10*, 157–176.

Lummer, K., Riecks, A., Galunsky, B., Kasche, V., pH-dependence of penicillin amidase enantioselectivity for charged substrates, *Biochim. Biophys. Acta* **1999**, *1433*, 327–334.

McLaren, A.D., Packer, L., Some aspects of enzyme reactions in heterogeneous systems, *Adv. Enzymol.* **1970**, *33*, 245–303.

Monti, D., Carrea, G., Riva, S., Baldaro, E., Frare, G., Characterization of an industrial biocatalyst: immobilized glutaryl-7-ACA acylase, *Biotechnol. Bioeng.* **2000**, *70*, 239–244.

Palomo, J.M., Muñoz, G., Fernándes-Lorente, Mateo, C., Fuentes, M., Guisan, J.M., Fernándes-Lafuente, R., Modulation of *Mucor miehei* lipase properties via directed immobilization on different hetero-functional epoxy resins. Hydrolytic resolution of *(R,S)*-2-butyroyl-2-phenylacetic acid, *J. Mol. Catal. B: Enzym.* **2003**, *21*, 201–210.

Park, S.W., Choi, S.Y., Chung, K.H., Hong, S.I., Kim, S.W., Characteristics of GL-7-ACA acylase immobilized on silica gel through silanization, *Biochem. Eng. J.* **2002**, *11*, 87–93.

Rashevsky, N., *Mathematical Biophysics*, The University of Chicago Press, Chicago, **1940**.

Renken, E., Signalentstehung und Signalentwicklung in fluorimetrischen dynamischen und Affinitäts-Durchfluß-Biosensoren. Dissertation, TU Hamburg-Harburg, **1993**.

Sherwood, T.K., Pigford, L.R., Wilke, R., *Mass transfer*. McGraw-Hill, Kogakusha Ltd., Tokyo, **1975**.

Spieß, A., Schlothauer, R, Hinrichs, J., Scheidat, B., Kasche, V., pH gradients in heterogeneous biocatalysts and their influence on rates and yields of the catalysed processes, *Biotechnol. Bioeng.* **1999**, *62*, 267–277.

Spieß, A., Kasche, V., Direct measurement of pH profiles in immobilised enzymes during kinetically controlled synthesis using CLSM, *Biotechnol. Prog.* **2001**, *17*, 294–303.

Svedas, V.K., Margolin, A.L., Sherstiuk, S.F., Klyosov, A.A., Berezin, I.V., Inactivation of soluble and immobilized penicillin amidase from *E. coli* by phenylmethylsulphonylfluoride: kinetic analysis and titration of the active sites, *Bioorg. Khim.* **1977**, *3*, 546–553.

Thiele, E.W., Relations between catalytic activity and size of particle, *Ind. Eng. Chem.* **1939**, *31*, 916–920.

Tischer, W., Giesecke, U., Lang, G., Röder, A., Wedekind, F., Biocatalytic 7-Aminocephalosporanic acid production, *Ann. N. Y. Acad. Sci.* **1992**, *613*, 502–509.

Tischer, W., Kasche, V., Immobilized enzymes: crystals or carriers?, *Trends Biotechnol.* **1999**, *17*, 326–335.

Ulbrich, R., Golbik, R., Schellenberger, A., Protein adsorption and leakage in carrier-enzyme systems, *Biotechnol. Bioeng.* **1991**, *37*, 280–287.

Villadsen, J., Michelsen M.L., *Solution of differential equation models by polynomial approximation*, Prentice-Hall, Englewood Cliffs, N.J., **1978.**

Warburg, O., Experiments on surviving carcinoma tissue, *Biochem. Z.* **1923**, *142*, 317–333.

Weisz, P.B., Diffusion and chemical reaction, *Science* **1972**, *179*, 433–440.

Wiesemann, T., Enzymmodifikationen für analytische und präparative Zwecke: natürliche und künstliche Penicillinamidase-Varianten. Dissertation, TU Hamburg-Harburg, **1991.**

Youschko, M.I., van Langen, L.M., de Vroom, E., van Rantwijk, F., Sheldon, R.A., Svedas, V.K., Penicillin acylase-catalyzed ampicillin synthesis using a pH gradient: a new approach to optimization, *Biotechnol. Bioeng.* **2002**, *78*, 589–593.

Zeldovich, Ya. B., The theory of reactions on powders and porous substances, *Acta Physicochim. U.R.S.S.* **1939**, *10*, 583–592.

9
Reactors and Process Technology

Reactor types	Selection according to mass balances for the reaction
Standard reactors	Stirred tank and tubular reactor
Special reactor types Case study 1	Reactors with recirculation and fluidized-bed reactor for wastewater treatment. Enzymatic production of 7-Amino-cephalosporanic acid (7-ACA)
Process fundamentals	Measurement and calculation of residence time distribution; mixing; pressure drop; mass transfer; energy requirement
Process technology: upstream- (substrate-) and downstream operations (product isolation and purification)	Unit operations: mixing, heat exchange, separation techniques
Case study 2	Production of glucose-fructose syrup

The application of biocatalysts in industrial processes aims at the synthesis of valuable products, or at the conversion and/or degradation of compounds in environmental technology. The bioreactor represents the central part of the plant as a whole, where the reaction takes place under controlled conditions. Based on chemical reaction engineering with few basic reactor types, a range of special reactor modifications and configurations may be derived which serve the aim to adapt the conversion to optimal conditions. In order to minimize the costs, the choice of the reactor must be based on the following aspects:

Biocatalysts and Enzyme Technology. K. Buchholz, V. Kasche, U.T. Bornscheuer

ISBN: 3-527-30497-5

- High yield of product (in general >90 % of the theoretical yield), at high concentration and high purity – that is, a low concentration of byproducts which are difficult to separate, and additives, such as buffer.
- High catalyst productivity – that is, high effectiveness and good operational stability.
- Low manufacturing and maintenance costs, as well as low space requirement, which means preferentially one or a few reactors with small volumes and minimal equipment (e.g., fittings, valves, instrumentation).

As in most cases in technology development, these requirements tend to be contradictory: high conversion and catalyst productivity are more readily realized in reactor cascades as compared to a single reactor; a constant and stable – as well as, with reference to reaction conditions, optimal – operation requires robust and multiple analytical instrumentation and control units. Although an economically viable decision can only be made by the manufacturer, the chemist, engineer or biochemical engineer must elaborate appropriate basic information in terms of calculated and experimentally verified protocols and options.

Reactor types, and important properties of reactors are covered in Sections 9.1 and 9.3.

Enzyme processes are integrated as one part of a process chain, which is of major importance for their design (Section 9.4). To illustrate this two case studies have been included in this chapter. They are:

- As a newer enzyme process the production of 7-aminocephalosporanic acid (7-ACA), a precursor for semisynthetic cephalosporin antibiotics. The selection of the suitable reactor is essential to optimize the product yield and to minimize byproduct formation. As enzyme processes are important in the production of β-lactam antibiotics (cephalosporins and penicillins) a short overview of this is also given (Section 9.2).
- As a classical example the integration of the enzymatic step of high fructose corn sirup (HFCS) manufacture from starch (Sections 5.3.2 and 6.4.1) with up- and downstream operations is dealt with (Section 9.4.2)

Despite the reactor occupying the central position, the equipment prior and subsequent to it for substrate preparation, product isolation and purification up to confectioning can also contribute major roles in terms of the overall costs of the process. Likewise, fittings, pipes and notably instrumentation for measurement and control, as well as data collection and processing, require high expenditure. Optimal integration of all these components forms the basis of process technology, and this is described briefly in Section 9.4.

9.1 Types of Reactors

9.1.1 Basic Types and Mass Balances

Although there are only two basic types of enzyme reactors, namely the stirred tank and the tubular reactor, combinations to form cascades of reactors play a major role in these processes. In particular, they are applied to experiments and mathematical modeling for the identification of optimal reactor configurations. In industrial practice, the stirred tank reactor is used in both the instationary (batch operation) and stationary mode (continuous operation). The characteristic differences between the two types result from the correlations of concentrations (of educts and products) and space, as well as (residence) times. This relationship is shown schematically in Figure 9.1. It should be pointed out, that whilst ideal reactors are discussed here, "real" reactors in which behavior deviates from the ideal situation are dealt with in Section 9.3.

Mass balances allow for the calculation of correlations of reactor size, or reaction (residence) time, respectively, and the amount of catalyst and required conversion. The more simple cases are dealt with subsequently (Prenosil et al., 1987; Reuss, 1991; Cabral and Tramper, 2000). Thus, the mass balances of the two limiting cases – the continuous stirred tank reactor (CSTR) and the continuous tubular reactor (see Fig. 9.1) – are discussed at this point. The discontinuous stirred tank reactor (batch process) is included insofar as the mathematical treatment of its mass balance is equivalent to that of the continuous tubular reactor. For simplicity, it is assumed that parameters such as pH, temperature and catalyst activity are constant.

The substrate converted ($\Delta[S]$) results from the initial $[S]_0$ and final $[S]_E$ substrate concentrations, respectively ($\Delta[S] = [S]_0 - [S]_E$), and is dependent on:

- the amount and activity of the catalyst (which are given by the maximal reaction rate V_{max});
- the reaction time (t) or residence time (τ), respectively, in the reactor; and
- the substrate concentration [S], which depends on the reaction time (t) or residence time (τ), respectively (or the length in a tubular reactor).

Together with the substrate concentration [S], the reaction rate v and also the effectiveness factor η will vary, the latter referring to the ratio of the reaction rates of free and immobilized enzyme, respectively (see Chapter 8, Section 8.3).

For the *ideal continuous stirred tank reactor* (Fig. 9.1b) the substrate concentration is constant throughout space and time, and is equal to the final concentration. Hence, this type of reactor is less favorable in general, as the reaction rate depends on the final substrate concentration, which should be low. Notably, for reactions with product inhibition it is inappropriate, as the reactor should operate at high product concentration. It may be favorable however for reactions with substrate inhibition when operating at the substrate concentration which results in the maximal reaction rate.

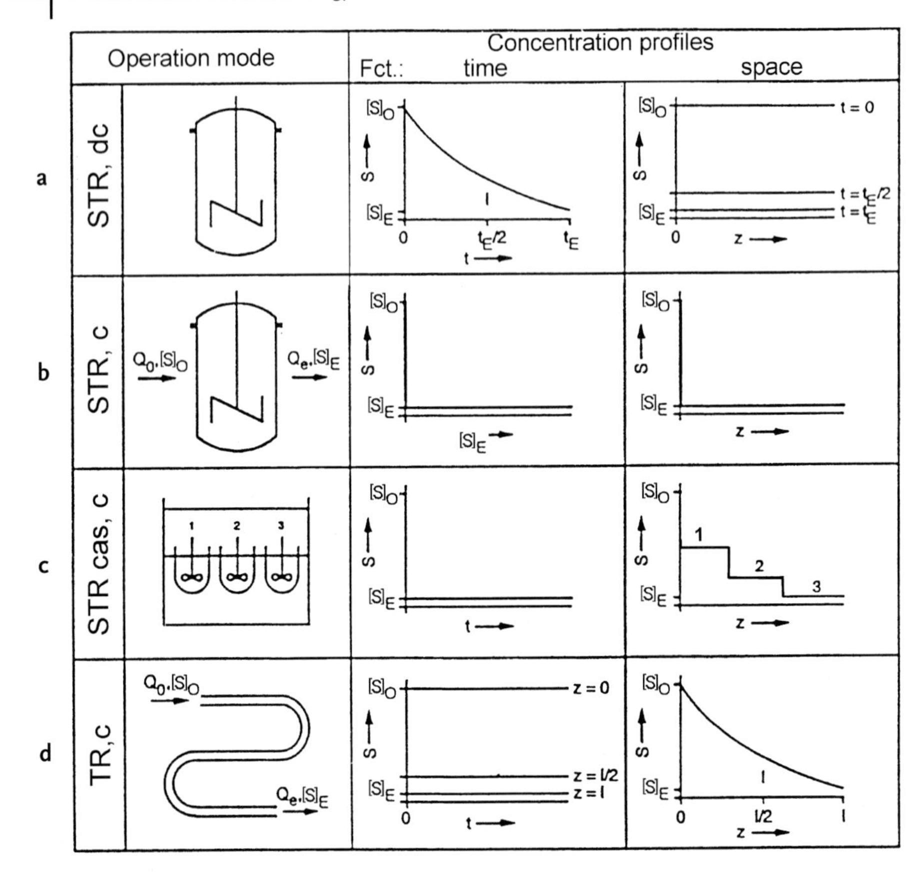

Fig. 9.1 Basic (ideal) reactor types and operational modes as well as the respective concentration profiles for a substrate S as a function of (residence) time (t), respectively. (a) STR, dc: stirred-tank reactor, discontinuous; (b) STR, c: continuous; (c) STR,cas,c: cascade, continuous; (d) TR,c: tubular reactor, continuous (Q = volumetric flow; z = length. Indices: entry = 0; exit = E).

To establish the mass balance of the CSTR in the steady state, the difference of the amounts of substrate introduced to and removed from the reactor (products of $[S]_0$ and $[S]_E$, respectively, and the volumetric flow rate Q (L s^{-1})) are taken as equal to the substrate converted in the reactor volume V. For the simple case of Michaelis–Menten kinetics, one obtains (Eq. 8.17, p. 344):

$$Q[S]_0 - Q[S]_E = v([S]) \cdot V = \eta \cdot \frac{V_{max} \cdot [S]_E}{K_m + [S]_E} \cdot V \tag{9.1}$$

In the case of the *discontinuous stirred tank reactor* and the continuous tubular reactor, the substrate concentrations vary with the reaction time (t) or residence time (τ), or length (z), respectively, in the reactor (see Figs. 9.1a and d). The equation for the reaction rate (Eq. 9.2) therefore must be integrated in order to obtain the conversion

$X = \Delta[S][S]_o^{-1}$ as a function of the residence time (τ) (or reaction time t) (Eqs. 9.3 and 9.4). If mass transfer plays a role, the operational effectiveness factor η_o must be introduced (see Section 8.3, Eq. 8.5 and Section 8.5).

$$v([S]) = -\frac{d[S]}{dt} = \eta_0 \frac{V_{max} \cdot [S]}{K_m + [S]} \tag{9.2}$$

$$-\int_{[S]_0}^{[S]_E} (K_m + [S]) \cdot \frac{d[S]}{[S]} = \eta_0 \cdot V_{max} \cdot \int_0^t dt \tag{9.3}$$

$$[S]_0 - [S]_E + K_m \cdot \ln([S]_0/[S]_E) = \eta_0 \cdot V_{max} \cdot t \tag{9.4}$$

The *mass* balance is the basis for the calculation of the reactor volume (or the residence time τ) or the amount of biocatalyst (or the activity V_{max}) required for a certain conversion of a given amount of substrate and reaction time (see Chapter 8, Section 8.5). Thus, it is possible to calculate from Eq. (9.4), for a given amount or activity of enzyme (with $V_{max} = k_{cat} \times [E]_0$) the reaction time required for the conversion of a certain amount of substrate (given by $[S]_0$, $[S]_E$). For a given reaction time t, it can also be used to calculate the enzyme activity per volume required for the conversion of a certain amount of substrate. For immobilized systems the operational effectiveness factor η_o for the range of concentration considered and the conversion ($X = ([S]_0 - [S]_E)/[S]_0$) must be taken into account.

The general equations for the product ($V_{max} \cdot \tau$) are summarized in Table 9.1, both for the continuous stirred tank and tubular reactors for three different kinetics (sim-

Table 9.1 Equations for $V_{max}\,\tau$ for the continuous stirred tank and tubular reactors, and for three different kinetics (substrate conversion $X = \Delta[S]\,[S]_0^{-1}$; P = Product = $[S]_0 - [S]_E$; K_I = inhibitor association constant).

Reaction rate v	***Correlation for:*** $\frac{V_{max} \cdot V}{Q} = V_{max} \cdot \tau$ ***Stirred-tank reactor***	***Tubular reactor***
Michaelis–Menten		
$V_{max} \frac{[S]}{K_m + [S]}$	$X \cdot \left(\frac{K_m}{1-X} + [S]_0\right)$	$[S]_0 \cdot X + K_m \cdot \ln(1-X)$
Competitive product inhibition		
$V_{max} \frac{[S]}{K_m \cdot (1 + [P]/K_I) + [S]}$	$[S]_0 \cdot X + K_m \frac{X}{1-X} + [S]_0 \frac{K_m}{K_I} \cdot \frac{X^2}{(1-X)}$	$K_m \cdot (1 + [S]_0/K_I) \cdot \ln(1-X) + (1 - K_m/K_I) \cdot [S]_0 \cdot X$
Substrate inhibition		
$V_{max} \frac{[S]}{K_m + [S] + [S]^2/K_I}$	$[S]_0 \cdot X \left[1 + \frac{K_m}{[S]_0 \cdot (1-X)} + \frac{(1-X) \cdot [S]_0}{K_I}\right]$	$[S]_0 \cdot X - K_m \cdot \ln(1-X) + \frac{[S]^2}{2K_i}(2X - X^2)$

ple Michaelis–Menten kinetics as well as practically relevant cases with inhibition by product and substrate, respectively). In this table and in the subsequent equations the effectiveness factor η is taken as 1. For immobilized biocatalyst however, the reaction rate v must be multiplied by the effectiveness factor η or η_o, respectively (Section 8.5).

The residence time τ required for the conversion of a component S at the reaction rate v can be calculated from the mass balance of the reactor under consideration.

For the CSTR, when using the mass balance (Eq. 9.1), the residence time is obtained from:

$$\tau = V/Q = \frac{[S]_0 - [S]_E}{v([S])} \tag{9.5}$$

For the continuous tubular reactor one obtains for the residence time, with

$$\tau = \int_0^t dt = V/Q \tag{9.6a}$$

the following correlation:

$$\tau = V/Q = -\int_{[S]_0}^{[S]_E} d[S]\,(v([S]))^{-1} \tag{9.6b}$$

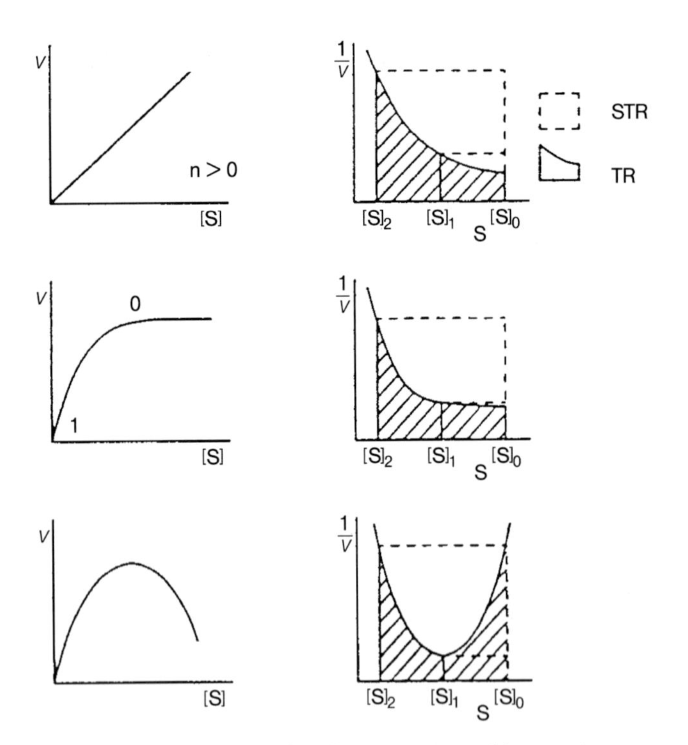

Fig. 9.2 Graphical analysis of residence time required for a continuous stirred tank reactor (STR) and a tubular reactor (TR) respectively, for different types of reactions. For explanation see text above and below.

In Figure 9.2, the two correlations for the residence time τ are interpreted by graphical means (according to Reuss, 1991). The upper part of the figure shows the case for a reaction of an order >0. The residence time required, or the reaction volume necessary for a given substrate amount and conversion is shown for the CSTR in terms of a rectangular plane with the coordinates v^{-1} and [S], with the boundary conditions $[S]_0$ and the outlet concentration $[S]_E$ required (two cases with different outlet concentrations $[S]_1$ and $[S]_2$ are shown). The corresponding residence time for the tubular reactor results from the integral below the correlation given for v^{-1} and [S], again with the boundary conditions $[S]_0$ and the outlet concentration $[S]_E$. It is obvious that the volume required for the tubular reactor is significantly smaller as compared to that of the CSTR, the difference becoming larger with increasing conversion.

The graphs in the center deal with Michaelis–Menten kinetics, which may often be applicable in biocatalysis. Here, it is clear that in the range of zero-order kinetics no difference exists as to the volume required with either a tubular or stirred tank reactor (both initial $[S]_0$ and final $[S]_E$ concentrations $>K_m$). If the final conversion should be higher (final concentration $[S]_2$), the reaction order is in a range of $n>0$ and the conditions are the same as discussed before.

The lower graphs in Figure 9.2 show the case of substrate inhibition. The correlation of v^{-1} and [S] point out that the optimal choice of the reactor depends again on the conversion required. At low conversion (final concentration $[S]_1$), the CSTR is superior, whereas at high conversion (final concentration $[S]_2$) the tubular reactor turns out to be the better choice. The optimal reactor configuration is a combination of a stirred tank as the first and a tubular reactor as the second one.

The following rule can be given for a general comparison of the efficiency of tubular and stirred tank reactors, respectively (both operating continuously): reactions with kinetics where the reaction rate decreases with increasing conversion (e.g., Michaelis–Menten kinetics) the tubular reactor provides for higher conversion as compared to the stirred tank reactor, when both operate with the same volume and catalyst activity. These conclusions are similar to those elaborated for chemical reaction engineering.

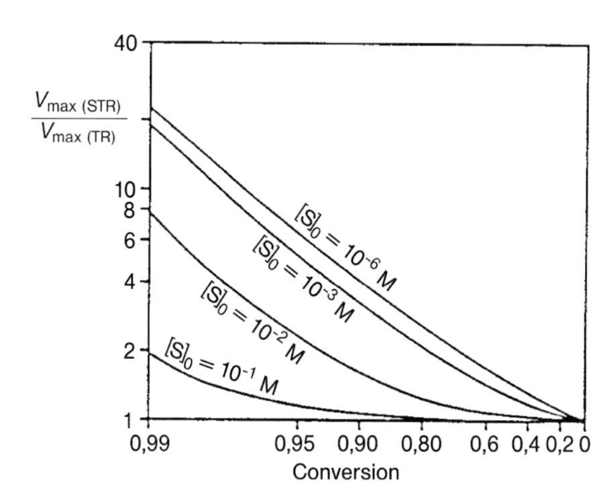

Fig. 9.3 Ratio of enzyme activities required for a continuous stirred tank reactor and a tubular reactor, respectively, for different initial substrate concentrations $[S]_0$, given by simple Michaelis–Menten kinetics and a Michaelis constant (K_m) of 1 mM (Lilly, 1978).

In order to compare the reactor volumes required for a different conversion and for different initial substrate concentrations, Figure 9.3 shows the enzyme activities required for a CSTR and a tubular reactor, respectively (see Chapter 8, Section 8.5).

9.1.2
Further Reactor Types and Configurations: Application Examples

Simple standard reactors offer the advantage of flexibility – that is, the possibility to serve different applications. Often, older stirred tank reactors are used for newly developed processes in order to save or reduce investment cost. In contrast, special modifications of basic reactor types offer the advantage of being adjusted to create an optimal design to meet the requirements of a specific process.

First, some modifications and combinations of stirred tank reactors will be presented (for reviews, see Lilly, 1978; Prenosil et al., 1987; Cabral and Tramper, 2000; Hempel, 2004).

Stirred tank reactors with internal loop – the so-called "loop reactors" (Fig. 9.4) – offer short circulation times and thus provide for intense mixing of the reaction solution. For continuous operation, the circulated mass (or volume) should be at least

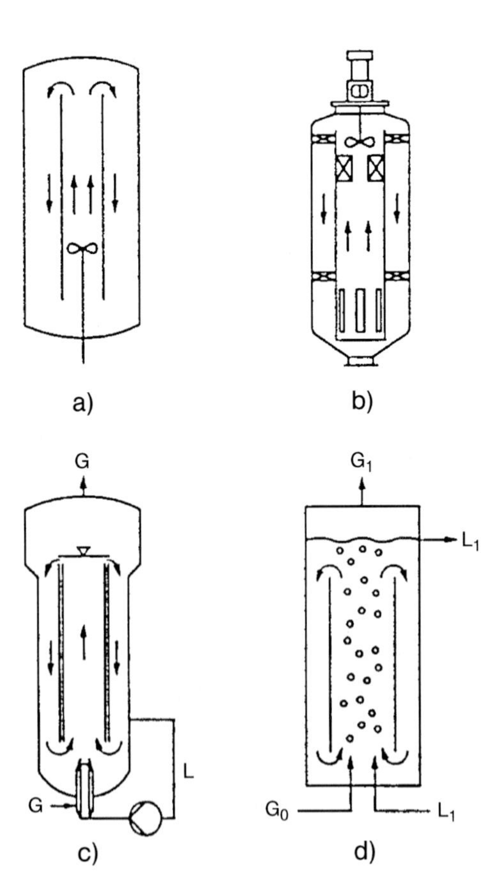

Fig. 9.4 Loop reactors (G = gas phase; L = liquid phase). (a) Simple principle with internal tube and stirrer; (b) slim type with baffles/breakers; (c) reactor with *enlarged* head (for three-phase-systems and degassing); (d) bubble column loop reactor.

5- to 10-fold that of the feed in order to provide for short mixing time. With scale-up over several orders of magnitude, the mixing time – and thus the control of essential parameters such as pH – can be problematic; this is notably the case with solutions of high viscosity. In environmental technology, for example in anaerobic wastewater treatment with immobilized biomass, scale-up must proceed from the laboratory scale with reactors of 1–10 L up to a reactor volume of several hundreds of m^3 – an increase by four to five orders of magnitude. The control of the mixing behavior (which is essential) in general requires special devices, for example an internal tube (Fig. 9.4a). The control of flow inside the reactor by an appropriate circulation rate ensures the control of the concentration distribution, which in turn depends on substrate inflow and reaction (Blenke, 1985).

A *cascade of stirred tank reactors* represents the most convenient method for continuous processes where the advantages of stirred tank reactors – simplicity and flexibility – can be utilized and the disadvantages – application of the biocatalyst at stationary conditions of low substrate and high product concentration – are avoided. The concentration levels of the cascade tend to approximate the profile of a tubular reactor (see Fig. 9.1 c vs. 9.1 d). The construction of the cascade can be rather simple in terms of a multistage tubular reactor equipped with sieve or perforated plates. A central shaft drive carries the stirrers in each chamber (Fig. 9.5a).

The mass balance for the cascade is based on Eq. (9.1), by taking as the inlet concentration $[S]_i$ of a reactor the outlet concentration of the preceding reactor. For Michaelis–Menten kinetics, it follows that:

$$v = \eta_{s,i} \frac{k_{cat} \cdot [E]_i \cdot [S]_i}{K_m + [S]_i} \tag{9.7}$$

For the inlet concentration related to the exit concentration in each reactor one may write (where X is the conversion $([S]_0 - [S]_E) \times [S]_0^{-1}$ and V the volume of a reactor; the residence time is $\tau = V \times Q^{-1}$):

$$[S]_i/[S]_{i-1} = 1 - V/Q(v_i/[S]_{i-1}) = 1 - X_i \tag{9.8a}$$

and

$$X_i = (V/Q) \cdot (v_i/[S]_{i-1}) = \tau \cdot v_i/[S]_{i-1} \tag{9.8b}$$

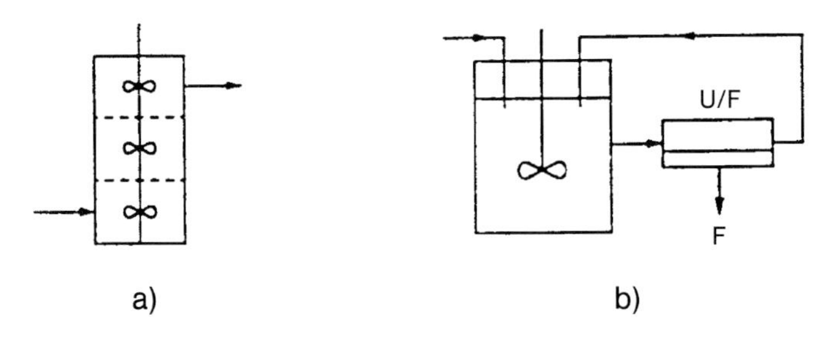

Fig. 9.5 Reactor configurations with continuous stirred tank reactor.
(a) Cascade of stirred tank reactors in the form of a tube subdivided into chambers.
(b) Stirred tank reactor with subsequent ultrafiltration unit.

Multiplying the singular equations gives the final outlet concentration of the cascade with *i* reactors related to the initial inlet concentration:

$$[S]_i/[S]_0 = \prod_{1}^{i} (1 + \tau \cdot v_i/[S]_i)^{-1} \tag{9.9}$$

For the final conversion X_i one obtains:

$$X_i = 1 - [S]_i/[S]_0 \tag{9.10}$$

(for the mass balances of other reactor modifications, e.g., of reactors with recycle, see Moser, 1985).

Continuous stirred tank reactors are used for starch hydrolysis, both for liquefying with α-amylase as well as for further hydrolysis with glucoamylase. For this technically important process, cascades of CSTRs are also used (see Chapter 5, Sections 5.3.2 and 9.4.2). For glucose manufacture, mainly cascades with six to 12 reactors are applied in order to avoid losses in yield of the final product (Uhlig, 1991; Olsen, 1995).

In continuous processes a soluble biocatalyst is lost which, nevertheless, is economically feasible when cheap enzymes (as in the case of starch processing) are used. If the cost of the biocatalyst is increased (about >2 % of the product price), reactors with recycle are required – that is, a separation unit for the biocatalyst must be installed downstream of the reactor. For soluble enzymes, ultrafiltration membrane systems are appropriate (see Chapter 5, Section 5.4 and Fig. 5.9).

Another reactor configuration to be used is the combined stirred tank reactor and differential fixed-bed reactor (Fig. 9.6). This is used in penicillin hydrolysis with immobilized penicillin amidase in a discontinuous process (see Section 9.1.3, Case study) (Liese et al., 2000, pp. 286, 287). Following a differential conversion of penicillin in the fixed bed of biocatalyst, mixing is required for pH control. The biocatalyst located in the fixed bed is not subject to variations of pH occurring in large reactors, nor is it subject to abrasion due to shear by the impeller (see Section 9.3.3).

Tubular reactors are applied mainly as *fixed-bed reactors* with immobilized biocatalyst (spheres or granules) (Fig. 9.7a), as for glucose isomerization and for kinetic res-

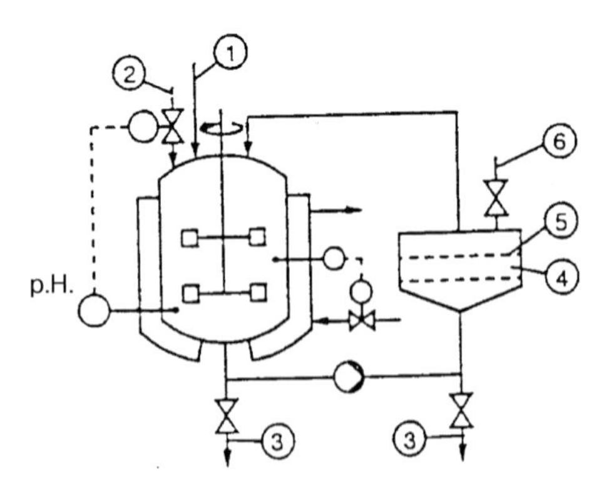

Fig. 9.6 Reactor for penicillin hydrolysis, with differential fixed bed with immobilized penicillin amidase and a stirred tank reactor for neutralization of 6-aminopenicillanic acid formed. ① addition of penicillin; ② addition of ammonia; ③ exit to extraction of 6-aminopenicillanic acid; ④ immobilized enzyme; ⑤ sieve plate; ⑥ degassing.

olution of racemic amino acids (see Chapter 6, Section 6.4). Specific requirements (such as a sediment-free substrate solution) require appropriate upstream operations, and these are detailed in Section 9.4. The essential advantage of fixed-bed reactors is simple continuous operation. When compared to CSTRs, a significantly higher catalyst productivity is obtained due to the *profiles* of substrate and product inside the tubular reactor, which are higher by a factor of about 1.7 in the cases of glucose isomerase and amino acylase (Chibata, 1987; Jensen and Rugh, 1987). Reactors with immobilized glucose isomerase have bed heights of 5–8 m, and the substrate feeding rate is 1–2 tonnes per m^3 and day.

Soluble enzymes can, in principle, be applied in simple tubular reactors. When scaling up to tubes of large diameter, however, back-mixing may occur (see Section 9.3), though this can be controlled by applying internals. Enzyme recycling can be accomplished with a subsequent ultrafiltration unit (see Section 5.4; Fig. 5.9d). Hollow-fiber membrane reactors comprise bundles of hollow fibers in which the membrane serves for the retention of the enzyme(s) (Fig. 9.7b).

Trickle bed reactors are applied in manifold varieties in biological exhaust gas purification (Fig. 9.7c). These are three-phase reactors containing a carrier with an adhering mixed bacterial culture degrading organic pollutants; the aqueous phase

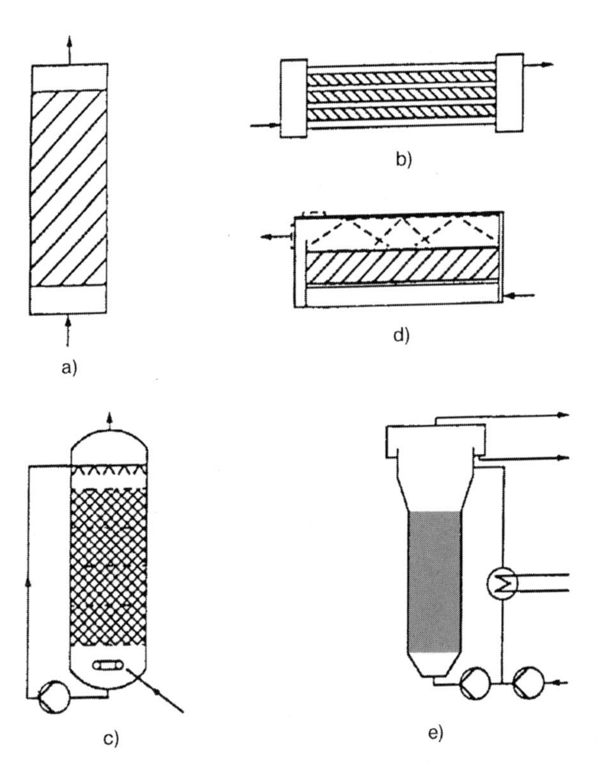

Fig. 9.7 Different configurations of tubular reactors. (a) Column with fixed bed of biocatalyst; (b) hollow-fiber membrane reactor; (c) trickle-bed reactor; (d) biofilter; (e) fluidized-bed reactor.

flows from the top down through the reactor, absorbing the organic components from the exhaust air, which flows countercurrent from the lower part of the reactor to the top. The biofilter represent another system for exhaust gas treatment (Fig. 9.7d; see also Chapter 7, Section 7.5.2 for details of both systems).

Fluidized-bed reactors (Fig. 9.7e) can be applied as either two- or three-phase systems. The liquid (substrate-) phase enters the reactor from the downside, while the upflow keeps the biocatalyst suspended inside the reactor tube. The regime of flow rates is limited on the one hand by the minimum rate at which the particles are suspended (minimum fluidization), and on the other hand by the maximum rate where the particles tend to leave the reactor. The flow rate required is generally controlled by an external loop. The advantages of such systems are the tolerance (no clogging) for fluid media with suspended fine particles, or precipitates (occurring in many cases in wastewater treatment), and the ease of introducing or removing a gas phase (e.g., biogas). A drawback is that mixing of the fluid phase, the residence time approaching that of a stirred tank reactor, if not prevented by internals or sieve plate installations.

Fluidized-bed reactors are applied in the large scale (up to 500 m^3) in anaerobic wastewater treatment (Fig. 9.8). Anaerobic mixed cultures (including methanogenic bacteria) grow slowly, thus requiring retention inside, or recycle to the bioreactor, in order to achieve good performance with high reaction rates. When immobilized by adhesion (e.g., on sand, pumice; see Chapter 7, Section 7.4.2), they can be applied notably in fluidized-bed reactors where the large amounts of biogas formed separate from the liquid phase and leave the reactor at the top. Mixing by recirculation via an external loop is necessary in order to control the pH (Fig. 9.7e). The mixed culture is able to adapt to low stationary substrate concentrations and to operate efficiently under these conditions.

Fig. 9.8 Industrial fluidized-bed reactor for anaerobic wastewater treatment; working volume 500 m^3, height 30 m. (Illustration reproduced with kind permission of Nordzucker AG, Braunschweig; Jördening et al., 1996).

The examples of industrial application mentioned show that, in the food and pharmaceutical industries, large reactors as well as sophisticated reaction engineering are important and standard techniques. The largest systems have been installed in environmental technology (from 100 m^3 but also up to several 1000 m^3). Here, reaction engineering plays an essential role in order to achieve process intensification and increase the efficiency of these systems.

9.2 Case Study: The Enzymatic Production of 7-ACA from Cephalosporin C

9.2.1 Enzyme Processes in the Production of β-Lactam Antibiotics

Penicillins and cephalosporins belong to the class of β-lactam antibiotics that are formed from the common precursor tripeptide isopenicillin N (Fig. 9.9). The β-lactam structure is formed by ring-closure reactions between Cys and Val, where (*S*)-Val is isomerized to (*R*)-Val. Penicillins and cephalosporins are the main antibiotics

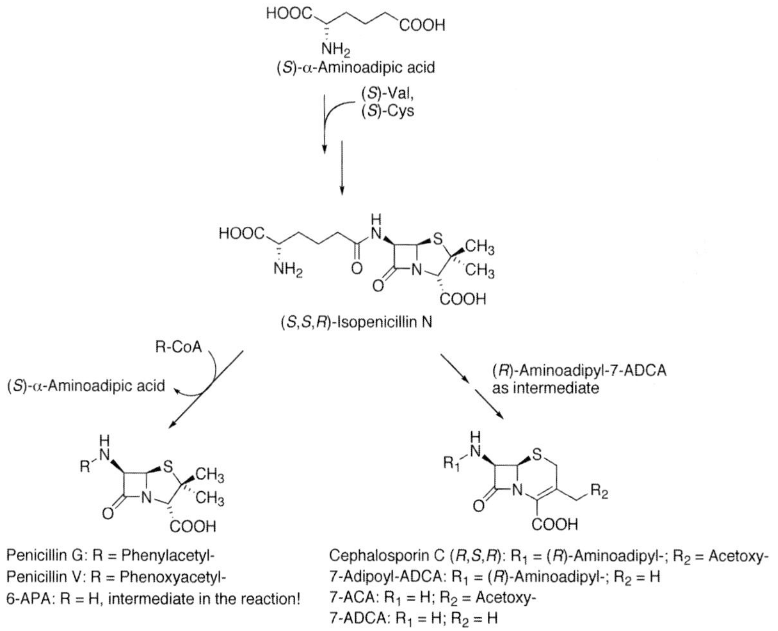

Penicillin G: R = Phenylacetyl-
Penicillin V: R = Phenoxyacetyl-
6-APA: R = H, intermediate in the reaction!

Cephalosporin C (*R,S,R*): R_1 = (*R*)-Aminoadipyl-; R_2 = Acetoxy-
7-Adipoyl-ADCA: R_1 = (*R*)-Aminoadipyl-; R_2 = H
7-ACA: R_1 = H; R_2 = Acetoxy-
7-ADCA: R_1 = H; R_2 = H

Fig. 9.9 Biosynthesis of penicillins and cephalosporins used for the production of semi-synthetic β-lactam antibiotics. Note the change in isomer structure of the amino acids given in brackets (Ingolia and Queener, 1989; Crawford et al., 1995; Weil et al., 1995).

1) In order to avoid the unnecessary selection and transmission (via food) of resistant bacterial strains, the same antibiotics should not be used for humans and animals. Antibiotics should therefore also not be used to prevent infections, that is, in animal feed.

for human use,[1)] with a market share of ≈65 % (Elander, 2003). The β-lactam precursors of all penicillins and cephalosporins are produced by fermentation in fermentors of up to ≈1000 m^3. The concentration of the products in the medium on completion of fermentation that takes between five and seven days, is up to 100 g L^{-1} (penicillin) and 20 g L^{-1} (cephalosporin C). The lower yields for the cephalosporins explains why they are more expensive than penicillins. The yields have been continuously increased by strain improvement, and this further reduces the processing costs.

The enzymatic and chemical production processes for semisynthetic penicillins and cephalosporins derived from the fermentation products in Figure 9.9 are summarized in Figure 9.10. The market value of the products is approximately 20–30×10^9 € (2002). A considerable proportion of the penicillins and cephalosporin C produced by fermentation is hydrolyzed to obtain 6-aminopenicillanic acid (6-APA) and 7-aminocephalosporanic acid (7-ACA), both of which are used in the production of semisynthetic β-lactam antibiotics such as ampicillin, amoxicillin, cephalexin, and claforan (Fig. 9.10). Until recently 7-ADCA was prepared chemically from penicillin G (Bruggink, 2001). Alternatives to produce 7-ADCA by fermentation include the following; in the biosynthesis of cephalosporin C, (*R*)-aminoadipyl-7-ADCA is an intermediate (Fig. 9.11). When the enzyme that catalyzes the oxidation of this compound is deleted, 7-ADCA can be obtained from (*R*)-aminoadipyl-7-ADCA using the same enzyme-catalyzed processes as for the production of 7-ACA from cephalosporin C. This two-step procedure can be replaced by a one-step procedure. An adipoyl group is bound to 6-APA instead of a phenylacetyl group in the conversion of isopenicillin to penicillin G.

The adipoyl-6-APA produced can be converted to adipoyl-7-ADCA in the cells with the enzyme that catalyzes the ring expansion in Figure 9.11. The adipoyl-7-ADCA

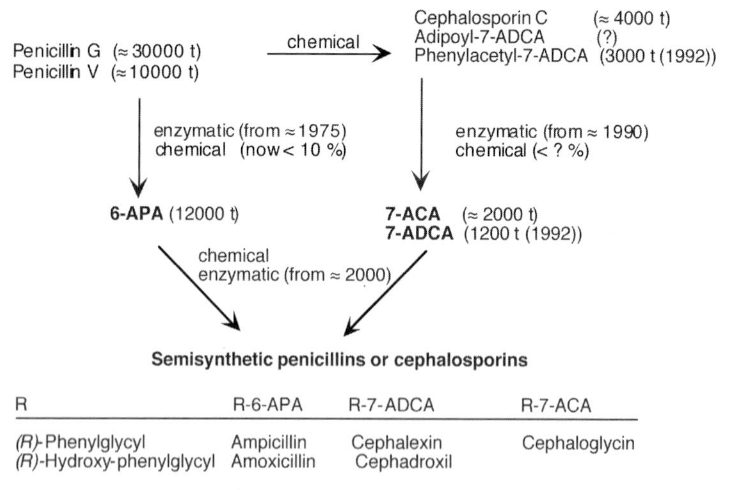

R	R-6-APA	R-7-ADCA	R-7-ACA
(*R*)-Phenylglycyl	Ampicillin	Cephalexin	Cephaloglycin
(*R*)-Hydroxy-phenylglycyl	Amoxicillin	Cephadroxil	

Fig. 9.10 Enzymatic and chemical production of semisynthetic penicillins and cephalosporins from hydrolysis products (6-APA, 7-ACA, 7-ADCA) of the fermentation products given in Figure 9.9. The byproducts phenylacetate and adipate can be recycled in the fermentations. The amounts produced are estimated from literature data (Bruggink, 2001; Elander, 2003).

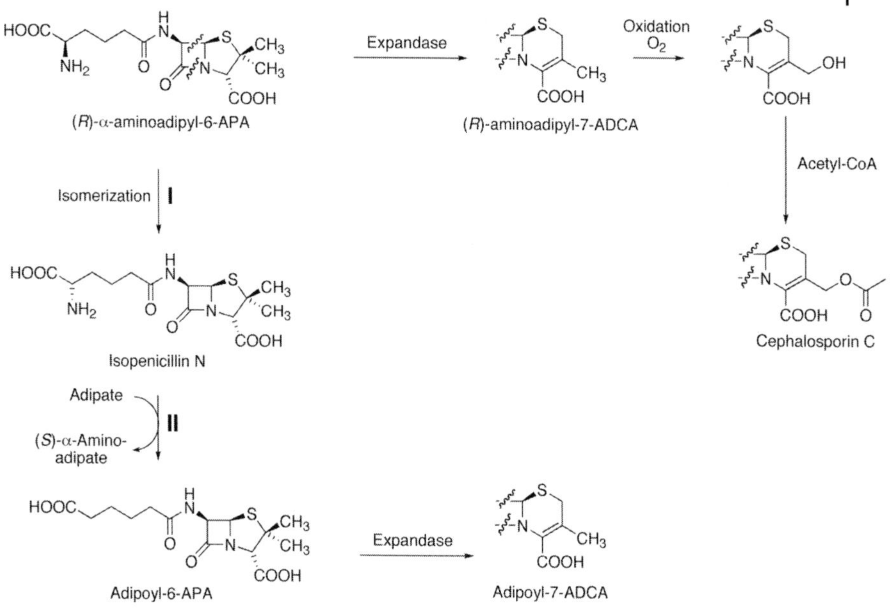

Fig. 9.11 Alternative processes for the production of (*R*)-7-ADCA from isopenicillin N by fermentation. (I) In cells where the oxidation of (*R*)-aminoadipyl-7-ADCA has been blocked. Only the part of the ring that is expanded is shown (Adrio et al., 2002). (II) In cells with expandase but without the racemase that catalyzes isomerization of the aminoadipyl-group in isopenicillin N (Crawford et al., 1995; Bruggink, 2001).

can then be hydrolyzed by the glutaryl amidase already used for the hydrolysis of Glu-7-ACA. This elegant example for a successful metabolic engineering has recently been applied on an industrial scale (Bruggink, 2001). In this process, much less waste is produced than in the chemical process formerly used.

The β-lactam ring is essential for the biological function as an antibiotic. β-Lactam antibiotics act as a practically irreversible inhibitor of an enzyme that catalyzes formation of the glycoprotein cell wall in Gram-positive bacteria. Similar enzymes do not exist in mammals, and this explains the organism-specific function and fewer adverse side effects of β-lactams compared to other antibiotics. It is possible that β-lactams may acylate proteins, leading to an allergic reaction as they act as haptens and cause antibody formation, but this effect has been reduced by improving antibiotic purity (Schneider and de Weck, 1981). When given in large doses for a long time (e.g., to treat bone infections), β-lactam antibiotics have been shown to influence the biosynthesis of nucleic acids (Do et al., 1987).

The hydrolysis of penicillins to produce 6-APA or 7-ADCA, or of cephalosporin C to produce 7-ACA, can be carried out as either a chemical or biotechnical (enzyme) process. The latter is now the dominating process due to lower costs and minimal production of wastes. In the equilibrium-controlled hydrolysis of natural penicillins, phenylacetyl-7-ADCA, and cephalosporin C, the enzyme used as a biocatalyst cannot influence the end-point of the reaction (see Chapter 2, Section 2.6) as the latter is a function of pH, T and ionic strength (see Chapter 2, Section 2.8). To obtain a high yield, it is essential to utilize conditions where the hydrolysis yield is near 100 %. Under these conditions, the product or enzyme may be unstable, and the enzyme kinetic properties not optimal. To design an optimal enzyme process it is necessary to consider the steps shown in Figure 1.10. Finally, other side chains are added to the amino group of 6-APA, 7-ACA and 7-ADCA to produce semisynthetic β-lactam antibiotics with better properties than penicillin G or V and cephalosporin C (acid-stable for oral use, a wider range of antibiotic activity, etc.) (Fig. 9.10). Although chemical synthesis remains the main procedure, synthesis is also possible using the same enzymes as are used for the hydrolysis of penicillin G as biocatalysts. Equilibrium-controlled procedures are not possible here as the possible products yields are too low, and much higher yields can be obtained in kinetically controlled processes (see Chapter 2, Sections 2.6 and 2.8, and Fig. 2.20).

Enzyme-catalyzed kinetically controlled processes for the synthesis of β-lactam antibiotics have recently been introduced on an industrial scale (Bruggink, 2001). In the following section, the design of an enzyme processs for the production of 7-ACA from cephalosporin C will be discussed in detail based on the scheme given in Figure 1.10 (p. 22).

9.2.2
Overall Process for the Production of 7-ACA

The process is given as a flow-sheet in Figure 9.12. Here, the design of the last three process steps will be discussed. For this aim it is essential to know the concentration and purity of the cephalosporin C solution after the chromatographic isolation procedure. The cephalosporin C and byproducts (Figs. 9.9 and 9.11) in the filtrate from the fermentation medium are adsorbed to a hydrophobic adsorbent and separated from all ions and uncharged polar molecules in this filtrate. Cephalosporin C is then desorbed with an isopropanol:water mixture. The pH in the desorbed solution is increased to ≈5 and Cephalosporin C adsorbed to an anion-exchange column. In the desorption step with an acetic acid containing eluent, it can be isolated with a purity

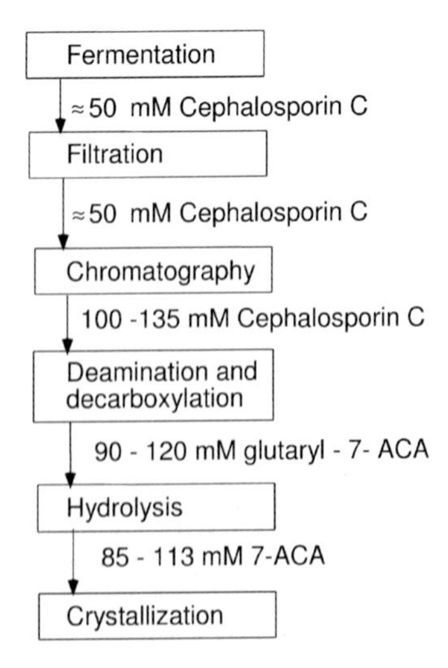

Fig. 9.12 Flowsheet for the production of 7-ACA, with concentrations of the precursors after each processing step. The increase in concentration after the chromatographic step is due to either the use of displacement chromatography or to nanofiltration.

of >90–95 % (concentration ≈0.1 M). This chromatographic separation procedure was originally developed during the 1970s for the subsequent chemical hydrolysis reaction (Fig. 9.12) (Voser, 1973).

9.2.3
Conversion of Cephalosporin C to 7-ACA

The possible chemical and enzyme processes for this conversion are shown in Figure 9.13. The main reason for developing the enzyme process was to reduce the amount

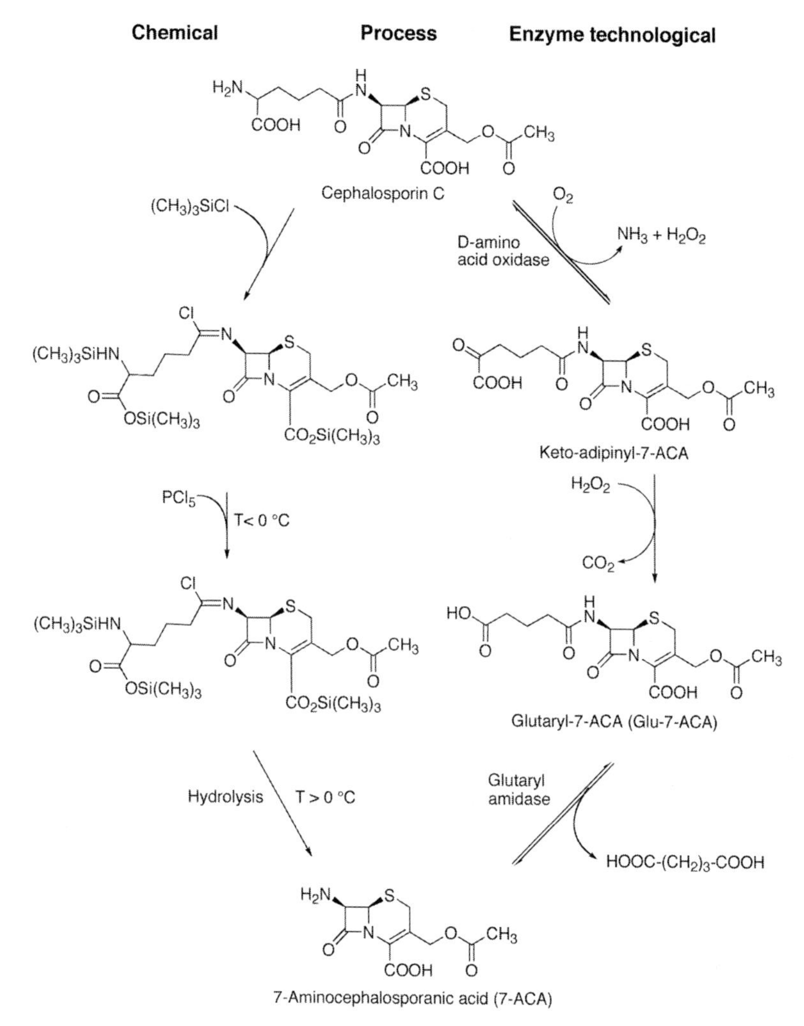

Fig. 9.13 Chemical (left) and enzyme technological (right) processes to produce 7-aminocephalosporanic acid (7-ACA) from cephalosporin C. (Note that the charges of the acidic and basic functional groups of the compounds at the process pH are not given.) The enzyme process is catalyzed by the enzymes (*R*)-amino acid oxidase (oxidative deamination) and glutaryl amidase (hydrolysis). Keto-adipinyl-7-ACS is decarboxylated in an uncatalyzed reaction in which the hydrogen peroxide formed in the first enzymatic reaction is consumed.

of problematic waste produced in the chemical process. This was successful, and waste production was reduced typically from 31 to ~0.3 tons per ton 7-ACA synthesized. Consequently, the enzyme process has now replaced the chemical process.

The two-step enzyme process could be improved considerably if a single enzyme (a hydrolase) were available that would catalyze a one-step hydrolysis of cephalosporin C to 7-ACA. Although many investigations have been conducted to solve this problem, a solution has not yet been found (see Chapter 2, Table 2.15). This enzyme must have similar properties as the glutaryl amidase.

9.2.4 Reaction Characterization and Identification of Constraints

9.2.4.1 Enzymatic Deamination and Chemical Decarboxylation

The first reaction is an equilibrium-controlled process, the rate of which depends critically on the O_2-content, which is only ≈0.2 mM in air-saturated aqueous solutions. As the product is converted further, a 100 % product yield is possible. This requires that all H_2O_2 formed in the enzymatic reaction is used in the chemical reaction. Any loss of H_2O_2 will reduce the yield of Glu-7-ACA. One reaction where this can occur is the reaction of H_2O_2 with the enzyme that may reduce its activity. In practice, it has been found that not all H_2O_2 is used to produce Glu-7-ACA. Literature values for the yield (mol Glu-7-ACA per mol cephalosporin C) are in the range 0.85 to 0.93 (Tischer et al., 1992; Conlon et al., 1995). No thermodynamic data are available for these reactions. Due to the NH_3 and CO_2 formed in the enzymatic and chemical reactions the pH will not be changed to any great extent, although the pH during the enzyme process must be kept constant.

9.2.4.2 Hydrolysis of Glu-7-ACA

The pH- and temperature-dependence of the apparent equilibrium constant of this reaction has been determined (Spieß et al., 1999). For more than 95 % yield, the reaction must be carried out at pH ≥ 8 (Fig. 9.14). The yield increases with the ionic strength. The solution from the deamination and decarboxylation reaction has a higher ionic strength than the original cephalosporin C solution, due to the formation of CO_2 and NH_3. The ions formed from these (HCO_3^-, CO_3^{2-} and NH_4^+) increase the buffer capacity in the hydrolysis reaction. In turn, this reduces the pH change due to the formation of a free carboxyl group of glutarate and the free amino group of 7-ACA that have the pK-values ≈4 and 4.6, respectively. Without buffer this would result in a pH of 4–5 in the reaction mixture during the hydrolysis.

During the hydrolysis, the 7-ACA yield is reduced due to the following monomolecular reactions:

(1) 7-ACA $\rightarrow$ desacetyl-7-ACA + HAc
(2) 7-ACA $\rightarrow$ 7-ACA-lactone

with a total rate constant k_1, and a bimolecular reaction (Tischer et al., 1992):

(3) 7-ACA + 7-ACA $\xrightarrow{k_2}$ Dimers

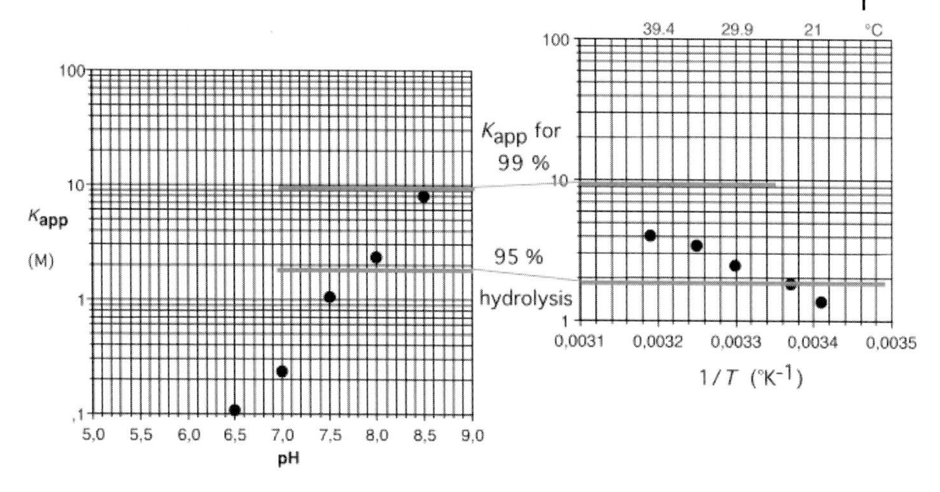

Fig. 9.14 Temperature and pH dependence of the apparent equilibrium constant (K_{app} = [glutarate][7-ACA]/[Glu-7-ACA]) for the hydrolysis of 100 M Glu-7-ACA. Left: pH dependence at 30 °C with phosphate buffers with NaCl (I = 1.5 M). Right: temperature dependence at pH 8.0 in phosphate buffer (I = 0.2 M) (Spieß et al., 1999).

The rate of reaction (2) is increased in the presence of compounds with free amino groups (Yamana and Tsuji, 1976; V. Kasche et al., unpublished results). The pH- and temperature-dependence of reaction rates in the ranges where the hydrolysis is carried out are given in Figure 9.15. From these data it follows that the loss of 7-ACA due to the reactions (1) to (3) in a solution of 100 mM at pH 8 will be 6, 3 and 1 %

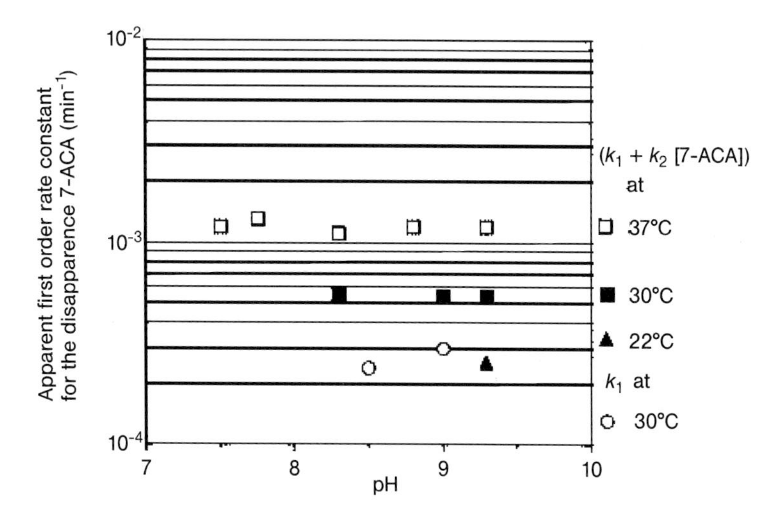

Fig. 9.15 pH- and temperature-dependence of the apparent first-order rate constant ($k_1 + k_2$[7-ACA]) for the reactions (1) to (3) that reduce the yield of 7-ACA.

after 1 h at 37, 30 and 22 °C, respectively. Thus, to avoid this loss of 7-ACA the endothermal hydrolysis reaction should be carried out at temperatures ≤25 °C with short hydrolysis times (≤30 min).

9.2.5
Enzyme Characterization and Identification of Constraints

9.2.5.1 (*R*)-Amino Acid Oxidase

This is an intracellular enzyme that is mainly used in an isolated form. The enzyme has been selected and its production optimized by companies that now use the enzyme process (Tischer et al., 1992; Sauber, 1992). With cephalosporin C as a substrate, the enzyme has a pH optimum of 8.3–9.5. More detailed enzyme kinetic data are not available. The enzyme is expensive to produce, and must be removed from the product. It is therefore used in immobilized form (covalently bound to macroporous particles as Eupergit C (Katchalski-Katzir and Krämer, 2000)). It can be used for approximately 150 reaction cycles each of 60 min (Conlon et al., 1995). Its stability is not affected by the H_2O_2 produced in the deamination reaction (Sauber, 1992).

9.2.5.2 Glutaryl Amidase (also named Acylase)

This is an extracellular (periplasmic) enzyme obtained from different microorganisms, mainly *Pseudomonas* sp. It can be produced in large quantities (up to 5 g L^{-1}) as a heterologous enzyme in *E. coli* (Koller et al., 1998). The enzyme is processed in a similar manner as some penicillin amidases (PA) where a linker peptide is removed in the pro-peptide (Li et al., 1999). The first reaction is an intramolecular autoproteolysis reaction, where a N-terminal Ser is formed that is essential for the catalytic activity (see Section 4.4). The enzyme's three-dimensional structure has been determined (Fritz-Wolf et al., 2002). This, together with an alignment with some penicillin amidases, indicates the functional similarity of the glutaryl (GA) and penicillin (PA) amidases. Their substrate specificities and stability do, however, differ considerably.

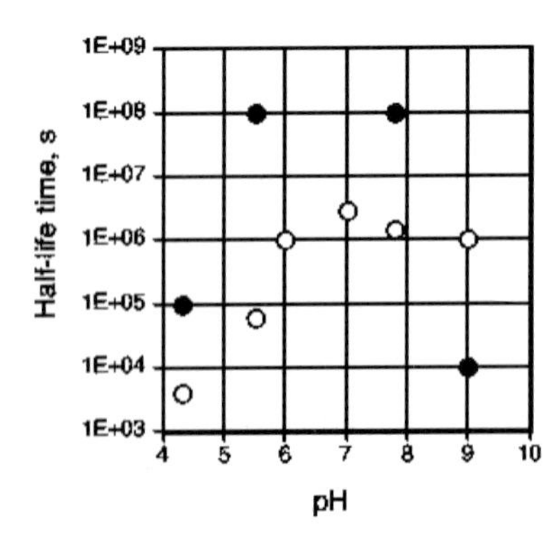

Fig. 9.16 The half-life of glutaryl amidase (○) and penicillin amidase (●) produced in *E. coli* as function of pH (25 °C, buffer with I = 0.2 M).

The k_{cat}-value of GA for Glu-7-ACA is ≈10 s^{-1}, whereas k_{cat} of PA for penicillin G is ≈80 s^{-1} at pH 8 and 30 °C. The K_m-values for these substrates differ by three orders of magnitude, namely 10 mM for GA and 10 μM for PA (Galunsky et al., 2000).

The half-life of the enzyme as a function of pH is shown in Figure 9.16. This value is constant, but is much lower than for PA in the pH range 6–8. This may be due to a lack of a tightly bound Ca^{2+}-ion found in PA (see Chapter 2, Section 2.4 and Fig. 2.6). The enzyme must be removed from the product, and is therefore immobilized, using similar carriers as for the (*R*)-amino acid oxidase. Immobilized glutaryl amidase which had been used for 300 cycles of 90 min was seen to have lost ≈50 % of its initial activity (Monti et al., 2000).

9.2.6 Evaluation of Process Options

9.2.6.1 Process Windows

The pH-*T* process window for >95 and 99 % hydrolysis of Glu-7-ACA is pH >8.0 or >8.6, as shown in Figure 9.14. The upper limit both for the enzyme and to reduce byproduct formation is around pH 9. In order to reduce product losses due to byproduct formation the temperature should be <25 °C. The lower temperature limit is given by requirements on the minimal reaction rate. This process window is shown in Figure 9.17, and is also suitable for the deamination and decarboxylation reactions. Thus, a pH change for the two consecutive reactions is not necessary.

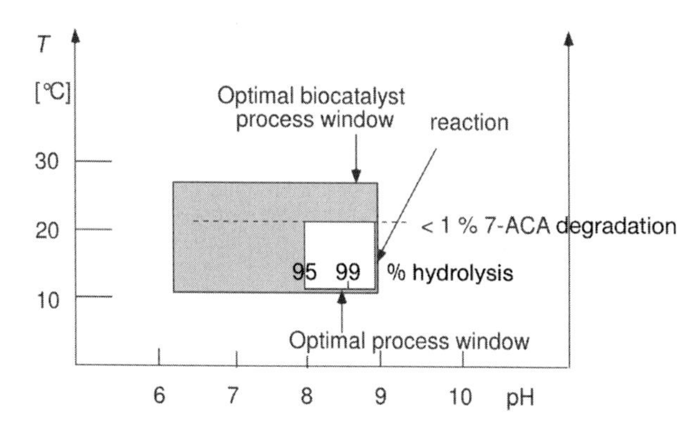

Fig. 9.17 The pH-T process window for the hydrolysis of 100 mM Glu-7-ACA is the overlapping part of the optimal biocatalyst and reaction windows (≥ 95 % hydrolysis).

9.2.6.2 Suitable Reactors and pH-controlling Buffers

A 7-ACA plant as discussed here produces about 1000 tons per year. Each day, a certain amount of cephalosporin C must be converted to 7-ACA, and this can be carried out as either a continuous or a discontinuous process. Due to product inhibition, a CSTR is unfavorable, as much more enzyme is required for a given space–time yield than for a packed-bed reactor. The latter is not suitable in the deamination step as it is difficult to supply the necessary O_2, and the pH will change along the length of the reactor. The same applies to the hydrolysis reactor. Thus, both processes must be carried out as discontinuous reactions.

Attempts have been made to perform both enzyme reactions simultaneously in the same reactor, but this has not been successful as the concentration of Glu-7-ACA will be low and will lead to an increased influence of product inhibition compared to two separate reactors.

The hydrolysis of Glu-7-ACA must be carried out rapidly (within 30 min) in order to avoid losses in 7-ACA. This requires large quantities of enzyme in the immobilized enzyme reactor, and consequently a batch reactor is not suitable. Also, the loss of immobilized enzyme due to abrasion caused by particle–particle collisions at immobilized enzyme contents >10 % (v/v) cannot be neglected. As a fixed-bed reactor cannot be used, a recirculation reactor consisting of a fixed bed and a mixing vessel where the pH is adjusted is a suitable reactor configuration (Fig. 9.18). The substrate is recirculated through the fixed bed until the end-point of the reaction has been reached. In such a reactor, 20–30 % of the total reactor volume (fixed bed + mixing vessel) can be occupied by immobilized enzyme (see also Fig. 9.6).

Such a reactor can also be used for the oxidative deamination and decarboxylation reaction, instead of the currently used batch reactors. As the O_2-content is critical here these reactions are carried out at higher pressures (3–5 bar) in order to obtain higher oxygen concentrations.

The reactors must be designed to ensure a minimal space–time yield given by the amount of cephalosporin C that must be converted to 7-ACA per day. The enzyme will undergo inactivation and must be replaced by new immobilized enzyme. This can be done when the fixed-bed reactor consists of several parallel fixed beds where

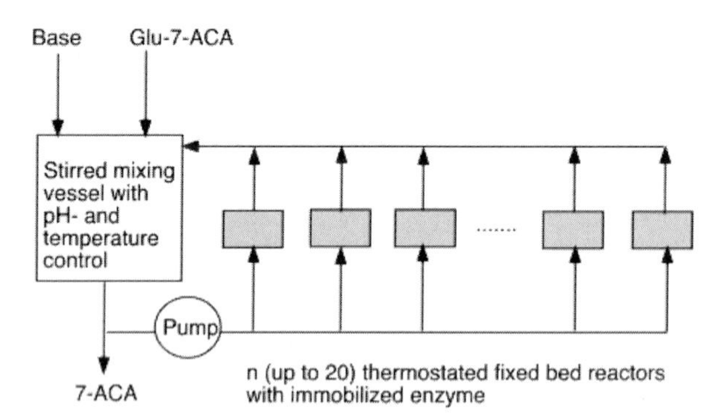

Fig. 9.18 A differential reactor with high content of immobilized enzyme and minimal abrasion. The system consists of a mixing vessel for pH control (and oxygenation of the substrate solution for the deamination of cephalosporin C) and one or more parallel fixed-bed reactors through which the substrate solution is pumped until the end-point of the reaction is reached. The temperature in both the vessel and the packed beds must be controlled. The above system, though developed for the hydrolysis of penicillin G, can also be used for the conversion of cephalosporin C to 7-ACA.

the inactivated biocatalyst is replaced by a new one in order to maintain the minimum space–time yield (Fig. 9.18).

The pH in the mixing vessels of both enzyme processes must be kept constant at pH 8–9. The most suitable bases to be used here are those with p*K*-values in this range, for example NH_3 and CO_3^{2-}. The latter is the better choice as it reduces costs for the denitrification of the wastewater produced. NaOH is not a suitable buffer as the local pH in the mixing chamber may increase to >10, thereby decreasing the stability of the biocatalyst, substrates and products.

9.2.6.3 Reaction End-Point and Immobilized Enzyme Requirement for Minimum Space–Time Yield

The critical reaction is the hydrolysis of Glu-7-ACA, which must be carried out within 30 min in order to reduce losses due to the degradation of 7-ACA to less than 1 %. As shown above (Figs. 9.14 and 9.17), 99 % hydrolysis is possible. This is a suitable end-point. With $[S]_0 = 100$ mM and $t = 30$ min, the required space–time yield is 3.3 mM min^{-1}.

The amount of biocatalyst per unit volume $V_{max,0}$ required to obtain this result can be calculated as shown in Chapter 8 (see Chapter 8, Sections 8.5 and 8.7). For a 30-min hydrolysis the deactivation of the enzyme can be neglected. For 99 % hydrolysis the degradation of 7-ACA, and the reverse reaction can be neglected. Then, $V_{max,0}$ can be calculated using Eq. (8.31), written to include biocatalyst inactivation.

$$V_{max,0} = \frac{K_m (1 + [S]_0/K_i) \ln([S]_0/[S]_t) + (1 - K_m/K_i)([S]_0 - [S]_t)}{\eta_0 \cdot (1 - e^{-k_i t})/k_i} \tag{9.11}$$

This equation was derived for a similar process, namely the hydrolysis of penicillin G. Without biocatalyst inactivation the denominator equals t. The operational effectiveness factor then is ≈ 1 . With $K_m = 7$ mM in the pH interval 8–9, both glutarate and 7-ACS are competitive inhibitors, with K_i-values of 80 and 200 mM in this pH interval, they are considered to be one inhibitor with the K_i-value $(200 \times 80/280)$ mM (Spieß et al., 1999), and $V_{max,0} = 6000$ U L^{-1}. Immobilized glutarylamidase preparations have activities of $\approx 50\,000$ U L^{-1}, and can be reused up to 300 times until their activity is 50 % of the initial value (Monti et al., 2000). From this it follows that ~10 % of the enzyme reactor volume must be taken up by immobilized biocatalyst.

For the deamination and decarboxylation reaction the end-point should be 99 % substrate conversion. It should be reached in the same time as for the subsequent hydrolysis reaction. Only about 90 % of the substrate is converted to Glu-7-ACA, as not all keto-adipoyl-7-ACA reacts with H_2O_2, and some Glu-7-ACA is deacylated (Fig. 9.13) (Conlon et al., 1995). Thus, $[S]_0$ and the space–time yield for this reaction must be about 10 % higher than for the hydrolysis of Glu-7-ACA. No kinetic data are available to estimate the amount of immobilized enzyme required to reach the desired space–time yield. Rather, this must be determined from laboratory- and pilot-scale experiments.

For the size of reactors used for the industrial enzyme process, see Exercises, Section 9.5.1.

9.2.6.4 Product Isolation

7-ACA has a low solubility (<10 mM) around pH 4 and is crystallized by a rapid pH change to this value immediately after the hydrolysis reaction. The byproduct glutarate is discarded. In the hydrolysis of adipoyl-7-ADCA the byproduct adipate is recycled into the fermenter, where it is reused (Bruggink, 2001).

9.3 Residence Time Distribution, Mixing, Pressure Drop and Mass Transfer in Reactors

9.3.1 Scale-up, Dimensionless Numbers

Scale-up is a difficult task with conflicting requirements, such as constant power input, or constant stirrer tip speed, both of which have been proposed with reference to solid particle suspension, resulting in evident disparity. Transport processes are very dependent on scale, and this is the main reason that scale-up problems exist. Whereas a process is determined by kinetic phenomena at small scale, it is determined by transport phenomena at large scale in general. The main problems are the mixing, suspending and abrasion of particles in stirred tank reactors, and pressure drop in tubular fixed-bed reactors. Optimal conditions for a biocatalytic process often ask for conflicting process conditions. Further, the accuracy of various scale-up correlations is questionable (Kossen and Oosterhuis, 1985; Middleton and Carpenter, 1992).

In the laboratory – and in part also in the pilot scale – reactors may be constructed and operated at near-ideal conditions so that concentration profiles as shown initially (see Fig. 9.1) are correct in good approximation. This includes ideal mixing – that is, identical concentrations at any position in ideal stirred-tank reactors for a given time, or independent of time at continuous operation under steady state. In large reactors however, significant deviations occur due to the problems of scale (Fig. 9.19) (Reuss and Bajpai, 1991; Schmalzried et al., 2003).

In addition, several parameters do not play any role at the laboratory scale, or they may be neglected; examples include the pressure drop in packed-bed reactors or the abrasion of particles in stirred vessels. A considerable number of parameters must be taken into account which require inter-correlation. In order to reduce these, groups of parameters are summarized in dimensionless numbers in physically meaningful ways, which in turn can be correlated, and this considerably reduces the number of correlations necessary to describe functions. These functions have been derived, for example, for external mass and heat transfer.

Dimensionless numbers, such as Re, Ne, Sh, and Sc are commonly used in order to correlate power consumption, and mass transfer with hydrodynamics – that is, fluid flow at different flow regimes. Their definitions are as follows:

- For characterizing fluid flow: Reynolds number
  $\mathrm{Re} = (\rho\, n \cdot d_\mathrm{i}^2 / \eta)$ (inertia forces/viscous forces) for stirred-tank reactors
  $\mathrm{Re} = (\rho\, u \cdot d_\mathrm{p} / \eta)$ for tubular reactors with fixed bed of particles

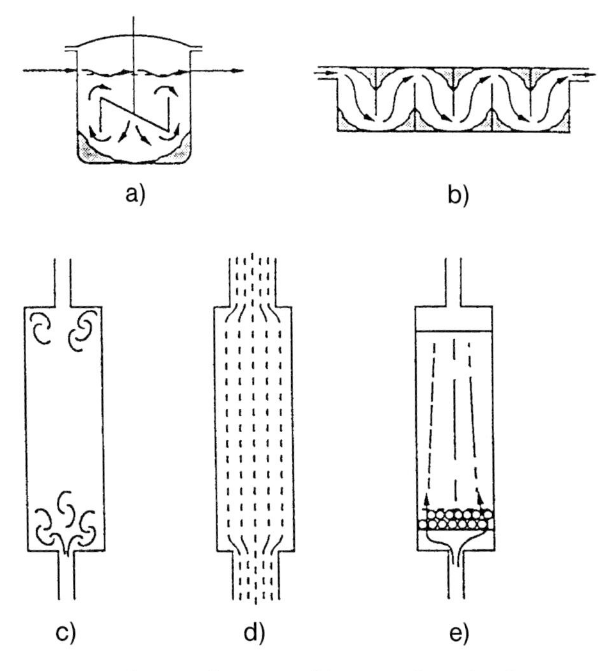

Fig. 9.19 Schematic illustration of deviations from ideal flow in stirred vessels (a) and tubular reactors (b–e). (a) Stagnant zones and short-circuit flow/channeling; (b) stagnant zones; (c) back-mixing in local turbulent field; (d) insufficient axial mixing; (e) channeling in fixed bed with catalyst.

- For characterizing power consumption: Newton number
  $\mathrm{Ne} = (P_R / (\rho\, n^3 \cdot d_i^5))$
- For characterizing external mass transfer: Sherwood number and Schmidt number
  $\mathrm{Sh} = (k_L \cdot d_p / D_s)$ (total mass transfer/mass transfer by diffusion)
  $\mathrm{Sc} = (\nu / D_s)$ (hydrodynamics boundary layer/mass transfer boundary layer)
  ν = kinematic viscosity
- Another most relevant parameter (not dimensionless) for scale-up is the power input:
  $P_R = 6\, \rho\, n^3 \cdot d_i^5$

9.3.2
Residence Time Distribution

Three parameters are important for describing fluid flow in reactors: (1) the (mean) residence time; (2) residence time distribution; and (3) mixing time. The residence time provides information about the time for which a volume element stays inside a reactor. In real reactors – and notably in CSTRs – different volume elements may have greatly differing residence times; this is described by the residence time distribution $E(t)$. This function gives the probability for a volume element (or a tracer pulse with the concentration c), that has entered the reactor at time $t=0$, leaves the reactor at time t (this also means the molar flow n' referring to the amount n_0 at time $t = 0$; Fig. 9.20):

$$E(t) = \frac{\dot{n}}{n_0} = \frac{Q \cdot c(t)}{\int_0^\infty Q \cdot c(t)\, dt} \quad \text{where} \quad \int_0^\infty E(t)\, dt = 1 \tag{9.12}$$

The mean hydraulic residence time τ_h can be calculated from the ratio of reactor volume and volumetric flow. It is often advantageous to relate the real-time distribution to this residence time, and to introduce a dimensionless time according to the following definition:

$$\tau_h = V \times Q^{-1} \quad \text{and} \quad \theta = t \times \tau_h^{-1} \tag{9.13}$$

The *integral* of the residence time distribution (Eq. 9.14) gives the total amount $n(t)$ of a substance which has left the reactor at time t (see Eq. 9.12). It may be described by the ratio of the actual concentration of a tracer $c(t)$ and the constant concentration at the inlet c_0.

$$F(\Theta) = \int_0^\Theta E(\Theta)\, d\Theta = \frac{c(\Theta)}{c_0} \tag{9.14}$$

For the experimental determination of the residence time function, a tracer is introduced to the reactor inlet and its concentration at the outlet is measured as a function of time. The tracer substance should not influence the reactor content physically (by density, viscosity) and it should be chemically inert; depending on the tracer, the electrical (addition of a salt solution) or thermal conductivity, or the absorption spectrum (e.g., a *lithium* salt with specific atomic absorption) can be measured. For the tracer addition two different methods may be applied:

- In the first of a pulse with a certain amount of tracer is added over a time interval which is as short as possible. Continuous measurement of the tracer at the reactor outlet gives the residence time distribution as a result of the pulse. When comparing the different reactor types (Fig. 9.20a), the ideal CSTR exhibits the broadest residence time distribution: when adding the tracer at time $t = 0$ it is immediately distributed equally throughout the reactor and leaves it with the concentration c_0 ($t = 0$) at the outlet; at the hydraulic residence time $t = \tau_h$ about 25 % of the tracer remains inside the reactor. In the ideal tubular reactor, the whole amount of the tracer leaves the reactor at the hydraulic residence time τ_h at the outlet.

- In the second method, a sudden change of the tracer solution is applied at the inlet of the reactor, for example with a sudden increase of a tracer concentration. The signal at the outlet corresponds to the integral of the residence time distribution $F(t)$ (Fig. 9.20b).

The residence time functions for cascades of CSTRs are intermediate between the extremes of the ideal stirred tank and tubular reactors, respectively. For practical

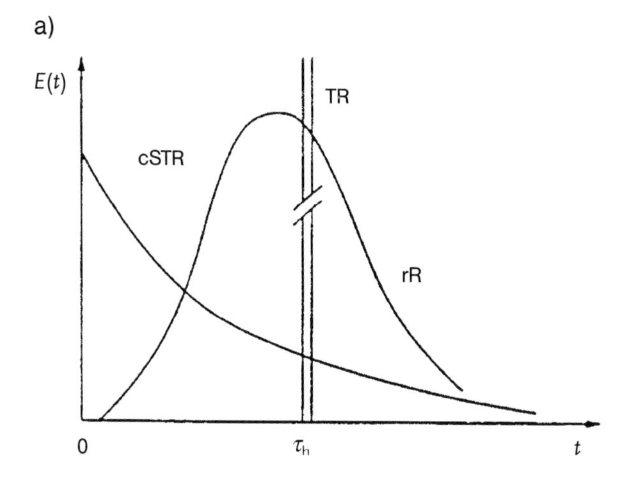

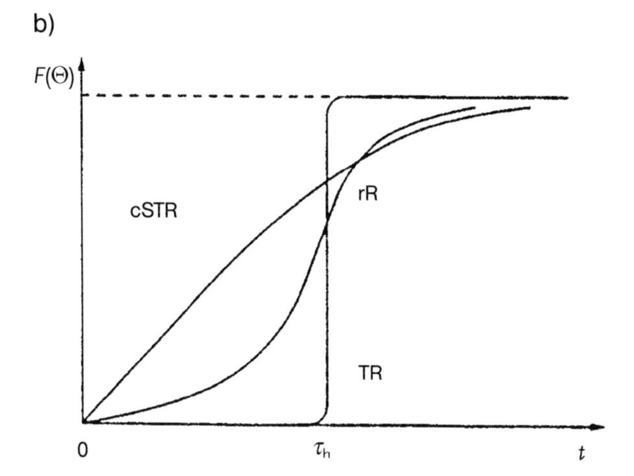

Fig. 9.20 (a) Residence time distribution $E(t)$ for ideal continuous stirred tank (cSTR) and tubular (tR) and real reactors (rR); (b) *Integral* of the residence time distribution $F(\Theta)$ for ideal and real reactors.

purposes, three to five vessels are sufficient in order to provide for a favorable residence time distribution (e.g., with reference to the catalyst efficiency). The residence time distributions and their integrals for cascades of different numbers of vessels are presented in Figure 9.21.

The residence time distribution of a reactor is the most important criterion for its efficiency, and notably for the efficiency of the biocatalyst. Deviations are common in practice, and in general they have unfavorable consequences, especially for high conversion and/or high selectivity. The reasons for deviations of the residence time distribution from ideal behavior are shown schematically in Figure 9.19. For a stirred vessel, deviations may be due to unfavorable installation of inlet and outlet (e.g., at opposite positions) so that bypass or short-circuit flow may occur, and part

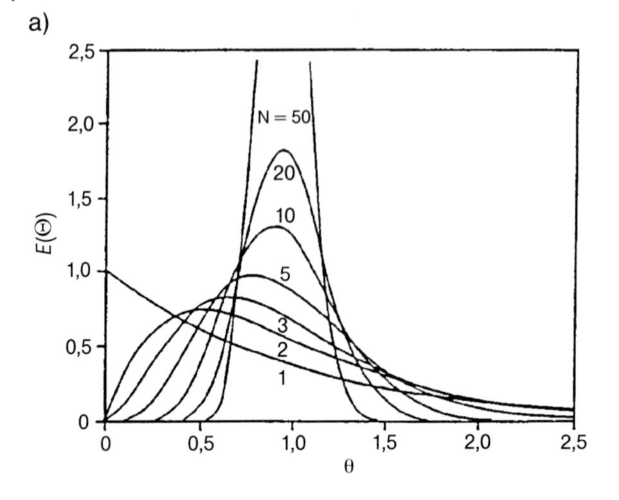

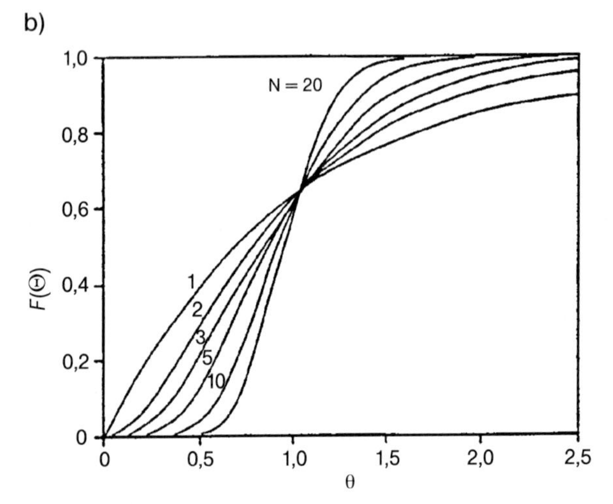

Fig. 9.21 (a) Residence time distribution in cascades of continuous stirred-tank reactors; parameters: number of vessels; (b) Integral of the residence time distribution for cascades; N: number of vessels.

of the substrate introduced flows directly towards the outlet, essentially without being mixed with the solution inside the reactor (see Fig. 9.19 a). Poorly mixed zones (dead zones) are likely to occur in viscous media (e.g., in starch hydrolysis, and generally at high substrate concentrations; Fig 9.19 a and b). Figure 9.22 shows a model calculation for a residence time distribution exhibiting a large deviation from the behavior of an ideal reactor, which was simulated with the aid of a model with bypass. In practical cases secondary maxima – both before and after the main maximum of the residence time distribution curve – may be observed for reactor cascades with insufficient mixing.

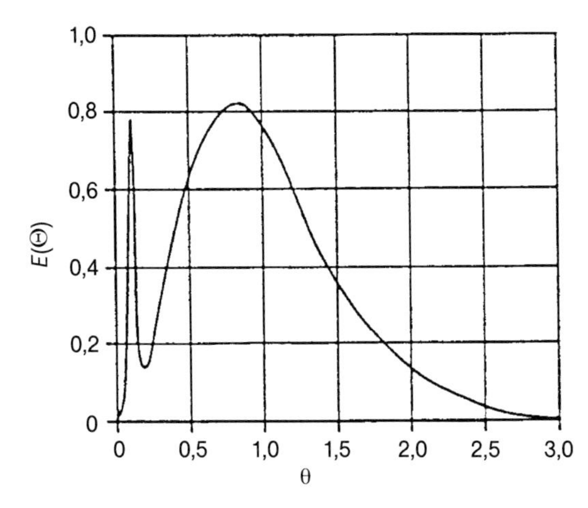

Fig. 9.22 Residence time distribution with short-circuit (simulated).

9.3.3
Mixing in Stirred-Tank Reactors

In the laboratory, where energy demand can be neglected, stirred-tank reactors are used experimentally and in model calculations as ideally mixed reactors. This situation is different in technical reactors, where electrical energy demand (stirrer motor) must be considered for economic reasons. Furthermore, the impeller dimension and speed must be limited according to the aspects of shear force and mechanical abrasion of the (immobilized) biocatalyst. On the other hand, intense mixing with short mixing time is essential for the transport of substrates with low solubility (e.g., oxygen) to the catalyst, or for the quick and efficient mixing of reactants (e.g., for neutralization as in penicillin hydrolysis).

The stirred-tank reactor has been treated extensively for these reasons as it represents the dominant type of reactor in practice. The mixing time, which signifies a certain extent of distribution of a tracer pulse in the reaction volume, is taken as a characteristic parameter for the quality of mixing. Figure 9.23 shows schematically the flow patterns for fluid elements in reactors with one and two impellers, respectively. For the measurement of mixing time, the conductivity of the solution may serve as a signal, for example when salt is added as a tracer. Different models and their mathematical treatment have been presented by Reuss and Bajpai (1991) and Schmalzried et al. (2003).

The example illustrates the relevance of the mixing time t_m as a key parameter for stirred-tank reactors. It gives the definition for the time at which a certain degree of mixing is obtained. It refers to an amplitude A_m of a signal (corresponding to the concentration of a tracer) which is taken as acceptable for a specific problem (Fig. 9.24) (Reuss and Bajpai, 1991) It can be calculated from the following correlation (with the amplitude decay rate constant K_A):

$$t_m = K_A^{-1} \ln (2 \cdot A_m^{-1}) \quad (9.15)$$

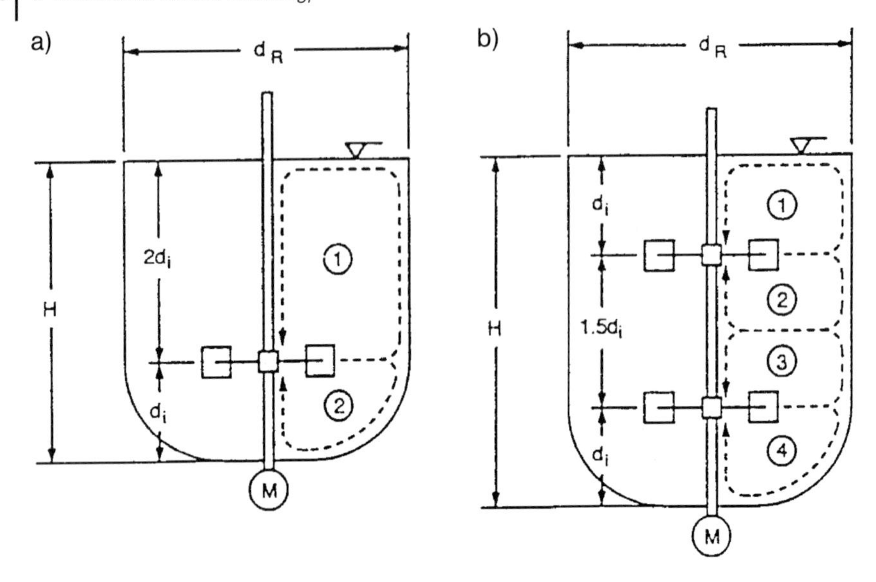

Fig. 9.23 Flow patterns with circulation flow in stirred-tank reactors. (a) Reactor with impeller in unsymmetrical position; (b) reactor with two impellers (H = height; d_R = reactor diameter; d_i = impeller diameter; 1 to 4 = circulation pathways) (Reuss and Bajpai, 1991).

It must be considered that the position of the impeller (either in the middle or in the lower part of the reactor) as well as the impeller speed n (via the Reynolds number Re) have an influence on this correlation; the mixing time can therefore be estimated only. K_A can be calculated approximately from the following equation:

$$\frac{n}{K_A}\left(\frac{d_i}{d_R}\right)^{2,3} \cong 0{,}5 \quad \text{for Re} = \frac{n\,d_i^2}{\mathrm{v}} > 2 \cdot 10^3 \tag{9.16}$$

The mean circulation time τ' (c.f. Fig. 9.23) is correlated with the systems parameters by the correlation (Eq. (9.17)) given subsequently; here, asymmetries of the stirrer position inside the vessel are also neglected (Reuss and Bajpai, 1991) (mean circulation times are in the range of several seconds up to several minutes for small or large reactors, respectively):

$$n\,\tau' = 0.76\left(\frac{H}{d_R}\right)^{0,6} \cdot \left(\frac{d_R}{d_i}\right)^{2,7} \tag{9.17}$$

For an advanced insight, mixing and mixing time in a stirred-tank reactor may be illustrated by figures taken from simulations based on computational fluid dynamics (CFD) (Schmalzried et al., 2003). Such simulations can help and accelerate the process of scale-up. The Rushton turbine (see Figs. 9.23 and 9.26 a-1) used in the experi-

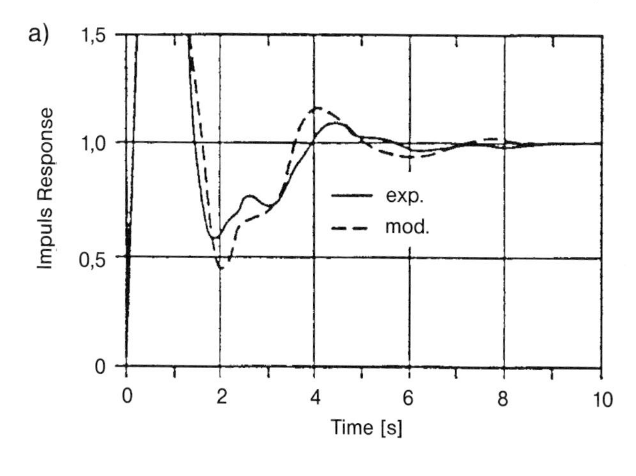

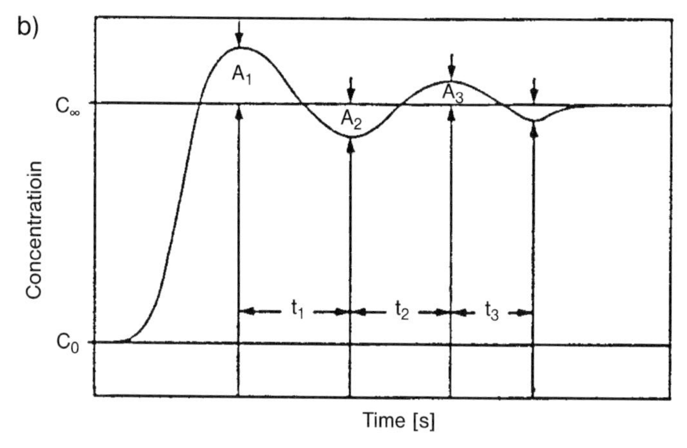

Fig. 9.24 Experimental and simulated signal from mixing after a tracer pulse.

ments and simulation is well established for mixing of liquids with low viscosity. It generates a flow which leaves the impeller in radial and tangential directions and recirculates back from the wall into the impeller region, this being the main reason for the mixing capability. Figure 9.25 shows the simulated dynamics of the tracer distribution at different times after a pulse onto the liquid surfacc in a stirred-tank reactor. It is clear from this example that mixing requires an extended time scale (over 8 s in this example) which might cause significant side reactions in large reactors. An example of practical relevance is the enzymatic hydrolysis of penicillin and Cephalosporin C (Section 9.2), where decomposition of the substrate and inactivation of the biocatalyst in a stirred tank can occur in fields of high pH near the point of addition of the base, which must be added at high concentration in order to avoid dilution of the product solution (Carleysmith et al., 1980). Similarly, the substrate distribution in large stirred vessels depends strongly on the feeding position

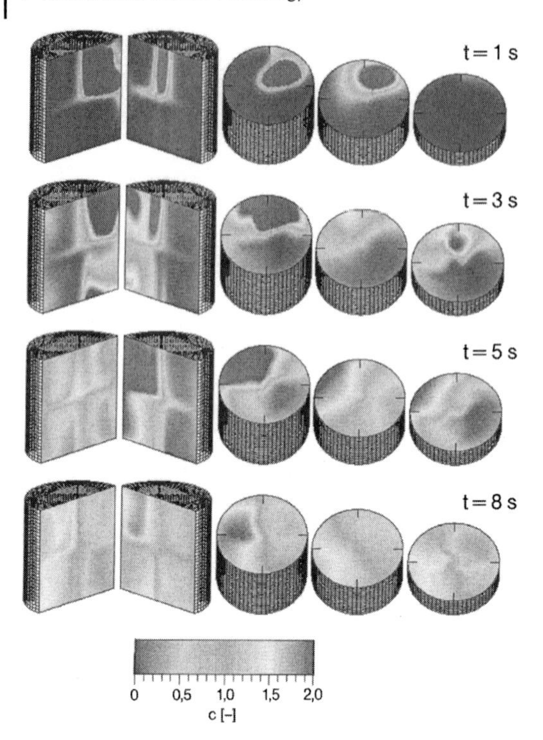

Fig. 9.25 Simulated dynamics of the tracer distribution at different times after a pulse onto the liquid surface in a stirred-tank reactor (height/tank diameter ratio = 1.0, Rushton turbine, impeller/tank diameter ratio = 0.3125, impeller clearance/height of liquid = 0.31). (Illustration taken from Figure 9 in Schmalzried et al., 2003.)

(Schmalzried et al., 2003; Fig. 14). Therefore, distributed points for the addition of reactands near the stirrer tip (field of highest turbulence) should be used.

Several types of impeller are shown in Figure 9.26. The mixing time for a certain impeller can be taken from empirical correlations, as shown in Figure 9.26b for several types of impeller. The product of stirrer speed multiplied by the mixing time t_m is given as a function of the Reynolds number (Re).

The power consumption of a stirrer can also be taken from empirical correlations; therefore, the power number (Newton number, Ne) is correlated with the Reynolds number (Re) for different types of impeller (Brauer, 1985)

The Newton number (Ne) is, as is the mixing time, for a given type of stirrer and a certain geometrical arrangement a function of Re only:

$$Ne = \frac{P_R}{\rho \cdot n^3 \cdot d_i^5} = f(Re) \tag{9.18}$$

Three different ranges of flow pattern are to be considered (Brauer, 1985):

1. The laminar region, where the following correlation holds:

$$Ne \approx Re^{-1}, \quad \text{or} \quad Ne \times Re = \text{const.} \tag{9.19}$$

2. The region of transition from laminar to turbulent fluid flow:

$$10^1 < Re < 10^2 \tag{9.20}$$

3. The turbulent flow region, where the influence of the baffles (which are essential) results in a ten-fold increase of the impeller power consumption, where, depending upon the references:

$$\mathrm{Ne} = \text{const.}; \; P_R = \text{const.} \cdot n^3 \times d_i^5 \times \rho \text{ (Zlokarnik, 1972), or} \tag{9.21}$$

$$\mathrm{Ne} \approx \mathrm{Re}^{-0.28} \text{ (Brauer, 1985)} \tag{9.22}$$

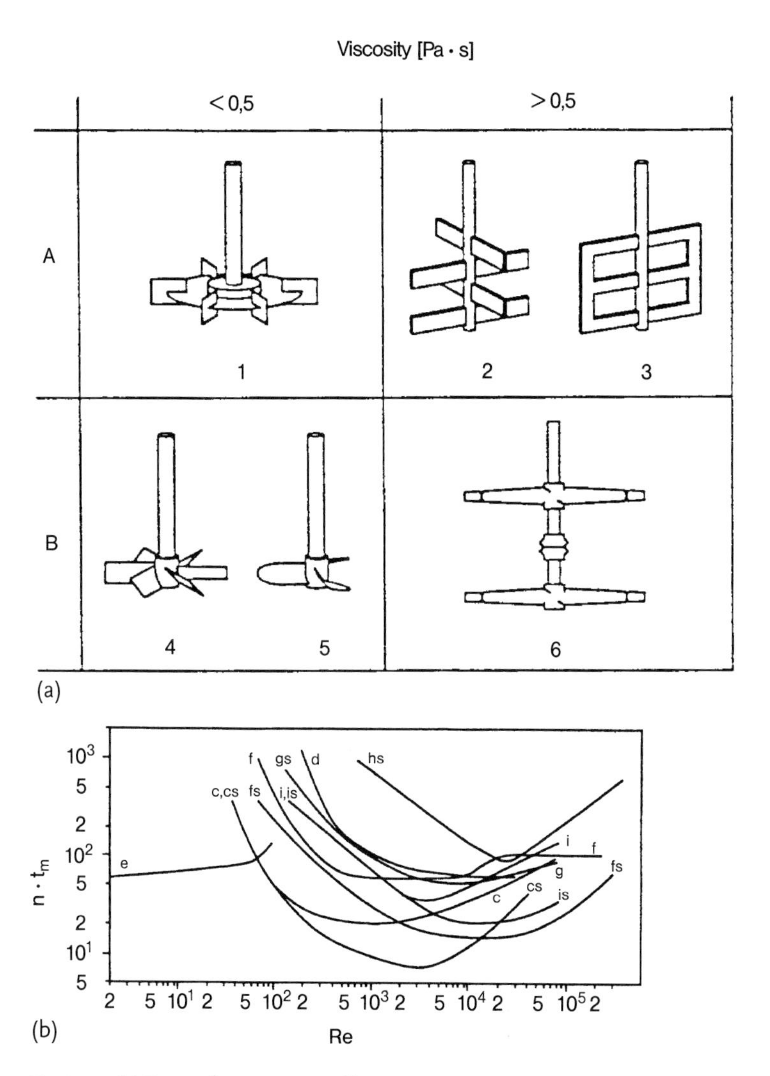

Fig. 9.26 (a) Types of common impellers, in the sequence of application for media of different viscosity and the type of fluid flow; A = tangential to radial; B axial fluid flow; 1 = Rushton turbine; 2, 3 = special types for high viscosity; 4 = pitched-blade turbine; 5 = propeller; 6 = MIG impeller (Ekato). (b) Correlations for estimating the mixing time for different types of impellers: s = with baffles; c = type 2; d = anchor; e = helical ribbon; f, fs = MIG impeller; gs = turbine; i = impeller; hs = propeller (Zlokarnik, 1972).

The *shear stress* increases in large-volume reactors, depending on the type of impeller, with $(N \times d_i^2)$, or with $(N \times d_i)^2$ (given by different sources) (Kossen and Oosterhuis, 1985); with the scale-up from a laboratory reactor of about 10 L volume (with $n = 5\ s^{-1}$, $d_i = 0.03$ m) to a technical reactor of 10 m^3 size (with $n = 2\ s^{-1}$, $d_i = 0.5$ m) and similar power input (proportional to $n^3 \times d_i^5$) the shear stress increases by a factor of 50, which may cause severe problems of abrasion for biocatalyst particles.

For *suspending particles* the general scale-up rule is to keep P/V = const. (Kossen and Oosterhuis, 1985).

9.3.4
Mass Transfer in Reactors

External mass transfer depends on hydrodynamics in reactors; it is in general not rate-limiting in stirred tank reactors under turbulent conditions, whereas it may be rather slow and eventually rate-limiting in fixed-bed reactors, depending on the flow rate. The phenomena involved are conveniently discussed with the aid of the simplified film model, assuming a quasi stagnant layer of fluid thickness (δ) which adheres to the outer catalyst surface (Buchholz, 1982). Substrates and products must penetrate this film by diffusion according to their concentration profile (see Fig. 8.29). The rate of this mass transfer step is described by:

$$k_L\,(c_S(\infty) - c_S(R)) = (D_S/\delta)(c_S(\infty) - c_S(R)) \tag{9.23}$$

In the steady state, it is equal to the overall reaction rate:

$$v = k_L\,A_P\,(c_S(\infty) - c_S(R))\;V_P^{-1} \tag{9.24}$$

The importance of the external particle surface A_P is obvious. k_L can be determined experimentally (whereas δ does not have a real physical significance for most experimental conditions). It is strongly correlated with the hydrodynamics of the reactor system, notably the flow rate in fixed-bed and fluidized-bed reactors, and the viscosity of the fluid.

The correlations for mass transfer and experimental conditions are commonly expressed in terms of dimensionless groups (Section 9.3.1):

$$\mathrm{Sh} = c\,\mathrm{Re}^a\,\mathrm{Sc}^b \tag{9.25}$$

Data compiled from the literature for packed-bed and trickle-bed reactors are (Atkinson and Mavituna, 1991):

$\mathrm{Sh} = 0.99\,\mathrm{Re}^{1/3}\,\mathrm{Sc}^{1/3}$; $\mathrm{Re} < 1$;
$\mathrm{Sh} = 0.95\,\mathrm{Re}^{1/2}\,\mathrm{Sc}^{1/3}$; $10 < \mathrm{Re} < 10^4$
$\mathrm{Sh} = 2 + 0{,}73\,\mathrm{Re}^{1/2}\,\mathrm{Sc}^{1/3}$; overall approximation of various correlations.

A correlation reported for isotropic turbulence is:

$$\mathrm{Sh} = 0.13\,\mathrm{Re}^{3/4}\,\mathrm{Sc}^{1/3} \tag{9.26}$$

9.3.5
Pressure Drop and Fluidization in Tubular Reactors

In *fixed-bed reactors*, the pressure drop is of major importance for the application of biocatalysts. These may be several meters in height, with pressure drops of several bar. A general correlation for the pressure drop Δp as a function of bed height H, flow rate u, voidage ε and particle diameter d_p is given by Eq. (9.27) for laminar flow conditions (when Re < 10):

$$\Delta p\, H^{-1} = 150\,(1-\varepsilon)^2\,\eta\,u\,(\varepsilon^3\,d_p^2)^{-1} \tag{9.27}$$

For turbulent flow, which is not likely to occur in large packed-bed reactors, this equation must be extended (Buchholz, 1982; Eigenberger, 1992).

The correlation given before makes the high relevance of both bed porosity ε and particle diameter d_p obvious. In the choice of particle size of immobilized biocatalysts for technical applications, two opposite requirements must be considered: (1) high efficiency at high enzyme loading in order to provide for high space–time yield, with optimal size in the range of 50–100 µm; and (2) technical handling and performance in filtration, or sieve retention in batch, or fixed-bed applications. Filtration time and pressure drop limit the particle size to a minimum range of 200–300 µm (see also Chapter 8, Section 8.4.1). To limit the pressure drop in large, fixed-bed reactors the range of particle size applied is 400–1000 µm (Antrim and Auterinen, 1986; Pedersen and Christensen, 2000; Jensen and Rugh, 1987; Busse, 2004). The particle size distribution, which determines the void volume, is also essential, and should be narrow.

In *fluidized-bed reactors*, the terminal settling velocity for a single particle is the limiting case where the particles tend to leave the reactor. It may be estimated from classical correlations (Jördening and Buchholz, 1999; Werther, 1992). One important point here is the fluidized bed expansion behavior, which may be calculated using the correlation of Richardson and Zaki (1954), most often as:

$$u = u_T\,\varepsilon^n \tag{9.28}$$

where u_T is the terminal velocity of isolated single particles. The exponent n is given as follows:

$n = 4.65$	$0 < Re < 0.2$
$n = 4.4\,Re^{-0,03}$	$0.2 < Re < 1$
$n = 4.4\,Re^{-0.1}$	$1 < Re < 500$
$n = 2.4$	$500 < Re$

where $Re = u_T\,d_p\,\nu^{-1}$

Typical operation ranges for fluidization are, for particles of low density such as pumice, 5–10 m h^{-1}, and for sand about 10–30 m h^{-1} (particle diameter 0.5 mm). The fluid bed pressure loss for low-density media is significantly lower than, for example, sand (Jördening and Buchholz, 1999).

9.4
Process Technology

Reaction engineering relevant to the technology of the bioconversion of educts (substrates) into products – that is, the reaction – has been dealt with in Sections 9.1 and 9.2. Here, the so-called upstream, and the subsequent so-called downstream steps, will be detailed. In practice, the reaction – as well as the final product quality – essentially depend on these operation steps, termed unit operations. First, a short survey will be provided, followed by an actual example, namely the process to manufacture glucose-fructose-syrup, with a range of generally relevant unit operations.

9.4.1
Survey

Process technology cannot be treated here in depth, as this would require the treatment of all unit operations including the engineering fundamentals with mass transfer principles, as well as the design and construction of devices and equipment.

Practical work requires an in-depth knowledge of the fundamentals of these engineering aspects. Even if the solution of technical problems in many cases proceeds in empirical and experimental approaches, the selection of optional methods and devices would require many experimental investigations to be conducted, or the rational selection of approaches would not be possible. This requires the knowledge of unit operations and reaction engineering (for relevant reports on this topic, see Hempel, 2004; *Ullmann's Encyclopedia of Industrial Chemistry*, 1988, 1992). Biochemical engineering and key unit operations in biotechnology have been compiled in comprehensive special monographs (Moo-Young, 1985; Atkinson and Mavituna, 1991), as well as in special reviews (see the series of *Advances in Biochemical Engineering*, e.g., Schmalzried et al., 2003; Brauer, 1985; Kossen and Oosterhuis, 1985).

The conversion represents the central step of a biochemical or biotechnological process. However, in general it corresponds to less than half of the equipment, operational cost and investment required. The preparation of substrates and additives, the isolation and purification of products, and the utilization or recycling of byproducts and treatment of wastes mostly require more equipment, expenditure of work and investment cost than the bioconversion itself. All individual steps of the overall process must be coordinated carefully in order to guarantee economic processing. The process as a whole must be controlled by appropriate measurement and instrumentation. The results from laboratory experiments – including separation operations such as chromatography – must be translated to the technical scale by careful scale-up.

A survey of the most important steps of a process, including the corresponding unit operations and equipment, is provided in Table 9.2. All individual steps must be coordinated corresponding to mass and heat flow and submitted to measurement and control. This comprises the volume of tanks, the horsepower of pumps and stirrers, the cross-section of pipes and columns (e.g., of ion exchangers), and the capacity of filters and heat exchangers (for the calculation of dimensions and capacity of pumps and pipes, etc., see Dunlap, 1985). Planning must also include the aspects of substrates and additives quality.

The aim of process technology and biochemical process design is to establish a process of:

- high product quality and
- high productivity under economic conditions.

Table 9.2 Elements of process technology.

Process step		***Unit operation***	***Equipment***
1. Pretreatment of educts (substrates, additives)	Separation/Purification	Filtration Sterile filtration Adsorption and Ion-exchange chromatography	Tanks for storage Filter Columns
	Pumping Concentrating	 Evaporation Ultrafiltration Reverse osmosis	Pumps Evaporators Membrane systems
	Dilution (pH-adjustment) Heating Cooling	Mixing	Stirred tank
2. Biochemical conversion	Reaction engineering[a)]		
3. Product isolation and purification	Separation/Purification	See above and Crystallization Precipitation	
	Stabilization	Sterilization Drying	 Dryer
	Concentrating	See above	Spray dryer
4. Recycling of auxiliary material (e.g., solvents)			
5. Treatment of residual material and wastes (e.g., regeneration water from ion exchangers)			
6. Confectioning (packaging, etc.) Concentrating			

a) See Sections 9.1 and 9.3.

This is possible only as a compromise, since special aspects and individual steps often exhibit conflicting requirements. Scale-up imposes specific problems and challenges which have been discussed in brief in the preceding section. The general treatment is outlined in the specialized literature (Atkinson and Mavituna, 1991; Kossen and Oosterhuis, 1985).

One example of the many questions arising in the context of biochemical process design is the manufacture of glucose-fructose-syrup. The quality, value and price of such a sweetener syrup depends on several parameters, including its color, which in turn depends on the formation of byproducts, the so-called browning products. Not only the process parameters (e.g., pH, temperature) but also non-sugar compounds in the substrate solution have significant influences on the formation of color. The question remains as to whether a purification step preceeding the enzymatic conversion is required and is economically feasible, or if appropriate process conditions can sufficiently avoid color formation. This decision must be made in the context of process technology.

9.4.2
Case Study 2: The Manufacture of Glucose-Fructose Syrup

Fructose exhibits a significantly (ca. two-fold) higher sweetening power as compared to glucose, which is produced in large amounts from corn (maize) starch. Consequently, glucose is isomerized in part to fructose (see Section 6.4). This is the largest process performed, with ca. 1500 t of immobilized enzyme producing a glucose-fructose syrup (high-fructose corn syrup, HFCS) with more than 10 million tonnes of product (dry matter) per year (Antrim and Auterinen, 1986; Misset, 2003; Pedersen and Christensen, 2000; Reilly and Antrim, 2003; Swaisgood, 2003). At least five major companies provide and offer the process as a whole, and this is a condition of successful marketing (c.f. Section 6.4, Table 6.8). The enzyme-catalyzed conversion is the central step of the process, but a considerable number of further operations must be carried out in a correct technical manner in order to ensure an economically feasible overall process.

The most relevant steps and unit operations for an example of the overall process are detailed in Table 9.3. Further variations with modified process conditions are applied practically (Uhlig, 1991). During one part of the process, starch is hydrolyzed with at least two soluble enzymes to yield a glucose syrup of dextrose equivalents (DE) 95–97 % (see Section 5.3.2, Table 5.6). In the second part, glucose is isomerized to fructose by immobilized glucose isomerase. The reaction can only approach chemical equilibrium with about 50 % glucose and 50 % fructose (depending upon the temperature employed).

In an initial step, gelatinization of suspended ground starch is performed by thermal treatment (105–110 °C) in a so-called jet cooker (tubular reactor(s)) in order to make the polysaccharide particles accessible to α-amylase. The first hydrolysis step – partial hydrolysis by an endoglucanase (α-amylase) – follows at about 95 °C for 90 min residence time in cascades of stirred-tank reactors (Fig. 9.27) (Olsen, 1995). The conditions for optimal enzyme activity are listed in Table 9.3. For classical

Table 9.3 Scheme for the manufacture of glucose-fructose syrup[a] (Antrim et al., 1989; Jørgensen et al., 1988; Uhlig, 1991; Olsen, 1995).

Process step	*Equipment/unit operation*	*Reaction parameters*
Educt: **starch** (30–40 % d.m. suspension)[b]	Storage tanks Heat exchangers	
1. Gelatinization[c]	Tubular reactor	105–110 °C, 5–10 min
2. Partial hydrolysis (Liquefaction by α-amylase[d])	Cascade of stirred-tank reactors	85–95 °C, 1–2 h, pH 6.5, 35 % d.m.[b], 3–50 ppm Ca^{2+}
3. Hydrolysis (with glucoamylase[d] to glucose)	Cascade of 6–12 stirred-tank reactors	55–60 °C, pH 4.5, 24–72 h
4. Downstream processing	Filtration, or centrifugation Adsorption (activated carbon) Cation-, anion exchange	
5. Evaporation	Evaporators	
6. Isomerization	Tubular fixed-bed reactors	55–60 °C, pH 7.5–8.2, 40–50 % d.m.[b], 45 ppm Ca^{2+}
7. Product purification	Adsorption (activated carbon) Ion exchange	
8. Evaporation, Cooling	Evaporators Heat exchanger	71 % d.m.[b], fructose ≥42 % (on d.m.), pH 3.5–4.5, 0.05 % ash
Product: **glucose-fructose-syrup**		

a) For simplicity, pumps and mixing operations (pH adjustment) are omitted.
b) d.m. = dry matter.
c) Option: addition of thermostable α-amylase.
d) The used enzymes are mostly mixtures of enzymes with different activities and specificities.

α-amylases, appropriate concentrations of Ca^{2+} ions are essential for stabilization of the enzyme, whereas genetically modified variants function without added Ca^{2+} ions.

A process variant utilizes bacterial hyperthermophilic amylases. One part of the thermostable α-amylase is added in the gelatinization step (at T≥105 °C) in order to utilize the synergism of thermal gelatinization and partial hydrolysis. The second part of the α-amylase is added in the second step (partial hydrolysis) at 85–95 °C. The degree of hydrolysis (for obtaining dextrins of a specific DE value) is determined by rapid inactivation of the enzymes (acidification and/or thermal treatment). Additional use of isoamylase and/or pullulanase as debranching enzymes makes the starch conversion more efficient (see Chapter 5, Section 5.3.2).

The hydrolysis to produce glucose syrup (DE 96–97 %) is performed subsequently at a lower temperature and pH, with a requirement for heat exchange and mixing

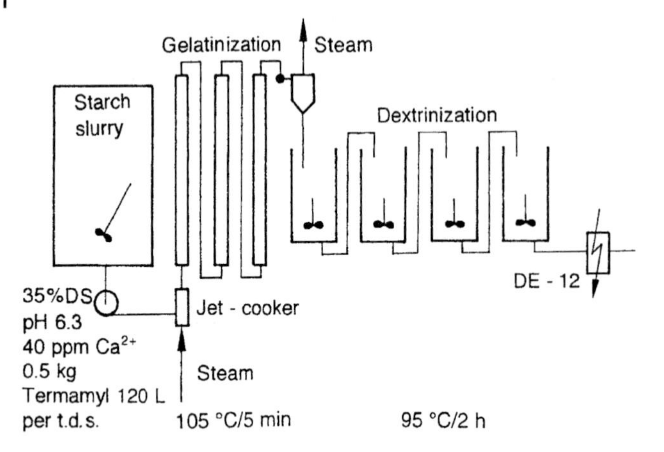

Fig. 9.27 Reactor configuration for the starch liquefaction process. (Termamyl is a bacterial thermophilic α-amylase) (Olsen, 1995).

to adjust the pH. The conversion takes place in large stirred tanks at a high residence time. Yields for glucose syrups are in the range of 96–97 %, and the byproducts are 2–3 % disaccharides (maltose and isomaltose) and 1–2 % higher oligosaccharides.

In the downstream processing, several unit operations must ensure the optimal conditions for the subsequent isomerization step. Thus, the Ca^{2+} ions required for the stability of amylases must be removed by ion exchange as they inhibit glucose isomerase, which in turn requires Mg^{2+}. Furthermore, organic substances such as browning products, acids, peptides and amino acids, must be eliminated by adsorption onto activated carbon. The dry matter content is adjusted by evaporation. The last part of the process – the isomerization of glucose – takes place in a series of tubular reactors with fixed beds of the immobilized glucose isomerase run in parallel (c.f. reactors for 7-ACA hydrolysis, Section 9.2). The steps involved, from preparation of the substrate (glucose solution) until the final product treatment (glucose-fructose syrup) are detailed in Table 9.4.

The fixed-bed reactors for isomerization have diameters in the range of 0.6 and 1.5 m, and heights between 2 and 5 m; thus, volumes in the range of 0.5 to 9 m^3. The pressure drop may be up to 3 bar. Biocatalyst particle sizes are in the range of 0.3 to 1.0 mm, with an activity of about 300 U g^{-1}. A reactor with 1.5 m diameter and 5 m height contains about 4000 kg of the biocatalyst. In a plant with 20 reactors, about 1000 t of product (with 42 % fructose, d.m.) can be produced each day (Olsen, 1995; Misset, 2003).

Most commercial immobilized glucose isomerases show an exponential activity decay as a function of residence time. Typically, the catalyst is replaced when the activity has fallen to 12.5 % of the initial value. Half-lives are in the range of 80–150 days, with the most stable commercial glucose isomerases exhibiting a half-life of around 200 days. Biocatalyst productivities are in the range of 12–20 t (product) per

Table 9.4 Unit operations for the isomerization of glucose (Olsen, 1995; Misset, 2003).

Process step	*Unit operation/equipment*	*Reaction parameters*
Substrate: **glucose solution**		40–50 % d.m.[a], ≥ 95 % glucose, peptides, organic substances
1. Substrate preparation	Adsorption Anion exchange	Elimination of: Browning products NO_3^-; (amino-) acids
2. Addition of (solution) $Na_2S_2O_8$, or SO_2 $MgSO_4$		SO_2 50 ppm Mg^{2+} 45 ppm (Ca^{2+} <1 ppm)
3. Adjustment of pH NaOH	Mixing vessel Storage tank	pH 7.5–8.2
4. Sterilization Steam Cooling water	Heat exchangers	
5. Isomerization	Fixed-bed reactors	55–60 °C, pH 7.5–8.2 40–47 % d.m.[a]
6. Product treatment, finishing	Filtration Adsorption (activated carbon) Cation exchange Anion exchange	Browning products Elimination of ionic compounds
Product: **glucose-fructose-syrup**		≈45 % d.m.[a], fructose: 42 % (ref. to d.m.), or fructose: 55 %

[a] d.m. = dry mass.

kg (biocatalyst). In order to maintain a constant fructose concentration, the feed flow rate is adjusted according to the enzymes actual activity. The production rate can be kept nearly constant while operating several reactors with different age of biocatalyst (Olsen, 1995; Swaisgood, 2003).

The costs of the individual steps have been described by Olsen (1995). As in most processes with a low product price (often the case in the food industry), the raw material (starch) dominates the overall cost at 37 % (8.8 € per 100 kg HFCS), followed by milling and feed preparation at 29 % (not included in the scheme discussed here). As suggested previously, the enzyme cost must be kept very low, and consequently liquefaction enzymes (α-amylases) comprise 0.7 %, saccharification enzymes (glucoamylases and pullulanases) 1.5 %, and glucose isomerase 1.6 % of the overall cost. Total production costs have been reported as about 24 € per 100 kg HFCS.

The product of the enzymatic isomerization must meet established quality standards, and the most important of these are listed in Table 9.5. These standards are ensured by the introduction of several downstream unit operations (see Table 9.4).

Table 9.5 Quality requirements for glucose-fructose syrups (HFCS, three specifications) (Aschengreen, 1984; Bonse, 1986; Misset, 2003).

Parameter	*A*	*B*	*C*
Fructose content [%]	42	42	55
Glucose content [%]	52–53	52–53	41
Oligosaccharides [%]	4		
Dry mass [%]	71	95–96	
Density [kg L^{-1}, 38 °C]	1.34		
pH	3.5–4.5		
Viscosity [Centipoise, 27 °C]	160		
Concentration of Ca^{2+} [ppm]		<1	
Ash content [%]	0.05		
Turbidity	no	no	no
Extinction (UV and visible)	0.5[a] (UV 280 nm)	low	low
Relative sweetness (relating to sucrose = 100)	100	100	110

[a] Value given for substrate.

9.4.3 Reactor Instrumentation

Reactor equipment is shown in Figure 9.28 for another example, namely the discontinuous hydrolysis of penicillin, together with details of measurement and control units for automated (processing during 24 h in batch mode). The supplementation of concentrated and thermostated substrate solution, as well as deionized water, is controlled by computer, as is the temperature of the reactor. The most important factor is pH control, which is achieved by neutralization with ammonia (see Case study 9.2). The rate of the amount of ammonia consumed indicates the progress of the reaction – that is, its reaction rate. These data are compiled continuously by a computer. A high standard of process control is required in order to ensure a stable processing and a high standard of quality.

Reactors as well as unit operations require further peripheral equipment and, in particular, instrumentation for measurement and control. The optimal and common solution is a central process control unit which provides all of the services required: acquisition of measurements from on-line data analysis, computerized data analysis and data reduction, representation of results, including display in tabular and graphical formats, storage of results, documenting, thereby providing a process database. Detailed information with regard to instrumentation, equipment for measurement, reliability and control of failure has been reported in specialized publications (Atkinson and Mavituna, 1991b; Zabriskie, 1985). The economics of specific singular process steps, and also of whole process, has been described by Atkinson and Mavituna (1991c). For bulk products it should be pointed out that enzyme cost must be kept very low in order that the process is economically feasible; for example, typical enzyme costs should be <1 % of the product price (cf. glucose-fructose syrup).

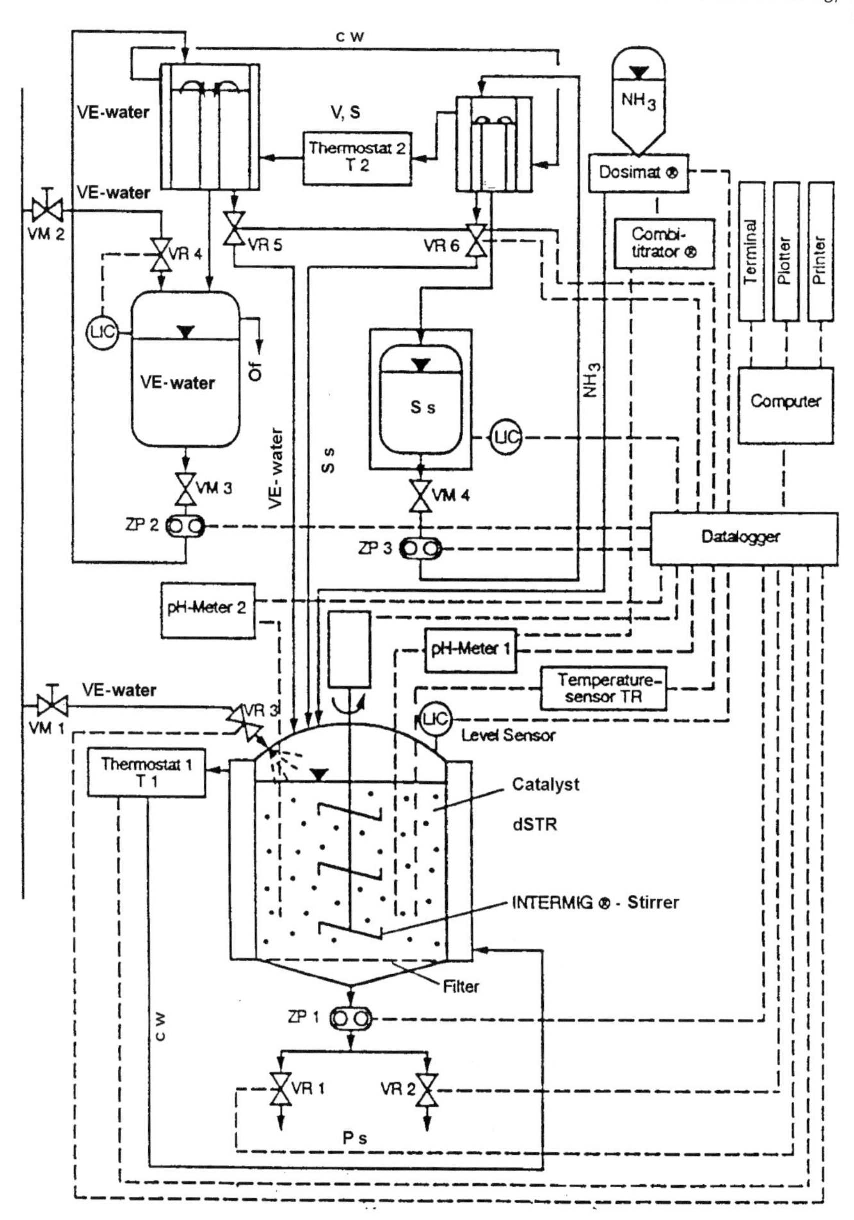

Fig. 9.28 Measurement and control units of a laboratory reactor for automated discontinuous penicillin hydrolysis. VM, VR = mechanically or electrically controlled valves; VE = storage tank; ZP = pump (Tischer, 1990).

9.4.4
Outlook: Perspectives for Integrated Process Engineering

One strategy to reduce down-stream processing cost is to integrate product isolation in the bioreactor. Some references will be given here, which may be taken as a suggestion for further reading and a challenge for the design of concepts for integrated processing. The topic has not been developed on a broad basis, but might open routes to easier and more economic production.

One option for *in situ* product removal (ISPR) is the selective adsorption of one product within the reactor during its synthesis, if a specific adsorbent is available. For instance, the synthesis of isomaltose by immobilized dextransucrase using glycosyltransfer from sucrose (substrate) to glucose (acceptor) requires an immediate selective separation of the product in the reactor, because isomaltose can be glucosylated in turn, leading to a sequence of higher oligosaccharides. A Multiphase Fluidized Bed Reactor has been developed for the selective separation of the product by *in situ* adsorption. As biocatalyst the enzyme dextran-sucrase is used immobilized in a calcium alginate matrix (d_p 1 mm) with enclosed sand particles to increase its density to retain the immobilized biocatalyst inside this reactor. The integrated separation of isomaltose in the reactor is achieved by adsorption on the suspended zeolites (< 50 µm), which are fed to and leave the reactor following the fluid phase (Fig. 9.29) (Berensmeier et al., 2004).

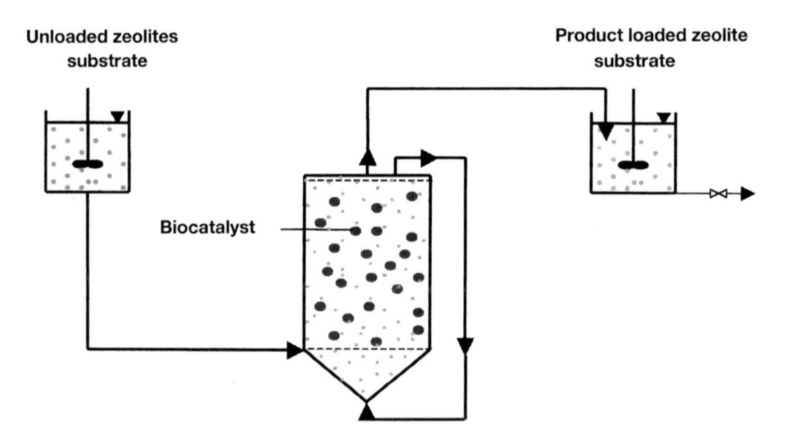

Fig. 9.29 Multiphase fluidized bed adsorber; substrate solution (sucrose, glucose), including zeolite for product adsorption, is introduced in the reactor; immobilized dextransucrase catalyses glucosyltransfer from sucrose to glucose to give isomaltose, fructose being formed as a byproduct; isomaltose is adsorbed selectively to the zeolite, which leaves the reactor with the product solution (including residual sucrose, glucose, and fructose as the byproduct); the product loaded zeolite is separated, the product isomaltose desorbed by water/ethanol and recovered by evaporation/rectification of solvents.

Two phase systems and membrane modules (see Chapter 5, Section 5.4) can facilitate product recovery from the enzyme reactor. Thus the production of unsaturated fatty acids form low cost saturated substrates has been performed by a desaturase from a *Rhodococcus* sp. *in* an oil/water emulsion. The product (and residual substrate) are recovered by a hydrophobic membrane based filtration (Liese et al., 2000, pp. 165–168).

SPR is also useful to improve yields in equilibrium (thermodynamically) controlled processes. Examples are water removal in lipase-catalyzed esterifications by membrane systems, or using a low-boiling-point azeotrope (Bartling et al., 2001; Yan et al., 2002).

The kinetic resolution of racemic naproxen thioesters in isooctane, coupled with a continuous *in situ* racemization of the remaining (*R*)-thioester substrate, was performed with immobilized lipase. A hollow-fiber membrane was integrated in the reactor system to reactively extract the desired (*S*)-naproxen out of the reaction medium (Lu et al., 2002).

Another alternative is product precipitation. This not only can shift the equilibrium towards product formation, but also can avoid product inhibition. Examples are the synthesis of Aspartame, in which the Z-protected sweetener precipitates from the reaction mixture (see Chapter 3, Section 3.2.2.3 and Chapter 6, Section 6.5.1), or the synthesis of the β-lactam antibiotic amoxicillin (Spieß and Kasche, 2001).

9.5 Exercises

9.5.1 Exercises Relating to Case Study 9.2

1. How can glutarate (adipate) formed in the hydrolysis of glutaryl-7-ACA or adipoyl-7-ADCA (Fig. 9.11) be recycled?
2. Discuss how the frequency in the selection of β-lactam resistant bacteria can be reduced.
3. How can the results in Figure 9.14 be determined? Is the hydrolysis of 7-ACA endo- or exothermal? Determine the free energy change at 25 °C, and the enthalpy and entropy change of the hydrolysis reaction in the studied temperature interval.
4. Under what conditions can the loss of 7-ACA be kept at <1% in the hydrolysis of glutaryl-7-ACA (Fig. 9.15)?
5. How could glutaryl amidase be improved so that it also can hydrolyze cephalosporin C? Hints: What properties must be improved? Then it must hydrolyze a (*R*-*S*) peptide bond in a (*R*-*S*-*R*) tripeptide, where the first (*R*)-isomer is a charged residue. Another enzyme that can do this is penicillin amidase.
6. Could the improved enzyme in Question 5 be used for racemate resolutions or to deprotect peptides?

7. Plants exist for the production of 1000 tons 7-ACA per year. In such a plant, 3.8 - tons of Glu-7-ACA must be hydrolyzed per day. This is contained in 100 m^3 with 100 mM Glu-7-ACA. Each hydrolysis cycle is 30 min, with 99 % substrate conversion, and the enzyme content per liter immobilized biocatalyst is 50000 U. Other conditions are given in Section 9.2.6.3).
   (a) What is the minimal total packed-bed reactor size (Fig. 9.18) to hydrolyze this per day without biocatalyst inactivation? What is the biocatalyst productivity after 300 cycles?
   (b) Try to estimate the total packed-bed reactor size, the productivity, and how the reactors must be replaced to keep the required space–time yield when biocatalyst inactivation is considered (half-life ≈15 days).

9.5.2 Exercises Relating to Mixing and Shear

1. Calculate the mean circulation times for the following reactors (consider dimensions! Eq. 9.17).
   (a) $n = 5\ s^{-1}$; $H = 0.5$ m; $d_R = 0.3$ m; $d_i = 10$ cm
   (b) $n = 5\ s^{-1}$; $H = 2.0$ m; $d_R = 1.0$ m; $d_i = 0.3$ m
   For what stirrer rates would equal mixing times be obtained? Compare the result for the value which can be taken from Figure 9.26 b for a turbine and propeller, respectively, in aqueous solution.

2. Calculate the increasing shear in stirred tank reactors with scale up of a 10-L reactor to
   (a) pilot scale with 0.5 m^3
   (b) production scale with 15 m^3
   in each case with constant energy input $P_R = 6\rho n^3 \cdot d_i^5$, with the stirrer diameter being one-third of the reactor diameter and the dimensions of the stirred tank being $H{:}d_R = 3{:}1$. The shear is (under turbulent conditions) proportional to $\rho\,(n \cdot d_i^2)$.

Literature

General

Atkinson, B., Mavituna, F., *Biochemical Engineering and Biotechnology Handbook*, 2nd edn., MacMillan Publishers Ltd., New York, 1991

Brauer, H., Stirred vessel reactors, in: Brauer, H. (Vol. Ed.), *Biotechnology*, Vol. 2, pp. 395–444, Wiley-VCH, Weinheim, 1985

Hempel, D.C., Bioverfahrenstechnik, in: Dubbel, Taschenbuch für den Maschinenbau, 21st edn., Springer-Verlag, Berlin, Heidelberg, New York (English version in preparation), 2002

Kossen, N.W.F., Oosterhuis, N.M.G., Modeling and scaling-up of bioreactors; in: Brauer, H., (Vol. Ed.), *Biotechnology*, Vol. 2, pp. 571–605, VCH, Weinheim, 1985

Moo-Young, M., *Comprehensive Biotechnology*, Vols. 1 and 2, Pergamon Press, Oxford, 1985

Ullmann's Encyclopedia of Industrial Chemistry (a and b): *Unit Operations I and II*, 5th edn., Vol. B2 and B3, VCH, Weinheim, 1988

Ullmann's Encyclopedia of Industrial Chemistry (1992): Principles of Chemical Reaction Engineering and Plant Design, 5th edn., Vol. B4, VCH, Weinheim, 1992

References

Adrio, J.L., Hintermann, G.A., Demain, A.L., Piret, J.M., Construction of hybrid bacterial deacetoxycephalosporin C synthases (expandases) by in vivo homologous recombination, *Enzyme Microb. Technol.*, **2002**, *31*, 932–940

Antrim, R.L., Auterinen, A.-L., A new regenerable immobilized glucose isomerase, *Stärke*, **1986**, *38*, 132–137.

Antrim, R.L., Lloyd, N.E., Auterinen, A.-L., New isomerization technology for high fructose syrup, *Stärke*, **1989**, *41*, 155–159.

Aschengreen, N.H., Technical isomerization of glucose, in: Immobilisierte Biokatalysatoren, DECHEMA-Kurs, Braunschweig, 1984.

Atkinson, B., Mavituna, F., Biochemical Engineering and Biotechnology Handbook, 2. edn., pp. a: 706, 707; b: 1023–1057; c: 1059–1109, MacMillan Publisher, New York, 1991.

Bartling, K., Thompson, J., Pfromm, P.H. et al., Lipase-Catalyzed Synthesis of Geranyl Acetate in *n*-Hexane with Membrane-Mediated Water Removal, *Biotechnol. Bioeng.*, **2001**, *75*, 676–681.

Berensmeier, S., Ergezinger, M ., Bohnet, M., Buchholz, K., Design of Immobilised Dextransucrase for Fluidised Bed Application, *J. Biotechnol.*, **2004**, in press.

Berensmeier, S., Buchholz, K., Separation of Isomaltose From High Sugar Concentrated Enzyme Reaction Mixture by Dealuminated Beta-Zeolite, *Sep. Purific. Technol.*, **2004**, *38*, 129–138.

Blenke, H., Biochemical loop reactors, in: Rehm, H.-J., Reed, G., Brauer, H., (Eds), *Biotechnology*, Vol. 2, pp. 465–517, VCH, Weinheim, 1985.

Bonse, D., Schindler, H., Immobilisierte Glucoseisomerase, in: Immobilisierte Biokatalysatoren, DECHEMA-Kurs, Braunschweig, 1986.

Bruggink, A., *Synthesis of ß-lactam antibiotics*, Kluwer Academic Publishers, Dordrecht, 2001

Buchholz, K., Reaction engineering parameters for immobilized biocatalysts, *Adv. Biochem. Eng.*, **1982**, *24*, 39–71.

Busse (Amino) (2004), personal communication: standard particle size of ion exchangers: 300–1200 μm; optimized particles: 600 μm.

Cabral, J., Tramper, J., Bioreactor design, in: Straathof, A.J.J., Adlercreutz, P., (Eds), *Applied Biocatalysis*, pp. 339–378, Harwood Academic Publishers, Amsterdam, 2000.

Carleysmith, S.W., Dunnil, P., Lilly, M.D., Kinetic behavior of immobilized penicillin acylase, *Biotechnol. Bioeng.*, **1980**, *22*, 735–756.

Chibata, I., Tosa, T., Sato, T., Application of immobilized biocatalysts in pharmaceutical and chemical industries, in: Rehm, H.-J., Reed, G., (Eds), *Biotechnology*, Vol. 7a, pp. 653–684, VCH, Weinheim, 1987.

Conlon, H., Baqai, J., Baker, K., Shen, Y.Q., Wong, B.L., Noiles, R., Rausch, C.W., Two-step immobilized enzyme conversion of cephalosporin C to 7-aminocephalosporanic acid, *Enzyme Microb. Technol.*, **1995**, *46*, 510–513.

Crawford, L., Stepan, A.M., McAda, P.C., Ramsbosek, J.A., Conder, M.J., Vinci, V.A., Reeves, C.D., Production of cephalosporin intermediates by feeding adipic acid to recombinant *Penicillium chrysogenum* strains expressing ring expansion activity, *Bio/Technology*, **1995**, *13*, 58–62.

Do, U.H., Neftel, K.A., Spadari, S., Hübscher, U., Beta-lactam antibiotics interfere with eucaryotic DNA-replication by inhibiting DNA polymerase, *Nucleic Acid. Res.*, **1987**, *15*, 10945.

Dunlap, C.E., Solids and Liquids Handling. In: Moo-Young, M. (Ed.), *Comprehensive Biotechnology*, Vol. 2, pp. 237–271, Pergamon Press, Oxford, 1985.

Eigenberger, G., Fixed bed reactors, in: *Ullmann's Encyclopedia of Industrial Chemistry (1992): Principles of Chemical Reaction Engineering and Plant Design*, 5th edn., Vol. B4, pp. 199–238, VCH, Weinheim, 1992.

Elander, R.P., Industrial production of β-lactam antibiotics, *Appl. Microbiol. Biotechnol.*, **2003**, *61*, 385–392

Fritz-Wolf, K., Koller, K.P., Lange, G., Liesum, A., Sauber, K., Schreuder, H., Aretz, W., Kabsch, W., Structure based prediction of modifications in glutarylamidase to allow single-step enzymatic production of 7-aminocephalosporanic acid from cephalosporin C, *Protein Sci.*, **2002**, *11*, 92–103.

Galunsky, B., Lummer, K., Kasche, V., Comparative study of substrate- and stereospecificity of penicillin G amidases from different sources and hybrid isoenzymes, *Chem. Monthly*, **2000**, *131*, 623–632.

Ingolia T., Queener, S.W., Beta-lactam biosynthetic genes, *Med. Res. Rev.*, **1989**, *9*, 245–264.

Jensen, V.J., Rugh, S., Industrial scale production and application of immobilized glucose isomerase, in: Mosbach, K., (Ed.), *Meth. Enzymol.*, Vol. 136, pp. 356–370, Academic Press, Orlando, 1987.

Jördening, H.-J., Buchholz, K., Fixed film stationary-bed and fluidized-bed reactors. In: Winter, J. (Vol. Ed.), *Biotechnology*, Vol. 11a, pp. 493–515, WILEY-VCH, 1999.

Jördening, H.-J., Mösche, M., Küster, W., Entwicklung und Betrieb von Fließbettreaktoren für die anaerobe Abwasserreinigung, *Chem. Ing. Techn.*, **1996**, *68*, 1152–1153; and *Zuckerindustrie*, **1996**, *121*, 847–854.

Jørgensen, O.B., Karlsen, L.G., Nielsen, N.B., Pedersen, S., Rugh, S., A new immobilized glucose isomerase with high productivity produced by a strain of *Streptomyces murinus*, *Stärke*, **1988**, *40*, 307–313.

Katchalski-Katzir, E., Krämer, D., Eupergit C, a carrier for immobilization of enzymes of industrial potential, *J. Mol. Catal. B: Enzymatic*, **2000**, *10*, 157–176.

Koller, K.P., Riess, G.J., Aretz, W., Process for the preparation of glutarylacylase in large quantities, US Patent 5.830.743, 1998.

Liese, A., Seelbach, K., Wandrey, C., *Industrial Biotransformations*, Wiley-VCH, Weinheim, 2000.

Li., Y., Chen, J., Jiang, W., Mao, X., Zhao, G., Wang, E., *In vivo* post-translational processing and subunit reconstitution of cephalosporin acylase from *Pseudomonas* sp. 130, *Eur. J. Biochem.*, **1999**, *262*, 713–719.

Liese, A., Seelbach, K., Wandrey, C., (Eds), *Industrial Biotransformations*, Wiley-VCH, 2000.

Lilly, M.D., Immobilized enzyme reactors, in: *Biotechnology*, Dechema-Monographs, Vol. 82, pp. 165–180, Verlag Chemie, Weinheim, 1978.

Lu, C.-H., Cheng, Y.-C., Tsai, S.-W., Integration of Membrane Extraction, *Biotechnol. Bioeng.*, **2002**, *79*, 31–34.

Middleton, J.C., Carpenter, K.J., Stirred-tank and loop reactors, in: *Ullmann's Encyclopedia of Industrial Chemistry*, Vol. B 4, pp. 167–180, Wiley-VCH, Weinheim, 1992.

Misset, O., Xylose (glucose) isomerase, in: Whitaker, J.R., Voragen, A.G.J., Wong, D.W.S., (Eds), *Handbook of Food Enzymology*, pp. 1057–1077, Marcel Dekker, New York, 2003.

Monti, D., Carrea, G., Riva, S., Baldaro, Frare, G., Characterization of an industrial biocatalyst: immobilized glutaryl-7-ACA acylase, *Biotechnol. Bioeng.*, **2000**, *70*, 239–244.

Moser, A., Imperfectly mixed bioreactor systems, in: Moo-Young, M., (Ed.), *Comprehensive Biotechnology*, Vol. 2, pp. 77–98, Pergamon Press, Oxford, 1985.

Olsen, H.S., Use of enzymes in food processing, in: Reed, G., Nagodawithana, T.W., (Eds), *Biotechnology*, Vol. 9, pp. 663–736, VCH, Weinheim, 1995.

Pedersen, S., Christensen, M.W., Immobilized biocatalysts, in: Straathof, A.J.J., Adlercreutz, P., (Eds), *Applied Biocatalysis*, pp. 213–228, Harwood Academic Publishers, Amsterdam, 2000.

Prenosil, J.E., Dunn, I.J., Heinzle, E., Biocatalyst Reaction Engineering; in: Rehm, H.-J., Reed, G. (Eds.), *Biotechnology*, Vol. 7a, pp. 489–545, VCH, Weinheim, 1987.

Reilly, P., Antrim, R.L., Enzymes in grain wet milling, in: *Ullmann's Encyclopedia of Industrial Chemistry*, Wiley-VCH, Weinheim, 2003.

Reuss, M., Bioreaktionstechnik; Skriptum zur Vorlesung, Universität Stuttgart, Institut für Bioverfahrenstechnik, 1991.

Reuss, M., Bajpai, R., Stirred tank models, in: Rehm, H.-J., Reed, G., (Eds), *Biotechnology*, Vol. 4, pp. 299–348, VCH, Weinheim, 1991.

Richardson, J.F., Zaki, W.N., Sedimentation and fluidization: Part 1, *Trans. Inst. Chem. Eng.*, **1954**, *32*, 35–53.

Sauber, K., Lessons from industry, in: Van den Tweel, W., Harder, A. Buitelaar, R.M., (Eds), *Stability and stabilization of enzymes*, pp. 145–151, Elsevier Science Publishers, Amsterdam, 1992

Schmalzried, S., Jenne, M., Mauch, K., Reuss, M., Integration of physiology and fluid dynamics, *Adv. Biochem. Eng.*, **2003**, *80*, 19–68.

Spieß, A., Kasche, V., Enzymatic synthesis of amoxicillin. Process integration using multiphase systems, *in Novel Frontiers in the Production of Compounds for Biomedical Use* (A. van Broekhoven et al., Eds), Kluwer Academic Publishers, Amsterdam, 169–192, 2001.

Spieß, A., Schlothauer, R, Hinrichs, J., Scheidat, B., Kasche, V., pH gradients in heterogeneous biocatalysts and their influence on rates and yields of the catalysed processes, *Biotechnol. Bioeng.*, **1999**, *62*, 267–277.

Swaisgood, H.J., Use of immobilized enzymes in the food industry, in: Whitaker, J.R., Voragen, A.G.J., Wong, D.W.S. (Eds), *Handbook of Food Enzymology*, pp. 359–366, Marcel Dekker, New York, 2003.

Tischer, W., Immobilisierte Enzyme in der Anwendung, in: Präve, P., Schlingmann, M., Crueger, W., Esser, K., Thauer, R., Wagner, J., (Eds), *Jahrbuch Biotechnologie*, pp. 251–275, Carl Hanser Verlag, München, 1990.

Tischer, W., Giesecke, U., Lang., Röder, A., Wedekind, F., Biocatalytic 7-aminocephalosporanic acid production, *Ann. N.Y. Acad. Sci.*, **1992**, *612*, 502–509.

Uhlig, H., *Enzyme arbeiten für uns*, pp. 214–241, Hanser-Verlag, München, 1991.

Uhlig, H., *Industrial enzymes and their applications*, John Wiley & Sons, Inc., New York, 1998.

Voser, W., Process for the recovery of hydrophobic antibiotics, US Patent 3,725,400, 1973.

Werther, J., Fluidized bed reactors, in: *Ullmann's Encyclopedia of Industrial Chemistry (1992): Principles of Chemical Reaction Engineering and Plant Design*, 5th edn., Vol. B4, pp. 239–274, VCH, Weinheim, 1992.

Yamana, T., Tsuji, A., Comparative stability of cephalosporins in aqueous solution: kinetics and mechanism of degradation, *J. Pharm. Sci.*, **1976**, *65*, 1563–1574.

Yan, Y., Bornscheuer, U.T., Schmid, R., Efficient Water Removal in Lipase-Catalyzed Esterifications Using a Low-Boiling-Point Azeotrope, *Biotechnol. Bioeng.*, **2002**, *78*, 31–34.

Zlokarnik, M., Rührtechnik, in: *Ullmann's Encyclopädie der technischen Chemie*, 4. Aufl., Bd. 2, S259–281, Verlag Chemie, Weinheim, 1972.

Zabriskie, D.W., Data analysis, in: Moo-Young, M., (Ed.), *Comprehensive Biotechnology*, Vol. 2, pp. 175–190, Pergamon Press, Oxford, 1985.

Appendix I
The World of Biotechnology Information: Eight Points for Reflecting on Your Information Behavior

Prepared with the assistance of Thomas Hapke

> *"Information literacy is the adoption of appropriate information behaviour to obtain, through whatever channel or medium, information well fitted to information needs, together with critical awareness of the importance of wise and ethical use of information in society."* (B. Johnston and S. Webber, 2003, p. 336)

1
Thinking About Your Information Behavior

Information literacy is a crucial key skill for self-directed learning in scholarly and professional everyday life. In addition to efficient retrieval and navigation strategies, it includes – above all – the creativity to organize and shape one's own information process in a conscious and demand-oriented way. For the searcher, it is no longer questionable to find some information, but rather to filter reliable information from many similar offers.

Like every subject, biotechnology has its own special information media, in addition to particular retrieval strategies, to meet the subject-related information needs. Which of the available databases match your specific needs and are reliable? The so-called "invisible web" or "deep web" contains information sources which are not collected by most search engines such as Google(TM) – that is, it includes the content of special databases, for example for patents, websites secured by password access or only available in an intranet and script-based websites, which offer for example dynamic content.

There is a whole range of reasons for reading and informing for research: To provide you with ideas and enhance your creativity; to understand and be able effectively to criticize what other researchers have done in your subject; to broaden your perspective and view your work in context to others (direct personal experience is never enough); to legitimize your arguments; to avoid double efforts in research; to learn more about research methods and their application in practice; and to find new areas for research (Blaxter et al., 2001). Before beginning to search information, first reflect on your topic and specific information needs, gather background information – and then focus your research.

Biocatalysts and Enzyme Technology. K. Buchholz, V. Kasche, U.T. Bornscheuer

ISBN: 3-527-30497-5

2
Playing with Databases and Search Terms

When searching a database it is important to use appropriate key words which allow retrieval of the desired information. Too general key words lead to too many hits from which often only a fraction is useful; when using too specific key words, important information might not be found. It is also recommend to use logical, so-called Boolean operators (AND, OR, NOT) to link search terms and to use wildcard (joker) symbols ('?' or '*' or '$', which one depends on the search interface). For example, searching with 'biodegr?' retrieves documents containing 'biodegradation or biodegradable or biodegraded or biodegradability or ...'.

A search term worksheet can help to structure your query. For this, the topic must be divided into components and key words chosen for every component. For searching, terms in each of the worksheet's columns have to be combined with the "OR", the resulting sets with "AND":

Topic: Microbial degradation of aromatic compounds in soil

Component 1	Component 2	Component 3
microbi? degrad?	aromat?	soil?
biodegrad?	polyaromat?	clay?
bioremed?	Benzene	compost?
microbi? decompos?	PAH	sediment?

3
Tutorials, Subject Gateways and Literature Guides

Tutorials, subject gateways on the net and literature guides help to inform yourself about searching information. DISCUS (Developing Information Skills & Competence for University Students) is an example of a web-based bilingual (German, English) learning tutorial for information literacy in engineering which can be used independent of time and space. It contains also a module about biotechnology.

DISCUS was developed at the University Library of the Hamburg University of Science and Technology and is offered at http://discus.tu-harburg.de. Another example is the Texas Information Literacy Tutorial (TILT) at http://tilt.lib.utsystem.edu.

So-called "subject gateways" are good starting points for relevant web sites containing collections of subject-specific links. Two examples are "EEVL (Enhanced and Evaluated Virtual Library) – the Internet Guide to Engineering, Mathematics and Computing" at http://www.eevl.ac.uk and the Engineering Subject Gateway of the TIB/UB, the German National Library of Science and Technology at http://vifatec.tib.uni-hannover.de/index.php3?L=e. Special link collections in biotechnology you find in web catalogs such as Yahoo™ (http://dir.yahoo.com/science/biology/biotechnology/) or the Open Directory Projekt (http://www.dmoz.org/Science/

Biology/Biotechnology/) as well as at web sites like the "U.S. National Center for Biotechnology Information" (http://www.ncbi.nlm.nih.gov).

Literature guides provide a comprehensive overview about all forms of primary and secondary literature of the treated subject. The most actual guide to information sources in chemistry, which is also of value for the process and biochemical engineer, is written by Maizell (1998). See also *Information Sources in Engineering* (3rd edition, from 1996) which will be updated soon by MacLeod and Corlett (2004). Guides for the life sciences come from Schmidt (2002) and Wyatt (1997), and for business information from Moss (2004).

4
Using Your Local Library

Even in the Internet age, a visit to the local university library can ease information retrieval. If they do not possess the item you are interested in, library union catalogs offer a wide range of library materials which can be ordered through interlibrary loan or document delivery. In many countries special libraries functions as a National Library for Science and Technology, (e.g., in Germany the TIB/UB in Hannover; http://www.tib.uni-hannover.de). Databases available in the local intranet also provide references to further information (e.g., journal articles) not necessarily housed by the library itself. Subject librarians can also provide information consulting.

5
Using Encyclopedias

A range of encyclopedias in chemical and process engineering has been published, including volumes dedicated to biotechnology. These are listed in this book at the end of each chapter. Encyclopedias contain a detailed view of evaluated knowledge, in addition to references for further reading. Libraries offer a selected range of such reference works in printed form in their reading rooms. Electronic versions may be available in the local intranet.

6
Searching Journal Articles, Patents, and Data

Today, most recent research results are published in scientific journals and subject-specific text books. In addition, patents are an important and often less frequently used source by academia. A literature search (see above) provides rapid identification of specific journal articles, reviews and recent books. The original publications can then either be downloaded through the University library homepage as gateway or are available as printed versions in the library (see also: "German Electronic Journals Library" at http://www.bibliothek.uni-regensburg.de/ezeit/ or the "DOAJ Directory of Open Access Journals" at http://www.doaj.org/).

A range of databases for information retrieval in chemistry, biotechnology and related fields is listed in Table A.1.

Table A.1 Internet databases useful for biocatalysis (selection).

Searching for	Database name and website	Comments
Articles in journals	CEABA-VTB (Chemical Engineering And Biotechnology Abstracts – Verfahrenstechnische Berichte) at http://www.dechema.de/ceaba-lang-en.html	First choice for process engineering and biotechnology (produced by the Dechema)
	Chemical Abstracts Service at http://www.cas.org	For all areas of chemistry, and related sciences like the materials sciences and the environmental sciences (with user interface SciFinder perhaps in your local intranet)
	COMPENDEX (COMPuterized ENgineering InDEX) at http://www.ei.org/eicorp/compendex.html	Most important and comprehensive database for general engineering
	INSPEC (Information Service in Physics, Electrotechnology, Computer and Control) at http://www.iee.org/publish/inspec/	Of importance because information technology plays a considerable role in all areas of engineering today
	FIZ Technik (Fachinformationszentrum, Specialized Information Center of Technology) at http://www.fiz-technik.de/en/	Databases contain also German resources
	"Science Citation Index" in the "Web of Science" from the Institute of Scientific Information at http://www.isinet.com/	In interdisciplinary citation databases you can search with documents as "search terms" and answer questions such as: Who have cited a specific document? How often is a document cited?
	PubMed at http://www.ncbi.nlm.nih.gov/entrez/query.fcgi	Interdisciplinary for medicine, also of great importance for biotechnology
	Toxline at http://toxnet.nlm.nih.gov	Toxicology and hazardous substances
	Agricola at http://agricola.nal.usda.gov	For agricultural sciences
	Ulidat at http://doku.uba.de	German database for the environmental sciences

Table A.1 (Continued)

Searching for	Database name and website	Comments
Patents	DEPATISnet at http://depatisnet.dpma.de	The German patent information system contains for free the fulltext of every German and American patent in pdf format, also patents from other countries. You have to know the exact patent number. Searching in other database fields – e.g., title, patent inventor, or abstract field – is possible from a distinct year. So you can search for German patents in the title or inventor field from the year 1981
	Esp@cenet, at http://ep.espacenet.com	European patents (European Patent Office)
	US Patent and Trademark Office at http://www.uspto.gov/patft/	Example for fulltext access to national patents
Chemicals	ChemFinder at http://chemfinder.camsoft.com	Meta-search engine for chemical substances information
	NIST Webbook at http://webbook.nist.gov/chemistry/	Detailed data for many common substances
Hazardous substances	GESTIS at http://www.hvbg.de/e/bia/fac/stoffdb/index.html	Free of charge information system of the German institutions for statutory accident insurance and prevention (english version available)
	TOXNET of the at http://toxnet.nlm.nih.gov/	U.S. National Library of Medicine
	International Chemical Safety Cards (ICSC) at http://www.bfr.bund.de/cd/444	German data base in english
Enzyme manufacturers (see also Table 5.4, p. 209)	http://www.amfep.org	Homepage of the Association of Manufacturers and Formulators of Enzyme Products. It contains information on enzymes, safety rules for their use, and links to similar organizations, the companies, organizations of importance for the regulation of enzymes (EU, FAO, FDA, WHO etc.)

Table A.1 (Continued)

Searching for	Database name and website	Comments
Enzyme classification and structure	http://www.chem.qmul.ac.uk iubmb/enzymes	(enzyme classification)
	http://www.expasy.ch/	Extensive information on all aspects of proteins/enzymes. Includes enzyme nomenclature database and links to more specialized databases on enzymes
	http://www.rcsb.org/pdb	A database for 3-D structures of proteins/enzymes and cofactors important for structure and function
	http://biochem.ucl.ac.uk/bsm/cath	A database on enzyme classification based on 3-D structures
Enzyme properties	http://www.brenda.uni-koeln.de	A comprehensive database on enzyme properties (k_{cat}, K_m for different substrate; cofactors; inhibitors; stability etc.)
Enzyme catalyzed reactions	http://xpdp.nist.gov/enzyme thermodynamics/enzyme1.pl	Thermodynamic data for enzyme catalyzed reactions
	http://genome.ad.jp/kegg/ligand.-html	A database to find an enzyme that can catalyze the biotransformation of a compound (ligand). Links to metabolic charts that show the enzyme that catalyzes the metabolic reactions
Enzymes, specific	http://merops.sanger.ac.uk	A database for peptides
	http://www.led.uni-stuttgart.de	A database for lipases
	http://afmb/cnrs-mrs.fr/CAZY	A database on enzymes catalyzing reactions with carbohydrates

Table A.1 (Continued)

Searching for	Database name and website	Comments
Bioinformatics	http://www.ncbi.nlm.nih.gov/education/index.html	Education page on bioinformatics
	http://www.ebi.ac.uk http://www.hnbioinfo.de	Two extensive databases from the European Bioinformatics Institute and the German Helmholtz Research Organization on the bioinformatics of proteins/enzymes with links to the main databases and tools on sequence and structure analysis (gene or protein), alignment of enzyme sequences etc.
Biocatalysis	http://www.accelrys.com/chem_db/biocat.html	An extensive biocatalysis database with >38 000 reactions using biocatalysts (requires subscription fee)
	http://umbbd.ahc.umn.edu/	Biocatalysis/biodegradation database

7
After Searching: Evaluating and Processing Information

After searching successfully, you have to evaluate your search findings with respect to relevance. How to be sure, that all the potentially important documents are included in your resulting set? How to modify your query to reach this goal? But it is also important to evaluate critically the quality of the documents you have found. In case the document is published in a scholarly peer-reviewed journal, the article has been evaluated by independent experts before acceptance/publication. Who is the author and what is his or her background? Why is the document being provided? How current is it?

To keep yourself up-to-date, several publishers of journals offer free access at least to the contents of the journal issues, or offer a free table of contents via e-mail whenever an issue is published. Another way to stay current is to subscribe to subject-specific mailing lists or to read subject-specific weblogs such as the bioinformatics weblog at http://www.nodalpoint.org.

8
Information and the World

What is publication – what is an author, a document, a journal, a collection, or a library? In the electronic world of the Internet all of these terms have changed their meaning and use. Thinking about information is particularly of interest in biotechnology (Braman, 2004). At a time when historians of science describe "... biology's metamorphosis in an information science" (Lenoir, 1999), it is necessary to reflect about information and its communication and use (Feather and Sturges, 2003).

In spite of information overload, only a limited part of information is freely available on the Internet. Access to commercial information sources for scholarly research such as reference databases and the fulltext of a specific journal is usually subject to a license fee and controlled by password. However, they are often offered within the intranet of universities or companies. Open access activities try to free access to scholarly publications at least for research and educational purposes. An example is the journal *PLOS Biology* at http://www.plosbiology.org.

Issues in intellectual property and copyright increase in a "cut-and-paste" environment. Why is it important to cite sources of information? What is the right way to cite? Questions of information ethics (plagiarism) as well as information policy (ownership, access, privacy) become important. Does there exist a digital divide? Even think of the preservation of information. What will be happening with electronic records or data in 30 or 50 years' time?

Literature

Blaxter, L., Hughes, C., Tight, M., *How to research*, 2nd edn., Open University Press, Buckingham, 2001

Braman, S., (Ed.), *Biotechnology and Communication: The Meta-Technologies of Information*, Lawrence Erlbaum, Mahwal, NJ, 2004

Feather, J., Sturges, P., (Eds), *International Encyclopedia of Information and Library Science*, 2nd edn., Routledge, London, 2003

Johnston, B., Webber, S., Information literacy in higher education: a review and case study, *Studies in Higher Education* **2003**, *28*, 335–352

Lenoir, T., Shaping biomedicine as an information science, in: Bowden, M.E., Hahn, T.B., Williams, R.V., (Eds), *Proceedings of the 1998 Conference on the History and Heritage of Science Information Systems*, pp. 27–45, Information Today, Medford, NJ, 1999. Available online at www.chemheritage.org/explore/ASIS_documents/ASIS98_Lenoir.pdf

Maizell, R.E., *How to Find Chemical Information: A Guide for Practicing Chemists, Educators, and Students*, 3rd edn., Wiley, New York, 1998

MacLeod, R.A., Corlett, J., (Eds), *Information Sources in Engineering*, 4th edn., Saur, Munich, 2004

Moss, R.W., *Strauss's Handbook of Business Information: A Guide for Librarians, Students, and Researchers*, 2nd edn., Libraries Unl., Westport, Conn., 2004

Schmidt, D., Davis, E.B., Jacobs, P.F., *Using the Biological Literature: A Practical Guide*, 3rd edn., Dekker, New York, 2002

Wyatt, H.V., (Ed.), *Information Sources in Life Sciences*, 4h edn., Bowker-Saur, London, 1997

Appendix II
Symbols

a	Surface area per unit volume	$[m^{-1}]$
A_p	Particle outer surface area	$[m^2]$
A_m	Amplitude in concentration during mixing	$[mol\ L^{-1}]$
c	Extent of a reaction in racemate resolutions/Conversion	
c	Concentration	$[mol\ L^{-1}]$
$c_S(r)$	Substrate concentration at distance r from the particle center	$[mol\ L^{-1}]$
$c_S(\infty)$	Bulk substrate concentration	$[mol\ L^{-1}]$
$c_S(z)$	Dimensionless substrate concentration	
d_i	Stirrer diameter	[m]
d_p	Particle diameter	[m]
d_R	Reactor diameter	[m]
D_S	Diffusion coefficient in solution	$[cm^2\ s^{-1}]$
D'_S	Diffusion coefficient in the immobilized biocatalyst particle	$[cm^2\ s^{-1}]$
DS	Dry substance content (weight-%)	
ee	Enantiomeric excess	
E	Enantio- or stereoselectivity/ Enantiomeric ratio	
E_{eq}	Enantio- or stereoselectivity for an equilibrium controlled process	
E_{kin}	Enantio- or stereoselectivity for a kinetically controlled process	
$E(t)$	Residence time distribution	$[s^{-1}]$
h	Hour	

Biocatalysts and Enzyme Technology. K. Buchholz, V. Kasche, U.T. Bornscheuer

ISBN: 3-527-30497-5

H	Height	[m]
I	Ionic strength	[mol L^{-1}]
k_{cat}	Turnover number	[s^{-1}]
k_H	Apparent hydrolysis rate constant in a kinetically controlled process	[L mol^{-1} s^{-1}]
k_i	Apparent first order rate constant for the inactivation of enzymes	[s^{-1}]
k_L	Mass transfer coefficient	[m s^{-1}]
k_T	Apparent transferase rate constant in a kinetically controlled process	[L mol^{-1} s^{-1}]
K	Equilibrium constant (always as dissociation constants)	[mol L^{-1}]
K_A	Amplitude	[s^{-1}]
K_{app}	Apparent dissociation constant for a hydrolysis reaction	[mol L^{-1}]
K_i	Dissociation constant for the binding of inhibitor	[mol L^{-1}]
K_m	Michaelis-Menten constant	[mol L^{-1}]
K_{ref}	Reference equilibrium constant for a hydrolysis reaction	[mol L^{-1}]
M	Concentration	[mol L^{-1}]
MW	Molecular weight	[Da]
n	Stirrer speed	[s^{-1}]
n_E	Concentration of active immobilized enzyme	[mol L^{-1}]
n_0	Molar amount of a compound at t=0	[mol]
$\dot{n}$	Molar flow rate	[mol s^{-1}]
Ne	Newton number	
p	Partition coefficient	
p	Pressure	[Pa]
P	Productivity (kg product/kg biocatalyst)	
P	Permeability	[m s^{-1}]
P_R	Power input	[W s^{-1}]
[P]	Product concentration	[mol L^{-1}]
Q	Volumetric flow rate	[m^3 s^{-1}]

r	Distance from particle center	[m]
Re	Reynolds number	
R	Particle or cell radius	[m]
Sh	Sherwood number	
Sc	Schmidt number	
$[S]_E$	Substrate concentration at the end-point of a reaction	[mol L^{-1}]
$[S]_0$	Substrate concentration at t=0	[mol L^{-1}]
$[S]_i$	Substrate concentration after the reactor i	[mol L^{-1}]
$[S]_t$	Substrate concentration at t	[mol L^{-1}]
STY	Space-time yield	[mol L^{-1} s] or [g L^{-1} s]
t	Time	[s]
t_{12}	Time required for the substrate conversion from $[S]_1$ to $[S]_2$	[s]
$t_{1/2}$	Half-life time	[s]
t_m	Mixing time	[s]
T	Temperature	[K, °C]
TTN	Total turnover number	
u	Linear flow rate	[m s^{-1}]
U	Unit for enzyme activity	[µmol min^{-1}]
v, V_{max}	Reaction rate, maximal reaction rate of an enzyme catalyzed reaction	[mol L^{-1} s]
v_{obs}	Observed rate of a reaction catalyzed by a biocatalyst	[mol L^{-1} s]
V	Reactor volume	[m^3]
V_p	Particle (cell) volume	[m^3]
V_{CST}, V_{PB}	Volume of a continuous stirred tank (CST) or packed bed (PB) reactor	[m^3]
x	Cells per volume unit	[L^{-1}]
X	Degree of substrate conversion $\Delta[[S]/[S]_0 = ([S]_0-[S]_t)/[S]_0)$	
z	Dimensionless distance from particle center (r/R_p)	
α	Volumetric fraction (biocatalyst, solid)	
γ	Relative substrate concentration $= [S]/K_m$	

δ	Thickness of diffusion layer	[m]
φ	Thiele modulus	
ε	Porosity of a packed bed	
η	Stationary effectiveness factor	
η_o	Operational effectiveness factor	
η	Viscosity	[kg m^{-1} s^{-1}]
ρ	Density	[kg m^{-3}]
σ^2	Variance of the circulation distribution	
τ	Residence time	[s]
τ'	Circulation time	[s]
τ_h	Hydrodynamic residence time	[s]
ν	Kinematic viscosity	[m^2 s^{-1}]

Subject Index

Biocatalysts and Enzyme Technology. K. Buchholz, V. Kasche, U.T. Bornscheuer

ISBN: 3-527-30497-5

h

i